LOGIC COLLOQUIUM 2000

LECTURE NOTES IN LOGIC

A Publication of

THE ASSOCIATION FOR SYMBOLIC LOGIC

LECTURE NOTES IN LOGIC 19

LOGIC COLLOQUIUM 2000

Proceedings of the Annual European Summer Meeting of the Association for Symbolic Logic, held in Paris, France July 23–31, 2000

Edited by

René Cori
Équipe de Logique Mathématique
Université Paris 7 – CNRS

Alexander Razborov
Institute for Advanced Study and
Steklov Mathematical Institute

Stevo Todorčević
Équipe de Logique Mathématique
Université Paris 7 – CNRS

Carol Wood
Department of Mathematics and Computer Science
Wesleyan University

ASL

ASSOCIATION FOR SYMBOLIC LOGIC

A K Peters, Ltd. • Wellesley, Massachusetts

Addresses of the Editors of Lecture Notes in Logic and a Statement of Editorial Policy may be found at the back of this book.

Sales and Customer Service:
A K Peters, Ltd.
888 Worcester Street, Suite 230
Wellesley, Massachusetts 02482, USA

Association for Symbolic Logic:
C. Ward Henson, Publisher
Mathematics Department
University of Illinois
1409 West Green Street
Urbana, Illinois 61801, USA

Library of Congress Cataloging-in-Publication Data

Logic Colloquium (2000 : Paris, France)
Logic Colloquium 2000 : proceedings of the Annual European Summer Meeting of the Association for Symbolic Logic, held in Paris, France, July 23-31, 2000 / edited by René Cori ... [et al.].
p. cm. – (Lecture notes in logic ; 19)
Includes bibliographical references.
ISBN 1-56881-251-5 (acid-free paper) – ISBN 1-56881-252-3 (pbk. : acid-free paper)
1. Logic, Symbolic and mathematical–Congresses. I. Cori, René. II. Title. III. Series.

QA9.A1L64 2000
511.3–dc22 2004066230

Publisher's note: This book was typeset in LaTeX, by the ASL Typesetting Office, from electronic files produced by the authors, using the ASL documentclass `asl.cls`. The fonts are Monotype Times Roman. This book was printed in Canada by the Friesens Corporation, on acid-free paper. The cover design is by Richard Hannus, Hannus Design Associates, Boston, Massachusetts.

14 13 12 11 10 09 08 07 06 05 5 4 3 2 1

INTRODUCTION

Logic Colloquium 2000, the European Meeting of the Association for Symbolic Logic, took place in La Sorbonne, Paris, France, on July 23–31, 2000, on the site of Hilbert's presentation of his famous list of problems in the summer of 1900. LC2000 was part of World Mathematical Year 2000 and was organized to have a special impact on the future of logic, in recognition of the new millennium. The main themes of the meeting were: Proof theory and logical foundations of computer science, Set theory, Model theory, Computability and complexity theory, History of 20th century logic, Philosophy and applications of logic to cognitive sciences. The tutorials, philosophy symposium, historical talks and certain other lectures also comprised the European Logic Summer School 2000, emphasizing the importance of the training of young researchers in logic.

The Program Committee consisted of D. Andler, C. Berline, B. Cooper, D. van Dalen, A. Kanamori, C. Parsons, A. Razborov, H. Schwichtenberg, J. Steel, S. Todorčević, A. Wilkie, and C. Wood (Chair). The Local Organizing Committee included C. Berline, Z. Chatzidakis, R. Cori (Chair), M. Dickmann, J. Dubucs, J.-B. Joinet, D. Lascar, Y. Legrandgérard, J. Mosconi, M.-H. Mourgues, C. Muhlrad-Greif, L. Pacholski, J.-P. Ressayre, B. Veličković, and F. Ville. The program consisted of 4 tutorials and 24 invited plenary talks. There were also over 200 contributed papers, presented during 11 blocks of time, each block having 7 parallel sessions. A few papers were presented by title only.

This Logic Colloquium was certainly the largest in recent memory, with 426 registered participants from 52 countries. Through grants from the ASL, the European Community, the French and the US governments, financial support was provided to 98 participants, many of them junior researchers and students attending their first international meeting.

The following tutorials were given, and all but the first of these are included in the volume:

Lenore Blum and Stephen Smale (Carnegie Mellon University, Pittsburgh), *Complexity and real computation.*

Elisabeth Bouscaren (Université Paris 7 – CNRS), *Model theory and geometry.*

John Longley (University of Edinburgh), *Realizability and higher-type computability.*

W. Hugh Woodin (University of California, Berkeley), *The continuum hypothesis.*

The following invited plenary talks were presented: Peter Aczel (University of Manchester), *Can ZF classical mathematics be understood from a constructive standpoint?* Sam Buss (University of California, San Diego), *Definability and complexity in bounded arithmetic.* Martin Davis (New York University), *Computability theory in the twentieth century.* Ilijas Farah (Rutgers University), *Continuous liftings.* Leo Harrington (University of California, Berkeley), *Automorphisms of the computably enumerable sets.* Wilfrid Hodges (Queen Mary and Westfield College), *Model theory in its first century.* Martin Hofmann (University of Edinburgh), *On the decision problem for typed lambda calculus with coproducts.* Gerhard Jäger (Universität Bern), *Metapredicative and explicit Mahlo: a proof-theoretic perspective.* Jean-Louis Krivine (Université Paris 7), *Realizability and generalized forcing in set theory.* Yves Lafont (CNRS), *Proof theory and computer science: from typographical engineering towards geometry.* Richard Laver (University of Colorado), *Large cardinal results and problems.* Angus Macintyre (University of Edinburgh), *Model theory of Frobenius on Witt vectors.* David Marker (University of Illinois at Chicago), *Differential equations and logarithmic-exponential series.* Donald Martin (University of California, Los Angeles), *Set theory in the 20th century.* Daniel Osherson (Rice University), *A psychologist looks hopefully to logic.* Peter Shor (AT&T Labs, USA), *Quantum computing.* Theodore A. Slaman (University of California, Berkeley), *Aspects of the Turing jump.* Otmar Spinas (Christian-Albrechts-Universität zu Kiel), *Canonical behaviour of Borel functions on Polish planes.* S. Starchenko (University of Notre Dame), *On the Zil'ber principle for structures definable in o-minimal models.* Martin Zeman (Universität Wien, Austria), *Combinatorial principles in inner models.* Boris Zil'ber (University of Oxford, U.K.), *Analytic and pseudo-analytic structures.*

Three invited plenary talks were organized into a Philosophy Symposium *The foundation of mathematics around* 1900, chaired by Charles Parsons: Philippe de Rouilhan (Université Paris I-IHPST), *Russell's logics.* William Ewald (University of Pennsylvania), *The philosophical significance of Hilbert's problems* and Richard Heck (Harvard University), *Frege's contributions to foundational studies.*

In addition, a session on *the history of logic in France* (*Petite histoire de la logique en France*) was held in late afternoon on July 28, with participants M. Demazure, S. Grigorieff, M. Guillaume, B. Jaulin, J.-L. Krivine, D. Lacombe, K. McAloon, and Ph. de Rouilhan. This session was held in fond memory of our recently deceased colleague Jean Coret, who played a fundamental role in logic in Paris.

The editors owe a debt of gratitude to the speakers and authors and committees named above, but also to many other individuals and institutions who made the meeting possible. We mention here Université Paris 7 Denis Diderot (Michel Delamar, President), Université Paris I – Panthéon-Sorbonne (Michel Kaplan, President), Université Paris IV – Paris-Sorbonne (Georges Molinié, President), Académie de Paris (René Blanchet, Recteur), Direction de la Recherche du Ministère de l'Éducation Nationale, Cité des Sciences et de l'Industrie, the European Community, the National Science Foundation, the Association for Symbolic Logic, Centre National de la Recherche Scientifique, Institut Universitaire de Formation des Maîtres de Créteil, Société Mathématique de France, and France Télécom.

It is our pleasure to take this opportunity to acknowledge with thanks several individuals who were directly involved in organizing the meeting but not mentioned above: Benjamin Ainardi, Odile Ainardi, Laure Coret, Yvonne Girard, David Lucas, Pascal Manoury and Suzanne Thiriot.

Our joy on the occasion of this conference was tempered by the sad loss of an invited speaker, a logician of unusual breadth and insight as well as a kind man.

Only a few days before his death did K. Jon Barwise abandon hope of participating in LC2000. We dedicate this volume to his memory.

The Editors
René Cori, *Paris*
Alexander Razborov, *Princeton and Moscow*
Stevo Todorčević, *Paris*
Carol Wood, *Middletown*

TABLE OF CONTENTS

In Memoriam
K. Jon Barwise (1942–2000)

TUTORIALS

MODEL THEORY AND GEOMETRY

ELISABETH BOUSCAREN

This paper is based on a series of three lectures that I gave during the LC 2000, in the context of the "tutorials" which have now become a tradition at the European meetings of the ASL. I have kept fairly close to the actual format and style of the talks.

It is always difficult to identify precisely the audience such a tutorial should address. A fair number of broad and ambitious surveys have already been published on the subject of the applications of model theory to algebraic geometry (see section 4.4). I did not, during this tutorial, choose to address the specialists of the subject. The audience I had in mind consisted of both young "inexperienced" researchers in model theory and more "mature" logicians from other parts of logic. Rather than attempting one more broad survey, I tried to present some of the main concerns of "geometrical model theory" by looking at concrete examples and this is what I will try to do also in the present paper.

We will discuss three algebraic examples, algebraically closed fields, differentially closed fields and difference fields (fields with automorphisms). The geometric application we will take up as illustration is Hrushovski's approach to the Manin-Mumford conjecture. This is based on a fine study of the model theory of difference fields and is quite emblematic of the method. Perhaps the key technical notion is that of "local modularity"(or "one-basedness"), which arises in a purely model theoretic setting. We will see that the Diophantine conjectures of the Manin-Mumford type can be rephrased in terms of this notion. Furthermore, as one thinks through the rephrasing process, one realizes the need for the introduction of auxiliary algebraic theories such as the theory of difference fields.

I would like to thank the anonymous referee, despite my temporary shock at the initial suggestion that the paper be totally rewritten and turned into a survey of a completely different type. Fortunately, he/she also provided a long list of detailed comments and less draconian suggestions, in case I did not choose to follow this first drastic piece of advice. I have found these comments very helpful and have followed most of these suggestions.

Logic Colloquium 2000
Edited by R. Cori, A. Razborov, S. Todorčević, and C. Wood
Lecture Notes in Logic, 19

§1. "Geometric" model theory. The most striking applications of model theory to algebra or number theory in the course of the last 40 years have typically been obtained using only the most basic tools of model theory, notably the compactness theorem and the technique of quantifier elimination, though the algebraic and analytic ingredients have been considerably deeper and more varied. This applies for example to the work of Ax-Kochen-Ershov on valued fields (1965), with later applications by Denef to the computation of p-adic integrals, to Ax's work on the elementary theory of finite fields (1968), to work of Denef, van den Dries and Macintyre on the p-adics (1970's and 1980's). It applies also to some of the more recent work on o-minimal structures, such as Wilkie's results on the theory of the reals with exponentiation and it's subsequent generalization to broad classes of analytic functions (see [10]).

In parallel pure model theory flourished at the same period, beginning with Morley's characterization of uncountably categorical theories (1965) and then with Shelah's monumental work on classification theory. Applications to algebra of these more sophisticated notions and results were at first rather few: existence and uniqueness of the differential closure of a differential field of characteristic zero (Blum 1977), application to the theory of modules (started by Garavaglia around 1978). It was soon apparent that the tools of the theory of stability were particularly well suited to the model theoretic analysis of groups and fields. Around 1980, Poizat introduced the notion of generic of a stable group which was directly inspired by the corresponding notion for algebraic groups and which became one of the main tools in the subject.

Then in the mid-eighties, under the influence of Zilber first and then of Hrushovski, stability theory started evolving and moved back to the study of the fine local behavior of structures of finite dimension. This was the beginning of what has been for some years known as "geometric stability" or more generally "geometric model theory".

Stability theory à la Shelah, developed a theory of abstract independence and dimension. Although this generalized the classical algebraic notions of independence (linear independence, algebraic independence), the methods used were often those of infinite combinatorics. One of the main aspects of the theory for example is the classification of structures according to which infinite combinatorial objects they interpret: orderings, trees ...

Geometric stability, as its name indicates, took much of its inspiration from geometry, both in the sense of combinatorial geometries (or matroids) and of algebraic geometry. This relationship turned out to go both ways: the abstract notions developed in model theory were applied to the disciplines of their origins in order to give new proofs or new results there. We will not discuss here combinatorial geometries nor any of the results that were proved in this domain by applying model theoretic tools or ideas (results of Zilber on homogeneous finite geometries for example [46] or results of Evans and Hrushovski about "algebraic matroids" [7, 8]), but we will focus on the relationship with algebraic geometry.

Geometric stability investigates the geometric properties of the abstract independence relation introduced by Shelah. One of the main focus points is the study of the algebraic structures coded via this relation (groups, fields). These questions and the results obtained can be considered to be, at a higher level of generality, in the direct line of two "classical" and well-known theorems:

– the old theorem of geometry which says that a Desarguesian projective geometry of dimension at least 3 is the projective geometry over some division ring;

– the theorem of Weil which constructs an algebraic group from a generically rational associative operation on an algebraic variety.

One of the central notions in the subject is that of "one-basedness" or local modularity, which was introduced into the subject independently from several different points of view. For sets of "dimension one", local modularity corresponds exactly to the cases where the combinatorial geometry associated to the dependence relation is affine, projective or trivial. The *Zilber Trichotomy principle* states that if D is a set of dimension one, there are only three possibilities:

– either the geometry is trivial and there is no group definable in D (the geometry associated to D is then the infinite set with no structure, example (1) in section 3.3);

– or the geometry is affine or projective and every group definable in D is of linear type (see the precise definition in section 2.2). The structure D then behaves very similarly to a vector space (example (2) in section 3.3);

– or there is an algebraically closed field definable in D.

This principle was shown to be false in general by Hrushovski [12]. But it holds with extra assumptions, namely in the context of abstract *Zariski geometries*, defined by Hrushovski and Zilber [20, 21]. This trichotomy, or more precisely this dichotomy in the case of a group of dimension one plays an essential role in the applications to the Manin-Mumford type of conjectures.

The general "abstract" framework in which this material was originally developed, namely stability theory, was eventually seen as part of a broader one, "simplicity theory", which has now become a very active aera in model theory. This is the point of view we will adopt for the presentation of the abstract notions involved.

This ends our introductory sketch of geometric stability theory. In the next section we will discuss the theory of algebraically closed fields, which is the model theoretic context for classical algebraic geometry, and explain how the Manin-Mumford type of conjectures fit within the model theoretic framework.

In the third section, we will give the abstract definition of independence and state the definition and main results about local modularity. These notions will be illustrated by four basic examples, presented at the end of the section (in 3.3). In the fourth and last section, we present two of the theories of fields which are used in Hrushovski's proofs of the algebraic geometry results and finish in 4.3 with a brief sketch of the actual strategy for the proof of Manin-Mumford.

§2. Algebraically closed fields and the Mordell-Lang conjecture.

2.1. The theory of algebraically closed fields. We consider fields K as first-order structures in the usual language of rings: $L_R = \{0, 1, +, -, .\}$.

The theory of algebraically closed fields ACF is axiomatized by axioms which say:

(i) K is a field

(ii) K is algebraically closed, that is, every polynomial in one variable with coefficients in K has a solution in K. This can be axiomatized by the following scheme: for every $n > 1$

$$\forall y_1, \dots \forall y_n \, \exists x \; x^n + y_1 x^{n-1} + \cdots + y_n = 0.$$

Every field L embeds into an algebraically closed field; there is a smallest such algebraically closed field containing L, the algebraic closure of L, which we denote by L^{alg} and which is unique up to isomorphism over L. The theory ACF is not complete but it suffices to specify the characteristic of the field to obtain a complete theory. For $p \geq 0$, we let ACF_p denote the (complete) theory of algebraically closed fields of characteristic p. In fact the theory ACF_p is categorical in every uncountable cardinality, that is, has a unique model up to isomorphism in every uncountable cardinality. Indeed if K, K' are two models of ACF_p, then K and K' are isomorphic if and only if they have the same transcendence degree over the prime field of characteristic p.

From now on, for the sake of simplification, we consider only the theory ACF_0 of algebraically closed fields of characteristic zero.

2.1.1. Definable subsets. Let K be a model of ACF_0 of infinite transcendence degree over $\mathbb{Q}$.

In first-order logic, we study the subsets defined by first-order formulas. We start with the basic or atomic subsets, defined using the basic operations and relations in the language. In this particular context, our basic sets will be:

• The Zariski closed sets: $E \subseteq K^n$ is Zariski closed if E is the zero-set of a finite number of polynomials over K, that is, if $E = \{(a_1, \dots, a_n) \in K^n;\ f_1(a_1, \dots, a_n) = \cdots = f_r(a_1, \dots, a_n) = 0\}$, for $f_1, \dots, f_r \in K[X_1, \dots, X_n]$.

The Zariski closed sets define a Noetherian topology on $\bigcup_n K^n$, the classical Zariski topology.

• The Zariski constructible sets: the finite boolean combinations (closure under finite intersection, finite union and complement) of the Zariski closed sets. They are exactly the sets definable by quantifier-free formulas in the language L_R.

• QUANTIFIER ELIMINATION: the theory ACF_0 has quantifier elimination, which means exactly that the projection of a constructible set is also constructible and hence that THE DEFINABLE SETS ARE EXACTLY THE CONSTRUCTIBLE SETS.

REMARK. The model theoretic notion of algebraic closure (see section 3.2) coincides with the usual field notion of algebraic closure.

The theory ACF_0 does not only eliminate quantifiers but it also ELIMINATES IMAGINARIES: for every definable equivalence relation E on $K^n \times K^n$, there is a definable map f_E from K^n to some K^m such that for all a, b in K^n $f_E(a) = f_E(b)$ if and only if a and b are E-equivalent.

2.1.2. VARIETIES AND ALGEBRAIC GROUPS ARE DEFINABLE. For those who already know their way around algebraic varieties and algebraic groups, the aim of this section is to explain how these objects can be considered as definable objects in the theory of algebraically closed fields. Those unfamiliar with the subject can consider them directly as definable subsets and definable groups with some specific properties and this should be sufficient for them to understand the statement of the Mordell-Lang conjecture in the next section.

For a more complete and elaborate introduction to the model theoretic approach to algebraic varieties see [35]. For basic definitions and results in algebraic geometry, see for example [25] and [26].

AN AFFINE VARIETY over K is a Zariski closed subset of K^n, for some $n \geq 1$, endowed with the induced Zariski topology from K^n. A QUASI-AFFINE VARIETY is a Zariski open subset of an affine variety, also endowed with the induced topology. Quasi-affine sets are special cases of Zariski constructible sets.

Let $V \subseteq K^n$ and $W \subseteq K^m$ be two quasi-affine varieties, a MORPHISM from V to W is a map f from V to W which is locally rational (or regular): for every $a \in V$, there is an open subset U of V containing a and polynomials $P_1, \ldots, P_m, Q_1, \ldots, Q_m$ in $K[X]$ such that on U, the Q_i's are non zero and

$$f(x) = (P_1/Q_1(x), \ldots, P_m/Q_m(x)).$$

By the compactness theorem, f is a definable map from V to W, i.e., the graph of f is definable: there are open subsets $U_1, \ldots, U_k$ of V such that on each U_i f is given by a fixed tuple of rational fractions.

An ISOMORPHISM is a bijective morphism whose inverse is also a morphism.

So far, we can see directly that we are dealing with definable sets and maps. It is a little more difficult in the case of an abstract variety which is obtained by gluing together a finite number of affine varieties.

A VARIETY V over K is a set V covered by a finite number of subsets $V_1, \ldots, V_k$ together with some maps $f_1, \ldots, f_k$, where each f_i is a bijection between V_i and some affine variety U_i, such that:

(i) for each i, j the set $U_{ij} := f_i(V_i \cap V_j)$ is open in U_i
(ii) the map $f_{ij} := f_i \circ f_j^{-1}$ is an isomorphism from U_{ji} into U_{ij}.

The U_i's are called the affine charts of V.

The Zariski topology on V is defined by declaring that $S \subseteq V$ is open if and only if for each i, $f_i(S \cap V_i)$ is open in U_i. A morphism from a variety $V = (V_i, f_i, U_i)$ to a variety $W = (W_j, g_j, Z_j)$ is a map h from V to W which is a morphism when read in the charts, i.e., h is continuous and for any i, j, the map $g_j \circ h \circ f_i^{-1}$ restricted to (the quasi-affine variety) $f_i(h^{-1}(W_j) \cap V_i))$ is a morphism.

There are different possible ways to identify a variety V, given by a fixed system of affine charts, (V_i, f_i, U_i), to a definable set. One way is to consider V to be the disjoint union of its affine charts $U_1, \ldots, U_k$, modded out by the definable equivalence relation which identifies U_{ij} and U_{ji} via the definable maps f_{ij}. By elimination of imaginaries this will indeed be (definably isomorphic to) a definable subset and the morphisms will be definable maps.

Notation: If V is a variety defined over K, we denote by $V(K)$ the set of K-rational points of V, or equivalently, if V is seen as a definable set in K^n, the subset of tuples in K^n which belong to the definable subset V.

We need two more definitions: An ALGEBRAIC GROUP G is a variety G equipped with a group multiplication $.: G \times G \mapsto G$ and an inverse $^{-1}: G \mapsto G$ which are morphisms for the variety structures on G and $G \times G$. So in particular, an algebraic group is a definable group, that is, a group which lives on a definable set and such that the group multiplication is a definable map.

The additive and multiplicative groups of the field K, $(K^n, +)$ and $((K^*)^n, .)$ are affine algebraic groups (algebraic groups which are isomorphic to affine varieties). So are all the linear groups, i.e., all the closed subgroups of $GL_n(K)$.

We will be interested in very different groups, the ones which have no affine subgroup at all. An ABELIAN VARIETY is an algebraic group G which is a complete irreducible variety, where complete means that, for any variety Y, the projection map $\pi: G \times Y \mapsto Y$ is closed (i.e., takes closed sets to closed sets).

The Abelian varieties of dimension one are exactly the elliptic curves, in fact the fundamental examples of Abelian varieties are the Jacobians of curves. Over $\mathbb{C}$ the Abelian varieties are complex tori, that is they are of the form $\mathbb{C}^n/\Lambda$ where Λ is a discrete subgroup of rank $2n$ (but not every complex torus is an Abelian variety).

Abelian varieties are commutative divisible groups. They have a certain number of other rather strong properties of which I will only mention one: in an Abelian variety G, for every $n > 0$, the number of torsion elements of order n is finite but the torsion subgroup of G, $Tor(G)$ is infinite and Zariski dense in G.

So we can consider Abelian varieties over K as a specific class of commutative divisible definable groups with a certain number of additional "nice" properties.

2.2. The Mordell-Lang conjecture. Recall that a commutative group Γ is said to be of *finite rank* if there is a finitely generated subgroup Γ_0 such that for every $\gamma \in \Gamma$, for some integer $n \geq 1$, $n\gamma \in \Gamma_0$. In any commutative group G, the group of torsion elements $Tor(G)$ is of course of finite rank.

We now have all the necessary elements in order to give the statement of the Mordell-Lang conjecture for Abelian varieties over a field of characteristic zero.

THE MORDELL-LANG CONJECTURE. *Let K be an algebraically closed field of characteristic zero, let A be an Abelian variety over K, X a closed irreducible subset of A and Γ a finite rank subgroup of $A(K)$. Then $X \cap \Gamma$ is a finite union of translates of subgroups of Γ, that is, there are $m \geq 1$, $H_1, \ldots, H_m$ subgroups of Γ and elements $b_1, \ldots, b_m$ in Γ, such that*

$$X \cap \Gamma = \bigcup_{i=1}^{m} b_i + H_i.$$

There are two different cases, "the number field case" when K is the algebraic closure of $\mathbb{Q}$, that is when A is in fact defined over a number field (a finite algebraic extension of $\mathbb{Q}$), and "the function field case" when A is not defined over $\mathbb{Q}^{alg}$.

The Mordell conjecture follows from the case when X is a curve defined over a number field k, A is the Jacobian of X and Γ is the (finitely generated) group of k-rational points of A.

The Manin-Mumford conjecture is the particular case when K is the algebraic closure of $\mathbb{Q}$ and the group Γ is the group $Tor(A)$. By taking Zariski closures, one can give the equivalent statement:

THE MANIN-MUMFORD CONJECTURE. *Let K be an algebraically closed field of characteristic zero, let A be an Abelian variety over $\mathbb{Q}^{alg}$ and X a closed irreducible subset of A. Then for some integer $m \geq 0$,*

$$X \cap Tor(A) = \bigcup_{i=1}^{m} b_i + Tor(B_i)$$

where for each i, B_i is an Abelian sub-variety of A (an irreducible closed subgroup of A) and $b_i + B_i$ is contained in X.

The Manin-Mumford conjecture was first proved by Raynaud in 1983, and the full Mordell-Lang conjecture was finally proved by Faltings in 1993. For more history and annotated bibliographies, one can look at [11] or [33]. Hrushovski gave a new proof of the function field case of the Mordell-Lang conjecture in 1994 [14], inspired by a previous proof of Buium's [3]. At the same time he also gave the first full proof of the characteristic $p > 0$ version of the Mordell-Lang conjecture. Then, in 1995, he gave a new proof of Manin-Mumford. One of the interesting aspects of these proofs is that they all fit in a common framework which was developed a priori in model theory, as I hope will become apparent very soon. Another interesting aspect of his Manin-Mumford proof is that it yields rather easily some effective bounds for the number m of translates involved. In fact I believe that at the time, in 1995, this was the first proof giving effective bounds which did not depend on the field of definition of the variety X.

Now let us consider again the statement of the Mordell-Lang conjecture and try to understand its meaning.

The first thing to remark is that it deals with two different kinds of objects: we have on one hand A and X, which are algebraic or from our point of view, definable objects, and the group Γ on the other hand, which is not definable. In algebraic geometry, one has tools to deal with algebraic or geometric objects, like varieties; similarly in model theory we have tools to deal with definable objects. So the first basic idea in the proof is going to be to replace the group Γ by a definable group.

The second remark is that the Mordell-Lang conjecture is usually considered as saying something about curves, or about closed subsets of A, but one can also consider that it is in fact a statement about the group Γ and the topology induced on it by the closed subsets of A. It says that this induced topology is determined by the subgroups and their translates. This is not the case for the topology on A itself: consider for example a curve X of genus strictly bigger than one, and A its Jacobian. It is classical that a curve of genus strictly bigger than one cannot be a group (or the coset of a group). In fact more generally, the topology on an algebraic group is never determined by its closed subgroups (see section 3.2). In model theory we are familiar with this type of questions about the "induced structure" on a subset. If M is a first-order structure and if $E \subset M^n$ is a definable subset, the *induced structure* on E is the new first-order structure consisting of the set E, together with all relatively definable subsets of M: $(E, D \cap E^m; m \geq 1,\ D \text{ definable subset of } M^{nm})$.

In the case of definable groups the following notion is crucial. A definable group is of *linear type* if the induced structure on it is similar to a module, precisely:

Definition. Let M be a first-order structure, and $G \subseteq M^n$ be a definable group. We say that G is of **linear type** if for every integer $m \geq 1$ and every definable $D \subseteq M^{nm}$, $D \cap G^m$ is equal to a finite boolean combination of cosets of definable subgroups of G^m.

The Mordell-Lang conjecture fits into this framework. There is a formal equivalence between the Mordell-Lang conjecture and the following statement:

The model theoretic version of Mordell-Lang. *Let K be an algebraically closed field of characteristic zero, let A be an Abelian variety over K and Γ a finite rank subgroup of $A(K)$. Let $L_K = \{+, ., S, \{c_a : a \in K\}\}$ be the usual* *language for rings with an extra unary predicate S (and also constants for each element of K, for technical reasons). Then in the theory of the L_K-structure $(K, +, ., \Gamma, a)_{a \in K}$, where the new predicate S is interpreted by the group Γ, the definable group Γ is of linear type.*

To see that the above statement implies Mordell-Lang, one only needs to check that if X is a Zariski irreducible closed subset of A and if $X \cap \Gamma$ is a finite boolean combination of translates of subgroups of Γ, then in fact it is a finite union of translates of subgroups. This is fairly straightforward, using the properties of the Zariski topology on groups. For the other direction, note first that Mordell-Lang says something not only about A but also about Cartesian products of A: just consider A^n which is also an Abelian variety hence also satisfies the conclusion of Mordell-Lang. Then there remains only to pass from information about the intersections with Γ^n of all *closed irreducible* subsets of A^n to the intersections with Γ^n of *all* definable subsets of K^n, in the new language.

Model theory has developed abstract criteria in terms of independence which characterize, among the definable groups, those which are of linear type. We will see this in the next section with the definitions of stable and one-based. But the problem is that, with this very brutal way of making the group Γ definable, by just adding a name for it, it is not easier to show that Γ is now of linear type than it was to show the original statement. So the strategy is going to be to add some new structure to the field K, in order to add new definable subsets but in a way we can control, for example in such a way that yields a good dichotomy between groups of linear type and the others. This is what will be achieved, for the group $Tor(A)$, by adding a field automorphism, as we will explain in the last part of this paper. We will not be able to actually make $Tor(A)$ itself definable but will find a new definable subgroup of A, containing $Tor(A)$, and which we will be able to show is of linear type—and this will suffice.

This extension process, in which the original theory of algebraically closed fields is replaced by an enriched theory, is characteristic of the model theoretic approach to such questions. It should be noted that this was also the approach taken by Buium in [3]. As Hrushovski did after him, in the function field case, Buium added a derivation, denoted δ, and confined the group Γ within a δ-closed subgroup of finite rank. He then proceeded to use the tools of differential algebra and jetspaces, in combination with analytic methods, in order to reach the desired result.

In the case of the model theoretic approach, there are two good reasons that make this extension necessary. This approach is based on the powerful abstract tools that were previously developed around the dichotomy linear type/non linear type for definable (or infinitely definable) groups. In the original theory of algebraically closed fields, the smallest definable group containing $Tor(A)$

is the Zariski closure of $Tor(A)$ in A, that is A itself. Even more relevant is the fact, already mentioned above, that no infinite group definable in an algebraically closed field (in the pure language of fields) is of linear type.

2.3. Independence and rank. We have just seen how to fit the Mordell-Lang conjecture into the model theoretic framework of the theory of algebraically closed fields. But algebraically closed fields, together with vector spaces, are also the main examples which motivated many of the definitions essential to stability theory. Before giving the actual abstract definitions of forking, independence and rank, we will consider them in this concrete context.

We keep the same conventions and K is still an algebraically closed field of characteristic zero and of infinite transcendence degree over the rationals.

The abstract notion of independence from model theory coincides with the classical notion of algebraic independence. Recall that if $K_0 \prec K_1 \prec K$ and $K_0 \prec K_2 \prec K$, we say that K_1 and K_2 are algebraically independent over K_0 if any finite set of elements of K_2 algebraically independent over K_0 remains independent over K_1. When K_0 is algebraically closed, this is equivalent to K_1 and K_2 being linearly disjoint over K_0, i.e., such that every finite set of elements of K_2 which is linearly independent over K_0 remains linearly independent over K_1.

Definition. Let A, B, C be subsets of K; we say that A and B are **independent** over C if the two fields $(\mathbb{Q}(AC))^{alg})$ and $(\mathbb{Q}(BC))^{alg})$ are algebraically independent over $(\mathbb{Q}(C))^{alg}$.

There are many different notions of rank that one uses in model theory. In the case of algebraically closed fields, they all coincide with the classical algebraic notion of dimension.

Definition. Let $E \subseteq K^n$ be a definable subset of K. Let $K_0 \prec K$ be an algebraically closed subfield containing the parameters necessary to define E. We define the **rank** or **dimension** of E over K_0, $Dim(E/K_0)$, to be the maximum of the transcendence degrees of the fields $K_0(e)$ over K_0, when e varies in E.

For $E \subseteq K^n$, the dimension of E is at most equal to n, which is the dimension of K^n itself.

Note that for a finite tuple $e \in K^n$, if $K_0 \prec K_1 \prec K$, then e is independent from K_1 over K_0 if and only if the transcendence degree of $K_1(e)$ over K_1 remains equal to the transcendence degree of $K_0(e)$ over K_0.

The next two properties will tell us that the theory of algebraically closed fields is stable and is not one-based:

Properties. 1. Let $K_0 \preceq K_1 \preceq K$, be algebraically closed subfields of K. Suppose that a, b finite tuples in K are such that $(K_0(a))^{alg}$ and $(K_0(b))^{alg}$ are K_0-isomorphic and that K_1 is linearly disjoint from each of $(K_0(a))^{alg}$ and $(K_0(b))^{alg}$ over K_0. It is then classical algebra that $(K_1(a))^{alg}$ and $(K_1(b))^{alg}$

are isomorphic over K_1. This is the uniqueness of "independent extensions" over models.

2. There exist K_1, K_2, algebraically closed subfields of K, which are not independent over their intersection. Take a, b, c three transcendental independent elements in K. We claim that $\mathbb{Q}(a,b)^{alg}$ and $\mathbb{Q}(c, ac+b)^{alg}$ are not algebraically independent over $L := \mathbb{Q}(a,b)^{alg} \cap \mathbb{Q}(c, ac+b)^{alg}$. First we check that $L = \mathbb{Q}^{alg}$. Indeed, suppose there is some $d \in L \setminus \mathbb{Q}^{alg}$; then $ac+b \in \mathbb{Q}(d,c)^{alg}$. Let $P(X,Y)$ be an irreducible polynomial with coefficients in $\mathbb{Q}(d)^{alg}$ such that $P(c, ac+b) = 0$. The polynomial $P(X,Y)$ remains irreducible over $\mathbb{Q}(a,b)^{alg}$, hence up to multiplication by an element of $\mathbb{Q}(a,b)^{alg}$ it must be equal to $(Y - aX - b)$. But this implies that both a and b are in $\mathbb{Q}(d)^{alg}$ which is impossible. It is now clear that $\mathbb{Q}(a,b)^{alg}$ and $\mathbb{Q}(c, ac+b)^{alg}$ are not algebraically independent over $\mathbb{Q}^{alg}$ as $\mathbb{Q}(a,b,c,ac+b)^{alg} = \mathbb{Q}(a,b,c)^{alg}$ has transcendence degree three over $\mathbb{Q}^{alg}$ and each of $\mathbb{Q}(a,b)$ and $\mathbb{Q}(c, ac+b)$ has transcendence degree two.

§3. **Independence, simplicity, stability, modularity.** We are first going to define what we mean when we talk about an *abstract relation of independence*. In model theory, or more precisely in stability or in geometric model theory, we often explain that we are working in structures where one can define a "good" notion of independence and then proceed directly to classical examples which are particular instances of such an abstract independence, without actually giving the precise abstract definition. I will give here a precise axiomatic definition because I find it quite remarkable that there is a fairly "simple" axiomatic way to define what a relation of independence should be. On the other hand one should be aware that this definition is not a good practical tool: in practice when given a structure, if one wants to see if there is a good relation of independence, one will use other definitions such as the original definition of "forking" of Shelah. One should also be aware that I will present here as definitions (of simplicity and of stability in particular) properties which were in fact theorems established a posteriori from the original definitions.

In section 3.3, I present four easy examples of independence relations which illustrate the various definitions and properties given in sections 3.1 and 3.2.

Conventions: We have a complete theory T in a countable first-order language L. In order to avoid heavy notation, we suppose that we are working inside a *monster model* $\mathfrak{M}$ of T: this means that all sets of parameters we consider, usually denoted $A, B, C \ldots$ are subsets of $\mathfrak{M}$, of cardinality strictly smaller than the cardinality of $\mathfrak{M}$, and all models of T, usually denoted $M, N \ldots$ are elementary sub-models of $\mathfrak{M}$, also of cardinality strictly smaller than the cardinality of $\mathfrak{M}$. Definable sets will be usually denoted $D, E, F \ldots$, for example, E is a definable set in $\mathfrak{M}$ with parameters from A, will mean that $E \subseteq \mathfrak{M}^n$ for some n and that E is the set of n-tuples in M satisfying a particular formula (in n free variables) with parameters from the set A. We do not make any difference in notation between elements and finite tuples.

Furthermore we suppose that this monster model $\mathfrak{M}$ is saturated, which has the following consequences:

– any infinite conjunction of formulas of cardinality strictly smaller than $|\mathfrak{M}|$ which is finitely consistent is realized in $\mathfrak{M}$.

– any two n-tuples a and b satisfy exactly the same formulas over some set C if and only if there is an automorphism of $\mathfrak{M}$ which takes a to b and fixes C point-wise. In that case we write that $a \equiv_C b$ and say that a and b *have the same type over* C.

One brutal way to do this is to suppose that the cardinality of $\mathfrak{M}$ is an inaccessible cardinal. But one should not worry about this, everything that is done using using these properties of $\mathfrak{M}$, could be done otherwise, with much more cumbersome notation, by constantly changing the model we are working with to an ad hoc sufficiently big one.

3.1. Abstract independence. An *independence relation* in $\mathfrak{M}$ is a relation (or a collection of triples) $\mathfrak{I}(c, B, A)$ where c ranges over finite tuples of $\mathfrak{M}$ and A, B over subsets of $\mathfrak{M}$, with $A \subseteq B \subset \mathfrak{M}$, which satisfies the following conditions:

1. (invariance) $\mathfrak{I}$ is invariant under automorphisms of $\mathfrak{M}$,
2. (local character) for any c, B there is some countable $A \subseteq B$ such that $\mathfrak{I}(c, B, A)$,
3. (finite character) $\mathfrak{I}(c, B, A)$ if and only if for every finite tuple b from B, $\mathfrak{I}(c, A \cup \{b\}, A)$,
4. (extension) for any c, A and $B \supseteq A$, there is some d such that $c \equiv d$ over A and $\mathfrak{I}(d, B, A)$,
5. (symmetry) for any b, c, A $\mathfrak{I}(c, A \cup \{b\}, A)$ if and only if $\mathfrak{I}(b, A \cup \{c\}, A)$ also,
6. (transitivity) suppose that $A \subseteq B \subseteq C$, then $\mathfrak{I}(e, C, B)$ and $\mathfrak{I}(e, B, A)$ if and only if $\mathfrak{I}(e, C, A)$.

These properties make it legitimate to say, for any B,C and A subsets of $\mathfrak{M}$, that B and C are $\mathfrak{I}$-*independent* over A if for every finite subset c of C, $\mathfrak{I}(c, B \cup A, A)$.

There is a first trivial example, where one puts in $\mathfrak{I}$ all possible triples (c, B, A), $A \subseteq B$. In a (monster) algebraically closed field K, if one sets $\mathfrak{I}$ to be the set of triples (e, K_2, K_1) where $K_1 < K_2$ are algebraically closed subfields of K and e and K_2 are independent over K_1 in the sense of section 2.3, then $\mathfrak{I}$ is an abstract independence relation. We give four more examples in section 3.3. In addition, we will see the two theories of enriched fields presented in section 4, differentially closed fields of characteristic zero and algebraically closed fields with automorphisms.

The independence relations in these different examples do not all behave similarly. For many years, the crucial dividing line was between stable theories and unstable theories. In the past few years, this line has shifted to include a much larger class of theories in which the tools of "geometric stability" apply, the simple theories.

Simple theories were originally introduced by Shelah in 1980, but it was only after work of Hrushovski on specific examples and then of Kim, and Kim and Pillay, that the following property and its consequences was isolated:

THE INDEPENDENCE THEOREM. We say that the independence relation $\mathfrak{I}$ satisfies the **independence theorem** (over models) if,

For any model M, and any a, b, c, d finite tuples such that

- a and b are $\mathfrak{I}$-independent over M,
- c and a (resp. d and b) are $\mathfrak{I}$-independent over M,
- $c \equiv d$ over M,

there is some e such that e and $\{a, b\}$ are $\mathfrak{I}$-independent over M, $e \equiv c$ over $M \cup \{a\}$ and $e \equiv d$ over $M \cup \{b\}$.

The independence theorem says that one can "amalgamate" types in an independent way.

DEFINITION. We say that T is **simple** if there is a notion of independence $\mathfrak{I}$ in T which satisfies the independence theorem over models.

We can already remark (which is rather reassuring) that the first trivial example, that is the relation $\mathfrak{I}$ consisting of all triples, does not satisfy the independence theorem (take $a \neq b$, $a = c$ and $b = d$).

The independence theorem is in fact a very strong condition, as it forces the independence relation to be uniquely determined:

PROPOSITION 3.1. *If T is simple then the relation $\mathfrak{I}$ for which T satisfies the independence theorem is uniquely determined (and is the notion of non-forking as originally defined by S. Shelah).*

DEFINITION. We say that T is **stable** if there is a notion of independence $\mathfrak{I}$ in T which satisfies the following property (**stationarity over models**): for any model M of T, for any a, b finite tuples such that $b \equiv a$ over M, and for any $C \supseteq M$, if a and C (resp. b and C) are $\mathfrak{I}$-independent over M, then $a \equiv b$ over C.

Stability means that, if $M \subseteq C$, there is (up to isomorphism) only one way C and a can be independent over M.

IF T IS STABLE, THEN T IS SIMPLE. Given a, b, c, d and M as in the independence theorem, by the extension property, we know that there is some c' (resp. some d') which looks like c (resp. like d) over $M \cup a$ and is independent from $\{a, b\}$ over M. By stability, as $c \equiv d$ over M , then $c' \equiv d'$ over $M \cup \{a, b\}$, so we also have $c' \equiv d$ over b.

One of the main consequences of stability, which is used in an essential way for example in the group configurations type of constructions, is that certain subsets turn out to be definable: given a model M, a formula $\phi(x, y)$ and some tuple b in $\mathfrak{M}$ (the monster model), the set of tuples a in M such that $\phi(a, b)$ holds is a definable subset of M, definable with parameters from M.

Examples (1) and (2) from section 3.3 are stable, (3) is simple but not stable and (4) is not simple. Algebraically closed fields (ACF_p) are stable, as shown by Property 1 in 2.3. Differentially closed fields of characteristic zero (DCF_0, section 4.1) are stable, algebraically closed fields with an automorphism (ACFA, section 4.2) are simple but not stable.

Finally, we will need an essential notion which was originally introduced by Shelah in the context of stable theories, namely orthogonality:

Definition. Let T be a simple theory, $M \preceq \mathfrak{M}$, and E and F two definable subsets in $\mathfrak{M}$. We say that E and F are **orthogonal** over M if for every finite sequence of elements e from E, and for every finite sequence of elements f from F, e and f are independent over M.

3.2. Modularity. First we are going to need a local version of stability; there may be stable definable subsets inside a model whose theory is not stable, as we will see in the next section when looking at algebraically closed fields with an automorphism.

From now on we suppose that T is a simple theory, hence that there is a (unique) notion of independence which satisfies the independence theorem.

We also suppose that T has elimination of imaginaries (this is relevant for the definition we give here of modularity). Recall that T has **elimination of imaginaries** if for every definable equivalence relation E on $\mathfrak{M}^n \times \mathfrak{M}^n$, there is a definable map f_E from $\mathfrak{M}^n$ to some $\mathfrak{M}^k$ such that, for all a, b in $\mathfrak{M}^n$, $f_E(a) = f_E(b)$ if and only if a and b are E-equivalent. We mentioned in the previous section that algebraically closed fields had elimination of imaginaries.

Definition. Let $F \subseteq \mathfrak{M}^n$ be a definable subset with parameters from A. We say that F is **stable** if, for all model $M \preceq \mathfrak{M}$, $A \subseteq M$, for all a, b tuples from F and all $C \supseteq M$, if $a \equiv b$ over M, a and C are independent over M and b and C are independent over M, then $a \equiv b$ over $M \cup \{C\}$.

Keeping in mind that we wish to study the induced structure on some definable subsets, we are also going to need:

Definition. Let $F \subseteq \mathfrak{M}^n$ be a definable subset with parameters from A. We say that F is **stably embedded** in $\mathfrak{M}$ if for every k and every definable subset $D \subseteq \mathfrak{M}^{nk}$, there is some definable $D' \subseteq \mathfrak{M}^{nk}$, definable *with parameters from F*, such that $D \cap F^k = D' \cap F^k$. In a stable theory, any definable set is both stable and stably embedded. In an unstable theory, a set can be stably embedded without being stable (it will be the case for example of the fixed field in a model of ACFA_0, see section 4.2.1) or stable without being stably embedded.

THE MODEL THEORETIC ALGEBRAIC CLOSURE. Recall that we say that a is **algebraic** over A ($a \in acl(A)$) if there is a finite set F, definable with parameters from A, such that $a \in F$; equivalently if a has a finite number of conjugates by the automorphisms of $\mathfrak{M}$ which fix A point-wise.

DEFINITION. Let F be a definable subset of $\mathfrak{M}^n$. We say that F is **locally modular** or **one-based** if for all C, all a, b finite tuples of elements from F, a and b are independent over $acl(C \cup \{a\}) \cap acl(C \cup \{b\})$. We say that the theory T is one-based if the formula "$x = x$" (i.e., $F = \mathfrak{M}$) is one-based.

The notion of modularity, in presence of stability, gives information of an algebraic type about the structure. We will not use this result here but in particular, any non trivial relation between three elements has to come from the action of an Abelian group. If we have a stable theory T and a definable group $(G, .) \subseteq \mathfrak{M}^n$, then there are a, b independent elements of G such that a, b and $a.b$ are pairwise independent but not independent ($a.b$ is not independent from $\{a, b\}$). I am not going to prove this here but it is easy to check that this is true for example in algebraically closed fields for both addition and multiplication (take a, b two algebraically independent transcendental elements over $\mathbb{Q}$). So the existence of three such elements is necessary for the existence of a stable definable group. Local modularity implies that it is also a sufficient condition.

PROPOSITION. [2] *Suppose that T is stable and one-based and that there are a, b, c finite tuples in $\mathfrak{M}$ which are pairwise independent but not independent, i.e., a and b, c are not independent. Then there is an infinite Abelian group definable in $\mathfrak{M}$.*

In fact one can draw much stronger conclusions from the existence of such a, b, c; the above is just a very weak version of the existing results. We will not be using this "group construction" here anyways but in contrast the following proposition is fundamental for what we are going to do. It is interesting to note that it was proved in 1985, hence long before the relation with Diophantine questions of the Manin-Mumford or Mordell-Lang type was realized.

PROPOSITION. [17] *Let G be a definable group in $\mathfrak{M}^n$ which is stable, stably embedded and one-based. Then for any m and for any definable set X in $\mathfrak{M}^{nm}$, $X \cap G^m$ is a finite boolean combination of cosets of definable subgroups of G^m.*

It follows that G has a definable Abelian subgroup of finite index. In any theory of modules, by the quantifier elimination to positive primitive formulas, it is true that any definable subset is a boolean combination of cosets of (positive primitive) definable subgroups. What the above says is that if G is one-based, then the structure induced by $\mathfrak{M}$ on G reduces to that of a "generalized module", that is a module with predicates for some subgroups.

Property 2, in section 2.3, shows that algebraically closed fields are not one-based. The same argument will be used later in section 4 to show that the two theories of enriched fields we consider there are not one-based either. In fact, more generally, one-basedness rules out the existence of a definable field. But, as we will see, some of the definable subsets inside an enriched field can be one-based and this is at the heart of the applications to algebraic geometry. As we have mentioned earlier, in the theory of (non enriched) algebraically closed fields, this cannot happen, and no definable set can be one-based. This comes from the fact that this theory is "unidimensional", that is, any two definable subsets are not orthogonal.

The three stable examples from section 3.3, are one-based. In order to check this more easily, we will now introduce the notion of strongly minimal sets. This notion and its link to combinatorial geometries was essential to the development of geometrical stability theory.

STRONGLY MINIMAL SETS. As we have mentioned above, the use of imaginary elements in the definition of local modularity is crucial. There is a context though in which one can avoid using imaginaries in the definition (or avoid assuming that the theory eliminates imaginaries) namely that of strongly minimal sets.

We say that a definable set $D \subseteq \mathfrak{M}^n$ is **strongly minimal** if for any other definable $F \subseteq \mathfrak{M}^n$, $F \cap D$ is finite or $D \setminus (F \cap D)$ is finite. We say that the theory T is strongly minimal if the formula "$x = x$" is strongly minimal. The theory ACF_0 of algebraically closed fields of characteristic 0 is strongly minimal: a Zariski closed subset of K is the zero set of a finite number of polynomial equations in one variable, and, by quantifier elimination, any definable subset K is a boolean combination of Zariski closed sets. Our first three examples below in 3.3 are also strongly minimal.

In a strongly minimal theory, (model-theoretic) algebraic closure gives rise to the unique independence relation satisfying the independence theorem, which is also stable: e and C are independent over B if e does not belong to $acl(C) \setminus acl(B)$. Moreover, considered as a closure operator, algebraic closure in a strongly minimal set satisfies the exchange principle and gives rise to a pregeometry in the classical sense (see for example [30]). Then one-basedness, or local modularity, corresponds to the local modularity of the associated pregeometry in the usual combinatorial use of the word and can be expressed in the following way:

Let T be a strongly minimal theory (with or without elimination of imaginaries). Then T is locally modular, or one-based, if and only if for all a, b finite tuples of elements from $\mathfrak{M}$ such that $acl(a) \cap acl(b) \neq acl(\emptyset)$, a and b are independent over $acl(a) \cap acl(b)$.

3.3. Some basic examples. We present here four basic examples. In these four examples, as well as in algebraically closed fields, the relation of independence is given by the relation of (model theoretic) algebraic closure. This means that we define A to be independent from B over C if and only if for no

$a \in A$, $a \in [acl((A \setminus \{a\}) \cup B \cup C)] \setminus [acl((A \setminus \{a\}) \cup C)]$. There are two important remarks to be made about this: first, this is a special situation, there are many examples where independence is not given directly by the algebraic closure, in particular the two examples of fields we will see in the next section. Secondly, it is not always the case that (model theoretic) algebraic closure gives rise to an independence relation in our sense. In particular the symmetry axiom is not always true (it corresponds to the fact that model-theoretic algebraic closure, considered as a closure operator, satisfies the exchange property, which is not always the case).

(1) Equality. Let L be the language consisting only of equality, and consider the theory in L which says that there are infinitely many distinct elements. This is a totally categorical theory, that is, it has exactly one model (up to isomorphism) in every (infinite) cardinality. It is clearly strongly minimal. Let E be an infinite set, hence a model. For $A \subseteq B \subset E$, and for $\bar{e} \in E^n$, say that $\bar{e} = (e_1, \ldots, e_n)$ is independent from B over A if for every i, $1 \leq n$, $e_i \in B$ iff $e_i \in A$. This is an abstract relation of independence which is stable and one-based (use the characterization of one-basedness in the case of strongly minimal sets at the end of the preceding section as this theory does not strictly speaking have elimination of imaginaries: one cannot eliminate for example the equivalence relation on n-tuples which define the same n element set).

Note that any set of pairwise independent elements is independent, hence (as one might expect) there is no definable group in any model.

(2) Vector spaces. Take a countable division ring S (finite or infinite) and V an infinite dimensional vector space over S. Consider V as an L_S-structure, where L_S is the language with addition, zero, and a unary function f_s for each element s of S, interpreted as scalar multiplication by s in V. The theory of infinite S-vector spaces, which we denote by T_S, is complete and admits quantifier elimination. If S is finite, T_S has one model up to isomorphism in every infinite cardinality; if S is infinite, T_S has countably many countable models and one model in each uncountable cardinality. This theory is strongly minimal. For $C \subseteq B \subset V$, and for $A \subseteq V$, say that A is independent from B over C if A and B are linearly independent over C: for every $a \in A$, a is in the subspace spanned by $B \cup (A \setminus \{a\})$ iff a is already in the subspace spanned by $C \cup (A \setminus \{a\})$. Then again this is a stable one-based theory. The fact that it is one-based corresponds exactly to the fact that vector spaces satisfy the classical dimension equality: for any finitely generated subspaces X, Y of V,

$$\dim(X) + \dim(Y) = \dim(X \cup Y) - \dim(X \cap Y).$$

There is a group of course in V and if v and w are independent, then the set $\{v, w, v + w\}$ is an example of a set which is pairwise independent but not independent.

(3) The random graph. Take the language $L = \{R\}$ with one binary relation R and consider the theory of the random graph E_R which is axiomatized by the following infinite scheme of axioms:

– R is symmetric irreflexive

– for every distinct $a_1, \ldots, a_n$ and $b_1, \ldots, b_m$, there exists x such that for all $i, 1 \leq i \leq n$, $R(x, a_i)$ and for all $j, 1 \leq j \leq m$, (not $R(x, b_j)$).

The theory of E_R admits quantifier elimination, has only one countable model (but has 2^κ non isomorphic models of power κ for every uncountable cardinal κ). Define independence as in example (1) above, i.e., for $A \subseteq B \subset E$, and for $\bar{e} \in E^n$, say that $\bar{e} = (e_1, \ldots, e_n)$ is independent from B over A if for every i, $1 \leq n$, $e_i \in B$ iff $e_i \in A$.

With this notion of independence, this theory is simple, as is easily checked. It follows that this is the unique possible way to obtain a relation of independence satisfying the independence theorem. But the theory is not stable; consider two models $M \prec N$ and two elements a and b such that a is not in relation via R to any element of N and b is related to exactly one element which is in $N \setminus M$. Then $a \equiv_M b$, a and b are each independent from N over M, but it is not the case that $a \equiv_N b$.

(4) Real closed fields. Consider the theory of the reals $\mathbb{R}$ in the language $L_{ord} = \{0, 1, +, -, ., <\}$ of ordered rings. The theory of $\mathbb{R}$, the theory of real closed fields, admits quantifier elimination and is o-minimal (i.e., every definable subset of $\mathbb{R}$ is a finite union of singletons and open intervals, allowing endpoints from $\mathbb{R} \cup \{\infty, -\infty\}$)). Take the relation of independence given by real closure (= algebraic closure in the model theoretic sense):

For $A \subseteq B \subset E$, and for $\bar{e} \in E^n$, say that $\bar{e} = (e_1, \ldots, e_n)$ is independent from B over A if for every i, $1 \leq n$, e_i is in the real closure of the field generated by $B \cup \{e_1, \ldots, e_{i-1}\}$ if and only if e_i is already in the real closure of the field generated by $A \cup \{e_1, \ldots, e_{i-1}\}$. This defines an independence relation which does not satisfy the independence theorem: in a big non standard model take a, b, c, d, such that $\mathbb{R} \ll c \ll a \ll b \ll d$, (where $x \ll y$ means that y is infinitely bigger than x), everything being independent over $\mathbb{R}$. No e can satisfy both $e \equiv c$ over $\mathbb{R} \cup \{a\}$ and $e \equiv d$ over $\mathbb{R} \cup \{b\}$.

The same kind of argument shows more generally that in the presence of a definable total ordering no independence relation can be simple.

3.4. Some references. Simple theories were first introduced by Shelah in 1980 in [42] as a class strictly containing stable theories. It was not known at the time if in simple theories, as defined there, forking was a symmetric relation. The interest for this class of theories was revived in the past few years for two reasons. First, it was realized by Hrushovski that many very interesting classes of algebraic structures were simple and that in these structures forking seemed to have very good properties (the independence theorem, symmetry etc). This was in particular the case of smoothly approximated structures ([16], for surveys see for example [6], [28]), pseudo-finite fields (see [18]) and of course a little later of algebraically closed fields with an automorphism which we describe in the next section. At around the same time, Kim proved that in simple theories forking was symmetric [22]. This changed the perspective on simple theories and also on what having a good relation of independence should mean. The definitions of independence, simplicity etc. which I gave in the preceding sections come from further work on the subject by Kim and Pillay [23]. For a survey on simple theories with the main results and open questions, there is [24]. A book by F. Wagner has recently appeared on this subject [44].

Concerning geometric stability, the main reference is A. Pillay's book "Geometric Stability Theory" [34]. More specifically on stable groups, see the books by B. Poizat (the original [37] or the recent english version [38]) and by F. Wagner [43].

§4. Fields with extra structure and the applications. All the present applications of model theory to classical Diophantine geometry questions fit into a common general framework. Each time, one uses a field with more definable sets than just the classical constructible ones and where a good dichotomy theorem is available which enables one to recognize when a group is one-based. Three theories have been used so far: (1) separably closed fields of characteristic $p > 0$ for the function field Mordell-Lang conjecture in characteristic p [14]; (2) differentially closed fields of characteristic zero for the function field Mordell-Lang conjecture in characteristic 0 [14]; (3) algebraically closed fields with an automorphism, in characteristic zero for the Manin-Mumford conjecture [15] and the Tate-Voloch conjecture [39], as well as in characteristic p for the case of Drinfeld modules [40].

We will present the two theories of fields used in the characteristic zero cases, differentially closed fields and algebraically closed fields with an automorphism, and then finish with a short sketch showing how to apply the model theoretic results in the case of a field with an automorphism in order to obtain the Manin-Mumford conjecture. At the end (section 4.4) we give a selection of references for surveys or introductory papers to all of these applications.

Both the theories we are going to discuss are expansions of algebraically closed fields by a unary function.

4.1. Differentially closed fields of characteristic zero. (see [32] or [1]).

The language is the usual language of rings L_R, which we already used for algebraically closed fields, together with a map δ. The theory (DCF_0) consists of the following scheme of axioms (i) to (iii):

(i) K is a field of characteristic zero

(ii) (K, δ) is a differential field, that is, δ is a derivation: $\delta\colon K \mapsto K$, such that, for all x, y in K, $\delta(x + y) = \delta(x) + \delta(y)$ and $\delta(xy) = x\delta(y) + y\delta(x)$.

Before stating the third set of axioms, we need some definitions. Given a differential field (K,δ), we define the ring $K_\delta[X]$ of differential polynomials (in one variable) over K to be the ring of polynomials in infinitely many variables

$$K[X,\delta(X),\delta^2(X),\dots,\delta^n(X),\dots].$$

The *order* of the differential polynomial $f(X)$ in $K_\delta[X]$ is -1 if $f \in K$ and otherwise the largest n such that $\delta^n(X)$ occurs in $f(X)$ with non zero coefficient. For example the differential polynomial equation $\delta(X) = 0$ which defines the *constants* for the derivations δ has order 1.

(iii) K is existentially closed. In this context, this can be axiomatized by saying (an infinite scheme): for any non-constant differential polynomials $f(X)$ and $g(X)$, where the order of $g(X)$ is strictly less than the order of $f(X)$, there is a z such that $f(z) = 0$ and $g(z) \neq 0$.

Basic Results. DCF_0 is a complete theory which admits quantifier elimination and elimination of imaginaries. We call the models of DCF_0 the differentially closed fields. It is the model completion of the theory of differential fields of characteristic zero, so, in particular, any differential field (K,δ) embeds into a differentially closed field (L,δ). Differentially closed fields are algebraically closed fields and one can show that they have infinite transcendence degree over $\mathbb{Q}$.

From now on (K,δ) is a monster model of DCF_0.

4.1.1. Definable sets in DCF_0. We saw earlier that in a "pure" algebraically closed field, the basic definable sets are the zero sets of polynomials. Here we start with the zero sets of differential polynomials. For any n let

$$K_\delta[X_1,\dots,X_n] = K[X_1,\dots,X_n,\delta(X_1),\dots,\delta(X_n),\delta^2(X_1),\dots,\delta^2(X_n),\dots].$$

We say that $F \subseteq K^n$ is a **δ-closed set** if there are $f_1,\dots,f_r \in K_\delta[X_1,\dots,X_n]$ such that $F = \{(a_1,\dots,a_n)\in K^n; f_1(a_1,\dots,a_n) = \cdots = f_r(a_1,\dots,a_n) = 0\}$. The ring $K_\delta[X_1,\dots,X_n]$ is of course not Noetherian but the δ-closed sets (which correspond to radical differential ideals) form the closed sets of a Noetherian topology on K, the δ-topology.

We now consider the δ-constructible sets, that is, the finite boolean combinations of δ-closed sets. The elimination of quantifiers for DCF_0 means that this class is closed under projection hence that all definable sets (we call them δ-definable sets) are δ-constructible.

Examples: First, if $D \subset K^m$ is a set definable in the language L_R, without using δ, as K is algebraically closed, D is constructible. This is a particular case of a δ-constructible set. Exactly as in the case of algebraically closed fields, if V is a variety defined over K, we can consider V as a δ-definable set.

The **field of constants of** K, $Cons(K) = \{a \in K;\delta(x) = 0\}$ is a δ-closed set which is not constructible; it is an algebraically closed subfield of K.

The induced structure on $Cons(K)$ is that of a **pure algebraically closed field**: if D is a δ-definable subset of K^n, $D \cap Cons(K)^n$ is a constructible subset (in the language of rings L_R) of $Cons(K)^n$, definable with parameters from $Cons(K)$.

We define the **δ-algebraic closure** of A, $acl_\delta(A)$, to be equal to the algebraic closure (in the usual sense of fields) of the differential field generated by A, i.e., the algebraic closure of the field $(A)_\delta := \mathbb{Q}(\delta^i(a); a \in A, i \geq 0)$ (this is exactly the algebraic closure of A in the usual model theoretic sense).

4.1.2. INDEPENDENCE AND RANK. If $C \subset A, B \subset K$, we say that A and B are **δ-independent** over C if $acl_\delta(A)$ and $acl_\delta(B)$ are algebraically independent (or equivalently linearly disjoint) over $acl_\delta(C)$. This δ-independence is a notion of independence in the sense of section 3.1 and DCF_0 is stable. One can check the stability easily thanks to the quantifier elimination: let $K_0 < K$ be a submodel and let a and b be such that $a \equiv b$ over K_0. So in particular, the ideal $I(a/K_0)$ of the differential polynomials f in $K_{0\delta}[X]$ vanishing on a is equal to the corresponding ideal for b, $I(b/K_0)$. By definition of δ-independence, if $K_0 < K_1 < K$ and if a (resp. b) and K_1 are δ-independent over K_0, then the ideal $I(a/K_1)$ is generated by $I(a/K_0)$, and similarly for b, $I(b/K_1)$ is the ideal generated by $I(b/K_0)$. It follows that $I(a/K_1) = I(b/K_1)$, and by quantifier elimination this implies that $a \equiv b$ over K_1.

In fact the theory DCF_0 is more than stable, it is what is called ω-stable, which means that it is possible to assign a rank (taking possibly infinite ordinal value) to each definable set. We are only going to consider definable sets with finite rank and give the definition of one rank, which will be sufficient for our purpose. The reader should be aware though that there are many different notions of rank available in model theory and that it is now known that no two of them coincide everywhere in DCF_0 (the Lascar rank, the Morley rank, the δ-degree we are going to define below …).

If E is a differential subfield of K and if a is a finite sequence of elements of K, we define the **δ-degree** of a over E, $d_\delta(a/K)$, to be the transcendence degree of the field $(E(a))_\delta$, the differential field generated by E and a, over E. If $D \subseteq K^n$ is a δ-definable set, we define the δ-degree of D to be the maximum of the δ-degrees of the elements of D.

The field $Cons(K)$ has δ-degree equal to one: for any differential subfield E, for any a element of $Cons(K)$, the differential field generated by E and a is equal to the field $E(a)$. Moreover, and this is fundamental for the application to Diophantine geometry, up to definable isomorphism, *$Cons(K)$ is the unique δ-definable field with finite δ-degree.*

In contrast, if V is any variety (of positive dimension) defined over K, as a δ-definable set, V has infinite δ-degree; this is in particular the case of K itself. When it is finite the δ-degree is a good notion of rank, in particular, if $d_\delta(a/E)$ is finite, then a and $B \supset E$ are δ-independent over E if and only if $d_\delta(a/E) = d_\delta(a/B)$.

4.1.3. MODULARITY AND THE DICHOTOMY THEOREM. The results below come from [19] and [14].

The field (K,δ) is not one-based, but neither is the definable subfield $Cons(K)$, by exactly the same argument as for the theory $ACFA_0$: consider a, b, c in the field $Cons(K)$ which are transcendental over $\mathbb{Q}$ and algebraically independent. In order to be able to do this, we have to suppose that $Cons(K)$ has big enough transcendence degree over $\mathbb{Q}$, but we can always suppose that by going to some big model K' extending K. Then $acl_\delta(a,b) = \mathbb{Q}(a,b)^{alg}$ (the field algebraic closure) and $acl_\delta(c, ac+b) = \mathbb{Q}(c, ac+b)^{alg}$ intersect in $\mathbb{Q}^{alg}$, but they are not algebraically independent over $\mathbb{Q}^{alg}$.

For our purpose, the interesting feature of differentially closed fields of characteristic zero, is that really, the constant field is the "unique" definable set of δ-degree one which is not one-based. Let us make this statement more precise. Let D be a definable set, we have defined in 3.1 the notion of orthogonality. In this particular context, D and $Cons(K)$ are orthogonal if, for every finite sequence of elements d from D, for every finite sequence of elements b from $Cons(K)$, and for every subfield $E = acl_\delta(E)$, $acl_\delta(Ea)$ and $acl_\delta(Eb)$ are algebraically independent over E.

Recall that a δ-definable set $D \subseteq K^n$ is **strongly minimal** if, for any δ-definable $F \subseteq K^n$, $F \cap D$ is finite or cofinite in D. A strongly minimal set has finite δ-degree. The constant field is strongly minimal.

THE DICHOTOMY THEOREM FOR DCF_0. *Let $D \subseteq K^n$ be a strongly minimal δ-definable subset. Then D is one-based if and only if D and the field of constants, $Cons(K)$, are orthogonal.*

Non-orthogonality between two strongly minimal sets is a very strong relation. In particular, if D is a δ-definable group which is non-orthogonal to the field $Cons(K)$, then D will be δ-definably isomorphic to $G(Cons(K))$, where G is an algebraic group defined over the field $Cons(K)$. The dichotomy theorem then means that the only strongly minimal groups which are not one-based are exactly the ones arising from algebraic groups over the constants.

Hrushovski's proof of the dichotomy theorem in [14] uses the fact that strongly minimal sets in DCF_0 are abstract Zariski geometries in the sense of Hrushovski-Zilber [21]. One can then apply their abstract dichotomy theorem which says that if a strongly minimal set D is a non locally modular Zariski geometry, there is a strongly minimal field definable in D. Then one uses the fact that the field $Cons(K)$ is, up to definable isomorphism, the unique strongly minimal field δ-definable in K. For introductory surveys to Zariski geometries, see [20] or [31]. A direct proof of the dichotomy theorem for DCF_0 was given very recently (two years after this tutorial actually took place) in [36].

4.2. Algebraically closed fields with an automorphism. An exposition of the basic properties (axiomatizability, decidability etc.) of ACFA, can be found in Macintyre's introductory paper [27]. The in-depth model theoretic analysis was carried out first by Chatzidakis and Hrushovski in [4], and continued in [5].

The way we are going to present this theory will make it seem very similar to the previous one, differentially closed fields. But although the results are very similar, the actual proofs need not be. One should note though that again in [36], a new proof of the dichotomy theorem for ACFA in characteristic zero is given, along similar lines as the one for the differential case.

A **difference field** is a field K together with an automorphism σ, which we consider as an $L_R \cup \{\sigma\}$-structure.

The class of existentially closed models for difference fields turns out to be axiomatizable (this fact needs a proof of course). Here we restrict ourselves to the case of characteristic zero.

The axioms (ACFA_0) say that:

(i) K is a an algebraically closed field of characteristic zero
(ii) (K, σ) is a difference field, i.e., σ is an automorphism of K.
(iii) K is existentially closed: every difference equation which has a solution in some extension of K has a solution in K.

ACFA_0 is not a complete theory and in order to make it complete one needs to describe the action of the automorphism σ on the algebraic closure of $\mathbb{Q}$. This theory does not have elimination of quantifiers, but it does have elimination of imaginaries. Every difference field of characteristic zero embeds into a model of ACFA_0.

Let us mention a striking recent result about ACFA [13] answering the long open question: what is the theory of a nonstandard Frobenius automorphism or more precisely, what is the theory of an ultraproduct of the difference fields $(\mathbb{F}_p^{alg},\ \sigma : x \mapsto x^p)$ for all p prime numbers ? The answer is that ACFA is exactly the theory of all nonprincipal ultraproducts of $(\mathbb{F}_q^{alg},\ \sigma_q : x \mapsto x^q)$, when q varies on the set of powers of prime numbers.

From now on (K, σ) is a monster model of ACFA_0.

4.2.1. DEFINABLE SETS IN ACFA_0. Here the basic sets are the zero sets of difference polynomials: for any n let

$$K_\sigma[X_1, \dots, X_n] = K[X_1, \dots, X_n, \sigma(X_1), \dots, \sigma(X_n), \sigma^2(X_1), \dots, \sigma^2(X_n), \dots].$$

We say that $F \subseteq K^n$ is a **σ-closed set** if there are $f_1, \dots, f_r \in K_\sigma[X_1, \dots, X_n]$ such that $F = \{(a_1, \dots, a_n) \in K^n; f_1(a_1, \dots, a_n) = \cdots = f_r(a_1, \dots, a_n) = 0\}$. The σ-closed sets form the closed sets of a Noetherian topology on K, the σ-topology. Consider now the σ-constructible sets. It is not true that every σ-definable set is σ-constructible (the theory does not eliminate quantifiers). Here is one example of a σ-definable set which is not σ-constructible: pick a in some extension of K, and extend σ to the field $K(a)$ by setting $\sigma(a) = a$. In order to extend σ to the algebraic closure of $K(a)$, there are choices to be made, in particular one can either choose to have σ fix point-wise the two square roots of a , or to have σ exchange them. This means that the set $\{x; \sigma(x) = x \wedge \exists t\ (t^2 = x \wedge \sigma(t) \neq t)\}$ is not σ-constructible.

The class of σ-definable sets is the closure under finite boolean operations and projections of the σ-closed sets.

The field $Fix(K) = \{a \in K; \sigma(a) = a\}$, the fixed field of σ in K, is σ-closed. It is not algebraically closed but it is pseudo-finite, i.e., it is an infinite model of the theory of all finite fields. It is also a "pure" field: if D is any σ-definable subset of K^n, $D \cap Fix(K)^n$ is a definable subset (in the language L_R) of $Fix(K)^n$ definable with parameters from $Fix(K)$.

We define the **σ-algebraic closure** of A, $acl_\sigma(A)$, to be equal to the algebraic closure (in the usual sense of fields) of the difference field generated by A, i.e., the algebraic closure of the field $(A)_\sigma := \mathbb{Q}(\sigma^i(a); a \in A, i \in \mathbb{Z})$.

4.2.2. Independence, stability and modularity. If $C \subset A \subset K$ and $C \subset B \subset K$, we say that A and B are **σ-independent** over C if $acl_\sigma(A)$ and $acl_\sigma(B)$ are algebraically independent (or equivalently linearly disjoint) over $acl_\sigma(C)$. We define the **σ-degree** of a definable set exactly like the δ-degree; if $D \subseteq K^n$ is a σ-definable set, the σ-degree of D is the maximum of the transcendence degrees of the difference fields generated by elements of D. The fixed field of σ, $Fix(K)$ has σ-degree one.

This gives a notion of independence which satisfies the independence theorem over models, which we will not prove here. Hence the theory is simple. But it is not stable, because the field $Fix(K)$ is not stable: one can find $E = acl_\sigma(E) \subset K$ and $a, b, c \in Fix(K) \setminus E$, such that a and c on the one hand, b and c on the other hand, are σ-independent over E, but such that $\sqrt{a-c} \in Fix(K)$ and $\sqrt{b-c} \notin Fix(K)$ (note that this is the same example which shows that quantifier elimination does not hold). This contradicts the uniqueness of independent extensions.

Exactly as in the case of the field of constants in DCF_0, the field $Fix(K)$ is not one-based and there is also a very powerful dichotomy theorem.

The dichotomy theorem for $ACFA_0$. *Let $D \subseteq K^n$ be a σ-definable subset of finite σ-degree. Then D is stable, stably embedded and one-based if and only if D and the fixed field, $Fix(K)$, are orthogonal.*

4.3. Application to the Manin-Mumford conjecture. Recall the statement of the conjecture from section 2.2. Let A be an Abelian variety defined over $\mathbb{Q}^{alg}$ and let X be a sub-variety of A; then $Tor(A) \cap X$ is a finite union of translates of subgroups of $Tor(A)$.

We have explained already that this is the same as showing that $Tor(A)$ is of linear type (section 2.2), and hence, by section 3.2 "stable, stably embedded and one-based", except that $Tor(A)$ is not definable in the algebraically closed field K. Indeed, as we remarked earlier, there are *no* definable one-based subsets in a "pure" algebraically closed field , so to make this approach work one must put additional structure on the field.

So the strategy is going to be: go to some bigger algebraically closed field L and add new structure on L, hence getting new definable sets, in such a way that there is some new definable subgroup of A, denoted H, which contains $Tor(A)$, and which we can prove is stable, stably embedded and one-based.

It is not immediately obvious that this is enough: this would say that $Tor(A) \cap X$ is contained in $H \cap X$, which itself is a boolean combination of translates of subgroups of H (definable in the bigger field with the extra structure). But it is then fairly straightforward to check, using the fact that X is Zariski closed, that this does imply that $X \cap Tor(A)$ is a finite union of translates of subgroups of $Tor(A)$.

Let $k < \mathbb{Q}^{alg}$ be a finite extension of $\mathbb{Q}$ such that A is defined over k.

We want to find an algebraically closed field L and an automorphism σ of L such that (L, σ) is a model of ACFA_0 and such that there is some σ-definable subgroup of $A(L)$ (the group of L-rational points of the Abelian variety A) containing $Tor(A)$ and which is stable, stably embedded and one-based.

What kind of group H are we looking for in (L, σ)? How can we be sure that this H will indeed be stable, stably embedded and one-based, i.e., by the dichotomy theorem, will be orthogonal to $Fix(\sigma)$? Let us consider groups defined by rather simple difference equations. First $H_1 = \{a \in A(L);\ \sigma(a) - a = 0\}$. This is $A(Fix(\sigma))$, so of course H_1 is not orthogonal to $Fix(\sigma)$ and hence is not stable one-based. Similarly if $H_n = \{a \in A(L);\ \sigma^n(a) - a = 0\}$, this is $A(Fix(\sigma^n))$. The field $Fix(\sigma^n)$ is a finite extension of $Fix(\sigma)$ and it follows that there is a σ-definable map (with finite fibers) from $(Fix(\sigma))^r$ (for some $r > 0$) onto H_n which is hence also not orthogonal to $Fix(\sigma)$.

Now these groups are particular cases of groups defined by polynomial equations. Let $P(T) = m_n T^n + \cdots + m_1 T + m_0$, where the m_i's are in $\mathbb{Z}$. Then define

$$H_P = \{a \in A(L); m_n \sigma^n(a) + \cdots + m_1 \sigma(a) + m_0 a = 0\}$$

where $+$ denotes addition in A, and for $a \in A(L)$ and $m \in \mathbb{N}$, ma denotes as usual $a + \cdots + a$, *m times*.

Then H_P is a σ-definable subgroup of $A(L)$ of finite σ-degree. If, for some $n \geq 1$, the polynomial $P[T]$ is not prime to $X^n - 1$, i.e., if $P[T]$ has a root which is also a root of unity, then H_P is contained in $Ker(\sigma^n - 1)$ and the argument given just above implies that H_P is not stable one-based. The remarkable result at the heart of Hrushovski's proof of the Manin-Mumford conjecture for number fields is that the converse is true:

PROPOSITION 4.1. *The group H_P is orthogonal to the field $Fix(\sigma)$ if and only if $P[T]$ has no root which is also a root of unity.*

The proof of this result goes through an analysis of the ring of σ-definable endomorphisms of $A(L)$ when A is a simple Abelian variety and then various reductions to minimal cases, using in particular the following fact: if $0 \mapsto A_1 \mapsto A_2 \mapsto A_3 \mapsto 0$ is an exact sequence of σ-definable homomorphisms, where the A_i's are σ-definable groups, then A_2 is one-based if and only if both A_1 and A_3 are one-based.

So from the dichotomy theorem for ACFA_0 one now knows that if $P[T]$ has no root which is also a root of unity, then H_P is stable, stably embedded and one-based.

Now in order to apply this, one needs to show that there is an automorphism σ of $\mathbb{Q}^{alg}$, fixing the number field k, and a polynomial $P[T]$ with integer coefficients such that no root of $P[T]$ is a root of unity and H_P contains $Tor(A)$. This part of the proof involves no model theory and consists of two steps. First, one fixes a prime p (of good reduction for A) and one considers only the p'-torsion of A, denoted $Tor_{p'}(A)$, that is, the torsion elements of order prime p. By applying a classical result of Weil [45] one gets such an automorphism σ_1 and a polynomial $P_1(T)$ with H_{P_1} containing $Tor_{p'}(A)$. Then using two different primes p and q, and a result of Serre ([41], pages 33–34 and 56–59), one gets the required automorphism working for the full torsion subgroup.

Fix such an automorphism σ, and extend the difference field $(\mathbb{Q}^{alg}, \sigma)$ to a model (L, σ) of ACFA_0. In (L, σ), the group H_P is of linear type, hence $X \cap H_P$ is a finite boolean combination of translates of (σ-definable) subgroups of H_P. And we can conclude that $X \cap Tor(A)$ is a finite union of translates of subgroups of $Tor(A)$.

An important remark: this sketch of the proof is correct but does not yield effective bounds for the number of translates involved in the representation of $X \cap Tor(A)$ as a finite union. In fact Hrushovski shows that one can bound the number of translates involved by a function of the degree of the polynomial $P[T]$ and of the size of its coefficients. But if one is not careful, one looses track of any effective bounds on the degree and coefficients of the polynomial $P[T]$ during the passage from the p'-torsion to the full torsion via the Serre result.

So Hrushovski in fact, in order to deal with the full torsion group, gives a more complicated proof, which uses model theory and yields sharper information. What I have described above is exactly his proof for the case of the elements of p'-torsion , $Tor'_p(A)$. In that case, the classical result of Weil mentioned above, (a result about the characteristic polynomial of the Frobenius in an Abelian variety defined over $\mathbb{F}_p$), provides directly a polynomial $P(T)$ such that its degree and the size of its coefficients are bounded by a function of p, and of invariants of A (dimension, degree). In order to deal with the full torsion and keep effective bounds, one needs to work simultaneously with two different automorphisms, σ and τ, hence two distinct models of ACFA_0, and two different polynomials, $P[T]$ and $Q[T]$, such that in $(\mathbb{Q}^{alg}, \sigma)$, H_P contains the torsion elements of order prime to p, and in $(\mathbb{Q}^{alg}, \tau)$, H_Q contains the torsion elements of order a power of p.

One last remark: in fact Hrushovski's result in [15] is more general than the one I quoted. He proves the result for all commutative algebraic groups, and not only Abelian varieties.

4.4. A selection of references on the model theory of fields and the applications to Algebraic Geometry. Some general surveys on geometric model theory and applications:

- A. Pillay, *Model Theory, Differential Algebra and Number Theory*, in Proceedings of the ICM 94, Zürich, Birkhäuser 1996.
- A. Pillay, *Model Theory and Diophantine geometry*, Bull. Am. Math. Soc. 34 (1997), 405–422.
- D. Marker, *Strongly minimal sets and geometries*, Tutorial, LC '95, in [29].
- E. Hrushovski, *Geometric model theory*, in Proceedings of the ICM 98, Berlin, Vol. I, Doc. Math., 281–302, 1998.
- A. Pillay, *Aspects of Geometric Model Theory*, Tutorial, LC '99, preprint.
- T. Scanlon, *Diophantine geometry from model theory*, Bulletin of Symbolic Logic 7 (2001), 37–57.

For surveys on algebraically closed fields with an automorphism (ACFA) and the Manin-Mumford conjecture or on the Mordell-Lang conjecture:

- J. B. Goode (B. Poizat) *H.L.M. (Hrushovski-Lang-Mordell)*, Séminaire Bourbaki, exposé 811, Février 1996.
- Z. Chatzidakis *A survey on the model theory of difference fields*, in Model Theory, Algebra and Geometry, D. Haskell and C. Steinhorn ed., MSRI Publications 2000, 65–96 [10].
- E. Bouscaren *Théorie des Modèles et Conjecture de Manin-Mumford [d'après E. Hrushovski]*, Séminaire Bourbaki, Exposé 870, Mars 2000.

Books.

- One can find an introduction to the model theory of fields with special emphasis on differentially closed fields of characteristic zero and a survey on separably closed fields in *Model theory of fields*, D. Marker, M. Messmer and A. Pillay, Lecture Notes in Logic 5, Springer 1996 [32]. (The Lecture Notes in Logic are now published by the ASL; a new edition of this book is planned).
- For a reasonably self-contained introduction to Hrushovski's proof of the Mordell-Lang conjecture, based on the lectures given at a summer-school held in Manchester in 1994, see *Model Theory and Algebraic Geometry*, Lecture Notes in Mathematics 1696, E. Bouscaren Ed., Springer, 1998 [1].

- In *Algebraic Model Theory*, B. Hart, A. Lachlan and M. Valeriote eds., NATO ASI Series, Kluwer Academic Publishers 1997 [9], one can find introductory lectures with proofs (by Z. Chatzidakis and A. Pillay) to Hrushovski's proof of the Manin-Mumford Conjecture [9].
- In *Model Theory, Algebra and Geometry*, D. Haskell, A. Pillay and C. Steinhorn Eds., MSRI Publications 2000, one can find the proceedings of the introductory workshop of the MSRI semester on "Model theory of fields" (January 98–June 98) [10].

REFERENCES

[1] E. Bouscaren (editor), *Model theory and algebraic geometry*, Lecture Notes in Mathematics, no. 1696, Springer, 1998.

[2] E. Bouscaren and E. Hrushovski, *One-based theories*, ***The Journal of Symbolic Logic***, vol. 59 (1994), pp. 579–595.

[3] A. Buium, *Intersections in jet spaces and a conjecture of Serge Lang*, ***Annals of Mathematics***, vol. 136 (1992), pp. 583–593.

[4] Z. Chatzidakis and E. Hrushovski, *The model theory of difference fields*, ***Transactions of the American Mathematical Society***, vol. 351 (1999), pp. 2997–3071.

[5] Z. Chatzidakis, E. Hrushovski, and K. Peterzil, *The model theory of difference fields II: periodic ideals and the trichotomy in all characteristics*, ***Proceedings of the London Mathematical Society***, vol. 85 (2002), pp. 257–311.

[6] G. Cherlin, *Large finite structures with few type*, In Hart et al. [9].

[7] D. Evans and E. Hrushovski, *Projective planes in algebraically closed fields*, ***Proceedings of the London Mathematical Society***, vol. 62 (1991), pp. 1–24.

[8] ———, *The automorphisms group of the combinatorial geometry of an algebraically closed field*, ***Journal of the London Mathematical Society***, vol. 52 (1995), pp. 209–225.

[9] B. Hart, A. Lachlan, and M. Valeriote (editors), *Algebraic model theory*, NATO ASI Series, Kluwer Academic Publishers, 1997.

[10] D. Haskell, A. Pillay, and C. Steinhorn (editors), *Model theory, algebra and geometry*, MSRI Publications, 2000.

[11] M. Hindry, *Introduction to abelian varieties and the Mordell-Lang conjecture*, In Bouscaren [1].

[12] E. Hrushovski, *A new strongly minimal set*, ***Annals of Pure and Applied Logic***, vol. 62 (1993), pp. 147–166.

[13] ———, *The first-order theory of the Frobenius*, preprint, 1996.

[14] ———, *The Mordell-Lang conjecture for function fields*, ***Journal of the American Mathematical Society***, vol. 9 (1996), pp. 667–690.

[15] ———, *The Manin-Mumford conjecture and the model theory of difference fields*, ***Annals of Pure and Applied Logic***, vol. 112 (2001), pp. 43–115.

[16] E. Hrushovski and G. Cherlin, *Finite structures with few types*, to appear in the Annals of Mathematical Studies, Princeton.

[17] E. Hrushovski and A. Pillay, *Weakly normal groups*, ***Logic Colloquium '85***, North-Holland, 1987, pp. 233–244.

[18] ———, *Groups definable in local fields and pseudofinite fields*, ***Israel Journal of Mathematics***, vol. 85 (1994), pp. 203–262.

[19] E. Hrushovski and Z. Sokolovic, *Minimal subsets of differentially closed fields*, to appear in the Transactions of the American Mathematical Society.

[20] E. Hrushovski and B. Zilber, *Zariski geometries*, ***Bulletin of the American Mathematical Society***, vol. 28 (1993), pp. 315–323.

[21] ———, *Zariski geometries*, ***Journal of the American Mathematical Society***, vol. 9 (1996), pp. 1–56.

[22] B. Kim, *Forking in simple unstable theories*, ***Journal of the London Mathematical Society***, vol. 57 (1998), pp. 257–267.

[23] B. Kim and A. Pillay, *Forking in simple unstable theories*, ***Annals of Pure and Applied Logic***, vol. 88 (1997), pp. 149–164.

[24] ———, *From stability to simplicity*, ***The Bulletin of Symbolic Logic***, vol. 4 (1998), pp. 17–36.

[25] S. Lang, ***Introduction to algebraic geometry***, Interscience tracts in pure and applied mathematics, Interscience Publishers, 1958.

[26] ———, ***Abelian varieties***, Interscience tracts in pure and applied mathematics, Interscience Publishers, 1959.

[27] A. Macintyre, *Generic automorphisms of fields*, ***Annals of Pure and Applied Logic***, vol. 2–3 (1997), pp. 165–180.

[28] D. Macpherson, *Homogeneous and smoothly approximated structures*, In Hart et al. [9].

[29] J. A. Makowsky and E. V. Ravve (editors), *Logic Colloquium '95, Proceedings of the Annual European Summer Meeting of the Association for Symbolic Logic, Haifa, Israel*, Lecture Notes in Logic, Springer, 1998.

[30] D. Marker, *Strongly minimal sets and geometry*, In Makowsky and Ravve [29].

[31] ———, *Zariski geometries*, In Bouscaren [1].

[32] D. Marker, M. Messmer, and A. Pillay, ***Model theory of fields***, Lecture Notes in Logic, vol. 5, Springer, 1996.

[33] B. Mazur, *Abelian varieties and the Mordell-Lang conjecture*, In Haskell et al. [10].

[34] A. Pillay, ***Geometric stability theory***, Oxford Logic Guides, vol. 32, Oxford University Press, 1996.

[35] ———, *Algebraically closed fields*, In Bouscaren [1].

[36] A. Pillay and M. Ziegler, *Jet spaces of varieties over differential and difference fields*, to appear in Selecta Math.

[37] B. Poizat, ***Groupes stables***, Nur al-matiq wal ma'rifah, 1987.

[38] ———, ***Stable groups***, Mathematical Surveys and Monographs, vol. 87, American Mathematical Society, 2001.

[39] T. Scanlon, *The conjecture of Tate and Voloch on p-adic proximity to torsion*, ***International Mathematical Research Notices Journal***, vol. 17 (1999), pp. 909–914.

[40] ———, *Diophantine geometry of the torsion of a Drinfeld module*, ***Journal of Number Theory***, vol. 97 (2002), pp. 10–25.

[41] J. P. Serre, ***Oeuvres, collected works 1985–1998***, vol. IV, Springer, 2000.

[42] S. Shelah, *Simple unstable theories*, ***Annals of Mathematical Logic***, vol. 19 (1980), pp. 177–203.

[43] F. Wagner, ***Stable groups***, London Math. Soc. Lecture Notes, vol. 240, Cambridge University Press, 1997.

[44] ———, ***Simple theories***, Mathematics and its applications, vol. 503, Kluwer Academic Publishers, 2000.

[45] A. Weil, ***Courbes algébriques et variétés abéliennes***, Hermann, 1971.

[46] B. Zilber, *Finite homogeneous geometries*, ***Proceedings of the sixth Easter conference on model theory (Wendisch-Rietz 1988)***, Humboldt Univ., Berlin, 1988, pp. 186–208.

CNRS-UNIVERSITÉ PARIS 7, UFR DE MATHÉMATIQUES, CASE 7012
2 PLACE JUSSIEU, 75251 PARIS CEDEX 05, FRANCE
E-mail: elibou@logique.jussieu.fr

NOTIONS OF COMPUTABILITY AT HIGHER TYPES I

JOHN R. LONGLEY

Abstract. This is the first of a series of three articles devoted to the conceptual problem of identifying the natural notions of computability at higher types (over the natural numbers) and establishing the relationships between these notions. In the present paper, we undertake an extended survey of the different strands of research to date on higher type computability, bringing together material from recursion theory, constructive logic and computer science, and emphasizing the historical development of the ideas. The paper thus serves as a reasonably comprehensive survey of the literature on higher type computability.

CONTENTS

Logic Colloquium 2000
Edited by R. Cori, A. Razborov, S. Todorčević, and C. Wood
Lecture Notes in Logic, 19

§1. Introduction. This article is essentially a survey of fifty years of research on higher type computability. It was a great privilege to present much of this material in a series of three lectures at the Paris Logic Colloquium.

In elementary recursion theory, one begins with the question: what does it mean for an ordinary first order function on ℕ to be "computable"? As is well known, many different approaches to defining a notion of computable function — via Turing machines, lambda calculus, recursion equations, Markov algorithms, flowcharts, etc. — lead to essentially the same answer, namely the class of (total or partial) *recursive* functions. Indeed, *Church's thesis* proposes that for functions from ℕ to ℕ we identify the informal notion of an "effectively computable" function with the precise mathematical notion of a recursive function.

An important point here is that many *prima facie* independent mathematical constructions lead to the same class of functions. Whilst one can argue over whether this is good evidence that the recursive functions include *all* effectively computable functions (see Odifreddi [1989] for a discussion), it is certainly good evidence that they represent a mathematically natural and robust class of functions. And since no other serious contenders for a class of effectively computable functions are known, most of us are happy to accept Church's thesis most of the time.

Now suppose we consider second order functions which map first order functions to natural numbers (say), and then third order functions which map second order functions to natural numbers, and so on. We will use the word *functional* to mean a function that takes functions of some kind as arguments. We may now ask: what might it mean at these higher types for a functional to be "computable"? (Some reasons why we might want to ask this will be discussed shortly.)

A moment's reflection shows that a host of choices confront us if we wish to formulate a definition of higher type computability. For example:[1]

- **Domain of definition.** Do we wish to consider *partial* or *total* computable functionals? Do we want them to act on partial functions of the next type down, or just on total functions? Should they act only on the "computable" functions of this type, or on some wider class of functions?
- **Representation of functions.** If we wish to perform "computations" on functions, how do we regard the functions as given to us? As infinite graphs? As algorithms or "programs" of some kind? Or as oracles or "black boxes", for which we only have access to the input/output behaviour?
- **Protocol for computation.** What ways of interacting with functions do we allow in computations? For example, do we insist that calls to func-

[1]Many of these points are also made in a survey article by Cook (Cook [1990]), whose point of view is very close to our own.

tions are performed sequentially, or do we allow parallel function calls? Do we insist that terminating computations are in some sense finite mathematical objects, as must be the case if we are seeking a genuinely effective notion of computability — or do we allow infinite computations in accordance with the infinitistic nature of the arguments?
- **Extensionality.** Do we want to restrict our attention to computable *functions* (as implicitly assumed in the preceding discussion)? Or do we want to consider computability for other, possibly non-extensional, operations of higher type? If the latter, what do we mean by an "operation"?

The spirit in which we are asking these questions is not to demand definitive answers to them, but to make the point that many choices are possible. Indeed, as we shall see, many different responses to the above questions are exemplified by the definitions of higher type computability that have been proposed in the literature. Moreover, the effects of all these choices escalate rapidly as we climb up the types. For example, if two definitions yield different classes of computable functions of type σ, it may be difficult even to compare these definitions at type $\sigma \to \mathbb{N}$, since the domains of the functions may differ. Indeed, we often find that a question needs to have a positive answer at type σ in order to be even meaningful at type $\sigma \to \mathbb{N}$.

It thus appears that very many approaches to defining higher type computability are possible, but it is not obvious *a priori* whether some approaches are more sensible than others, or which approaches lead to equivalent notions of computability. Moreover, in contrast to the first order situation, there does not seem to be a clear canonical pre-formal notion of "effective computability at higher types" to which we can refer for guidance. (This is hardly surprising, in view of the fact that there are several possible pre-formal conceptions of what a function is.) In short, it is unclear in advance whether at higher types there is really just one natural notion of computability (as in ordinary recursion theory), or several, or indeed no really natural notions at all.

This paper is the first of a planned series of three articles devoted to the conceptual problem of finding good, natural notions of higher type computability. Whereas previous work in the area has explored various *particular* notions of computability in some detail, our wish is to take a step back and look at the overall picture. Our main objectives are as follows:

- To discover what natural notions of computability exist at higher types, and to collect evidence for their naturalness.
- To develop some basic "recursion theory" for each of these notions, analogous to the elementary parts of ordinary recursion theory.
- To investigate how these notions of computability are related.
- To provide a coherent framework for pulling together and organizing the existing knowledge in the area.

We will be concerned mainly with objects of *finite type* (that is, nth order operations for some $n \in \omega$). In principle one can also consider *transfinite* types, though we will touch on these only occasionally.

Many ideas and results relevant to our project are already known, although they are rather widely scattered across the literature in recursion theory, constructive logic and computer science, and have never previously been presented together as contributions to a single subject. In this article we will give a fairly comprehensive survey of the work to date on different approaches to higher type computability. This will amass some raw material for our project. In two sequel papers, we will present a more systematic view of much of this material, proposing some simple general frameworks for discussing the "space of possible notions of computability", and showing that within these frameworks a reasonably cohesive picture does indeed emerge from the disparate strands.

To expand on our working philosophy a little further: It appears (to us) that *a priori* considerations are by themselves of limited use in determining what are the natural notions of higher type computability — any particular definition one can write down seems to involve some choices which might be felt to be arbitrary. We are therefore led to adopt a more empirical attitude: we can explore a range of possible definitions, and see what natural notions emerge and how they are related. Various criteria may be used to determine which notions of computability count as "natural", for instance:

- Whether they arise from some intuitively appealing informal concept of "computation".
- Whether they admit a wide range of independent mathematical characterizations — the more independent the better.
- Whether they occupy some special position within the space of possible notions of computability.

There has already been much research over the last fifty years exploring different approaches to higher type computability, and we feel the time has come for bringing this material together and trying to make sense of the big picture. In view of our goals, it is natural that we should favour an eclectic attitude — since we do not know in advance where to look for good notions of computability, we should cast our net as wide as possible and embrace the diversity of definitions that have been proposed.

Our enterprise will be justified in retrospect if it does in fact lead to a coherent and satisfying picture. We will then be in a strong position to attempt a more conceptual explication of those notions of computability which we suspect to be fundamental.

To anticipate the outcome of our project, we will argue that, for computable *functionals* at least, there is in fact a manageable handful of around six natural and robust notions of higher type computability, each with a variety of different characterizations and some pleasing intrinsic properties. Although it is possible that there are other equally natural notions of computable functional awaiting discovery, the fact that very many attempts at defining a notion of computable functional lead to one of the known notions suggests (in this author's opinion) that the current picture is probably by now reasonably complete. For non-functional notions of computability, the situation is at present much more open-ended, but we are at least able to unify much of what is currently known in a satisfying way.

1.1. Motivations. Before proceeding further, we should mention some of the reasons why computability at higher types is interesting. Besides its intrinsic mathematical and conceptual appeal, the subject lies at an intriguing juncture between several areas of mathematical logic and computer science, and has (actual or potential) connections with the following areas. For reasons of space, however, we will say relatively little about these applications in the rest of the paper, choosing to concentrate on clarifying the basic notions of the subject.

1.1.1. *Constructive logic and metamathematics.* Historically, the first applications of the ideas of higher type computability were to the metamathematics of constructive systems. Computable objects of finite type can often be used to give interpretations of logics — such as *realizability* interpretations — that endow formulae with some kind of constructive content. For instance, we might stipulate that a *realizer* for a formula $\phi \Rightarrow \psi$ is a computable function mapping realizers for ϕ to realizers for ψ; in this case, formulae with nested implications will naturally lead us to consider higher type objects.

On a technical level, such interpretations can be used to obtain consistency and independence results for constructive logics. On a conceptual level, they can be helpful for clarifying various constructive views of mathematics, often from a classical standpoint. A good early discussion of possible applications of this kind appears in Kreisel [1959, Sections 1,2]. The area was extensively developed by Troelstra and his school (Troelstra [1973]), and by Beeson (Beeson [1985]), who focused on realizability and related interpretations.

In a somewhat similar spirit is Feferman's use of computable higher type objects in connection with systems for explicit mathematics and theories of finite type (Feferman [1975], [1977b]). These systems are typically intended to reflect "semi-constructivist" standpoints that suffice for most of mathematical practice. For other recent applications of this kind, see Kohlenbach [2002].

1.1.2. *Descriptive and admissible set theory.* Logical quantifiers may be regarded as objects of higher type: for instance, existential quantification over the natural numbers can be seen as a (non-computable) object ${}^2\exists\colon 2^{\mathbb{N}} \to 2$, where $2 = \{0, 1\}$. There are interesting relationships between computability *relative to* such quantifiers and logical complexity: for instance, a function on $\mathbb{N}$ is computable relative to ${}^2\exists$ (in a certain sense) iff it is definable by a hyperarithmetic predicate (see Section 3.2.1 below). This aspect of higher type recursion theory was an important ingredient Kleene's early work in the area, and was developed further by Moschovakis and others (see Section 3.2).

The relationship between higher type (relative) computability and logical definability also manifests itself in connections with admissible set theory. These connections can be exploited to apply forcing techniques from set theory to the solution of degree-theoretic problems for higher types (see Sacks [1990], and Section 3.2.2 below). Furthermore, a natural generalization of certain ideas from higher type computability leads to a good notion of "computability" on arbitrary sets, closely related to Gödel's notion of constructibility in set theory (see Normann [1978b], and Section 3.2.4 below).

1.1.3. *Abstract computability theories.* There is an enormous literature on finding suitable notions of computability for various kinds of mathematical objects, such as rings, fields, topological spaces, Banach spaces, or ordinals. (Griffor [1999] provides a good starting-point for references on these topics.) In view of this, it is not surprising that attempts have been made to develop *abstract* theories that offer a uniform account of "computability" for a wide range of structures (see *e.g.*, Moschovakis [1969], Friedman [1971], Tucker [1980]). Several of these approaches are clearly described in Hinman [1999]. Some approaches to abstract computabiliy can themselves be seen as instances of a more general theory of inductive definability (Aczel [1977]).

Higher type computability, and especially Kleene's early work, has inspired many of the ideas in these abstract approaches, and has played a useful role as a motivating example. In turn, these abstract approaches have then suggested simpler, clearer ways of presenting higher type computability. Both directions of influence may be discerned in the work of Moschovakis (see Moschovakis [1974a], [1983], [1989]).

1.1.4. *Semantics and logic of programming languages.* Ideas from higher type computability have inspired the design of modern functional programming languages. This started with the theoretical work of Scott and Plotkin (Scott [1969], Plotkin [1977]), which led eventually to the design of fully fledged programming languages such as Standard ML (Milner, Tofte, Harper, and MacQueen [1997]).

In addition, much work in theoretical computer science has been concerned with the *semantics* of programming languages. Finding a well-matched semantic model for a programming language is often tantamount to finding a good mathematical characterization of the notion of computable operation it embodies. As argued in Longley and Plotkin [1997], Longley [1999a], this can often help us to design a good logic for proving properties of programs in the language. The finite types over the natural numbers are a good target for study here, because many other computational datatypes of importance can be obtained easily as *retracts* of these types (this will be explained in Part II).

Ideas from higher type computability also play a role in Feferman's approach to computation on abstract datatypes (see *e.g.*, Feferman [1996]). In the longer term, we expect that an understanding of higher type computability will contribute particularly to the study of object oriented languages, which naturally support higher order styles of programming.

1.1.5. *Subrecursion and complexity.* Notions of higher type computation can also be used to study the computational complexity of ordinary first order functions. Many interesting *subrecursive* notions of first order computability (that is, notions that do not allow the computation of all general recursive functions) can be conveniently characterized via systems for higher type recursion. For example, the $_0$-*recursive* functions (*i.e.*, the provably total functions of Peano arithmetic) are precisely those definable in Gödel's System T (see Section 3.1). Furthermore, an ordinal stratification of these functions, corresponding to the *extended Grzegorczyk hierarchy*, can be defined very elegantly using some simple higher type functionals (see Schwichtenberg [1975], [1999]). As argued in Schwichtenberg [1999], it seems that in giving definitions of functions there is some inherent connection or trade-off between type complexity and ordinal complexity.

Recently, characterizations of this kind have been achieved for much lower complexity classes, including even the polytime computable functions (see Bellantoni, Niggl, and Schwichtenberg [2000] and the papers cited therein).

In addition, of course, one can look for natural complexity classes at higher type. Ideas from higher type computability have informed the definition of complexity-theoretic notions (see *e.g.*, Cook [1990]), but it seems that many of the basic notions of higher type complexity are still open to discussion. Indeed, in order to formulate a notion of feasible computation at higher types (for instance), we will have to confront all the choices mentioned earlier, and many others besides. Clearly, a good understanding of higher type computability will stand us in good stead if we wish to clarify the fundamental notions of complexity at higher types.

The current state of the art regarding feasible computation at higher types is described in detail in Irwin, Kapron, and Royer [2001a], [2001b]. Other subrecursive notions of computability at higher types have been considered in Schwichtenberg [1991], Niggl [1993].

1.1.6. *Real number computability.* The connection between computability over the reals and higher type computability was recognized as far back as Lacombe [1955a], [1955b], [1955c] and Grzegorczyk [1957], and later manifested itself in the context of constructive interpretations of systems for analysis (*e.g.*, in Troelstra [1973]). Notions of computability over the reals and other metric spaces underpin constructive recursive analysis of the Markov school on the one hand (see Aberth [1980], Beeson [1985]), and classical computable analysis on the other (see Pour-El and Richards [1989], Weihrauch [2000b],

Bauer [2000]). Interest in real number computability has also recently been awakened within computer science (see Escardó [1996]), and exact real number computation appears attractive as a potential application area for higher type programming. For example, *integration* for real functions corresponds to a second order computable operation over the reals, and hence to a third order computable operation over ℕ (see Simpson [1998]); an operator for solving differential equations might therefore involve a fourth order operation over ℕ.

Higher types over the reals, and some associated notions of computability, have recently been considered *e.g.*, in Normann [2001], [2002], Bauer, Escardó, and Simpson [2002], Korovina and Kudinov [2001]. As with higher type complexity, it appears that the natural notions here are still open to discussion — though once again, a good understanding of the higher types over ℕ will surely help.

1.1.7. *Computability in physics.* Notions of computability in analysis can in turn be applied to questions of computability in various physical theories. Some of this territory is explored in Pour-El and Richards [1989] (see also Pour-El [1999]). Besides providing a foundation for this work, it seems conceivable that the study of higher type computability might suggest alternative definitions of computability for physical systems which may hold some philosophical interest. For some intriguing speculations along these lines, see Cooper [1999].

1.2. Overview of the series. This series of articles is intended as a fairly comprehensive account of what is currently known about notions of higher type computability over the natural numbers.

The present Part I is a historical survey of the work to date on higher type computability, tracing the various strands of research which have contributed to our present understanding. This is intended to serve several purposes: firstly, to offer a gentle introduction to the main ideas of the subject; secondly, to document the genesis of these ideas; thirdly, to facilitate comparison between different strands of work by placing them side by side in a uniform setting; and fourthly, to provide a reasonably complete map of the rather bewildering literature in the area. We have tried to include just enough technical detail to make the mathematical substance intelligible, without losing the broad sweep of the story. The point of view we wish to advocate is that all the strands of research that we describe can naturally be seen, with hindsight, as contributions to a single coherent subject. In the remaining papers in the series, we will attempt a more systematic and technically detailed exposition of this subject.

In Part II we will try to organize the material concerning notions of computable *functional* (that is, extensional operations of higher type), and the relationships between them. Here we will work within a general framework given by some simple definitions involving *finite type structures*. Although this framework is very simple and even somewhat crude, it suffices for clarifying much of the existing material. After developing the necessary general concepts, we will consider in turn the various good notions of *total* and *partial* computable functional. In both the total and partial settings, we will present arguments for the impossibility of a "Church's thesis for higher types". We will also discuss ways in which total and partial functionals can be related and combined.

In Part III we will consider, more generally, notions of computable *operation* (not necessarily extensional) — for example, the notions of computability embodied by various non-functional programming languages. In order to articulate these notions, we will use a more sophisticated general framework based on ideas of *realizability* — this will extend and refine the theory of Part II in a mathematically satisfying manner. Once again, we develop the general theory, then survey within this framework some of the notions of computability that appear to have some claim to naturalness.

Naturally, much of the material covered in Part I will be treated again from a different perspective in Parts II and III. We feel that this kind of overlap is justified in the interest of presenting a rounded view that takes account of both the historical and the purely logical aspects of the subject. Indeed, the historical and logical parts of the series are intended to be complementary, in the sense that each tries to emphasize ideas that receive scant attention in the other.

1.3. Outline of the present paper. As a glance at Figure 1 on page 50 will confirm, the study of higher type computability has not developed in a coherent, orderly fashion. Rather, it is mostly the result of the parallel activity of several research communities, each with their own set of motivations, and it is only in retrospect that the various strands can be seen as parts of a coherent whole. It is therefore not surprising that the history of the subject appears as somewhat chaotic.

The parallel nature of the subject's development, in particular, makes the history difficult to describe: some compromise between a strictly chronological presentation and a thematic one is necessary, and no linear ordering of the material seems completely satisfactory from an expository point of view. Here we have adopted the following course. In Section 2 we describe, more or less chronologically, the early work on computability at type 2, taking us up to about 1958. Around this time, several notions of computability at all finite types made their debut; at this point the subject effectively split into several streams, and it is only over the last few years that these have begun to converge again. For the main body of the paper (Sections 3 and 4) we therefore treat each of the main notions of higher type computability in turn, taking them (roughly) in order of their first appearance in the literature, and giving separate chronological accounts of the developments relating to each of them.

In Section 3 we discuss around four different notions of *total* computable functional, and in Section 4 we consider a similar number of notions of *partial* computable functional. (Fortunately for our scheme, all of the total notions made their first appearance before any of the partial ones!) As we shall see, the partial notions tend to be simpler and to have more pleasant properties than the total ones. Turning to more recent developments, in Section 5 we describe some ideas from realizability which cross-cut many of these streams. Finally, in Section 6 we briefly discuss some ideas relating to *non-functional* notions of "computable operation" at higher types.

For convenience, some diagrams summarizing the main structures of interest and the relationships between them have been included in an appendix.

We have tried to make our account as accurate and complete as possible. However, shortcomings are inevitable in a work of this kind, and the author can only offer his apologies to anyone whose work he has inadvertently misrepresented or overlooked. He would be glad to be informed of any significant errors or omissions so that these may be rectified in a future publication.

Throughout the paper we presuppose a good knowledge of elementary recursion theory, and some general background in logic. A few further prerequisites of a more specialized nature will be summarized in Section 1.5.

1.4. Notation. We first fix some general notational conventions. For any set X, we write $X_\perp$ for the set $X \sqcup \{\perp\}$, where $\perp$ is some distinguished element which intuitively will represent the *non-termination* of a process which is attempting to compute a value in X. We write $f : X \rightharpoonup Y$ to mean "f is a partial function from X to Y". Any partial function $f : X \rightharpoonup Y$ may be *represented* by a total function $X \to Y_\perp$, though formally we shall distinguish between the two entities.

We will use the following notational conventions in connection with potentially non-denoting expressions e, e' arising from the use of partial functions.

- $e\downarrow$ means "the value of e is defined" (that is, e denotes something).
- $e\uparrow$ means "the value of e is not defined".
- $e = e'$ means "the values of e and e' are both defined and they are equal" (strict equality).
- $e \simeq e'$ means "if either e or e' is defined then so is the other and their values are equal" (Kleene equality).

Note that these conventions relate to the definedness of mathematical expressions rather than the termination of computations. In particular, we have $\perp\downarrow$.

If e is an expression possibly involving the variable x, we write $\Lambda x.e$ to mean the set-theoretic (total or partial) function that maps x to e; the intended domain of this function will be determined by the context. Thus, the expression $\Lambda x.e$ names the function defined in more standard notation by $x \mapsto e$. However, the Λ notation seems more convenient in complex expressions, and it also provides a semantic counterpart to the formal syntax of the λ-calculus which we shall frequently use (see Section 1.5.2).

We write $\mathbb{N}$ for the set of natural numbers including 0. We also write:

- $\mathbb{N}^{\mathbb{N}}$ for the set of all (set-theoretic) total functions from $\mathbb{N}$ to $\mathbb{N}$,
- $\mathbb{N}_p^{\mathbb{N}}$ for the set of all partial functions from $\mathbb{N}$ to $\mathbb{N}$,
- $\mathbb{N}_{rec}^{\mathbb{N}}$ for the set of total recursive functions from $\mathbb{N}$ to $\mathbb{N}$,
- $\mathbb{N}_{p\,rec}^{\mathbb{N}}$ for the set of partial recursive functions from $\mathbb{N}$ to $\mathbb{N}$.

We write $\mathrm{Seq}(X)$ for the set of finite sequences over a set X; we will use the notation $[x_1, \ldots, x_n]$ to display such sequences. We will suppose $\langle - \rangle \colon \mathrm{Seq}(\mathbb{N}) \to \mathbb{N}$ is some fixed effective coding for finite sequences, and write $\langle x_1, \ldots, x_n \rangle$ in place of $\langle [x_1, \ldots, x_n] \rangle$. Given $f : \mathbb{N} \rightharpoonup \mathbb{N}$, we define its *course-of-values* function $\widetilde{f} : \mathbb{N} \rightharpoonup \mathbb{N}$ by

$$\widetilde{f}(n) \simeq \langle f(0), \ldots, f(n-1) \rangle.$$

We also suppose we have some effective indexing scheme for the partial recursive functions, given for example by an effective enumeration of Turing machines. We will write ϕ_m for the partial recursive function from $\mathbb{N}$ to $\mathbb{N}$ with recursive index m.

If X is any set and R is a partial equivalence relation (that is, a symmetric, transitive relation) on X, we write X/R for the set of R-equivalence classes over X.

We will follow certain conventions regarding variables and also in our use of certain typefaces. Ordinary mathematical fonts will be used for variables of all kinds; generally speaking, we will use

- i, j, k, l, m, n, r to range over $\mathbb{N}$;
- α, β to range over $\mathrm{Seq}(\mathbb{N})$;
- f, g, h to range over first order functions, or sometimes over functions of arbitrary type;
- F, G to range over second order objects (functions or operations);
- Φ, Ψ, Θ to range over higher order objects (*i.e.*, third order or above).

We will also use subscripted and superscripted variants of these symbols in the same way. We will occasionally depart from the above conventions when it is convenient to do so. Other variables (*e.g.*, x, y, z) will be used more flexibly as we have need of them.

We use boldface letters as abbreviations for *vectors*, or lists of variables. More precisely, a symbol such as $\boldsymbol{x}$, wherever it occurs, will textually abbreviate either $x_1 \ldots x_{l_x}$ or $x_1, \ldots, x_{l_x}$ (as demanded by the context), where $l_x \geq 0$.

We use Roman boldface (*e.g.*, ***Set***) for the names of particular categories of interest, and uppercase sans serif font (*e.g.*, HEO) for the names of particular type structures (see Section 1.5.1).

A few other typographical conventions associated with λ-calculi and types will be introduced in the course of the next section.

1.5. Prerequisites. Throughout this paper we will make incessant use of the idea of a *type structure*; we will also make reference to the *λ-calculus* and occasionally to *cartesian closed categories*. These notions will enormously facilitate our task of giving a unified presentation of our material. Our use of these concepts will sometimes mean recasting original definitions into a rather more modern form, but we believe this will not do too much violence to the historical point of view.

We now review the material that we shall need on these topics.

1.5.1. *Types and type structures.* The following basic concepts are central to the entire paper and will be used ubiquitously.

Given any set Γ of *basic type symbols* γ, the (*finite* or *simple*) *types* over Γ are the formal expressions σ built up according to the following grammar:

$$\sigma ::= \gamma \mid (\sigma_1 \to \sigma_2).$$

Informally, each symbol $\gamma \in \Gamma$ will represent some set of basic entities, and $(\sigma_1 \to \sigma_2)$ will represent the type of functions from σ_1 to σ_2. We use ρ, σ, τ as variables ranging over types. We will omit brackets wherever possible, and regard $\to$ as right-associative, so that $\rho \to \sigma \to \tau$ means $(\rho \to (\sigma \to \tau))$. We define the *level* of a type inductively by

$$\mathit{level}\,(\gamma) = 0, \quad \mathit{level}\,(\sigma \to \tau) = \max(1 + \mathit{level}\,(\sigma), \mathit{level}\,(\tau)).$$

Except where otherwise stated, we will be considering finite types over the single basic type symbol $\overline{0}$, which usually represents the type of natural numbers. We then define the *pure types* $\overline{n}$ inductively by $\overline{n+1} = \overline{n} \to \overline{0}$. If (X_σ) is a family of mathematical objects or relations indexed by types σ, we will often write just X_n in place of $X_{\overline{n}}$.

Our fundamental notion will be that of a *type structure*. For the purpose of the present paper, the following definition will suffice:

Definition 1.1 (Type structures).

(i) *A* partial type structure *A will consist of a family of sets A_σ (one for each type), together with "application" functions $\cdot_{\sigma\tau} : A_{\sigma\to\tau} \times A_\sigma \rightharpoonup A_\tau$.*
(ii) *A* total type structure, *or more simply a* type structure, *is a partial type structure in which all the application functions are total.*

We usually omit the type subscripts from the application functions, and treat $\cdot$ as a left-associative infix, so that $f \cdot x \cdot y$ means $(f \cdot x) \cdot y$. In accordance with the convention mentioned above, we often write A_n in place of $A_{\overline{n}}$. By a *type n object* of A we will mean an element of some A_σ where $\mathit{level}\,(\sigma) = n$. We say A is a (partial or total) type structure *over X* if $A_0 = X$.

Most of the examples we consider will be type structures over $\mathbb{N}$ or $\mathbb{N}_\perp$. An important example is the *full set-theoretic* type structure S over $\mathbb{N}$, in which $\mathsf{S}_0 = \mathbb{N}$ and $\mathsf{S}_{\sigma\to\tau}$ is the classical set of all functions from S_σ to S_τ.

A partial type structure A is *extensional* if for all types σ, τ and all $f, g \in A_{\sigma\to\tau}$ we have

$$(\forall x \in A_\sigma.\, f \cdot x \simeq g \cdot x) \Longrightarrow f = g.$$

It is easy to see that any extensional [partial] type structure A is isomorphic to a [partial] type structure B in which each set $B_{\sigma\to\tau}$ is a set of [partial] functions from B_σ to B_τ. We will therefore often refer to objects of type level 2 or more in extensional partial type structures as *functionals*.

There is a standard way to obtain extensional type structures from an arbitrary type structure:

DEFINITION 1.2 (Extensional collapse). *Let A be any partial type structure over X.*

(i) *Given any partial equivalence relation $\approx$ on X, define a partial equivalence relation $\approx_\sigma$ on each A_σ as follows:*
- *$x \approx_0 y$ iff $x \approx y$;*
- *$f \approx_{\sigma\to\tau} g$ iff for all $x, y \in A_\sigma$, $x \approx_\sigma y$ implies $f \cdot x\downarrow$, $g \cdot y\downarrow$ and $f \cdot x \approx_\tau g \cdot y$.*

Now let $\mathsf{EC}(A, \approx)$, *the* extensional collapse *of A with respect to $\approx$, be the total extensional type structure defined by* $\mathsf{EC}(A, \approx)_\sigma = A_\sigma/\approx_\sigma$, *with application inherited from A.*

(ii) *Define* $\mathsf{EC}(A)$, *the* extensional collapse *of A, to be* $\mathsf{EC}(A, =)$.

We will often speak, somewhat informally, of an element of a type structure A being *hereditarily P* for some property P. What we typically mean by this may be explained reasonably precisely by induction on types as follows:

- An element $x \in A_0$ is hereditarily P iff x has property P;
- An element $f \in A_{\sigma\to\tau}$ is hereditarily P if the restriction f' of f to the hereditarily P elements $x \in A_\sigma$ has property P, and for all such x, $f \cdot x$ is hereditarily P.

For example, if A is a type structure over $\mathbb{N}_\perp$, a hereditarily total functional of type $\overline{2}$ is a functional that acts totally on functions that act totally on elements of $\mathbb{N}$. Clearly, if all the elements of A have some property P, then they are all hereditarily P.

A more comprehensive armoury of definitions relating to type structures will be presented in Part II.

1.5.2. *The λ-calculus.* Next we review the basics of untyped and simply typed λ-calculi. For more details, see for example Barendregt [1984].

The λ-calculus is a convenient formal language for talking about functions and application. To define an *untyped* λ-calculus, one specifies a (possibly empty) set C of *constant* symbols, and we also assume we have available an unlimited supply of *variable* symbols. We will use teletype font for particular constant symbols that we shall introduce (*e.g.*, `0`, `succ`), and use c as a metavariable ranging over constant symbols. We will use ordinary Roman letters $x, y, z, f, g, h, \ldots$ as variable symbols. As usual in logic, we will not bother to distinguish notationally between particular variables and metavariables ranging over variable symbols.

The *untyped λ-terms* U over C are then built up according to the following grammar:

$$U ::= c \mid x \mid (\lambda x.U_1) \mid (U_1 U_2).$$

We use U, V, W as metavariables ranging over λ-terms. We will omit brackets wherever possible, taking application to be left-associative, so that UVW means $((UV)W)$. We will also write $\lambda x_1 \ldots x_n.U$ as an abbreviation for $\lambda x_1. \cdots .\lambda x_n.U$.

Informally we may read $U_1 U_2$ as "U_1 applied to U_2", and $\lambda x.U_1$ as "the function mapping any element x to U_1". We view λx as a *binder* analogous to the quantifiers $\forall x$ and $\exists x$ in logic; thus we have evident notions of free and bound variable occurrences, closed terms, and substitution. We write $U[V/x]$ for the result of substituting V for all free occurrences of x in U, renaming bound variables if necessary to avoid capture.

The untyped λ-calculus is a very fluid system: any term can be applied to any other term and even to itself. Often we wish to consider more restricted systems in which natural type distinctions between functions and arguments are enforced. For our purposes it will suffice to consider *simply typed λ*-calculi, in which the types σ are given precisely as in Section 1.5.1 above. To define a simply typed λ-calculus, one specifies a set C of constant symbols c^σ, each decorated with some type σ; we also assume we have an infinite supply of variable symbols x^σ for each type σ. We first define the set of untyped λ-terms U as above; we then define a relation $U : \sigma$ (read as "U is of type σ") inductively by means of the following clauses:

- $c^\sigma : \sigma$ for any constant c^σ,
- $x^\sigma : \sigma$ for any variable x^σ,
- if $U : \sigma \to \tau$ and $V : \sigma$ then $UV : \tau$,
- for any variable x^σ, if $U : \tau$ then $\lambda x^\sigma.U : \sigma \to \tau$.

We will frequently omit type superscripts where these can be inferred from the context. We say U is *well-typed*, or is a *simply typed λ-term*, if $U : \sigma$ for some (necessarily unique) σ.

We can think of a λ-calculus (whether untyped or simply typed) not just as a formal notation for functions, but also as a kind of programming language in which one can perform computations by symbolic manipulation. For instance, one may define a *reduction* relation $U \rightsquigarrow V$ on the terms of a λ-calculus, intended to capture the idea of a single computation step. The definition of $\rightsquigarrow$ will depend on the λ-calculus in question, but (for the purposes of this paper) it will always include at least the *β-rule*:

$$(\lambda x.U)V \rightsquigarrow U[V/x].$$

Frequently, the definition will also include the following *congruence rules*:

$$\text{if } U \rightsquigarrow U', \text{ then } UV \rightsquigarrow U'V,\ WU \rightsquigarrow WU', \text{ and } \lambda x.U \rightsquigarrow \lambda x.U'.$$

In addition, there may be other special reduction rules involving the constants of the language in question. In general we will write $\rightsquigarrow^*$ for the reflexive transitive closure of $\rightsquigarrow$. A term U is a *final value* with respect to $\rightsquigarrow$ if there is no V such that $U \rightsquigarrow V$. We write $U \Downarrow V$ if $U \rightsquigarrow^* V$ and V is a final value.

Simply typed λ-calculi provide good languages for defining objects of finite type. An *interpretation* of a simply typed λ-calculus $\mathcal{L}$ in a type structure A is a function assigning to each closed term $U : \sigma$ of $\mathcal{L}$ an element $[\![U]\!] \in A_\sigma$, in such a way that $U \rightsquigarrow V$ implies $[\![U]\!] = [\![V]\!]$. A λ-calculus that admits an interpretation in A is often a very convenient formal language for denoting elements of A. We may say an element x of A is *$\mathcal{L}$-definable* (for a particular language $\mathcal{L}$) if $x = [\![U]\!]$ for some closed term U of $\mathcal{L}$.

A *theory* on $\mathcal{L}$ will (for our purposes) be a type-respecting equivalence relation $\sim$ on closed terms of $\mathcal{L}$ which includes the reduction relation and which is a congruence with respect to application. Any interpretation $[\![-]\!]$ of $\mathcal{L}$ induces a theory on $\mathcal{L}$, given by $U \sim V$ iff $[\![U]\!] = [\![V]\!]$.

The set of closed terms of a simply typed λ-calculus $\mathcal{L}$ is itself a type structure, with application given by juxtaposition of terms. Furthermore, the quotient of this type structure modulo any theory on $\mathcal{L}$ is again a type structure. Type structures obtained from theories in this way are called *(closed) term models* for $\mathcal{L}$. Interesting term models can often be obtained by identifying terms that have equivalent behaviour in some sense.

1.5.3. *Cartesian closed categories.* We now give a brief sketch of the notion of a cartesian closed category. Only a general impression will be required in this paper. Further details may be found in Lambek and Scott [1986].

Informally, a cartesian closed category is one in which for any objects X and Y, we have an object Y^X playing the role of the space of functions from X to Y. A helpful motivating example is the classical category ***Set*** of sets and functions, in which Y^X is simply the set of all functions from X to Y. Note that for any set Z, functions from $Z \times X$ to Y correspond precisely to functions from Z to Y^X. Abstracting the essential features of this situation leads to the following definition:

DEFINITION 1.3 (Cartesian closed categories). *A* cartesian closed category *is a category $\mathcal{C}$ with finite products (including a terminal object 1), in which for any objects X and Y we have an object Y^X and a morphism ${}_{XY} : Y^X \times X \to Y$ with the following property: for all morphisms $f : Z \times X \to Y$ there is a unique morphism $\overline{f} : Z \to Y^X$ such that*

$$f = {}_{XY} \circ \langle \overline{f}, \mathrm{id}_X \rangle.$$

It is a consequence of the definition that the object Y^X is always uniquely determined up to isomorphism. The reader unfamiliar with the above definition should satisfy himself that in the case of ***Set*** it does indeed characterize the set of all functions from X to Y up to a canonical bijection.

Given any object X in a cartesian closed category $\mathcal{C}$, we may obtain a type structure as follows. First define an interpretation $[\![-]\!]$ of the finite types as objects of $\mathcal{C}$:

$$[\![\bar{0}]\!] = X, \quad [\![\sigma \to \tau]\!] = [\![\tau]\!]^{[\![\sigma]\!]}.$$

Now define a type structure $A = \mathsf{T}(\mathcal{C}, X)$ by taking $A_\sigma = \mathrm{Hom}(1, [\![\sigma]\!])$, with $\cdot_{\sigma\tau}$ the evident function induced by $_{XY}$. Many type structures of interest arise in this way from mathematically natural categories. Clearly, if $\mathcal{C} =$ ***Set*** and $X = \mathbb{N}$, this construction yields the full set-theoretic type structure S.

A cartesian closed category is *well-pointed* when for all $f, g \colon X \to Y$, if $f \circ x = g \circ x$ for every $x \colon 1 \to X$ then $f = g$. If $\mathcal{C}$ is well-pointed, then all type structures $\mathsf{T}(\mathcal{C}, X)$ will be extensional.

There is a very close relationship between cartesian closed categories and the simply typed λ-calculus. For instance, given a λ-calculus $\mathcal{L}$ and a suitable interpretation of its basic types and constant symbols in a cartesian closed category $\mathcal{C}$, one can define an interpretation of $\mathcal{L}$ in $\mathcal{C}$, in which a λ-term $U : \tau$ with free variables $x_1^{\sigma_1}, \ldots, x_n^{\sigma_n}$ is interpreted by a morphism

$$[\![U]\!] \colon [\![\sigma_1]\!] \times \cdots \times [\![\sigma_n]\!] \to [\![\tau]\!].$$

In the case of a single ground type $\bar{0}$, such an interpretation clearly gives rise to an interpretation of $\mathcal{L}$ in $\mathsf{T}(\mathcal{C}, [\![\bar{0}]\!])$ in the sense of Section 1.5.2.

We may say $\mathcal{C}$ is a *model* for $\mathcal{L}$ if we have an interpretation such that $U \rightsquigarrow V$ implies $[\![U]\!] = [\![V]\!]$. In fact, the β-rule and congruence rules mentioned in Section 1.5.2 are automatically validated by any interpretation of $\mathcal{L}$ in a cartesian closed category.

We will sometimes refer in passing to the notion of a *topos*. All that the reader will need to know is that a topos is a cartesian closed category with some strong additional properties, giving rise to a very rich categorical structure. In fact, any topos provides a model for higher order intuitionistic logic, and can therefore be viewed as a kind of "universe" for much of intuitionistic mathematics. The leading example of a topos is ***Set***; in this case, the corresponding interpretation of logic coincides with the familiar classical one. Again, more information can be found in Lambek and Scott [1986].

1.5.4. *Historical remarks.* The concept of functions of arbitrary finite type (and even of transfinite type) appeared in Hilbert [1925], in a discussion of Cantorian set theory. The language of simple types, and the simply typed λ-calculus, were introduced in Church [1940], though the question of interpretations of the system was deliberately left open. Definitions that more or less resemble our notion of type structure have appeared very many times in the literature: the first such was given in Henkin [1950], who considered interpretations of Church's system in the type structure S.

Particular instances of the extensional collapse construction (in the form in which we have defined it) appear in Kreisel [1959] and Kleene [1959a]. The general construction was probably folklore from an early stage, though its debut in the literature seems to be in Zucker [1971].

The connections between λ-calculi and cartesian closed categories were established in the 1970s by Lambek and others (see Lambek and Scott [1986]).

§2. Early work: Computability at type 2.

2.1. Prehistory. The main ideas concerning computability for type 1 functions of course date back to the development of basic recursion theory in the 1930s (Gödel [1931], Church [1936], Turing [1937b], [1937a], Kleene [1936a], [1936b], Post [1936]). These early papers furnished several characterizations of the class of (partial) recursive functions. The first explicit formulation of Church's thesis appears in Church [1936]. In Turing [1939, §4] Turing introduced the notion of a computing machine equipped with an oracle for deciding non-computable properties, but considered this only as a means of defining first order computability relative to a *fixed* oracle, so cannot truly be said to have introduced the concept of a computable type 2 function.

2.2. Banach-Mazur functionals. A very early definition of a class of type 2 functionals involving a notion of computability is due to Banach and Mazur (Banach and Mazur [1937]) (see also Mazur [1963]):

DEFINITION 2.1. *A total function* $F : \mathbb{N}^{\mathbb{N}}_{rec} \to \mathbb{N}$ *is* Banach-Mazur *if, for every total recursive function* $h : \mathbb{N} \times \mathbb{N} \to \mathbb{N}$, *the function* $\Lambda x.F(\Lambda y.h(x, y)) : \mathbb{N} \to \mathbb{N}$ *is total recursive.*

Notice that this condition says that, in some sense, F carries computable functions to computable functions, but it does not tell us how given g one might compute $F(g)$ in any sense. For this reason, the notion is rather tangential to the story we tell here — we do not regard it as a genuine candidate for a notion of computable functional, but rather as a property which computable functionals may possess. Early results showed that every computable functional (in the senses discussed below) is Banach-Mazur, but not *vice versa* (see Friedberg [1958a]), and some relationships to other properties of functionals were considered in Pour-El [1960]. Most of this material is helpfully summarized in Rogers [1967, §15.3].

The Banach-Mazur functionals later reappeared in the work of Lawvere and Mulry (Mulry [1982]), whose *recursive topos* provides a natural generalization of the notion to higher types (see also Section 4.2.4).

2.3. Computations on pure functions. The first explicit definition of a genuine notion of type 2 computability, as far as we are aware, was given by

FIGURE 1. History of higher type computability: a selective outline.

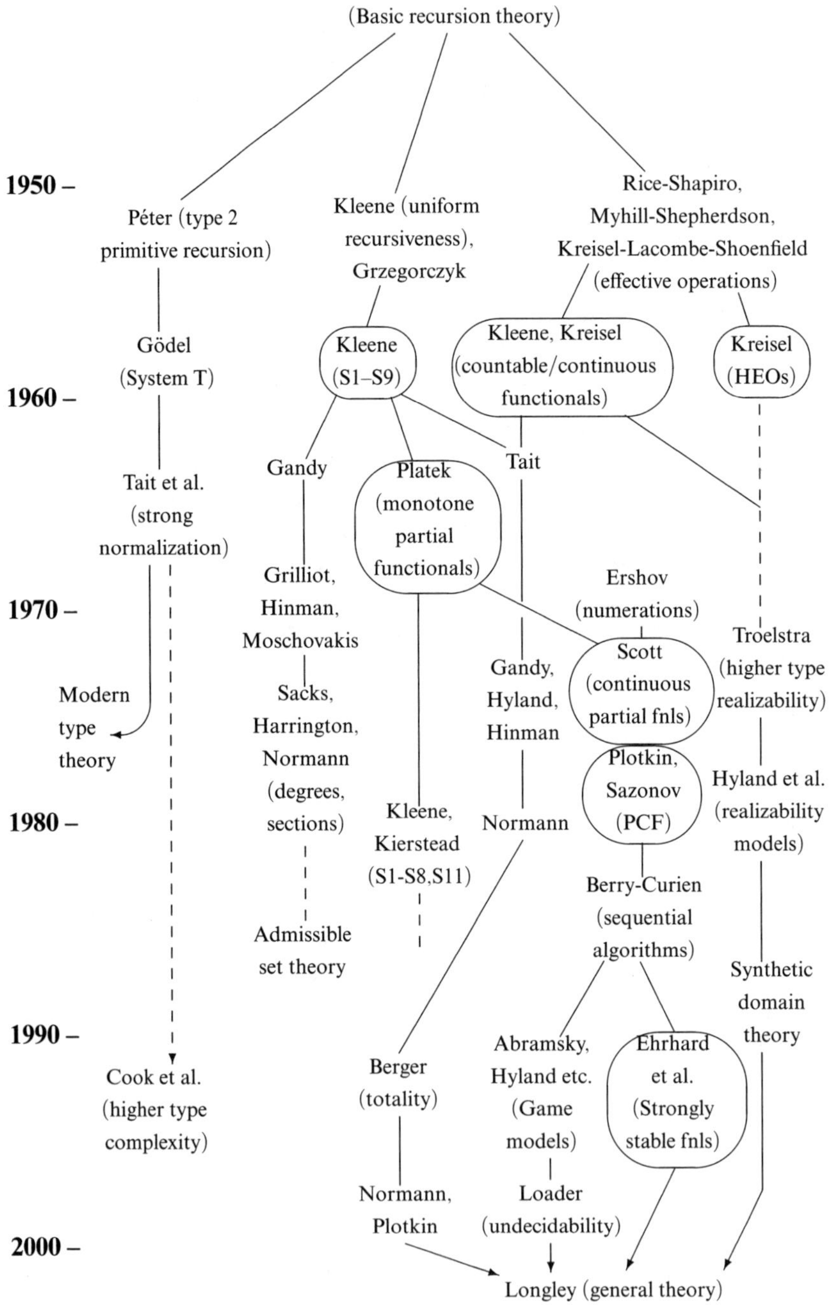

Péter in Péter [1951a], [1951b] (although a somewhat similar definition had been sketched in Hilbert [1925]). Péter considered a schema for "primitive recursion of the second degree" as a means of defining total functions with arguments of type $\mathbb{N}^{\mathbb{N}}$ as well as $\mathbb{N}$. Sacrificing some generality for the sake of clarity, the basic idea is as follows: given previously defined functionals G, H of suitable types, one may construct a new function $F : \mathbb{N} \times \mathbb{N} \to \mathbb{N}$ such that

$$\begin{aligned} F(0, m) &= G(m), \\ F(n+1, m) &= H(\Lambda y.F(n, y), n, m). \end{aligned}$$

By allowing additional parameters $g_1, \ldots, g_r$ of type $\mathbb{N}^{\mathbb{N}}$, we are able to construct new type 2 functionals from old ones. We may then define a class of *type 2 primitive recursive* functionals by starting from a suitable set of "basic" functionals and closing under substitution and primitive recursions of the above kind.

Péter showed that a certain class of "transfinite recursions" at type 1 could be systematically replaced by primitive recursions at type 2. Thus, for instance, the well-known *Ackermann* function, though not primitive recursive in the usual sense, could be defined by a type 2 primitive recursion.

In his famous book (Kleene [1952]), Kleene gave schemata for defining primitive, total and partial recursive functions *uniformly in* a finite list of functions $g_1, \ldots, g_l \in \mathbb{N}^{\mathbb{N}}$ (§47, §58, §63). Restricting for simplicity to the case $l = 1$, the definitions are as follows.

Definition 2.2 (Uniform computability).

(i) *A function* $f : \mathbb{N}^k \to \mathbb{N}$ *is* primitive recursive uniformly in g *if* f *can be defined from* g *(together with the usual repertoire of basic functions) by means of composition and (ordinary first order) primitive recursion.*
(ii) $f : \mathbb{N}^k \rightharpoonup \mathbb{N}$ *is* partial recursive uniformly in g *if* f *can be defined from* g *plus the basic functions via composition, primitive recursion and minimization. If in addition the definition of* f *results in a total function for all values of* $g \in \mathbb{N}^{\mathbb{N}}$, *we say* f *is* total recursive uniformly in g.

Kleene made explicit the possibility of regarding g as a variable and thus obtaining type 2 functionals $F : \mathbb{N}^{\mathbb{N}} \times \mathbb{N}^k \rightharpoonup \mathbb{N}$. Note that Kleene's notion of primitive recursive functional is more restrictive than Péter's, because whereas in Kleene's definition type 1 functions simply enter the computation as additional basic functions which remain fixed throughout, Péter's scheme in some sense allows infinitely many values of F to be collected into a new type 1 function during the course of a recursive computation. Kleene's definition yields just the usual class of primitive recursive functions at type 1.

Kleene also showed that his partial recursive functionals can be characterized equivalently as those computable by Turing machines with oracles (*op. cit.*, Chapter XIII).[2]

In two papers from 1954, Grzegorczyk (Grzegorczyk [1955b], [1955a]) considered a class of computable total functionals with arguments in $\mathbb{N}$ and $\mathbb{N}^{\mathbb{N}}$, defined as the smallest class containing some basic functionals and closed under substitution and minimization. It is fairly easy to see directly (and is immediate from Kleene's later results) that Grzegorczyk's notion coincides with Kleene's notion of uniform total recursiveness, though this does not seem to have been noted in the literature of the time. In Grzegorczyk [1955b], Grzegorczyk showed that his definition was equivalent to an Herbrand-Gödel style definition in terms of an equational calculus. In Grzegorczyk [1955a] he investigated some properties of this notion, showing that all computable functionals were *continuous*, and that certain *modulus of continuity* functionals were computable.

The relationship between computability and continuity was to become a recurring theme in the subject. In the case of Grzegorczyk's result (and others in the same vein), the intuition is simple: if we apply a computable type 2 function F to a type 1 function g, the "computation" of $F(g)$ can only interrogate g at finitely many arguments before terminating, so for any other function g' agreeing with g on this finite portion we would have $F(g) = F(g')$. Thus F is continuous with respect to the familiar Baire topology.

2.4. Computations on Kleene indices. All the approaches to type 2 computability mentioned above share the feature that the type 1 arguments are presented simply as oracles or "black boxes": the only way to interact with them is to feed them with an argument and observe the outcome. In parallel with this was another strand of research which considered notions of type 2 computability in which the type 1 arguments were presented as recursive indices (say, as Gödel numbers for Turing machines). Given a recursive index we can of course do everything that we can do with an oracle, since we can always apply the index to a type 0 argument. However, it would seem *a priori* that one might be able to do more with an index than with an oracle, since we are given extra *intensional* information — intuitively, we can "look inside" the box and examine how the machine or program is working.

The following definition will allow us to state the main results succinctly:

DEFINITION 2.3 (Effective operation). *Let R be either of the sets $\mathbb{N}^{\mathbb{N}}_{rec}$ or $\mathbb{N}^{\mathbb{N}}_{p\,rec}$. A partial function $F : R \rightharpoonup \mathbb{N}$ is a* partial effective operation *on R*

[2]Kleene introduced, somewhat peripherally, a notion of partial recursiveness uniformly in a list of *partial* functions $\mathbb{N} \rightharpoonup \mathbb{N}$ (Kleene [1952, §63]), but did not study it in detail. As pointed out in Platek [1966, pp. 128–130], Kleene's particular definition has some rather undesirable features; nevertheless, it turns out to give rise to the class of parallel-computable type 2 functions, see Section 4.2.

if there exists a partial recursive function $f : \mathbb{N} \rightharpoonup \mathbb{N}$ *such that for any* $m \in \mathbb{N}$ *with* $\phi_m \in R$ *we have* $F(\phi_m) \simeq f(m)$. *If additionally* F *is total, we say it is a* total effective operation *on* R.

Thus, an effective operation is given by a computable function f acting on recursive indices, which f may manipulate in any way it likes, subject only to the requirement of *extensionality*: f must give the same result on different indices for the same type 1 function.

In fact, all four cross-combinations of total and partial function spaces were investigated in the early literature. First, a theorem of Rice (Rice [1953]), originally phrased in terms of decidable properties of r.e. sets, essentially says the following:

THEOREM 2.4 (Total acting on partial). *Every total effective operation on* $\mathbb{N}^{\mathbb{N}}_{p\,rec}$ *is constant.*

This led, via a theorem of Rice and Shapiro (Rice [1956]), to the following important result, often known as the Myhill-Shepherdson theorem. It was first obtained in Myhill and Shepherdson [1955], and independently in Uspenskii [1955]. A related result also appeared in Nerode [1957]. Here we take $\theta \mapsto \widehat{\theta}$ to be an effective coding of the graphs of *finite* partial functions $\theta : \mathbb{N} \rightharpoonup \mathbb{N}$ as natural numbers.

THEOREM 2.5 (Partial acting on partial). *A function* $F : \mathbb{N}^{\mathbb{N}}_{p\,rec} \rightharpoonup \mathbb{N}$ *is a partial effective operation iff*

- F *is monotone and continuous (i.e.,* F *preserves existing least upper bounds of chains in* $\mathbb{N}^{\mathbb{N}}_{p\,rec}$*), and*
- F *acts effectively on finite elements (i.e., there is a partial recursive function* h *such that for every finite* $\theta : \mathbb{N} \rightharpoonup \mathbb{N}$ *we have* $F(\theta) = h(\widehat{\theta})$*).*

Moreover, every partial effective operation F *on* $\mathbb{N}^{\mathbb{N}}_{p\,rec}$ *extends uniquely to a monotone and continuous function* $\overline{F} : \mathbb{N}^{\mathbb{N}}_{p} \rightharpoonup \mathbb{N}$.

The other main positive result, often called the Kreisel-Lacombe-Shoenfield theorem, was obtained slightly later in Kreisel, Lacombe, and Shoenfield [1957], [1959], and independently in Tseitin [1959]. Our formulation here is somewhat less general than the original version.

THEOREM 2.6 (Total acting on total). *A function* $F : \mathbb{N}^{\mathbb{N}}_{rec} \to \mathbb{N}$ *is a total effective operation iff it is the restriction of a uniformly partial recursive function* $\overline{F} : \mathbb{N}^{\mathbb{N}} \rightharpoonup \mathbb{N}$ *(see Definition 2.2) whose domain contains* $\mathbb{N}^{\mathbb{N}}_{rec}$.

It follows that every total effective operation on $\mathbb{N}^{\mathbb{N}}_{rec}$ is continuous with respect to the usual Baire topology. Proofs of Theorem 2.6 may be found in many texts, *e.g.*, Rogers [1967], Beeson [1985], Aberth [1980], Odifreddi [1989]. A very short proof using Kleene's second recursion theorem was given by Gandy (Gandy [1962]), though it is perhaps too serendipitous for most ordinary mortals. A unified result subsuming both Theorems 2.5 and 2.6 may be found in Spreen and Young [1983].

Theorems 2.5 and 2.6 both provide very striking examples of the connection between computability and continuity. These results obviously say something deeper than the result of Grzegorczyk mentioned at the end of Section 2.3, since there we could interact with a type 1 function only as a black box, while here we have access to a recursive index for it. However, the final theorem of our quartet, due to Friedberg (Friedberg [1958b]), shows that the connection with continuity breaks down if totality and partiality are mixed:

THEOREM 2.7 (Partial acting on total). *There exists a partial effective operation on* $\mathbb{N}^{\mathbb{N}}_{rec}$ *which is not continuous (hence not the restriction of a Kleene partial recursive functional on* $\mathbb{N}^{\mathbb{N}}$*).*

The above theorems and their proofs are covered in Rogers [1967, §15.3].

Most of the remaining natural questions about type 2 computability turn out to have negative answers. For example, not every total effective operation on $\mathbb{N}^{\mathbb{N}}_{rec}$ is the restriction of a uniformly total recursive function on $\mathbb{N}^{\mathbb{N}}$. This can be shown using the celebrated *Kleene tree*, one of the most important counterexamples in the subject.

THEOREM 2.8 (Kleene tree). *There exists a binary tree* B *(i.e., a prefix-closed set of finite sequences over* $\{0, 1\}$*) such that*

- B *is primitive recursive (that is, there is a primitive recursive function* b *such that* $b\langle\alpha\rangle = 0$ *iff* $\alpha \in B$*).*
- B *contains finite paths of arbitrary length (so classically, by König's Lemma,* B *contains infinite paths).*
- B *contains no* recursive *infinite paths.*

The Kleene tree is so named after its appearance as Theorem LII of Kleene [1959b], although examples of the same phenomenon also appeared in Lacombe [1955d] and Zaslavskii [1955], Zaslavskii and Tseitin [1962]. We may now obtain a total effective operation Z that does not extend to a uniformly total recursive function on $\mathbb{N}^{\mathbb{N}}$, by defining

$$Z(f) = \mu n.\ [f(0), \ldots, f(n-1)] \notin B,$$

where μ is the minimization operator of ordinary recursion theory. This example also sheds light on the meaning of continuity in the foregoing results: Z is continuous on $\mathbb{N}^{\mathbb{N}}_{rec}$ by Theorem 2.6, but it cannot be extended even to a continuous function on the whole of $\mathbb{N}^{\mathbb{N}}$.

Two more negative results will help to complete the picture at type 2. Firstly, a plausible analogue of Theorem 2.5 fails for total functionals acting on total functions: not every continuous functional which acts effectively on a suitable basis of "finite" type 1 functions (namely, the eventually constant functions) is an effective operation. A simple counterexample was given in Pour-El [1960]; see also Rogers [1967, §15.3]. Secondly, the analogue of Theorem 2.6 for partial functionals acting on partial functions fails because the uniformly partial recursive functions $\mathbb{N}_p^{\mathbb{N}} \rightharpoonup \mathbb{N}$ are all *sequentially* computable, whereas the partial effective operations include *parallel* functions. This distinction was first made explicit in Platek [1966], and will be discussed at length in Section 4.

These early results show that the situation at type 2 is already quite complicated, and they serve to illustrate many of the general characteristics of the subject. Firstly, they exhibit a considerable diversity of approaches to defining a notion of "computable functional". Secondly, they show how quite different approaches sometimes lead to the same class of functionals, providing evidence for the intrinsic importance of this class (Theorem 2.6 is a good example of this). On the other hand, we have seen that not all definitions of computability conveniently collapse to a single notion, and that an important role is played by negative results and counterexamples. Finally, we have already seen several instances of the connection between computability and continuity.

§3. Total computable functionals. Once various notions of type 2 computability had been considered, the possibility of trying to extend them to higher types was obvious. (Péter speculated informally on this possibility in Péter [1951b], but did not develop it.) The years 1957–59 saw the appearance of no fewer than four important notions of higher type computability, represented by Gödel's System T, Kleene's schemata S1–S9, the Kleene-Kreisel countable or continuous functionals (and their effective substructure), and Kreisel's hereditarily effective operations.

It is worth noting that all these notions were concerned primarily with hereditarily *total* functionals — good notions of computability for hereditarily partial functionals of all higher types were a later development (see Section 4). A qualification is needed here in the case of S1–S9, since (as we shall see) this notion naturally gives rise to partial computable functionals, albeit acting on hereditarily total objects. However, the partial functionals that arise in this way are somehow second-class citizens, since they cannot themselves serve as arguments to functionals of higher types. It therefore seems natural to include S1–S9 in our discussion of hereditarily total notions of computability.

Roughly speaking, each of the four notions mentioned above gave rise to its own strand of research, so that the study of higher type computability appears somewhat fragmented from 1960 onwards. We now consider these notions in turn and the lines of research they gave rise to.

3.1. System T and related type systems. Gödel's System T, introduced in Gödel [1958], provides a higher type analogue of the notion of primitive recursive computation. It is one of a number of syntactic formalisms that define restricted classes of computable functionals, in the sense that the computational complexity of definable functionals is somehow limited. These formalisms fall somewhat outside our primary area of interest, since they make no claim to defining a "complete" class of computable functionals in any interesting sense, but they are close enough to our concerns to merit some attention.

3.1.1. *Gödel's T.* System T is essentially a simply typed λ-calculus over the ground type $\overline{0}$, with constants for zero, successor and primitive recursors of higher type:

$$\mathtt{0}\colon \overline{0}, \quad \mathtt{succ}\colon \overline{0} \to \overline{0}, \quad \mathtt{rec}_\sigma\colon \sigma \to (\sigma \to \overline{0} \to \sigma) \to \overline{0} \to \sigma.$$

These are equipped with the following reduction rules in addition to the usual β-rule and congruence rules given in Section 1.5.2:

$$\mathtt{rec}_\sigma UV\mathtt{0} \rightsquigarrow U$$
$$\mathtt{rec}_\sigma UV(\mathtt{succ}\, n) \rightsquigarrow V(\mathtt{rec}_\sigma UVn)n$$

(where n is a variable of type $\overline{0}$). It is fairly easy to see that System T naturally extends Péter's notion of second order primitive recursion to higher types (see Section 2.3).

One can consider this system either as a standalone syntactic formalism, or as a language for denoting functionals in some given type structure, as in Section 1.5.2. For instance, if S is the full type structure over $\mathbb{N}$ defined in Section 1.5.1, then any closed term $U\colon \sigma$ of System T denotes an element $[\![U]\!] \in \mathsf{S}_\sigma$ in an obvious way.

In fact, Gödel's original presentation of System T was as a logical system for reasoning about higher type objects — in place of our reduction rules $U \rightsquigarrow U'$ he gave equational axioms $U = U'$. Gödel's purpose was to give an interpretation of first order Heyting arithmetic (the so-called *Dialectica interpretation*) in which first order sentences involving complex nestings of quantifiers were replaced by logically simple sentences involving objects of higher type. Specifically, Gödel gave a translation from formulae $\phi(\boldsymbol{x})$ of Heyting arithmetic to formulae $\phi'(\boldsymbol{x})$ of the form $\exists \boldsymbol{y}.\forall \boldsymbol{z}.\phi^*(\boldsymbol{x},\boldsymbol{y},\boldsymbol{z})$, in which the variables in $\boldsymbol{y},\boldsymbol{z}$ may be of arbitrary finite type, and ϕ^* is just a propositional combination of equations between terms of System T. For the sake of completeness we give the definition of the translation here, though the details are not too important for our purposes:

Definition 3.1 (Dialectica translation). *For atomic formulae α, define $\alpha' \equiv \alpha$. Given $\phi'(\boldsymbol{x}) \equiv \exists \boldsymbol{y}.\forall \boldsymbol{z}.\phi^*(\boldsymbol{x},\boldsymbol{y},\boldsymbol{z})$ and $\psi'(\boldsymbol{u}) \equiv \exists \boldsymbol{v}.\forall \boldsymbol{w}.\psi^*(\boldsymbol{u},\boldsymbol{v},\boldsymbol{w})$, define*

$$(\phi \wedge \psi)' \equiv \exists \boldsymbol{y}\boldsymbol{v}.\forall \boldsymbol{z}\boldsymbol{w}.\phi^*(\boldsymbol{x},\boldsymbol{y},\boldsymbol{z}) \wedge \psi^*(\boldsymbol{u},\boldsymbol{v},\boldsymbol{w}),$$
$$(\phi \vee \psi)' \equiv \exists \boldsymbol{y}\boldsymbol{v}t.\forall \boldsymbol{z}\boldsymbol{w}.(t = 0 \wedge \phi^*(\boldsymbol{x},\boldsymbol{y},\boldsymbol{z})) \vee (t = 1 \wedge \psi^*(\boldsymbol{u},\boldsymbol{v},\boldsymbol{w})),$$

$$(\phi \Rightarrow \psi)' \equiv \exists \boldsymbol{V}\boldsymbol{Z}.\forall \boldsymbol{y}\boldsymbol{w}.\phi^*(\boldsymbol{x},\boldsymbol{y},\boldsymbol{Z}(\boldsymbol{y},\boldsymbol{w})) \Rightarrow \psi^*(\boldsymbol{u},\boldsymbol{V}(\boldsymbol{y}),\boldsymbol{w})^3,$$
$$(\forall s.\phi)' \equiv \exists \boldsymbol{Y}.\forall s\boldsymbol{z}.\phi^*(\boldsymbol{x},\boldsymbol{Y}(s),\boldsymbol{z}),$$
$$(\exists s.\phi)' \equiv \exists s\boldsymbol{y}.\forall \boldsymbol{z}.\phi^*(\boldsymbol{x},\boldsymbol{y},\boldsymbol{z}).$$

It can be shown that if ϕ is provable in HA then there are terms $\boldsymbol{Y}$ of System T such that $\phi^*(\boldsymbol{x}, \boldsymbol{Y}(\boldsymbol{x}), \boldsymbol{z})$ is provable in System T using just *quantifier-free* intuitionistic logic. The terms $\boldsymbol{Y}$ here can be thought of as embodying the constructive content of the proof of ϕ. We may now interpret the terms $\boldsymbol{Y}$, and the proof of $\phi^*(\boldsymbol{x}, \boldsymbol{Y}(\boldsymbol{x}), \boldsymbol{z})$, in any type structure satisfying the axioms of System T. We thus obtain a "functional interpretation" of Heyting arithmetic, which can be used to prove the consistency of Heyting arithmetic, and hence (in view of Gödel's double-negation translation) that of Peano arithmetic.[4]

However, the interest of System T seems to go beyond this original application. Gödel himself gave a proof-theoretic characterization of the expressive power of System T: the T-definable type 1 functions (in S, say) are precisely the functions provably total in first order Heyting or Peano arithmetic. Grzegorczyk (Grzegorczyk [1964]) gave some alternative characterizations of the System T definable functionals in the style of combinatory logic. Tait and several others (Tait [1967], Dragalin [1968], Hinata [1967], Hinatani [1966], Sanchis [1967], Shoenfield [1967]) proved independently that System T is *strongly normalizing* — that is, for any System T term, all reduction paths terminate yielding the same final value, or *normal form*. (A remarkable alternative proof of this was later given in Gandy [1980].) This important result implies that any term model for System T is itself a type structure with all the properties needed for the Dialectica interpretation. Thus, a functional interpretation of arithmetic can be given in terms of System T itself, without the need for an interpretation of System T in another type structure. (This idea was already implicit in a remark in Kreisel [1959, §3.4].)

The Dialectica interpretation also inspired work on interpretations of systems for analysis (or second order arithmetic) in a similar spirit (Kreisel [1959], Spector [1962]); however, these go beyond System T itself. For instance, Spector's system includes an additional operator for defining objects by means of

---

[3]Here, for instance, if $\boldsymbol{Z}$ abbreviates $Z_1 \ldots Z_r$, we write $\boldsymbol{Z}(\boldsymbol{y},\boldsymbol{w})$ to abbreviate $Z_1(\boldsymbol{y},\boldsymbol{w})$, $\ldots, Z_r(\boldsymbol{y},\boldsymbol{w})$. Gödel motivated the clause for implication as follows. We identify a proposition $(\exists \boldsymbol{y}.\cdots) \Rightarrow (\exists \boldsymbol{v}.\cdots)$ with the existence of computable functions $\boldsymbol{V}$ that to each sequence $\boldsymbol{y}$ making the antecedent true assign a sequence $\boldsymbol{v}$ making the consequent true. Moreover, we identify a proposition $(\forall \boldsymbol{z}.\cdots) \Rightarrow (\forall \boldsymbol{w}.\cdots)$ with the existence of computable functions $\boldsymbol{Z}$ that to each sequence $\boldsymbol{w}$ making the consequent false assign a sequence $\boldsymbol{z}$ making the antecedent false.

[4]All that is required for Gödel's argument is that *some* type structure satisfying the axioms of System T exists. In Gödel [1958], Gödel seemingly regarded the existence of a suitable notion of computable operation of finite type as immediately apparent. A footnote in a revised version of the paper (Gödel [1972]), however, suggests he had in mind the structure HRO (see Section 3.4).

bar recursion (an intuitionistic principle considered by Brouwer); this system admits a natural interpretation in the type structure of total continuous functionals (see Section 3.3).

Another set of questions concern *theories* for System T. For example, what equivalence relations on System T terms are induced by their interpretation in various type structures? The largest reasonable theory, which we shall call $\approx$, is the one given by the notion of "observational equivalence": $M \approx N$ at type σ iff for all closed $P\colon \sigma \to \overline{0}$, PM and PN reduce to the same normal form. An argument due to Kreisel (appearing as one of the final set of exercises in Barendregt [1984]) shows that the interpretation of System T in S induces a strictly smaller theory — in computer science terms, this interpretation is not *fully abstract*. By contrast, Loader has shown that the interpretation of System T in the type structure HEO (see Section 3.4) induces exactly the theory $\approx$ (Loader [1997]).

Other mathematical results pertaining specifically to System T include the theorems of Howard (Howard [1973]; see also Girard [1987, annex 7.A]) that all T-definable functionals are *hereditarily majorizable*, and hence that there is no T-definable *modulus of extensionality* functional. A further mathematical analysis of System T in terms of functors over the category of ordinals is given in Päppinghaus [1985] (see also Girard [1988]).

Further discussions of System T can be found in Kreisel [1959], Troelstra [1973, §3.5], Barendregt [1984, Appendix A.2], and Girard [1987, annex 7.A]. A good survey of later research related to System T can be found in Section 5 of Troelstra's introductory note to Gödel's original paper in Gödel [1990]. For more on the Dialectica interpretation, see Avigad and Feferman [1998].

3.1.2. *Kleene's S1–S8.* We have already observed that System T can be interpreted in various type structures, such as the full set-theoretic type structure S. Indeed, one reason why System T is interesting from our point of view is that it admits an interpretation of this kind in practically all the type structures we shall have occasion to consider. It therefore provides a kind of common "skeleton" for all these type structures, and gives us an effectively enumerable class of "total" functions that can play a role analogous to that of the primitive recursive functions in ordinary recursion theory.

However, in this regard there is nothing particularly unique about System T — many other systems would serve the same purpose. For instance, a weaker form of primitive recursion is given by the typed λ-calculus with constants 0, succ and

$$\widehat{\mathtt{rec}}_\sigma : \sigma \to (\overline{0} \to \overline{0} \to \sigma) \to \overline{0} \to \sigma$$

together with reduction rules

$$\widehat{\mathtt{rec}}_\sigma UV0\boldsymbol{W} \rightsquigarrow U\boldsymbol{W}$$
$$\widehat{\mathtt{rec}}_\sigma UV(\mathtt{succ}\, n)\boldsymbol{W} \rightsquigarrow V(\widehat{\mathtt{rec}}_\sigma UVn\boldsymbol{W})n\boldsymbol{W}$$

where $\sigma = \sigma_1 \to \cdots \to \sigma_n \to \overline{0}$ and $\boldsymbol{W}$ textually abbreviates $W_1 \dots W_n$.[5] This system, which we will call System $\widehat{\mathrm{T}}$, corresponds exactly in expressive power to Kleene's schemata S1–S8 (see Section 3.2 below); Kleene referred to the elements of S definable by these means as the *primitive recursive functionals*. Whereas System T offers a natural higher type generalization of Péter's notion of type 2 primitive recursiveness, System $\widehat{\mathrm{T}}$ offers a similar generalization of Kleene's notion (Definition 2.2(i)). Thus, System T can also define functions such as the Ackermann function, whereas the type 1 functions definable in System $\widehat{\mathrm{T}}$ are just the primitive recursive functions in the usual sense (see Kleene [1959b, §1]).

In some sense S1–S8 offers a better common core of "simple" functionals than System T, insofar as it can be interpreted in a wider class of type structures. It would seem, though, that overall Kleene's S1–S8 holds less mathematical interest than Gödel's T.

System T also provided a point of departure for two other strands of later work, namely modern type theory (which typically considers stronger systems than System T), and higher type complexity theory (which typically considers weaker systems). Both of these are significant areas of research in theoretical computer science. For detailed information on modern type theory, a good reference is Barendregt [1992]; for higher type complexity, we refer the reader to the surveys in the recent articles of Irwin, Kapron, and Royer [2001a], [2001b].

3.2. Kleene computability: S1–S9.

3.2.1. *Kleene's work.* The first serious attempt at a full-blown generalization of the notion of recursive function to all type levels was Kleene's definition of higher type computability via the schemata S1–S9 (Kleene [1959b], [1963]). This can be seen as generalizing his notion of partial recursiveness at type 2 (Definition 2.2). In Kleene's approach to higher type computation, we suppose we are given a type structure A of hereditarily total functionals over $\mathbb{N}$, and we define a class of computable *partial* functionals over A by means of certain computation schemata. However, as mentioned at the beginning of Section 3, the computable partial functionals appear to play a less central role than the total ones.

The spirit behind Kleene's definition is easily grasped: we are allowed to perform effective computations involving elements of A, in which the elements of function type are treated simply as *oracles* or *black boxes*. In particular, we may feed an element of function type with numbers or functions that are themselves computable (in the same sense), and observe the numerical results, but we are not granted access to information about such elements in any other way. Thus, Kleene's definition embodies the ideal of computing with

---

[5]The notation $\widehat{\mathtt{rec}}$ is adapted from Feferman [1977b], where the constants $\mathtt{rec}_\sigma$ are called $\mathbb{N}$-*recursion operators* and the $\widehat{\mathtt{rec}}_\sigma$ are called *elementary recursion operators*.

functions as pure extensions, without reference to how they are represented or implemented. Indeed, A will in general include non-computable objects, so that we have to imagine the corresponding oracles as working "by magic".

Kleene in fact concentrated his attention on computations over objects of *pure* type (see Section 1.5.1), though this was not an especially significant decision. Because of its historical importance, we reproduce Kleene's original definition here in all its glory, with slightly modified notation.[6] Given a type structure A, we write $X(A)$ for the disjoint union of all sets $A_{\sigma_1} \times \cdots \times A_{\sigma_r}$ $(r \geq 1)$ where the σ_i are pure types. We write $\#-$ for the coding of pure types as natural numbers given by $\#\overline{n} = n$, and take $\langle \cdots \rangle$ to be an effective coding of finite sequences of natural numbers. Throughout this section we will use $x : \sigma$ as an abbreviation for $x \in A_\sigma$, rather than as a formal expression. We also abbreviate $x_1, \ldots, x_r$ by $\boldsymbol{x}$, $x_1 : \sigma_1, \ldots, x_r : \sigma_r$ by $\boldsymbol{x} : \boldsymbol{\sigma}$, and $\langle \#\sigma_1, \ldots, \#\sigma_r \rangle$ by $\#\boldsymbol{\sigma}$.

The definition proceeds by introducing a system of *indexing* whereby natural numbers m encode definitions of computable partial functions $\{m\}$; this is reminiscent of the indexing of partial recursive functions in ordinary recursion theory.

DEFINITION 3.2 (Kleene computability). *Let A be an extensional type structure over $\mathbb{N}$.*

(i) *We inductively define a partial function $\{-\}(-) : \mathbb{N} \times X(A) \rightharpoonup \mathbb{N}$, called* index application, *by means of the following clauses (which we interpret as applying whenever they are well-typed).*

S1: Successor function: $\{\langle 1, \#\boldsymbol{\sigma} \rangle\}(\boldsymbol{x} : \boldsymbol{\sigma}) = x_1 + 1$.

S2: Constant functions: *For any $q \in \mathbb{N}$,* $\{\langle 2, \#\boldsymbol{\sigma}, q \rangle\}(\boldsymbol{x} : \boldsymbol{\sigma}) = q$.

S3: Projection: $\{\langle 3, \#\boldsymbol{\sigma} \rangle\}(\boldsymbol{x} : \boldsymbol{\sigma}) = x_1$.

S4: Composition: *For any $g, h \in \mathbb{N}$,*

$$\{\langle 4, \#\boldsymbol{\sigma}, g, h \rangle\}(\boldsymbol{x} : \boldsymbol{\sigma}) \simeq \{g\}(\{h\}(\boldsymbol{x}), \boldsymbol{x}).$$

S5: Primitive recursion: *For any $g, h \in \mathbb{N}$, if $m = \langle 5, \#\boldsymbol{\sigma}, g, h \rangle$ then*

$$\{m\}(0, \boldsymbol{x}) \simeq \{g\}(\boldsymbol{x}), \quad \{m\}(n+1, \boldsymbol{x}) \simeq \{h\}(n, \{m\}(n, \boldsymbol{x}), \boldsymbol{x}).$$

S6: Permutation of arguments: *For any $g \in \mathbb{N}$ and $1 \leq k < r$,*

$$\{\langle 6, \#\boldsymbol{\sigma}, k, g \rangle\}(\boldsymbol{x} : \boldsymbol{\sigma}) \simeq \{g\}(x_{k+1}, x_1, \ldots, x_k, x_{k+2}, \ldots, x_r).$$

S7: Type 1 application: *If $\sigma_1 = \overline{1}$ and $\sigma_2 = \overline{0}$, then*

$$\{\langle 7, \#\boldsymbol{\sigma} \rangle\}(\boldsymbol{x} : \boldsymbol{\sigma}) = x_1(x_2).$$

---

[6]Kleene's definition was restricted to the case $A = \mathsf{S}$. Moreover, Kleene adopted the peculiar convention that only the order of arguments within each type was material, so his version of S6 was slightly different from the one given here.

S8: Higher type application: *For any* $h \in \mathbb{N}$ *and* $t \geq 2$, *we have*

$$\{\langle 8, \#\boldsymbol{\sigma}, t, h\rangle\}(y : \bar{t}, \boldsymbol{x} : \boldsymbol{\sigma}) \simeq y(\Lambda z : \overline{t-2}.\, \{h\}(y, z, \boldsymbol{x}))$$

provided $\{h\}(y, z, \boldsymbol{x})$ *is defined for all* $z \in A_{t-2}$.

S9: Index invocation: $\{\langle 9, \#\boldsymbol{\sigma}, \#\boldsymbol{\tau}\rangle\}(n : \bar{0}, \boldsymbol{y} : \boldsymbol{\sigma}, \boldsymbol{z} : \boldsymbol{\tau}) \simeq \{n\}(\boldsymbol{y})$.

(ii) *We say* $F : A_{\sigma_1} \times \cdots \times A_{\sigma_r} \rightharpoonup \mathbb{N}$ *is* Kleene computable over A *if there is a number* m *(called an* index *for* F*) such that* $F(\boldsymbol{x}) \simeq \{m\}(\boldsymbol{x})$ *for all* $\boldsymbol{x}$.

One may wonder why Kleene, one of the pioneers of the λ-calculus, did not make greater use of the typed λ-calculus in formulating this notion of computability. It seems that his early attempts to give talks based on his λ-calculus work before the latter had achieved wide currency were less well-received than the above definition via indices. A definition of Kleene computability in a more modern spirit will be given in Part II.

As already noted, S1–S8 by themselves define a perfectly good class of *total* functionals — the schema S9 is the only one that introduces partiality into the definition. Notice that even if the indexing system were redesigned so that every natural number indexed precisely one function of each type, partial functions would still arise for reasons of circularity. Specifically, if m were an index for the function $F = \Lambda n.\, \{n\}(n)$ then $F(m)$ would be undefined according to the inductive definition above, since the clause for S9 would tell us merely that $\{m\}(m) \simeq \{m\}(m)$. That is, a value for $\{m\}(m)$ would appear at some stage in the inductive generation of $\{-\}(-)$ only if it had already appeared at some previous stage.

Kleene generally conceived computations in this setting as infinitely branching trees of transfinite depth — when invoking S8 one computes the entire graph of $\Lambda z.\{h\}(y, z, \boldsymbol{x})$ by means of auxiliary computations before presenting it to the oracle y. This is perhaps not the only possible conception: for example, one might imagine that the magic of the oracles included the ability to recognize functions from finitary descriptions of them, in which case these auxiliary computations would not be needed. But whatever conception one adopts, Kleene's schemata exhibit a curious tension between effective, finitary computational processes and calls to numinous or infinitistic oracles, so that the computational significance of Kleene's definition is rather difficult to gauge.

Kleene's motivation seems to have sprung from two distinct strands of his earlier work. On the one hand, he was certainly seeking an appropriate higher type generalization of the basic notions of ordinary recursion theory — including, if possible, some analogue of Church's thesis at higher types. On the other hand, he was seeking to explain his theory of logical complexity for classical predicate logic (embodied in the arithmetic and analytic hierarchies, as in Kleene [1955a], [1955b]) in terms of computability relative to certain higher type objects (hence the emphasis on *quantifiers* in his papers on higher type computability). This latter motivation goes some way towards explaining two features of Kleene's definition that sometimes appear puzzling to modern readers: the restriction to computations on *total* functionals, and the specialization to the type structure S rather than any more constructively given class of functionals.

Kleene developed some basic results for this notion of computability analogous to those of ordinary recursion theory: for example, higher type versions of the normal form theorem (Kleene [1959b, §5]), and a restricted version of the first recursion theorem (Kleene [1963, §10]). He also showed that the schema S9 is strictly more powerful than the μ-recursion (minimization) schema familiar from ordinary recursion theory (Kleene [1959b, §8]), and argued that the former gives rise to a more compelling notion of higher type computability. In addition, Kleene introduced the natural notion of relative computability in the setting of S1–S9:

DEFINITION 3.3 (Relative Kleene computability).[7]

(i) *A partial function* $g : A_{\sigma_1} \times \cdots \times A_{\sigma_r} \rightharpoonup \mathbb{N}$ *is* Kleene computable relative to $y \in A$ *if there is a Kleene computable partial function* f *such that* $f(y, \mathbf{z}) \simeq g(\mathbf{z})$ *for all* $\mathbf{z} : \boldsymbol{\sigma}$.

(ii) *Given* $x, y \in A$ *of type level* > 0, *we write* $x \preceq y$ *if* x *is Kleene computable relative to* y. *We say* x, y *are of the same* Kleene degree *if* $x \preceq y \preceq x$.

Other results made precise the connection with logical complexity. For instance:

THEOREM 3.4 (Kleene [1959b, §10]). *A total function* $F : \mathbb{N}^r \to \mathbb{N}$ *is hyperarithmetical (that is, its graph is definable by a* Δ^1_1 *formula) iff it is Kleene computable relative to the type* 2 *object* ${}^2\exists$, *given by*

$$ {}^2\exists(f) = \begin{cases} 0 & \text{if } \exists n.\, f(n) = 0, \\ 1 & \text{otherwise.} \end{cases} $$

Kleene also followed up his definition via S1–S9 with a clutch of papers (Kleene [1962c], [1962d], [1962b], [1962a]) showing that several alternative definitions of higher type computability — via Turing machines, λ-calculus,[8] and Herbrand-Gödel style recursive definitions — give rise to the same class of computable functionals. These results closely parallel the corresponding results of ordinary recursion theory, but they are not especially deep extensions of the familiar results, and all these definitions are based on essentially the same idea of computation with oracles. Nevertheless, these results go some way towards establishing the robustness of Kleene's class of computable functionals.

[7] Kleene's original definition did not take exactly this form, but the equivalence is trivial.

[8] In fact, in Kleene [1962b] Kleene used a system based on what is now known as Church's λI calculus.

Despite these reassuring results, however, Kleene's notion presents us with some strange anomalies. We mention two of these here, both arising somehow from the curious infinitary side-condition in S8.

The first of these is the notorious fact that Kleene's computable partial functionals are not closed under some basic substitution operations. For example, consider the functions h, Φ, G defined by

$$h(m\colon \overline{0}, n\colon \overline{0}) \simeq \begin{cases} 0 & \text{if } \neg T(m,m,n), \\ \text{undefined} & \text{if } T(m,m,n) \end{cases}$$

$$\Phi(F : \overline{2}, m\colon \overline{0}) \simeq F(\Lambda n.h(m,n))$$

$$G(f : \overline{1}) = 0$$

where T is the T-predicate of ordinary recursion theory. Then the partial functions Φ, G are both Kleene computable, but $\Lambda e.\Phi(G,e)$ is not. This is because $G(\Lambda n.h(m,n)) = 0$ iff $\Lambda n.h(m,n)$ is a total function, which is the case iff $\phi_m(m)$ is *not* defined. Hence, if $\Lambda e.\Phi(G,e)$ were computable, we could solve the halting problem.

This is obviously rather worrying — indeed, it even seems questionable in what sense we have defined a computable functional Φ if we are not even allowed to plug in the total computable functional G as its first argument. Essentially the same problem lies behind the failure of the natural generalization of the first recursion theorem. There is also other evidence that the notion of partial computable functional here is pathological: for instance, a set that is semi-recursive and co-semi-recursive need not be recursive (see Platek [1966, p. 131]), and the union of two semi-recursive sets need not be semi-recursive (see Grilliot [1969b, p. 233]).

One way of responding to this problem is to restrict one's attention to the *total* computable functionals. In most interesting cases, the total Kleene computable functions over A constitute a well-behaved class, so we may regard Kleene's definition as picking out an important substructure of A:

DEFINITION 3.5. *A type structure A is* closed under Kleene computation *if all the total Kleene computable functionals over A are themselves elements of A. We then write* $\mathsf{KC}(A)$ *for the type structure consisting of the total Kleene computable elements of A.*

In this situation, the elements of $\mathsf{KC}(A)$ behave well with respect to substitution and indeed constitute a cartesian closed category.

A second curious feature of Kleene's definition, less often noted, is its apparent *non-absoluteness*. The set of indices that define total functionals, for instance, may vary from one type structure A to another — intuitively, the fewer elements of type σ there are in A, the easier it is for an index to define a total functional of type $\sigma \to \overline{0}$. Even for the case of computability over S, it is unclear *a priori* whether the totality or otherwise of the function $\{m\}(-)$

is an absolute property of m, unless one is willing to accept the notion of "the" full set-theoretic type structure as absolute. Could there be Kleene-style computations whose termination is dependent on controversial principles of set theory? We will return to this question in Part II.

Other presentations of the Kleene computable functionals were later given in Gandy [1967a] and Platek [1966]. Both of these treatments sought to avoid some of the arbitrary features of Kleene's definitions and to emphasize the naturalness of this notion of computability. Gandy gave a more perspicuous definition involving register machines, and argued that one was led ineluctably to Kleene's notion of higher type computability if (a) one restricted attention to (hereditarily) total arguments, and (b) one treated functions as "pure extensions". Platek gave a characterization of Kleene's functionals within a general framework for recursion theory that stressed *definability* of functions by recursion rather than computability. In particular, he showed how a very natural theory of recursive definability could be developed for type structures of hereditarily *partial* functionals — here the troublesome side-condition in S8 is not needed, and one may easily recover Kleene's original notion via some simple relationships between partial and total type structures. (We will say more about this approach in Section 4.1.)

Platek's approach via inductive definability was streamlined by Moschovakis (Moschovakis [1976]), whose approach made use of ideas from his work on abstract computability theories (Moschovakis [1969], [1974a]), which in turn was inspired by Kleene's original definition of S1–S9. A key insight here was that higher type computability can in some sense be reduced to type 2 computability: if we think of each of the sets A_σ as a separate ground sort, then Kleene computability over A becomes simply a matter of what can in an abstract sense be computed relative to the (first order) application functions and the (second order) λ-abstraction operations. This treatment was followed in the expository article of Kechris and Moschovakis [1977]. A comparison of various approaches to Kleene computability was given in Fenstad [1978] (see also Feferman [1977a]).

Beyond the results already mentioned, however, the pure notion of Kleene computability over S appears to hold rather little mathematical interest. For this reason, the later study of recursion theory on S concentrated almost entirely on questions of relative computability or on certain kinds of non-computable object. Thus, at this point the subject largely lost contact with genuinely effective notions of computability. Nevertheless, the ideas involved can still be seen as plausible generalizations of "computation" to infinite objects, so an account of these developments is in order.

3.2.2. *Recursion in normal objects.* Most of the later work concentrated on the theory of *normal* functionals in S:

DEFINITION 3.6 (Normal functionals).

(i) *We write* ${}^k\exists$ $(k \geq 2)$ *for the object of* S_k *embodying existential quantification over* S_{k-2}.

(ii) *A functional* Θ *of type level k is said to be* normal *if* ${}^k\exists \preceq \Theta$.

We will be interested in notions of Θ-*computability* (that is, Kleene computability relative to Θ) where Θ is normal, and particularly in the notions of ${}^k\exists$-*computability*.

The study of normal objects and the associated notions of computability turns out to yield a very rich and beautiful theory. Roughly speaking, if ${}^k\exists$ is deemed computable, then the theory of computability is good up to and including type level k. One way to see intuitively why this should be so is to note that in a setting where we have the ability to quantify over S_{k-2}, it is consonant to think of the infinitary side-condition in S8 as somehow "semidecidable" (and hence harmless) for $t \leq k$; thus, the anomalies mentioned earlier for pure Kleene computability are washed out in this setting.

The significance of normal objects had already emerged from Kleene's early work, but their good properties were first seriously exploited by Gandy (Gandy [1967b]),[9] who introduced the key concepts of stage comparison and number selection. Informally, a *stage comparison function* for two functionals F and G is a partial function $\chi_{F,G}(\boldsymbol{x},\boldsymbol{y})$ that tells us which out of the Kleene computation trees for $F(\boldsymbol{x})$, $G(\boldsymbol{y})$ has the smaller ordinal depth, assuming that at least one of these trees is well-founded. In particular, it can tell us which of the two computations will terminate, given that at least one of them will. The following theorem was stated for type 2 in Gandy [1967b], and extended to higher types (independently, and by different methods) in Platek [1966] and Moschovakis [1967], [1976]:

THEOREM 3.7 (Stage comparison). *Let* Θ *be a normal functional of level* k. *If* F, G *are* Θ-*computable and of level* $\leq k$, *then* $\chi_{F,G}$ *is also* Θ-*computable.*

One can often use stage comparison where in ordinary recursion theory one might use interleaving arguments — for instance, in showing that (for Θ normal of level k) the union of two Θ-semidecidable subsets of S_k is Θ-semidecidable, or that a subset of S_k is Θ-decidable iff both it and its complement are Θ-semidecidable.

Given a predicate $P(\boldsymbol{x}, n\colon \overline{0})$, a *selection function* for P is a partial functional $F(\boldsymbol{x})$ that "selects" a value of n satisfying $P(\boldsymbol{x}, n)$ whenever there is one. Again, the basic result is proved for type 2 in Gandy [1967b], and generalized in Moschovakis [1967], [1976], Platek [1966]:

THEOREM 3.8 (Number selection). *Let* Θ *be normal of level* $k \geq 2$. *For any* Θ-*semidecidable predicate* $P(\boldsymbol{x}, n\colon \overline{0})$ *on* S_k, *there is a* Θ-*computable partial*

[9] Gandy first presented these results around 1962.

function $F(\boldsymbol{x})$ *such that*

$$\forall \boldsymbol{x}.\ ((\exists n.\, P(\boldsymbol{x}, n)) \Longrightarrow F(\boldsymbol{x})\downarrow \wedge P(\boldsymbol{x}, F(\boldsymbol{x}))).$$

It follows, for example, that the class of Θ-semidecidable subsets of S_k is closed under (uniform) $\mathbb{N}$-indexed unions, and that a partial function of level k is Θ-computable iff its graph is Θ-semidecidable. The basics of stage comparison and number selection up to this point are covered in Kechris and Moschovakis [1977].

A selection theorem of another kind is the following. It first appeared in Grilliot [1969b] but with an incorrect proof. A correct proof was given in Harrington and MacQueen [1976].

THEOREM 3.9 (Grilliot selection). *Let* Θ *be normal of level* $k+2$ $(k \geq 1)$. *For any inhabited* Θ*-semidecidable subset* P *of* S_{k-1}, *there is an inhabited* Θ*-decidable subset* Q *such that* $Q \subseteq P$.

Much attention has also been devoted to the study of *sections* as defined in Kleene [1963]. Given any $x \in \mathsf{S}$, the *section* of x is the set $\{y \in \mathsf{S} \mid y \preceq x\}$; likewise the k*-section* of x is $\{y \in \mathsf{S}_k \mid y \preceq x\}$.[10] One of the high points of the theory is the following result due to Sacks (Sacks [1974], [1977]):

THEOREM 3.10 (Plus-one theorem). *Suppose* $0 < k < n$. *Then for any normal object* $\Theta \in \mathsf{S}_n$ *there is a normal object* $\Psi \in \mathsf{S}_{k+1}$ *such that* Θ *and* Ψ *have the same* k*-section.*

The proof of this very substantial theorem exploits some beautiful connections with admissible set theory (Sacks [1971]). Let us consider the case $k = 1$ as an example. Observe that any hereditarily countable set X (that is, a set whose transitive closure is countable) can be coded up as a type 1 function: essentially, one just needs to code up the binary membership relation on the transitive closure of X. Under this correspondence, those sets of type 1 functions which arise as sections correspond precisely to those sets of hereditarily countable sets which constitute models for a certain weak fragment of set theory. Sacks was able to apply forcing techniques to such models in order to construct a suitable type 2 object Ψ for the above theorem.

Sacks also considered higher type analogues of problems from classical degree theory such as Post's problem (Sacks [1980], MacQueen [1972]; see also Section 3.2.4 below).

Sections are to total functionals what *envelopes* are to partial functionals. If $\sigma = \sigma_1 \to \cdots \to \sigma_r \to \overline{0}$, let us write $\mathsf{S}_{p\sigma}$ for the set of all partial functions $\mathsf{S}_{\sigma_1} \times \cdots \times \mathsf{S}_{\sigma_n} \rightharpoonup \mathbb{N}$. The k*-envelope* of $x \in \mathsf{S}$ is the set

$$\{f \in \mathsf{S}_{p\sigma} \mid \textit{level}\,(\sigma) \leq k \wedge f \text{ Kleene computable relative to } x\}.$$

[10] The k-section is sometimes taken to include all objects of type level $\leq k$ computable in x, but it makes little difference.

Moschovakis (Moschovakis [1974c]) showed that the analogue of the plus-one theorem for envelopes fails: indeed, if Θ is *any* normal type 3 object, the 1-envelope of Θ is not the 1-envelope of any normal type 2 object. However, we have the following result due to Harrington (Harrington [1973]):

THEOREM 3.11 (Plus-two theorem). *Suppose* $0 < k < n - 1$. *Then for any normal object* $\Theta \in \mathsf{S}_n$ *there is a normal object* $\Psi \in \mathsf{S}_{k+2}$ *such that* Θ *and* Ψ *have the same* k*-envelope.*

Regarding elements of the k-envelope of a normal Θ as representing Θ-semidecidable predicates, it is natural to ask what closure properties these enjoy. It is clear from earlier remarks that the 1-envelope of a normal type 2 object is closed under existential quantification over $\mathbb{N}$ (in an obvious sense). However, the situation at higher types is more interesting:

THEOREM 3.12. *Let* Θ *be normal of level* $m \geq 3$. *Then the* $(m-1)$*-envelope of* Θ *is closed under existential quantification over* S_j *for* $j < m - 2$ (*Harrington and MacQueen* [1976]), *but not under existential quantification over* S_{m-2} (*Moschovakis* [1967], *Grilliot* [1967]).

Both Sacks and Harrington worked with definitions of recursion in normal functionals that differed somewhat from Kleene's, but the equivalences are verified in Lowenthal [1976]. The proofs of many of the above results have been clarified by adopting the perspective of abstract recursion theory, which isolates the essential features of the domains (in this case, the S_σ) over which we are computing. This abstract approach is worked out in the books Moldestad [1977], Fenstad [1980].

3.2.3. *Hierarchies.* Another area of investigation has been the search for *hierarchies* for the functionals computable from a given object. The starting point for this endeavour is the fact that the 1-section of ${}^2\exists$ consists of exactly the hyperarithmetic or Δ^1_1 functions (Theorem 3.4). As shown in Kleene [1955b], these can be classified according to their logical complexity by means of an ordinal hierarchy of height ω_1^{CK}, known as the hyperarithmetic hierarchy. Kleene in Kleene [1963] introduced the general problem of seeking similar ordinal stratifications for the sections of other higher type objects.

Tugué (Tugué [1960]) considered in particular the normal type 2 object E_1 embodying the *Suslin quantifier*:

$$E_1(f : \overline{1}) = \begin{cases} 0 & \text{if } \exists g : \overline{1}.\, \forall n.\, f(\widetilde{g}(n)) = 0, \\ 1 & \text{otherwise,} \end{cases}$$

where $\widetilde{g}$ is as defined in Section 1.4. Tugué and later Richter (Richter [1967]) gave ordinal hierarchies for the 1-section of E_1. Kleene had conjectured that this class might coincide with the Δ^1_2 functions, but this was refuted by Shoenfield (Shoenfield [1962]).

Some of these hierarchies were considered in greater detail in Gandy [1967b], Grilliot [1969a]. In particular, Gandy showed how such hierarchies could be used to prove structural results such as the stage comparison and number selection theorems mentioned in Section 3.2.2.

A pleasing generalization of these hierarchy results to the 1-section of an arbitrary normal type 2 object was given independently by Shoenfield (Shoenfield [1968]) and Hinman (Hinman [1966], [1969]). The two versions are essentially the same; the key point in both cases is that the set of ordinal notations, rather than being fixed in advance, is generated simultaneously with the hierarchy itself. Later, Wainer (Wainer [1974]) showed how a somewhat more delicate construction yields a hierarchy for the 1-section even of an arbitrary non-normal type 2 object (see also Wainer [1975], [1978]).

A natural challenge was to obtain similar hierarchy results for 2-sections of ${}^3\exists$ and other type 3 objects. In Kleene [1963], Kleene outlined the construction of a *hyperanalytic hierarchy* for type 2 objects; this hierarchy was studied in detail by his student Clarke (Clarke [1964]), who conjectured that it does not exhaust the type 2 objects computable in ${}^3\exists$. The conjecture was confirmed by Moschovakis (Moschovakis [1967]), who however constructed an alternative hierarchy which does exhaust them, by using a much more powerful system for generating ordinal notations.

Other objects of interest are the *superjump* operators ${}^kS$ $(k \geq 2)$, defined by

$$ {}^kS(e\colon \overline{0}, x\colon \overline{k-1}) = \begin{cases} 0 & \text{if } \{e\}(x)\downarrow, \\ 1 & \text{if } \{e\}(x)\uparrow. \end{cases} $$

The operator ${}^2S$ is the ordinary jump operator of classical recursion theory, which is equivalent in strength to ${}^2\exists$. The superjump operator ${}^3S$, however, is strictly weaker than ${}^3\exists$ (and hence non-normal). Recursion in ${}^3S$ has been studied in Gandy [1967b], Platek [1971], Aczel and Hinman [1974]. Superjump operators of even higher types have been considered in Harrington [1973], [1974]. Much more recently, both ${}^3\exists$ and ${}^3S$ have been used in the construction of universes of domains closed under powerful type formation principles (Normann [1997], [1999b]).

Many of the above hierarchy results are covered in detail in Hinman [1978].

3.2.4. *Set recursion.* It was discovered by Normann (Normann [1978b]), and independently by Moschovakis, that the ideas of Kleene computability relative to all the ${}^k\exists$ could naturally be generalized to a theory of computability on arbitrary sets, in which the equality predicate on sets is deemed to be computable. This theory is called *E-recursion theory*; it has close connections with Gödel's notion of constructibility. (The connection had essentially been noted already in Sacks [1971].)

This much more general setting allowed the whole subject to break free from the shackles of the type structure S, and E-recursion was found to provide a natural framework for the later degree-theoretic investigations of Sacks and others (Sacks [1985], [1986], Griffor [1980], Slaman [1981], [1985]). Thus, at this point the study of higher type recursion became subsumed in admissible set theory, and specific references to the finite types are rare in the later papers.

For a full treatment of E-recursion theory see Sacks [1990]; for a survey see Sacks [1999].

3.3. The total continuous functionals.

3.3.1. *Early work: Kleene and Kreisel.* Whereas computations over the full set-theoretic type structure S have connections with the metamathematics of classical logic, the study of the *total continuous functionals* has largely been driven by interest in more constructive interpretations. The type structure C of total continuous functionals was obtained around the same time by Kleene (Kleene [1959a]) (who called them the *countable* functionals) and Kreisel (Kreisel [1959]) (who called them the *continuous* functionals). Both definitions were rather complicated and neither were *prima facie* very natural, but they both embodied the basic idea that any finite piece of information about the output from a functional F was determined by a finite amount of information about the input.

Kleene's definition centres around the observation that any total continuous type 2 function $F : \mathbb{N}^{\mathbb{N}} \to \mathbb{N}$ can be completely determined by a type 1 function — for instance, by the function $f_F : \mathbb{N} \to \mathbb{N}$ defined (classically) by

$$f_F\langle n_0, \dots, n_{r-1}\rangle = \begin{cases} m+1 & \text{if } F(g) = m \text{ for all } g \text{ such that} \\ & \quad g(i) = n_i \text{ for all } i < r, \\ 0 & \text{if no } m \text{ exists with this property.} \end{cases}$$

Reversing this idea, we can define a partial "application" operation $- \mid -$: $\mathbb{N}^{\mathbb{N}} \times \mathbb{N}^{\mathbb{N}} \rightharpoonup \mathbb{N}$ by

$$f \mid g \simeq f(\tilde{g}(r)) - 1,$$
$$\text{where } r \text{ is the least number such that } f(\tilde{g}(r)) > 0.$$

Thus, type 1 objects can be used to encode type 2 objects. Once this is in place, it easy to see how type 1 objects can be used to encode objects of arbitrary pure types:

DEFINITION 3.13 (Kleene's countable functionals). *For each n, we define a set C_n and a notion of* associate *for the elements of C_n as follows:*

- *Take $\mathsf{C}_0 = \mathbb{N}$, $\mathsf{C}_1 = \mathbb{N}^{\mathbb{N}}$, and declare each $f \in \mathsf{C}_1$ to be an associate for itself.*

- *For $n \geq 1$, say $f \in \mathsf{C}_1$ is an associate for the function $F: \mathsf{C}_n \to \mathbb{N}$ iff whenever g is an associate for $G \in \mathsf{C}_n$ we have $f \mid g = F(G)$; and take C_{n+1} to be the set of functions $\mathsf{C}_n \to \mathbb{N}$ that have an associate.*

This is essentially an extensional collapse construction (Definition 1.2; see also Section 5.1). Kleene extended the above definition from pure types to arbitrary types by means of standard type-changing techniques, giving rise to a type structure C.[11] This structure is called ECF in Troelstra [1973, §2.6].

In Kreisel's definition, continuity was built in via the concept of *neighbourhoods*. In any type structure A over $\mathbb{N}$, one may define for each type σ a system $\mathcal{N}_\sigma$ of subsets of A_σ as follows:

- $\mathcal{N}_0$ consists of all singletons $\{n\}$.
- $\mathcal{N}_{\sigma\to\tau}$ consists of all sets of the form
$$[U_1 \mapsto V_1] \cap \ldots \cap [U_r \mapsto V_r] \; (r \geq 1),$$
where $U_i \in \mathcal{N}_\sigma$, $V_i \in \mathcal{N}_\tau$, and $[U \mapsto V]$ means the set
$$\{f \in A_{\sigma\to\tau} \mid \forall x \in U.\, f \cdot x \in V\}.$$

The elements of $\mathcal{N}_\sigma$ are called *neighbourhoods* and correspond to finite pieces of information about elements of A_σ. Clearly, any basic neighbourhood may be denoted by a formal expression involving $\mapsto, \cap$, and singletons, and obviously one can define such "formal neighbourhoods" quite independently of any type structure.

Kreisel gave a syntactical construction of C in terms of these formal neighbourhoods and functions over them: elements of $\mathsf{C}_{\sigma\to\bar{0}}$ were first given by functions on formal neighbourhoods of type σ, and only then turned into functions on the elements of C_σ themselves. It was thus automatic that all the functionals in C were continuous in the sense determined by the neighbourhood systems. However, the precise details are a little complicated and we will omit them here.

The equivalence between the Kleene and Kreisel definitions, was recognized at the time of their discovery and is mentioned in both of the original papers. (However, the proof turned out to be less trivial than these authors initially suspected; see also Hinata and Tugué [1969].) Tait (Tait [1962]) later gave a somewhat cleaner, more axiomatic treatment of the continuous functionals in the spirit of Kreisel's definition.

Both Kleene and Kreisel identified a natural effective substructure RC of C, consisting of the *recursively countable* or *recursively continuous* functionals. In Kleene's terms, these are the functionals with at least one recursive associate;

[11] Actually, Kleene first defined the countable functionals to be certain elements of S, whose action on other countable functionals was as suggested by the above definition. However, the possibility of considering the countable functionals "by themselves" was also clear to him from the outset.

in Kreisel's terms, they are defined via recursive functions on bases of neighbourhoods. An important result, the *density theorem* (see Kreisel [1959, Appendix]), ensures that each $F \in \mathsf{C}_{\sigma\to\tau}$ is completely determined by its effect on the recursive (or even the "finite") elements of C_σ; it follows that RC is extensional when regarded as a type structure in its own right.

Kreisel, in particular, was interested in using C to give constructive interpretations for systems such as intuitionistic second order arithmetic. The idea is as follows: given a formula ϕ, first apply the Gödel Dialectica translation (see Section 3.1.1) to obtain a formula $\phi' \equiv \exists \boldsymbol{y}.\forall \boldsymbol{z}.\phi^*$ in which ϕ^* is quantifier-free and the $\boldsymbol{y}, \boldsymbol{z}$ may be of any finite type. We may now consider the interpretations of ϕ' arising by allowing the $\boldsymbol{y}$ and $\boldsymbol{z}$ to range over various kinds of finite type object. Kreisel suggested that we get an interpretation that fits well with the informal notion of constructive truth if we let $\boldsymbol{y}$ range over all functionals in RC and $\boldsymbol{z}$ over functionals in C. The intuition is that the constructive content of ϕ should be embodied by effectively given operations (the $\boldsymbol{y}$) which make ϕ^* true even when presented with arbitrary continuous data such as free choice sequences (the $\boldsymbol{z}$).

Kreisel also considered rival interpretations of this kind which are of interest for independence proofs: for instance, we may let $\boldsymbol{z}$ range over arbitrary functionals in S and $\boldsymbol{y}$ over Kleene computable ones; or we may let both $\boldsymbol{y}$ and $\boldsymbol{z}$ range over the hereditarily effective operations. In addition, he showed that the above interpretation in C could be combined with Gödel's double-negation translation to give a very satisfactory "no-counterexample interpretation" for *classical* second order arithmetic. The present author considers Kreisel's paper (Kreisel [1959]) to be a classic in the field, which remains well worth reading as a discussion of the metamathematical applications of higher type functionals, and which contains in seminal form an amazing number of ideas that were later to become major themes in the subject.

One can also consider the class KC(C) of Kleene computable functionals over C, as given by Definition 3.2. It was clear to Kleene and Kreisel that every total Kleene computable functional over C is a recursively continuous functional — intuitively, anything that is computable in Kleene's "extensional" sense is computable at the more intensional level of associates. Moreover, at type 2 the recursively continuous functionals clearly coincide with the Kleene computable functionals (over C or S), since *e.g.*, any $F \in \mathsf{C}_2$ is Kleene computable relative to an associate for F.

It was raised as an open problem in Kreisel [1959, §4.4] whether all elements of RC at types 3 and above were Kleene computable. This was answered negatively by Tait (Tait [1962]), who gave as a counterexample the type 3 *modulus of uniform continuity* functional Φ (also known as the *fan functional* because it embodies the constructive content of Brouwer's Fan Theorem). The existence of Φ depends on the fact that (classically) every continuous function

$F: 2^{\mathbb{N}} \to 2$ is uniformly continuous. To define Φ, let $binary(f : \overline{1})$ be the predicate $\forall x.\, f(x) = 0 \vee f(x) = 1$, and take

$$\Phi(F : \overline{2}) = \mu n. \forall f, g.\ (binary(f) \wedge binary(g) \wedge (\forall m < n.\, f(m) = g(m))) \Longrightarrow F(f) = F(g).$$

Tait showed that Φ is recursively continuous but not Kleene computable over C — nor, indeed, is it the restriction to C_2 of a function Kleene computable over S. (Published proofs of Tait's result may be found in Gandy and Hyland [1977, §4] or Normann [1999a, §5].) We therefore have two distinct notions of computability for the total continuous functionals, represented by RC and KC(C) respectively.

3.3.2. *Further characterizations.* Apart from Tait's work, there seems to have been little work on the total continuous functionals during the 1960s. The 1970s, however, saw a wealth of new developments which were mostly of two kinds: results providing further characterizations of C and RC, illuminating their basic character and confirming their natural status; and results concerning various notions of relative definability and degree structures on C.

Many of the later characterizations of C were much simpler and more immediately appealing than the original definitions. Ershov (Ershov [1974a], [1977a]) showed that the continuous functionals could be obtained from the type structure P of partial continuous functionals via an extensional collapse construction (see Theorem 4.13 below). A very similar characterization of C was obtained independently by Scott, using a type structure arising from the category of algebraic lattices (see Hyland [1975]).

Another pleasing characterization of C is that based on the idea of *sequential continuity*.[12] This definition may be presented very simply as follows. We use the notation $[x_i]$ for an infinite sequence $x_0, x_1, \ldots$.

DEFINITION 3.14 (Sequential continuity). *For each type σ, we define a set C_σ together with a relation $[x_i] \downarrow x$ (read as "$[x_i]$ converges to x") between infinite sequences and elements of C_σ:*

- $C_0 = \mathbb{N}$, *and* $[x_i] \downarrow x$ *if* $x_i = x$ *for all sufficiently large* i;
- $C_{\sigma\to\tau}$ *is the set of all* $f : C_\sigma \to C_\tau$ *such that* $[f(x_i)] \downarrow f(x)$ *whenever* $[x_i] \downarrow x$. *And* $[f_i] \downarrow f$ *in* $C_{\sigma\to\tau}$ *iff* $[f_i(x_i)] \downarrow f(x)$ *whenever* $[x_i] \downarrow x$.

This is tantamount to the definition of the type structure over $\mathbb{N}$ in the cartesian closed category of *L-spaces* (see Kuratowski [1952]). Scarpellini considered the type structure defined in this way as a model for bar recursion at higher types (Scarpellini [1971]), apparently without realizing that it coincided with the Kleene-Kreisel continuous functionals. The equivalence was shown by Hyland (Hyland [1975], [1979]), who collected together the

[12]The terminology, which was used in the literature of the time, refers to continuity based on sequences. It should not be confused with the more modern use of the term "sequential", which we discuss in Section 4.3.

known characterizations and clarified the relationships between them. He also discovered some new characterizations of C: for instance, as the type structure over $\mathbb{N}$ in the cartesian closed category of *filter spaces*, or in the cartesian closed category of compactly generated Hausdorff spaces. (A published proof of the latter fact appeared first in Normann [1980].) Normann later gave yet another construction via a hyperfinite type structure in the sense of non-standard analysis (Normann [1983], [1999a]).

Bergstra (Bergstra [1976], [1978]) gave a interesting characterization of C as a maximal type structure (subject to some basic closure requirements) in which all type 2 functions are continuous. Intuitively, one can construct C by taking, at each type level ≥ 3, all functions that can be added without inducing any discontinuous type 2 functions. Grilliot (Grilliot [1971]) had already shown that a type 2 functional F is continuous iff ${}^2\exists$ is *not* Kleene computable relative to F and some type 1 function; we therefore have a characterization of C as a maximal type structure closed under Kleene computation and not containing ${}^2\exists$.

By this stage it was very clear that C was the canonical choice of a full continuous type structure of total functionals over $\mathbb{N}$. Moreover, Hyland's work also showed that all the constructions of C that admitted a natural effectivization gave rise to the same effective substructure, namely RC.

3.3.3. *Degrees and relative computability.* Most of the work on degrees and relative computability for C has focussed on the notions arising from relative Kleene computability. Hinman (Hinman [1973]) gave an example of an *irreducible* element of C_2 — that is, one whose Kleene degree is not the Kleene degree of any type 1 function. Hyland in Gandy and Hyland [1977, §5] gave a simpler example of this phenomenon, making use of the Kleene tree. The same paper also contained an example, due to Gandy, of a type 3 object $\Gamma \in RC_3$ which is not Kleene computable relative to the fan functional.

In view of this last result, it was natural to ask whether any good "basis"of functionals in RC could be given, relative to which all functionals in RC were Kleene computable. (A closely related question, discussed by Feferman (Feferman [1977a]), was whether one could give a definition of RC as a substructure of C via monotone inductive schemata, in the spirit of generalized recursion theory.) It is fairly easy to specify an infinite basis for RC consisting of *partial* recursively continuous functions, one for each type level (see Bergstra [1978, §1]). Normann (Normann [1979b], [1981a]) gave a similar infinite basis consisting of total recursively continuous functionals (*i.e.*, elements of RC), but showed that none of the functionals in this were Kleene computable relative to any functionals of lower type. It follows that no *finite* basis of this kind for RC is possible. (See however Theorem 4.25 for a more satisfactory ending to this story.)

Another group of results from this period concerned properties of 1-sections for continuous type 2 objects. Early results in this vein were obtained by Grilliot (Grilliot [1971]), who pointed out that such a 1-section is never closed under the ordinary jump operator. Bergstra (Bergstra [1976]) showed that there are objects $F \in \mathsf{C}_2$ whose 1-section is not the 1-section of any $f \in \mathsf{C}_1$ (this improves on the result of Hinman mentioned above). Results relating to ordinal hierarchies for 1-sections of continuous objects are obtained in Bergstra and Wainer [1977], Normann [1978a], Normann and Wainer [1980].

A simple but beautiful result of Normann (Normann [1981b]) (who traces the idea to Kreisel [1959]) reveals a connection between the continuous functionals and some classical logical complexity classes for subsets of $\mathbb{N}^{\mathbb{N}}$. A proof of the theorem also appears in Normann [1999a].

THEOREM 3.15 (Projective hierarchy). *Let $k > 0$, $A \subseteq \mathbb{N}^{\mathbb{N}}$. Then*

(i) *A is Π^1_k iff there is a primitive recursive predicate R (e.g., definable by Kleene's $S1$–$S8$ over C) such that*

$$f \in A \iff \forall G \in \mathsf{C}_k.\exists n \in \mathbb{N}.\, R(f, G, n).$$

(ii) *A is Σ^1_k iff there is a primitive recursive R such that*

$$f \in A \iff \forall G \in \mathsf{C}_{k+1}.\exists n \in \mathbb{N}.\, R(f, G, n).$$

and moreover there is a uniform algorithm (e.g., a Kleene computable partial functional over C) which given any $f \notin A$ returns a G such that $\forall n.\neg R(f, G, n)$.

Normann showed that many facts about 1-sections and 2-envelopes flow from this theorem.

Finally, we mention that an alternative degree structure on C can be obtained by considering a more generous notion of relative computability: take $F \preceq_c G$ iff there is a recursive type 2 functional which transforms any associate for G into an associate for F. Most of the known results about degrees, sections and envelopes with respect to $\preceq_c$ are collected in Hyland [1978].

Normann's book (Normann [1980]) contains most of what was known about the total continuous functionals by 1980.

3.3.4. *Recent work.* Work on the continuous functionals abated again during the 1980s, but was renewed in the 1990s, partly owing to interest within the computer science community. Modern treatments have tended to favour versions of the Ershov-Scott construction of C (see Theorem 4.13) via domains or information systems (Berger [1993], Stoltenberg-Hansen, Lindström, and Griffor [1994], Schwichtenberg [1996], Normann [1999a]). A particular focus of recent work has been the search for abstract formulations of the concept of *totality* for elements of such domains. Berger (Berger [1993]) has demonstrated the significance of the dual notions of *density* and *codensity* (= totality) for subsets of domains, and given generalized versions of the density theorem and Kreisel-Lacombe-Shoenfield theorem in a domain-theoretic framework. A related approach to totality is pursued in Normann [1989], [1997]. More recent work has been concerned with extending the construction of C and the associated results on domains, density and totality to transfinite types and Martin-Löf style dependent types with universes (Berger [1997], Kristiansen and Normann [1997]). Bauer and Birkedal (Bauer and Birkedal [2000]) have shown how much of this material fits smoothly into the framework of Scott's *equilogical spaces* (see Section 5.2).

Some other recent results involving C (Normann [2000], Plotkin [1997]) will be mentioned in Section 4.3.4 below.

3.3.5. *Type two effectivity.* It is convenient at this point to mention briefly the work of Weihrauch and his colleagues (see *e.g.*, Weihrauch [1985], [2000b]), who have developed a framework for computable analysis known as *type two effectivity*. This framework was proposed independently of the work on C that we have described, but there are close connections. Weihrauch considers a model of computation consisting of Turing machines with infinite input and output tapes; these essentially compute uniformly partial recursive type 2 functionals on $\mathbb{N}^{\mathbb{N}}$ (Definition 2.2), subject to some minor caveats. For the most part, Weihrauch's theory is concerned with the use of this model to represent computability on other sets, such as spaces arising in analysis. A *representation* of a set S is a partial function from $\mathbb{N}^{\mathbb{N}}$ onto S — for instance, we might represent the reals by (coded) Cauchy sequences of rational approximations.

The key result which makes the theory work (presented in Weihrauch [1985]) is that the set of continuous functions on $\mathbb{N}^{\mathbb{N}}$ may itself be represented in this way by $\mathbb{N}^{\mathbb{N}}$. One obtains an effective version of the theory by restricting here to functions represented by $\mathbb{N}^{\mathbb{N}}_{rec}$. In essence, the idea here is the same as the idea behind Kleene's definition of associates (Definition 3.13), which provides representations of just this kind for the total continuous functionals. These and other connections have recently been made precise in Bauer [2001] (see also Section 5.2 below).

3.4. The hereditarily effective operations. As we have seen, one notion of total computable functional of higher type is given by the effective submodel RC of the total continuous functionals; here we have computable functionals acting on (possibly arbitrary) continuous data. One might also ask if there are interesting type structures based on the idea of computable functions acting only on computable data. Several ways of constructing such a type structure might suggest themselves. For instance:

- We might consider higher type generalizations of the definition of type 2 effective operations based on recursive indices (see Definition 2.3).
- We might consider effective *analogues* of various definitions of C: that is, we might mimic some construction of C taking only the effective functionals at each type level.

Both kinds of construction were considered in Kreisel [1959, §4.2]. As an instance of the first kind, Kreisel defined an extensional type structure HEO. Note that the definition is a straightforward extensional collapse construction:

DEFINITION 3.16 (Hereditarily effective operations). *For each type σ, define a partial equivalence relation $\equiv_\sigma$ on $\mathbb{N}$ as follows:*

- $x \equiv_0 x'$ *iff* $x = x'$;
- $x \equiv_{\sigma\to\tau} x'$ *iff for all* $y \equiv_\sigma y'$ *we have* $\phi_x(y) \equiv_\tau \phi_{x'}(y')$.

Now let HEO_σ *be the set of* $\equiv_\sigma$*-equivalence classes. Since there are well-defined total application operations* $\mathsf{HEO}_{\sigma\to\tau} \times \mathsf{HEO}_\sigma \to \mathsf{HEO}_\tau$ *induced by recursive index application, we have a total extensional type structure* HEO *over* $\mathbb{N}$.

Kreisel also introduced (in Kreisel [1958]) a closely related non-extensional type structure:

DEFINITION 3.17 (Hereditarily recursive operations). *For each type σ, define a subset* $\mathsf{HRO}_\sigma \subseteq \mathbb{N}$ *as follows:*

- $\mathsf{HRO}_0 = \mathbb{N}$;
- $x \in \mathsf{HRO}_{\sigma\to\tau}$ *iff for all* $y \in \mathsf{HRO}_\sigma$ *we have* $\phi_x(y)\downarrow$ *and* $\phi_x(y) \in \mathsf{HRO}_\tau$.

This defines a total non-extensional type structure HRO *over* $\mathbb{N}$, *with application given by recursive index application.*

The above names, and the notations HEO, HRO, were introduced later by Troelstra, who independently rediscovered these structures around 1970 (see Troelstra [1973, §2.4.17]). Note that the structure HEO is independent (up to isomorphism) of the choice of enumeration for the partial recursive functions, whereas this is not true for HRO — hence it is misleading, strictly speaking, to talk about "the" hereditarily recursive operations.

As an instance of the second kind of construction suggested above, Kreisel proposed a recursive analogue of his construction of C via formal neighbourhoods. We give here an equivalent definition based on a recursive analogue of Kleene's construction of C via associates (Definition 3.13).

DEFINITION 3.18 (Hereditarily recursively countable functionals). *For each n, define a set* HRC_n *and a notion of associate for elements of* HRC_n *as follows:*

- *Take* $\mathsf{HRC}_0 = \mathbb{N}$, $\mathsf{HRC}_1 = \mathbb{N}^{\mathbb{N}}_{rec}$, *and declare each* $f \in \mathsf{HRC}_1$ *to be an associate for itself.*
- *For* $n \geq 1$, *say* $f \in \mathsf{HRC}_1$ *is an associate for* $F : \mathsf{HRC}_n \to \mathbb{N}$ *iff whenever* g *is an associate for* $G \in \mathsf{HRC}_n$ *we have* $f \mid g = F(G)$; *and take* HRC_{n+1} *to be the set of functions* $\mathsf{HRC}_n \to \mathbb{N}$ *that have an associate.*

The definition may be extended from pure types to arbitrary types by standard methods, yielding a type structure HRC. This structure is called $\mathsf{ECF}(\mathcal{R})$ in Troelstra [1973, §2.6].

Kreisel noted the remarkable fact that these two approaches to constructing a class of hereditarily computable functionals coincide — that is:

THEOREM 3.19. HEO $\cong$ HRC.

At type 2 this is essentially the Kreisel-Lacombe-Shoenfield theorem (Theorem 2.6). The generalization to higher types was noted in Kreisel [1959, §4.2] without detailed proof. Some further details appeared in Tait [1962], but a complete published proof first appeared in Troelstra [1973, §2.6].

Some further remarks may help to clarify the relationship between HRC (the recursive *analogue* of C) and RC (the recursive *substructure* of C). To build RC, one first builds the whole of C and then extracts the effective elements. To build HRC, one considers at each type level only the effective total functionals on the set of effective objects of the next type down. Thus, for instance, the Kleene tree functional Z defined in Section 2.4 lives in HRC_2, but it is not (the restriction of) an element of RC_2 since it has no computable extension to the whole of $\mathbb{N}^{\mathbb{N}}$. We therefore have a proper inclusion of RC_2 in HRC_2. Conversely, there is no element in HRC_3 corresponding to the fan functional $\Phi \in \mathsf{RC}_3$, essentially because any computable functional $\Phi' \in \mathsf{HRC}_3$ that extended Φ (with respect to the above inclusion) would be undefined on Z. There is therefore no canonical inclusion from RC_3 to HRC_3 or *vice versa*.

The idea here is that Z and Φ cannot live together in the same universe of computable total functionals. Indeed, in Part II we will formulate and prove an "anti-Church's thesis for total functionals", to the effect that there is *no* type structure of computable functionals over $\mathbb{N}$ that subsumes both HRC and RC. This suggests that, for hereditarily total functions at least, the dichotomy between "computable acting on computable" and "computable acting on continuous" is a fundamental one.

We have seen that the recursive analogues of both Kleene's and Kreisel's definition of C yield the same type structure HRC. Later results confirmed the impression that whenever any natural definition of C admits a recursive analogue, this analogue turns out to be HRC. Several such equivalences are verified in Hyland [1975]. In addition, Ershov (Ershov [1976b], [1977a]) showed that just as the hereditarily total elements of P give rise to C (see Theorem 4.13), so the hereditarily total elements of $\mathsf{P}^{\mathit{eff}}$ give rise to HEO. Ershov also made explicit a higher type generalization of the Kreisel-Lacombe-Shoenfield theorem implicit in the fact that HEO $\cong$ HRC: namely, that all operations in HEO are continuous in the sense of the neighbourhood structure given in Section 3.3.1.

This contrasts with the surprising fact, discovered by Gandy around 1965, that not all the functionals in HEO are *sequentially* continuous in the sense of Definition 3.14. His ingenious construction of a counterexample at type 3 appears in Gandy and Hyland [1977, §8].

Later, Bezem (Bezem [1985a]) gave another characterization of HEO as the extensional collapse of HRO (see Definition 1.2):

THEOREM 3.20. HEO $\cong$ EC(HRO).

What is perhaps surprising is that this result is decidedly non-trivial. The proof essentially goes via the isomorphism with HRC as defined via Kleene associates (see Definition 3.18).

Another descendant of Kreisel's definition of HEO was Girard's category ***PER*** of partial equivalence relations on the natural numbers (Girard [1972]). We will give the definition of this category in Section 5; meanwhile, we simply remark that HEO can be naturally seen as the finite type structure over N in the cartesian closed category ***PER***.

§4. Partial computable functionals. The picture outlined in Section 3 exhibits various oddities which arise from the fact that we are restricting ourselves to hereditarily total objects. Intuitively, any computational paradigm powerful enough to generate all total computable functions will naturally generate partial ones as well, and so any restriction to total functionals will in some ways be artificial. Moreover, this artificiality is exacerbated as one passes to higher types. The most striking instance of this is the brood of difficulties associated with Kleene's S8 (see Section 3.2.1). Another tension arising from the insistence on totality can perhaps be discerned in the incompatibility of the Kleene tree functional and the fan functional (see Section 3.4). These observations might lead one to suspect that a theory of hereditarily partial objects of higher type would work more smoothly than one for total objects.

The idea that "partial is easier than total" at higher types was first aired in Kleene [1963, §9.3], and was frequently discussed in the later literature, *e.g.*, in Platek [1966], Gandy and Hyland [1977], Feferman [1977a], Ershov [1977a]. We will see in this section that the above suspicions are amply justified — that notions of partial computable functional do indeed lead to simpler theories, and enjoy more pleasant properties, than the total notions. (This accords with our experience in ordinary recursion theory: for example, the partial recursive functions are recursively enumerable but the total recursive functions are not.) This is not to say the total notions are necessarily of lesser interest than the partial ones — we have already seen that some total type structures have very good mathematical credentials — but as we shall see, a study of the partial notions greatly enriches our understanding of the total ones.

Nevertheless, there are some important conceptual questions to be addressed in formulating notions of hereditarily partial object of higher type. For instance, should our type 1 objects be partial functions $\mathbb{N} \rightharpoonup \mathbb{N}$ (*i.e.*, functions $\mathbb{N} \to \mathbb{N}_\perp$)? Or should our treatment be so thoroughly partial that we even interpret $\overline{0}$ as $\mathbb{N}_\perp$, in which case our type 1 objects might be (for instance) functions $\mathbb{N}_\perp \to \mathbb{N}_\perp$? Questions of this kind will proliferate as we pass to higher types, and at first sight it is unclear how we should respond to them. One can speculate that a lurking unease about these issues may partly account for the relatively late development of good notions of partial higher type object.

With the benefit of hindsight, we can see that these choices are not as significant as they might seem, since all reasonable choices turn out to be "equivalent" in some sense. We will discuss these issues in detail in Part II, where a precise statement to this effect will be given. For the purposes of our historical survey, however, it will be useful to introduce here two rival definitions of "partial type structure" that both figured in the development of the subject:

DEFINITION 4.1 (Call-by-name, call-by-value).

(i) *A* call-by-name (or CBN) structure *is an extensional total type structure over* $\mathbb{N}_\perp$.
(ii) *A* call-by-value (or CBV) structure *is an extensional partial type structure over* $\mathbb{N}$ (*in the sense of Section* 1.5.1).

This terminology is borrowed from computer science, and does not appear in the earlier recursion theory literature. The significance of the terms can be appreciated by considering the nature of type 1 objects under the two definitions. In a CBN structure, a type 1 object is a function $\mathbb{N}_\perp \to \mathbb{N}_\perp$, and the argument fed to such a function is not itself a natural number but a *name* for one — that is, a computational process which may or may not evaluate to yield a natural number. In a CBV structure, a type 1 object is a functions $\mathbb{N} \to \mathbb{N}_\perp$, which demands a genuine natural number as its argument — that is, the computation of the *value* of the argument must take place before the function is called.

As we shall see in Part II, in all cases of interest CBN and CBV structures are equivalent, in the sense that from any CBN structure we may recover a corresponding CBV structure and *vice versa*. We can therefore regard corresponding CBN and CBV structures simply as different concrete presentations of the same underlying notion of computability. However, this perspective was probably not available to the pioneers of the subject.

Throughout this section, if X is a poset, we write $X_\perp$ for the poset obtained by adding a new least element $\perp$.

4.1. Partial monotone functionals. As far as we are aware, the earliest formulation of a class of hereditarily partial computable functionals at all finite types is due to Davis (Davis [1959]). Davis (together with Putnam) realized that some restriction on set-theoretic partial functionals was needed to obtain a good theory, and proposed (essentially) a call-by-name structure A of *consistent* partial functionals ($f : A_n \rightharpoonup \mathbb{N}$ is consistent if whenever $v, w \in \mathrm{dom}(f)$, $t \in A_n$ and $v \subseteq t$, $w \subseteq t$, we have $f(v) = f(w)$), and a notion of computable element of A. Certain features of Davis's definition closely foreshadow Scott's later work on domain theory (see Section 4.2); however, the consistency condition appears to be too weak to be really fruitful, and Davis's type structure seems lacking in good properties at higher types.

4.1.1. *Platek's thesis.* A very satisfactory notion of partial functional was introduced and studied by Platek in his monumental Ph.D. thesis (Platek [1966]). His first fundamental insight was that a good theory could be obtained by restricting attention to *monotone* partial functionals. Essentially, he considered recursion theory over the following call-by-value structure:

DEFINITION 4.2 (Hereditarily monotone functionals). *For each type σ define a poset M_σ as follows: take $\mathsf{M}_0 = \mathbb{N}$ with the discrete order, and for $\sigma = \sigma_1 \to \cdots \to \sigma_r \to \overline{0}$ $(r \geq 1)$, let M_σ be the set of all monotone partial functions $\mathsf{M}_{\sigma_1} \times \cdots \times \mathsf{M}_{\sigma_r} \rightharpoonup \mathbb{N}$ endowed with the pointwise order. Application in M is defined by*

$$f \cdot x \simeq \Lambda z_2 \ldots z_r.\, f(x, z_2, \ldots, z_r).^{13}$$

(where $f \in \mathsf{M}_\sigma$ and $x \in \mathsf{M}_{\sigma_1}$).

The type structure S can be embedded in M via injections $\Psi_\sigma : \mathsf{S}_\sigma \to \mathsf{M}_\sigma$: take $\Psi_0(x) = x$, and for $\sigma = \sigma_1 \to \cdots \to \sigma_r \to \overline{0}$ we let $\Psi_\sigma(F)(\boldsymbol{g}) = y \in \mathbb{N}$ iff there exist $G_i \in \mathsf{S}_{\sigma_i}$ with $\Psi_{\sigma_i}(G_i) \sqsubseteq g_i$ and $F(\boldsymbol{G}) = y$. We say $f \in \mathsf{M}_\sigma$ *represents* $F : \mathsf{S}_{\sigma_1} \times \cdots \times \mathsf{S}_{\sigma_r} \rightharpoonup \mathbb{N}$ if $f(\Psi(\boldsymbol{G})) \simeq F(\boldsymbol{G})$ for all $\boldsymbol{G}$. If $F \in \mathsf{S}_\sigma$, clearly $\Psi(F)$ represents (the uncurried form of) F.

One can give a computational motivation for the restriction to monotone functionals: if f is any computable partial functional, we would expect an increase in the available information about the input to f to allow (if anything) an increase in the output information produced by f. Platek's main motivation, however, was to provide a setting in which definitions by recursion make sense. Given any recursion equation

$$f(x) = F(f, x)$$

where $x : \sigma$, $f : \sigma \to \tau$ are variables and $F \in \mathsf{M}_{(\sigma\to\tau)\to\sigma\to\tau}$, Tarski's fixed point theorem ensures that there is a unique least element $f \in \mathsf{M}_{\sigma\to\tau}$ satisfying the equation for all $x \in \mathsf{M}_\sigma$. This element f may be obtained via a transfinite iteration of F:

$$f_0(x) = \bot, \quad f_{\alpha^+}(x) = F(f_\alpha, x), \quad f_\lambda = \bigcup_{\alpha<\lambda} f_\alpha, \quad f = \bigcup f_\alpha$$

(where λ ranges over limit ordinals). As mentioned earlier, Platek was more concerned with questions of *definability* than of effective computability, so the transfinite nature of the computation here was not seen as a problem.

---

[13]Platek used the term *consistent* rather than *monotone*. The structure M can be construed as a call-by-value structure in our sense, though it does not quite coincide with the "natural" full monotone call-by-value structure over $\mathbb{N}$, for which one would take $\mathsf{M}_{\sigma\ \ \tau}$ to be the set of monotone partial functions $\mathsf{M}_\sigma \rightharpoonup \mathsf{M}_\tau$. However, for pure types the two definitions coincide exactly, and the minor difference can be glossed over.

Platek actually defined such a structure relative to an arbitrary set of objects in place of $\mathbb{N}$, since one of his aims was to give a uniform treatment of recursion theory which applied also to other domains, such as ordinals and transitive sets.

Platek's second fundamental insight was that definitions by recursion can be conveniently expressed by means of *fixed point operators*. Specifically, for each type $\rho = \sigma \to \tau$ there is an element $Y_\rho \in \mathsf{M}_{(\rho\to\rho)\to\rho}$ such that for all $F \in \mathsf{M}_{\rho\to\rho}$, $Y_\rho(F)$ is the unique least element $f \in \mathsf{M}_\rho$ such that $f(x) = F(f, x)$ for all x.

Other (much simpler) ways of explicitly defining new elements of M from old ones are encapsulated by the elements $I_\sigma, K_{\sigma\tau}, S_{\rho\sigma\tau}, D \in \mathsf{M}$ defined by

$$
\begin{aligned}
I_\sigma(x^\sigma) &= x, \\
K_{\sigma\tau}(x^\sigma, y^\tau) &= x, \\
S_{\rho\sigma\tau}(f^{\rho\to\sigma\to\tau}, g^{\rho\to\sigma}, x^\rho) &\simeq (fx)(gx), \\
D(x^{\bar{0}}, y^{\bar{0}}, f^{\bar{1}}, g^{\bar{1}}) &= \begin{cases} f & \text{if } x = y, \\ g & \text{otherwise.} \end{cases}
\end{aligned}
$$

(Here D stands for "definition by cases".) These lead us to the following definition of recursive definability over M:

DEFINITION 4.3. *Given any set $B \subseteq \mathsf{M}$, define the set $\mathcal{R}(B)$ of elements* recursively definable from *B to be the smallest subset of* M *containing B and all the elements Y, I, K, S, D, and closed under application.*

In the case of the monotone type structure over $\mathbb{N}$, we will usually be interested in the case where B contains "enough" basic computable elements (zero, successor and predecessor suffice[14]). We say an element of M is *recursively definable* if it is recursively definable from these basic elements alone.

Platek showed that the above definition can also be cast in terms of schemata in the spirit of Kleene's S1–S9, but with an important difference: *any* definition of a computation written using these schemata has a meaning in M. Thus the theory does not suffer from the difficulties arising from the infinitary side-condition in S8.

Moreover, one can recover Kleene's notion of computability over S as follows: any partial function $F : \mathsf{S}_{\sigma_1} \times \cdots \times \mathsf{S}_{\sigma_r} \rightharpoonup \mathbb{N}$ is Kleene computable iff it is represented by some recursively definable $f \in \mathsf{M}$. Using this as an alternative definition of Kleene computability, Platek was able to obtain clearer proofs of many of Kleene's results plus some new ones, such as the fact that the first recursion theorem holds subject to a type level restriction (see Platek [1966, p. 123]). However, the equivalence with Kleene's definition is non-trivial and depends on the difficult substitution theorems of Kleene [1959b]; this seems to reflect the unmanageability of the schemata S1–S9 (see discussions in Platek [1966, pp. 114–5, 168–9]).

Platek was also the first to draw attention to the famous *parallel or* function (called *strong or* by Platek) and the issues it raises (pp. 127–131). Let us

[14]Gandy pointed out that in fact zero and successor suffice in Platek's setting.

suppose $0 \in \mathbb{N}$ stands for "true" and any other natural number stands for "false". The parallel or function may then be defined as follows:

DEFINITION 4.4. *Let* $por \in \mathsf{M}_{\bar{0}\to\bar{0}\to\bar{0}}$ *be the element given by*

$$por(x, y) = \begin{cases} 0 & \textit{if } x \textit{ is true}, \\ 0 & \textit{if } y \textit{ is true}, \\ 1 & \textit{if } x \textit{ and } y \textit{ are both false}, \\ \bot & \textit{otherwise}. \end{cases}$$

Platek pointed out that *por* is not recursively definable, although it is definable in an Herbrand-Gödel style equation calculus, and is arguably computable if we allow a kind of non-determinism in computations with regard to the order of evaluation of arguments. As we shall see below, the dichotomy between these notions of computability was to become a major theme in later work.

Platek also gave a characterization of recursive definability in terms of a formal language based on the λ-calculus, closely foreshadowing PCF (see Section 4.2.4). Rather curiously, the language he introduced was an *untyped* λ-calculus — this allowed him to make use of the counterparts of Y and D as well as I, K, S in pure λ-calculus. Platek's calculus also featured some elaborate machinery for allowing a mixture of syntax and semantics in computations — *e.g.*, a constant $\lceil f \rceil$ for every element $f \in \mathsf{M}$ was admitted, and the definition of the evaluation relation $U \Downarrow n$ included rules of the following kind (related to Kleene's S8):

$$\begin{array}{rl} \text{if} & f \in \mathsf{M}_{(\sigma\to\bar{0})\to\bar{0}},\ g \in \mathsf{M}_{\sigma\to\bar{0}} \\ \text{and} & \text{for all } h \in \mathsf{M}_\sigma \text{ such that } g(h) \in \mathbb{N},\ U\lceil h \rceil \Downarrow g(h) \\ \text{then} & \lceil f \rceil U \Downarrow f(g). \end{array}$$

The infinitary character of such rules means that computations can be of transfinite length, although for terms containing no constants of type level 2 or above, computations are always finite.

One of the major results of Platek [1966] is the following, proved under mild conditions on B:

THEOREM 4.5. *An element* $f \in \mathsf{M}_{\sigma_1\to\cdots\to\sigma_r\to\bar{0}}$ *is recursively definable from* $B \subseteq \mathsf{M}$ *iff* f *is* λ-definable *from* B, *in the sense that there is a* λ*-term* U, *containing no constants except those corresponding to elements of* B, *such that for all* $g_i \in \mathsf{M}_{\sigma_i}$ *we have* $U\lceil g_1 \rceil \ldots \lceil g_r \rceil \Downarrow n$ *iff* $f(\mathbf{g}) = n$.

The forward implication here is easily shown by supplying lambda terms corresponding to I, K, S, Y, D. To show the converse, Platek proved the existence of recursively definable *enumerators* $E_\sigma \in \mathsf{M}_{\bar{0}\to\sigma}$ such that if U λ-defines $f \in \mathsf{M}_\sigma$ then $E_\sigma(\#U) = f$ (where $\#U$ is a Gödel-number for U). Essentially the same result and proof were rediscovered in Longley and Plotkin [1997].

In summary, Platek's thesis was a major achievement which both shed considerable light on Kleene's earlier work on S1–S9 computability, and introduced many of the fundamental ideas explored in later work on PCF. Indeed, his recursively definable elements of M are exactly the PCF-definable ones in the sense of Scott and Plotkin; and his work shows the existence of a programming language with a *finitary* notion of computation for expressing these elements. With a little charity, one can read into the results of his Chapter 4 a proof of adequacy for M as a model of combinatory PCF (cf. Definition 4.16 and Theorem 4.19 below).

Platek's thesis is unfortunately not generally available, but a detailed published account of much of the material appears (in a somewhat streamlined form) in Moldestad [1977].

4.1.2. *Kleene's later work.* In his later years, Kleene returned to the study of higher type computability and published a series of papers under the title "Recursive functionals and quantifiers of finite types revisited" (Kleene [1978], [1980], [1982], [1985], [1991]). Of these papers, Kleene [1985] provides the best introduction to the whole series.

Kleene's motivation was spelt out in Kleene [1978, §1.2]:

> I aim to generate a class of functions ... which shall coincide with all the partial functions which are "computable" or "effectively decidable", so that Church's 1936 thesis (IM §62) will apply with the higher types included (as well as to partial functions, IM p. 332).

(This has been called "Kleene's problem" in Hyland and Ong [2000].) Like Platek, Kleene sought to avoid the anomalies of the original S1–S9 theory by developing a theory of hereditarily partial objects. However, whereas Platek was interested primarily in questions of definability, Kleene was more interested in the concrete nature of computations at higher types.

Kleene concentrated mostly on the type levels $j = 0, 1, 2, 3$. In Kleene [1978] he introduced a collection of schemata for defining a class of computable functions: a group of schemata S0–S7 (similar in spirit to the earlier S1–S8 though different in a few details), together with a new schema S11 to take over the role of S9. (The label S10 was used in Kleene [1959b] for the schema of μ-recursion.) We adapt Kleene's formulation slightly, for consistency with the notation of Definition 3.2:

S11: *General recursion:*

$$\{\langle 11, \#\boldsymbol{\sigma}, g\rangle\}(\boldsymbol{x}\colon \boldsymbol{\sigma}) \simeq \{g\}(\Lambda \boldsymbol{x}.\, \{\langle 11, \#\boldsymbol{\sigma}, g\rangle\}(\boldsymbol{x}), \boldsymbol{x}).$$

(More briefly, $\{h\}(\boldsymbol{x}) \simeq \{g\}(\{h\}, \boldsymbol{x})$ where $h = \langle 11, \#\boldsymbol{\sigma}, g\rangle$).

Whereas in the earlier theory the schema S9 had incorporated the second recursion theorem as a defining principle, giving rise to restricted versions of the first recursion theorem as consequences, the new schema S11 has the effect of building in a natural formulation of the first recursion theorem, and closure under S9 can be derived as a consequence.

In order to give an interpretation of these schemata which makes sense of the apparent circularity in S11, Kleene gave in Kleene [1978] a syntactical description of computations as computation trees labelled by formal expressions. (Kleene uses the term *j-expression* for a formal expression of type $\overline{j}$.) The essential difference between these and the computation trees of Kleene [1959b], [1963] is that in order to evaluate an expression such as $f(\lambda x.g^1(x), y^0)$ (where f, g are themselves defined via the schemata), we are no longer required to launch a subcomputation to compute the whole graph of $\Lambda x.g^1(x)$, but instead may plough on with the main computation, carrying around the $\lambda x.g(x)$ as an unevaluated formal expression. Only if we later require a particular value for g, such as $g(5)$, do we engage in a computation for g. Computation therefore takes on a much more syntactic character: we are really computing with formal expressions rather than with the semantic objects they denote, and so it need not matter that not every formal expression denotes a semantic object.

In general, the formal expressions in Kleene's computations are allowed to contain free variables of type levels $0, 1, 2, 3$: computations then proceed relative to an interpretation of these free variables as semantic objects. For this reason, a successful computation tree for a 0-expression (that is, one that results in a numerical value) may be infinitely branching and hence of transfinite depth: for example, if we wish to compute $F^2(b)$ where F^2 is a free variable and b is a 1-expression, we really do need to compute $b(n)$ for every numeral n. However, a successful computation tree for a *closed* 0-expression will always be a finite object. Indeed, Kleene's definition of computations closely foreshadows the operational definition of PCF — though as in his earlier work on S1–S9, his seeming reluctance to make fuller use of the typed λ-calculus is puzzling.

In Kleene [1980] and the following papers, Kleene proceeded to develop a semantics for his formal expressions which allows us to give an interpretation for the objects considered at all stages during a computation. In effect, Kleene constructed (the first few levels of) a call-by-name type structure U consisting of what he termed the *unimonotone* functionals (*monotone* with a *un*ique and *i*ntrinsically determined basis). Here one takes U_0 to be $\mathbb{N}_\perp$ and U_{j+1} to be a certain set of monotone functions from U_j to U_0, ordered pointwise. As in Platek's work, the monotonicity reflects the computational intuition that if we are able to obtain some output value without seeing a certain piece of input information, this value should not be affected if we later see it. Furthermore, Kleene restricted attention to those monotone functionals that could be computed by an oracle who followed a strategy or protocol of a certain kind. Kleene gave detailed explicit descriptions of the possible behaviours of such oracles at types 2 and 3, which foreshadow later work on game semantics for PCF. However, Kleene's definitions, couched in terms of the colourful imagery of oracles who open envelopes containing other oracles, were somewhat lacking in clarity and mathematical crispness, and moreover the generalization to higher types was left unclear. It seems that the expression of Kleene's ideas might have been facilitated by some of the basic concepts of domain theory and even category theory which were then available.

Nevertheless, several interesting ideas emerged from Kleene's work. For instance, Kleene isolated an important feature of "sequential" strategies for oracles: a type 2 oracle O, unless she computes a constant type 2 function, must be able to produce of her own accord a type 0 object x (*i.e.*, an element of $\mathbb{N}_\perp$) to serve as the first object she feeds to her type 1 argument f (so that if $f(x) = \perp$ then the whole computation hangs up). Kleene originally conjectured that a similar property would hold at type 3: that a non-constant type 3 oracle would always be able to come up with a particular type 1 object to serve as the first object fed to its type 2 argument. However, this was refuted by Kleene's student Kierstead, who gave as a simple counterexample the type 3 functional defined by

$$\Psi(F^2) = F(\Lambda x^0.\, F(\Lambda y^0.x)).$$

Intuitively, the first object fed to the argument F here is the type 1 function $\Lambda x.F(\Lambda y.x)$, but this function is itself dependent on information about F which only emerges later in the dialogue. This example shows that the possible interactions between oracles and their arguments are somewhat more subtle than Kleene had at first envisaged.

In the subsequent papers, Kleene was able to draw his description of type 3 oracles to a satisfactory completion. An important feature of Kleene's semantics is its intensional character: his oracles work by acting on other oracles (who here are "computing agents" with a certain behaviour) rather than on pure function extensions. Indeed, the requirement that oracles behave extensionally (that is, give the same result when presented with two different oracles for the same function) has to be explicitly incorporated into the definition at certain points. This intensional character is also an important aspect of much of the work in computer science on the semantics of higher type computation: for instance, Berry and Curien's sequential algorithms model (Berry and Curien [1982]), or the game models of PCF (see Section 4.3.3).

Kleene's unimonotone semantics can be considered as one in which computable objects act only on computable objects, if "computable" here is understood to mean realized by some oracle. However, the behaviour of oracles is not required to be effectively given, and so unimonotone functionals need not be computable in a genuinely effective sense. Even more strikingly, they need not be continuous: in fact, Kleene seems to have made a quite deliberate decision to allow infinite computation trees, and moreover to allow the behaviour of oracles to depend on infinitely much information about their arguments. In this regard, Kleene's work contrasts sharply with most work in the computer science tradition; indeed, from a modern perspective, Kleene's decision to impose such tight and computationally motivated constraints in the direction of unimonotonicity but not in the direction of continuity stands as something of a curiosity.

In the meantime, Kierstead (Kierstead [1980], [1983]) developed an alternative semantics for Kleene's S0–S7 and S11, closer in spirit to Platek's work in that it makes use of the full monotone type structure over $\mathbb{N}_\perp$. Many of Kierstead's results are closely parallel to Platek's: for instance, he embeds the total set-theoretic type structure S in the monotone one in a way which respects Kleene computability, and from this is able to deduce certain substitution properties for the total Kleene computable functionals. One difference is that Kierstead considers the call-by-name monotone type structure whereas Platek considers the call-by-value one — this appears to be correlated to some difference in the conceptual status of the models in the two approaches, but from a purely technical point of view it does not seem to be a major difference. (See also Draanen [1995] for some further information in this area.)

The present author feels that Kleene's work in this area contributed some important ideas, although many of these have been more clearly expressed and more fully explored in the subsequent computer science literature. Some useful work by Bucciarelli (Bucciarelli [1993a], [1993b]) makes explicit the relationships between Kleene's work and subsequent developments in computer science, and in particular explores the connections between Kleene's oracles and Berry-Curien sequential algorithms (see Section 4.3). The influence of Kleene's ideas on the computer science tradition is also discussed in Hyland and Ong [2000].

4.2. Partial continuous functionals. Undoubtedly one of the most compelling notions of higher type computability was discovered independently by Scott (Scott [1969], [1993]) and Ershov (Ershov [1972], [1973b]). Coming from somewhat different directions, both these authors arrived at the same type structure P of *partial continuous functionals* and an effective substructure $\mathsf{P}^{\mathit{eff}}$ of partial computable functionals.

For expository convenience we will break slightly with chronological order, describing first the work of Scott, Ershov and others on purely mathematical characterizations of these type structures, and then returning in Section 4.2.4 to consider the connections with languages such as PCF, starting again with the work of Scott in 1969. This will allow us to avoid breaking up our account of the developments relating to PCF, which will be central to much of the rest of the paper.

4.2.1. *Scott's approach.* Scott's work brings together Platek's idea of using partial functionals in recursion theory, and the Kleene-Kreisel idea of using continuity to capture the finitary aspect of computations. This leads to a theory of computable functionals which is arguably simpler and more natural than either Platek's or Kleene and Kreisel's. Scott was also motivated by the idea of giving a mathematical theory of the meanings of computer programs — an abstract view of the mathematical functions they compute as distinct from a purely machine-oriented account of how they behave — and his work lies at the root of the modern computer science tradition in denotational semantics.

The simplest definition of the partial continuous functionals was formulated in Scott [1993][15] in terms of *complete partial orders*:

DEFINITION 4.6 (CPOs).

(i) *A* complete partial order (CPO) *is a poset* $(X, \sqsubseteq)$ *with a least element, in which every chain* $x_0 \sqsubseteq x_1 \sqsubseteq \ldots$ *has a least upper bound* $\bigsqcup x_i$.
(ii) *A function* $f : X \to Y$ *between CPOs is* continuous *if* f *is monotone and for every chain* $x_0 \sqsubseteq x_1 \sqsubseteq \ldots$ *in* X *we have* $f(\bigsqcup x_i) = \bigsqcup f(x_i)$.

DEFINITION 4.7 (Partial continuous functionals).
Define a type structure P *as follows: let* P_0 *be* $\mathbb{N}_\bot$ *(considered as a CPO with the usual partial ordering), and* $\mathsf{P}_{\sigma\to\tau}$ *is the CPO of continuous functions* $f : \mathsf{P}_\sigma \to \mathsf{P}_\tau$, *ordered pointwise.*

Thus, P is simply the natural type structure over $\mathbb{N}_\bot$ in the cartesian closed category of CPOs.

Scott showed in Scott [1969] that an intrinsic notion of computability in P can be given, by exploiting the fact that the CPOs P_σ are all *domains*. Intuitively, a domain is a CPO in which there is a good notion of "finite piece of information" about an element, and in which every element is determined by the set of finite pieces of information about it. The following definitions were introduced in Scott [1969] — for more details see any standard text on domain theory, *e.g.*, Stoltenberg-Hansen, Lindström, and Griffor [1994]. (A detailed understanding of these definitions will not be required for what follows.)

DEFINITION 4.8 (Scott domains).

(i) *A subset* D *of a poset* $(X, \sqsubseteq)$ *is* directed *if it is non-empty and for all* $x, y \in D$ *there exists* $z \in D$ *with* $x \sqsubseteq z$ *and* $y \sqsubseteq z$. *A* DCPO *is a poset* $(X, \sqsubseteq)$ *with a least element* $\bot$ *in which every directed subset* D *has a least upper bound* $\bigsqcup D$. *A function* $f : X \to Y$ *between DCPOs is a* continuous map *if it is monotone and preserves lubs of directed sets.*
(ii) *In a DCPO* X, *we say* x, y *are* consistent *(and write* $x \uparrow y$*) if they have an upper bound in* X. *A DCPO* X *is* consistently complete *if whenever* $x, y \in X$ *have an upper bound, they have a least upper bound* $x \sqcup y$.

[15]This paper was written in 1969 and circulated widely in manuscript form. It was eventually published in the Böhm Festschrift in 1993, along with illuminating retrospective comments by Scott himself.

(iii) *An element e of a DCPO X is* finite *if whenever D is directed and* $e \sqsubseteq \bigsqcup D$, *we have* $e \sqsubseteq d$ *for some* $d \in D$. *We write* F_x *for the set of finite elements* $e \sqsubseteq x$.

(iv) *A DCPO is* algebraic *if for every element* x, F_x *is directed and has lub* x. *A DCPO is* ω-algebraic *if it is algebraic and its set of finite elements is countable.*

(v) *A* domain *is a consistently complete,* ω*-algebraic DCPO.*

Scott showed that the CPOs P_σ are all domains, and moreover that in each P_σ one can give an *effective enumeration* $e_0, e_1, \ldots$ of the finite elements, in such a way that the relations $e_i \uparrow e_j$ and $e_i \sqcup e_j = e_k$ are semirecursive in i, j, k. We may then call an element $x \in \mathsf{P}_\sigma$ *effective* if the finite pieces of information about x are recursively enumerable — that is, if $\{i \mid e_i \sqsubseteq x\}$ is r.e. The effective elements of P are closed under application and so constitute a substructure $\mathsf{P}^{\mathit{eff}}$.

In subsequent papers (Scott [1970], [1972], [1976]), Scott shifted his attention from domains to *complete lattices*, which give rise a very similar theory but using a rather simpler and more familiar definition. Although this approach is very elegant, it can be argued that it takes us further away from computationally meaningful structures: for instance, even the type of natural numbers now needs to be represented by a poset with a top element, which does not typically correspond to the behaviour of any program (see Section 4.2.4). In the 1980s Scott returned to the original domain-theoretic ideas and gave a more concrete presentation of them in terms of *information systems* (Scott [1982]), in which the finite elements (finite pieces of information) were taken as primary.

4.2.2. *Ershov's approach.* In the meantime, Ershov had given a construction of $\mathsf{P}^{\mathit{eff}}$ of a quite different character, arising from his theory of *enumerated sets*. This theory offers a framework for studying a wide range of mathematical structures from an algorithmic or recursion-theoretic point of view (for a recent survey and further references, see Ershov [1999]). The basic definitions are as follows:

DEFINITION 4.9 (Enumerated sets).

(i) *An* enumerated set *is simply a set* X *equipped with a total function* $\nu \colon \mathbb{N} \to X$ *(called an* enumeration*). If* $\nu(n) = x$, *we may say that* n *is a* code *or* recursive index *for* x.

(ii) *A* morphism *from* (X, ν) *to* (X', ν') *is a function* $\phi \colon X \to X'$ *such that there exists a recursive function* f *satisfying* $\phi \circ \nu = \nu' \circ f$.

In Ershov [1971b], [1971a] Ershov introduced the category ***EN*** of enumerated sets, and considered the question of when the set of morphisms from (X, ν) to (X', ν') can be endowed with an enumeration making it into the category-theoretic exponential $(X', \nu')^{(X,\nu)}$. In particular, he defined certain classes $\mathcal{C}_2, \mathcal{C}_{20}$ of enumerated sets (with $\mathcal{C}_{20} \subseteq \mathcal{C}_2$), and proved:

THEOREM 4.10. *If $E \in \mathcal{C}_2$ and $E' \in \mathcal{C}_{20}$, then the exponential E'^E exists in* ***EN*** *and belongs to $\mathcal{C}_{20}$.*

We omit here the somewhat technical definitions of $\mathcal{C}_2$ and $\mathcal{C}_{20}$, which are reproduced *e.g.*, in Ershov [1999]. Since $\mathbb{N}$ (with the identity enumeration) is in $\mathcal{C}_2$ and $\mathbb{N}_\perp$ (with an obvious enumeration) is in $\mathcal{C}_{20}$, it follows that we can construct both call-by-name and call-by-value type structures by repeated exponentiation in ***EN***. Significantly, these type structures are constructed purely out of computable objects acting on computable objects — no notion of continuity is involved in the definition.

In Ershov [1972] Ershov gave a topological description of these structures using the concept of *f_0-spaces*:

DEFINITION 4.11 (f_0-spaces).

(i) *For any T_0 topological space X, let $\sqsubseteq_X$ be the partial order defined by*

$$x \sqsubseteq_X y \iff \textit{for every open set } V, \textit{ if } x \in V \textit{ then } y \in V.$$

Call an open non-empty set V an f-set *if it contains a least element with respect to $\sqsubseteq_X$.*

(ii) *An f_0*-space *is a T_0 space X in which the family of f-sets, together with the empty set, is closed under binary intersections and forms a basis for the topology on X, and moreover the whole of X is an f-set.*

(iii) *Let us write X_0 for the set of elements $x \in X$ for which $\{y \mid x \sqsubseteq_X y\}$ is open. We say $I \subseteq X_0$ is an* ideal *if it is downward closed under $\sqsubseteq_X$ and every pair of elements in I has a least upper bound in I. An f_0-space X is* complete *if $(X, \sqsubseteq_X)$ is canonically isomorphic to the set of ideals in X_0 ordered by inclusion.*

Ershov showed, in effect, that the category of complete f_0-spaces and continuous maps is cartesian closed, and moreover that a substructure of "effective elements" in the type structure over the complete f_0-space $\mathbb{N}_\perp$ coincides with the type structure over $\mathbb{N}_\perp$ in ***EN***. It is natural to think of this as a higher type generalization of the Myhill-Shepherdson theorem (Theorem 2.5).

In Ershov [1972, Section 8] Ershov noted in passing that this class of computable functionals had many pleasing recursion-theoretic properties, remarking that

> It is fully justified to consider [this class] as the most natural generalization (more precisely, extension) of the class of partially recursive functions.

Thus far Ershov's work had proceeded independently of Scott's, but the connections were made explicit in Ershov [1973b]. Indeed, it is not hard to show that domains and complete f_0-spaces are essentially equivalent, and that Ershov's type structures coincide exactly with P and P^{eff}. Ershov's generalization of the Myhill-Shepherdson theorem may therefore be recast as follows:

THEOREM 4.12. $\mathsf{T}(\boldsymbol{EN}, \mathbb{N}_\perp) \cong \mathsf{P}^{eff}$.

In general, Ershov's definitions made greater use of topological concepts whereas Scott's had emphasized the order-theoretic ideas, although in fact both authors made good use of the interplay between the two perspectives.

In Ershov [1974a], [1977a] Ershov obtained the following relationship between P and the Kleene-Kreisel type structure C:

THEOREM 4.13 (Hereditarily total elements). *Define a substructure* $A \subseteq \mathsf{P}$ *as follows:* $A_0 = \mathbb{N}$, *and* $A_{\sigma\to\tau} = \{f \in \mathsf{P}_{\sigma\to\tau} \mid \forall x \in A_\sigma . f \cdot x \in A_\tau\}$. *Define total equivalence relations* $\sim_\sigma$ *on* A_σ *as follows:* $x \sim_0 y$ *iff* $x = y$; *and* $f \sim_{\sigma\to\tau} g$ *iff* $\forall x \in A_\sigma . f \cdot x \sim_\tau g \cdot x$. *Then the sets* $A_\sigma / \sim_\sigma$ *constitute a type structure that is canonically isomorphic to* C.

Moreover, if $A_\sigma^{eff} = A_\sigma \cap \mathsf{P}_\sigma^{eff}$, *the sets* $A_\sigma^{eff} / \sim_\sigma$ *constitute a type structure isomorphic to* RC.

It follows easily that C is the extensional collapse of P with respect to the equality relation on $\mathbb{N}$. However, the above theorem says more than this, since it tells us that all hereditarily total elements are automatically hereditarily extensional. An analogous result for effective type structures, implying that the extensional collapse of P^{eff} gives rise to HEO, was proved in Ershov [1976b].

Ershov's main results on P and P^{eff} are conveniently summarized in Ershov [1977a]. Some further discussion of the relationship between the partial and total type structures also appears in Sections 9 and 10 of Gandy and Hyland [1977]. A streamlined presentation of the main results on f_0-spaces appears in Giannini and Longo [1984], together with some applications to the semantics of *untyped* λ-calculi.

4.2.3. *Later developments.* An important structural property of both P and P^{eff} was obtained in Plotkin [1978]. This makes use of the notion of a *coherent* Scott domain, that is, one in which every pairwise consistent subset has a least upper bound.

THEOREM 4.14. *The Scott domain* $\mathbb{T}^\omega$ *of functions from* $\mathbb{N}$ *to* $2_\perp$ *ordered pointwise is a* universal *coherent domain: that is, for any coherent Scott domain* X *there are continuous maps* $f_X : X \to \mathbb{T}^\omega$ *and* $g_X : \mathbb{T}^\omega \to X$ *such that* $g_X \circ f_X = \mathrm{id}_X$ *(we say* X *is a* retract *of* $\mathbb{T}^\omega$*). Moreover, if* X *is an effective domain then these maps are computable.*

Since all the domains P_σ are coherent, it follows that any domain that is rich enough to contain $\mathbb{T}^\omega$ as a computable retract (such as P_1) contains all the P_σ as computable retracts. Indeed, using some standard categorical techniques, the whole of P can be reconstructed from just P_1 together with its set of continuous endofunctions; likewise, P^{eff} can be reconstructed from the monoid of computable endofunctions on $\mathsf{P}_1^{\mathit{eff}}$. (For an exposition of these general techniques, see *e.g.*, Longley [2004].)

A few later results should also be mentioned. For instance, several people noted that Ershov's category ***EN*** has rather poor closure properties (in particular, it is not cartesian closed), and so proposed larger categories to remedy this. (Of course, the category ***EN*** is still of interest, since we know more about an object by knowing it belongs to ***EN*** than by knowing it lives in some larger category.) Mulry (Mulry [1982]) showed that ***EN*** could be embedded into his *recursive topos* preserving existing exponentials, so that the type structure over a suitable object $N_\bot$ in this topos coincides with $\mathsf{P}^{\mathit{eff}}$.[16] This is close in spirit to another characterization of $\mathsf{P}^{\mathit{eff}}$ given by Longo and Moggi (Longo and Moggi [1984b]), which gives the remarkable appearance of magically dispensing with any explicit computability requirement at higher types. Notice the analogy with Definition 2.1.

DEFINITION 4.15. *Let fst, snd*: $\mathbb{N} \to \mathbb{N}$ *be the projections associated with some recursive pairing function from* $\mathbb{N} \times \mathbb{N}$ *to* $\mathbb{N}$. *For each type* σ *we define a set* $\mathsf{P}_\sigma^{\mathit{eff}}$ *together with a set* $\mathsf{P}_\sigma^{\omega,\mathit{eff}}$ *of functions* $f : \mathbb{N} \to \mathsf{P}_\sigma^{\mathit{eff}}$ *as follows*:

$$\mathsf{P}_0^{\mathit{eff}} = \mathbb{N}_\bot.$$

$$\mathsf{P}_0^{\omega,\mathit{eff}} \cong \mathbb{N}_{p\,rec}^{\mathbb{N}}.$$

$$\mathsf{P}_{\sigma\to\tau}^{\mathit{eff}} = \{f : \mathsf{P}_\sigma^{\mathit{eff}} \to \mathsf{P}_\tau^{\mathit{eff}} \mid \forall g \in \mathsf{P}_\sigma^{\omega,\mathit{eff}}.\ f \circ g \in \mathsf{P}_\tau^{\omega,\mathit{eff}}\}.$$

$$\mathsf{P}_{\sigma\to\tau}^{\omega,\mathit{eff}} = \{f : \mathbb{N} \to \mathsf{P}_{\sigma\to\tau}^{\mathit{eff}} \mid \forall g \in \mathsf{P}_\sigma^{\omega,\mathit{eff}}.\ \Lambda n.\, f(\mathit{fst}\, n)(g(\mathit{snd}\, n)) \in \mathsf{P}_\tau^{\omega,\mathit{eff}}\}.$$

(This is a mild variation on the definition given in Longo and Moggi [1984b] — the latter relies on a theorem ensuring the existence of suitable pairing operations at higher types.) Inspired by this characterization, Longo and Moggi introduced another cartesian closed category extending ***EN***: the category of *generalized numbered sets* (Longo and Moggi [1984a]). Moggi later clarified the relationship between this category and the recursive topos (Moggi [1988]).

Subsequent work has shown how P and $\mathsf{P}^{\mathit{eff}}$ arise naturally from various realizability models. We will continue this part of the story in Section 5 below.

4.2.4. *PCF and parallelism.* We now return to discuss the important connections between P, $\mathsf{P}^{\mathit{eff}}$ and formal languages, beginning again with Scott's early work. In Scott [1993], Scott showed that P provided a model for a simple formal language inspired directly by Platek's work (Section 4.1.1). In the context of Scott's paper, this served as the term language of a logic intended for reasoning about computable functions, which later became known as LCF.

[16]It is worth noting, in passing, that the type structure over N in this topos does not coincide with any of our type structures of total computable functionals, but rather yields the *generalized Banach-Mazur functionals*: see Mulry [1982] and cf. Section 2.2.

Scott's language for functions may be presented as follows. As usual, we will work with the simple types over $\overline{0}$.[17]

DEFINITION 4.16 (PCF).[18] *Let* (combinatory) PCF *be the language consisting of the well-typed expressions built up via application from the constants*

$$\begin{aligned} \mathtt{K}_{\sigma\tau} &: \sigma \to \tau \to \sigma \\ \mathtt{S}_{\rho\sigma\tau} &: (\rho \to \sigma \to \tau) \to (\rho \to \sigma) \to (\rho \to \tau) \\ \mathtt{Y}_{\sigma} &: (\sigma \to \sigma) \to \sigma \\ \mathtt{if} &: \overline{0} \to \sigma \to \sigma \\ \mathtt{0} &: \overline{0} \\ \mathtt{succ} &: \overline{0} \to \overline{0} \\ \mathtt{pred} &: \overline{0} \to \overline{0}. \end{aligned}$$

The similarity to Platek's definition of recursive definability (Definition 4.3) is clear. The constant $\mathtt{if}$ takes over the role of Platek's D; $\mathtt{0}, \mathtt{succ}, \mathtt{pred}$ correspond to a set of basic computable elements; and Platek's I is redundant anyway, since one can define $I = SKK$.

Of course, there are some arbitrary choices involved in this definition, and many mild variants of it lead to essentially the same language. For instance, one might drop the constants $\mathtt{K}$ and $\mathtt{S}$, and instead consider the simply typed λ-calculus over the remaining constants (this is in fact the version that is most commonly considered today). Alternatively, one might define PCF as "System T plus general recursion": that is, we simply augment the definition of Gödel's System T (Section 3.1.1) with the constants $\mathtt{Y}_{\sigma}$. The possibility of such variants was already clear to Scott in Scott [1993]; they are all intertranslatable and for our purposes have the same expressive power, so we may freely refer to any of them as "PCF".

Scott gave an interpretation of PCF in P in the spirit of Platek's definition. The constants $\mathtt{K}, \mathtt{S}, \mathtt{0}, \mathtt{succ}, \mathtt{pred}$ are interpreted in the obvious way. The combinator $\mathtt{if}$ is interpreted by the element *if* of P is given by

$$\mathit{if}\,(0)(x)(y) = x, \quad \mathit{if}\,(n+1)(x)(y) = y, \quad \mathit{if}\,(\bot)(x)(y) = \bot.$$

The combinator $\mathtt{Y}_{\sigma}$ is interpreted by the function Y_{σ} which assigns to any $f \in \mathsf{P}_{\sigma\to\sigma}$ the least fixed point of f in P_{σ}. (Note that because all functionals in P are continuous, we may construct Y_{σ} simply by an iteration up to ω — we do not require transfinite iterations as in Section 4.1.1.) Finally, application

[17] Actually, Scott's system also had a ground type of truth-values as well as one of integers, and hence a slightly different selection of constants. Here, as elsewhere, we will content ourselves with representing "true" by zero and "false" by any other natural number. Otherwise, we have kept closely to Scott's original definition.

[18] The name 'PCF' first appeared in Milner [1977].

in PCF is interpreted by application in P. We thus have the notion of a *PCF-definable* element of P.

Scott's stated intention in introducing the above language was to provide "a restricted system that is specially designed for algorithms". However, he presented the system purely as a mathematical language for defining elements of P rather than as an executable "programming language", and initially it was not even immediately obvious whether all the PCF-definable elements of P were effectively computable in any reasonable sense (see Scott [1993, Section 4]).

In fact (as Scott quickly realized), it is easy to show that every PCF-definable element P is effective at least in the sense that it is an element of $\mathsf{P}^{\mathit{eff}}$. However, the converse is not true: the function *por* considered by Platek (Definition 4.4) is effective but not PCF-definable. It is clear that *por* is in some sense computable — one can evaluate $por\,(x)(y)$ by evaluating x and y "in parallel" — but like Platek, Scott noted that this kind of algorithm has a significantly different flavour from the means of computation that suffice to compute the PCF-definable elements (Scott [1993, Section 4]):

> Do we enjoy this new flavor enough to call it computable? Some people would say yes, but I wonder.

It thus began to appear that there might be two reasonable notions of computable element in P: the notion of effective element (given by $\mathsf{P}^{\mathit{eff}}$), and the more restrictive notion of PCF-definable element.

The next major results were obtained independently by Plotkin (Plotkin [1977]) and Sazonov (Sazonov [1976a]). (Closely related results were also obtained around the same time by Feferman (Feferman [1977a]).) These authors showed that the gap between PCF computability and effectivity could be bridged by augmenting PCF with just two operations of a "parallel" flavour:

THEOREM 4.17. *Let* PCF^{++} *be the language* PCF *extended with the two constants*

$$\texttt{por}\colon \overline{0} \to \overline{0} \to 0, \quad \texttt{exists}\colon (\overline{0} \to \overline{0}) \to \overline{0}.$$

Extend the interpretation of PCF in P *by interpreting* `por` *by the function por (see Definition* 4.4), *and* `exists` *by the functional given by*

$$exists(f) = \begin{cases} 0 & \text{if } f(n) = 0 \text{ for some } n \in \mathbb{N}, \\ 1 & \text{if } f(\bot) = k+1 \text{ for some } k \in \mathbb{N} \\ & \quad (\text{hence } f(x) = k+1 \text{ for all } x), \\ \bot & \text{otherwise.} \end{cases}$$

Then the elements of $\mathsf{P}^{\mathit{eff}}$ *are exactly the* PCF^{++}*-definable elements of* P.[19]

[19]In fact Plotkin considered a parallel conditional operator rather than parallel or, but it is easy to show that these are interdefinable over PCF (see Stoughton [1991a]).

Notice that the functional *exists* here is a genuinely effective "parallel search operator", quite different in flavour from the functionals ${}^k\exists$ considered in Section 3.2.2.

As pointed out in Feferman [1977a], this was a significant result in that it showed that this notion of computability on P could be captured by a finite collection of inductive schemata, somewhat in the spirit of Kleene's S1–S9. Interestingly, it seems that Kleene himself never considered parallel operations as a way of getting more expressive power, even though he was specifically interested in the problem of identifying a class of "all" computable functions at higher types.

4.2.5. *Operational semantics.* So far we have considered PCF and PCF^{++} simply as formal languages for defining elements of P. However, a crucial step was made by Plotkin (Plotkin [1977]), who showed how to give an *operational semantics* (that is, a set of symbolic evaluation rules) for these systems, turning them into executable programming languages. This meant that the "meaning" of programs (that is, their evaluation behaviour) could be defined without reference to a model such as P.

We illustrate the idea by giving one possible operational semantics for our version of PCF^{++}. Our definition is in a rather different style from Plotkin's, being closer in spirit to Milner [1977]. Specifically, we will define a transitive relation $\rightsquigarrow$ on terms corresponding to a notion of many-step reduction. We write $\widehat{k}$ as an abbreviation for $(\mathtt{succ}^k\, 0)$, and $\bot_\sigma$ for the term $\mathtt{Y}_\sigma \mathtt{I}$ where $\mathtt{I} = \mathtt{SKK}$.

DEFINITION 4.18 (Operational semantics for PCF^{++}). *Let* $\rightsquigarrow$ *be the smallest transitive relation on terms of* PCF^{++} *satisfying the following clauses, in which we assume all terms are well-typed.*

- $\mathtt{K}UV \rightsquigarrow U$, $\mathtt{S}UVW \rightsquigarrow (UW)(VW)$, $\mathtt{Y}U \rightsquigarrow U(\mathtt{Y}U)$.
- $\mathtt{pred}\, 0 \rightsquigarrow 0$, $\mathtt{pred}\, \widehat{k+1} \rightsquigarrow \widehat{k}$.
- $\mathtt{if}\, 0\, U\, V \rightsquigarrow U$, $\mathtt{if}\, \widehat{k+1}\, U\, V \rightsquigarrow V$.
- $\mathtt{por}\, 0\, V \rightsquigarrow 0$, $\mathtt{por}\, V\, 0 \rightsquigarrow 0$, $\mathtt{por}\, \widehat{j+1}\, \widehat{k+1} \rightsquigarrow \widehat{1}$.
- *If* $U\, \widehat{n} \rightsquigarrow 0$ *then* $\mathtt{exists}\, U \rightsquigarrow 0$.
- *If* $U\, \bot_\sigma \rightsquigarrow \widehat{k+1}$ *then* $\mathtt{exists}\, U \rightsquigarrow \widehat{1}$.
- *If* $U \rightsquigarrow U'$ *then* $UV \rightsquigarrow U'V$.
- *If* $U \rightsquigarrow U' : \overline{0}$ *then* $c\ U \rightsquigarrow c\ U'$ *for* $c = \mathtt{succ}, \mathtt{pred}, \mathtt{if}, \mathtt{por}, \mathtt{exists}$, *and* $\mathtt{por}\, V\, U \rightsquigarrow \mathtt{por}\, V\, U'$.

We say a term $U : \overline{0}$ evaluates to *some (necessarily unique) natural number k if* $U \rightsquigarrow \widehat{k}$. *A term* $U : \overline{0}$ converges *if it evaluates to some k; otherwise U* diverges.

Let us write $[\![\, U \,]\!]$ for the element of P denoted by a PCF^{++} term U. The following fundamental result (essentially Theorem 3.1 of Plotkin [1977]) says that the operational and denotational semantics agree.

THEOREM 4.19 (Adequacy of P). *For any closed term* $U : \overline{0}$, *U evaluates to* k *iff* $[\![U]\!] = k$.

The proof makes use of powerful ideas developed in Tait [1967] to prove strong normalization for calculi such as System T (see Section 3.1.1).

Plotkin's work on PCF marked something of a break between the older recursion theory tradition and what was to become the modern computer science tradition. Roughly speaking, whereas the former usually treated formal languages or inductive schemata principally as ways of picking out a class of computable elements from some predefined mathematical structure such as S or P, the tendency in computer science has been to take the programming languages as primary and then look for mathematical structures in which they can be interpreted. (One can think of exceptions to this, of course, but some cultural difference along these lines can be discerned in the literature and persists to this day.) The ideas of operational semantics, which allow us to give standalone syntactic definitions of programming languages, are what make the latter approach viable.

We can perhaps illustrate this point of view by using the operational definition of PCF^{++} to give an independent and purely syntactic construction of $\mathsf{P}^{\mathit{eff}}$ (up to isomorphism). For PCF^{++} terms $U, V : \overline{0}$, let us write $U \equiv V$ if U and V either both diverge or both evaluate to the same number k. Let A be the type structure of PCF^{++} terms, and let $\equiv_\sigma$ be the partial equivalence relations induced by the extensional collapse construction with respect to $\equiv$ (Definition 1.2). It is not too hard to show that $U \equiv_\sigma V$ iff $[\![U]\!] = [\![V]\!] \in \mathsf{P}_\sigma$. It follows by Theorem 4.17 that $(A_\sigma / \equiv_\sigma) \cong \mathsf{P}^{\mathit{eff}}_\sigma$; thus, $\mathsf{P}^{\mathit{eff}}$ is isomorphic to $\mathsf{EC}(A, \equiv)$.

4.3. PCF and sequential computability. By ignoring the references to `por` and `exists` in Definition 4.18, we obtain a standalone operational definition of PCF. Using such a definition, Plotkin was able to give rigorous proofs that *por* and *exists* are not PCF-computable, by an analysis of possible reduction behaviours. (A remarkably similar result had been obtained quite independently in Sasso [1971] — see Odifreddi [1989, p. 188].) Despite this incompleteness, the feeling persisted that the notion of computability embodied by PCF was a natural one and of interest in its own right.

Intuitively, computations in PCF proceed in a "sequential" manner in the sense that there is a single thread of computation — we never find two disjoint subterms of a term being evaluated at the same time. Since very many practical programming languages also have this sequential character, PCF appeared attractive from a computer science perspective as a prototypical programming language for suitable theoretical study.

4.3.1. *PCF versus S1–S9.* Before surveying the main body of research relating to PCF, we digress briefly to comment on its relationship to Kleene's S1–S9. It is sometimes remarked that PCF is essentially equivalent to S1–S9, but in our view this idea needs to be treated with caution, as there is a subtle issue here which seems never to have been clearly explained in the published literature.

Even if we restrict attention to the type structure P, taking a literal reading of S1–S9 as given in Kleene [1959b] (or our Definition 3.2), it is true that all the Kleene computable elements of P are PCF-definable, but not *vice versa*. This is because there is a difference between saying that $\{m\}(\boldsymbol{x})$ is not defined (by the inductive definition of the ternary relation $\{m\}(\boldsymbol{x}) = y$) and saying that $\{m\}(\boldsymbol{x}) = \bot$. Indeed, the fact that in P the notion of non-termination has been objectified by $\bot$ means that if we wish to have $\{m\}(\boldsymbol{x}) = \bot$, then this triple must be explicitly generated by the inductive procedure. Now suppose we attempt to construct an index m for, say, the fixed point operator $Y_1 \in \mathsf{P}_{(\overline{1} \to \overline{1}) \to \overline{1}}$. By invoking S9 and appealing to Kleene's second recursion theorem, we can certainly obtain an index m with the property that $\{m\}(F) \simeq F(\{m\}(F))$ for all $F \in \mathsf{P}_{\overline{1} \to \overline{1}}$. But now, even if we specialize F to the constant function $\Lambda g.\Lambda x.0$, we cannot conclude that $\{m\}(F) = 0$, since strictly speaking we cannot assign a meaning to $F(\{m\}(F))$ before $\{m\}(F)$ has been given a meaning! This is because F is just an ordinary mathematical function on the set $\mathbb{N} \sqcup \{\bot\}$. All we can deduce is that *if* $\{m\}(F)$ *means anything* (whether $\bot$ or a number) then $F(\{m\}(F)) = 0$ and so $\{m\}(F) = 0$. In fact, it can be shown that the element Y_1 is not Kleene computable under this strict interpretation.

One can overcome this problem by reinterpreting S1–S9 not as the inductive definition of a set of triples $(m, \boldsymbol{x}, y)$, but as the recursive definition of a total function $\{-\}(-) \colon \mathbb{N} \times X(\mathsf{P}) \to \mathbb{N}_\bot$ obtained as a gigantic simultaneous least fixed point. (Alternatively, one could perhaps recast it as the definition of a relation $\{m\}(\boldsymbol{x}) \sqsupseteq y$). It appears that this is often what people have in mind when referring to Kleene computability over P, and it *is* true that this more generous notion coincides with PCF-definability.

Even so, it seems to us questionable whether this latter interpretation is true to the *spirit* of S1–S9 as manifested in Kleene's papers. As remarked earlier, the schemata S1–S9 were introduced by Kleene to capture the ideal of computing with functions as pure extensions, whose characteristic behaviour is that when presented with a specified object of lower type they simply return an answer (or perhaps diverge). It was a conscious shift in perspective on Kleene's part to regard computations as acting on more intensional representations of functions, such as the formal expressions in Kleene [1978] or the oracles in Kleene [1980]; and for this purpose Kleene introduced S11. Now the strategy required to compute a functional such as Y_1 makes essential use of the idea that an object F of type $\overline{1} \to \overline{1}$ can not only provide answers when presented with specified arguments, but also respond with questions of its own when presented with unspecified (or incompletely specified) arguments. And if F can do this, it is (in our view) behaving as something more than a pure extension. It therefore seems to us more accurate, both in letter and in spirit, to say that PCF essentially corresponds to Kleene's S1–S8 plus S11 (this is indeed stated explicitly in Nickau [1994]). It is worth remarking that Kleene himself never used S9 in connection with hereditarily partial functionals.

One can, however, define a weaker language than PCF that does correspond in expressive power to the strict interpretation of S1–S9, essentially by using S1–S9 themselves as the basis of a language for *partial* functionals. The notion of computability embodied by this language seems to be a very natural one and deserves more attention than it has so far received. Such a language captures an intuitively appealing idea of "computing with pure extensions", and moreover seems to offer the most natural route to the study of Kleene computability even in total settings. We will say more about this notion of computability in Part II, where we will introduce some associated type structures K and K^{eff}.

Even weaker languages have also been considered. For instance, one can consider Gödel's System T extended with the minimization operator μ of ordinary recursion theory. Some preliminary expressivity results on this and related systems have recently been considered by Berger (Berger [2000]). Closely related issues have also recently been considered in Niggl [1999] and Normann and Rørdam [2002].

4.3.2. *The full abstraction problem.* The operational semantics of PCF gives rise to a notion of *observational equivalence* of PCF programs.

DEFINITION 4.20. *We say the PCF terms $U, V : \sigma$ are* observationally equivalent (*and write $U \approx_\sigma V$*) *if, for all term contexts $C[-]$ such that $C[U], C[V]$ are terms of type $\overline{0}$, $C[U]$ evaluates to k iff $C[V]$ evaluates to k.*

This is a natural notion from a programming point of view, since $U \approx V$ means that it is always safe to replace U by V in any larger program without affecting the overall result. It is therefore natural to ask whether one can give a mathematical characterization of observational equivalence (in PCF or in other languages).

An important result of Milner (Milner [1977]) shows that two PCF terms are observationally equivalent iff, intuitively, they induce the same functions on terms of lower type:

THEOREM 4.21 (Context Lemma). *For closed PCF terms*

$$U, V : \sigma_1 \to \cdots \to \sigma_r \to \overline{0},$$

we have $U \approx V$ iff, for all closed $W_1 : \sigma_1, \ldots, W_r : \sigma_r$, $UW_1 \ldots W_r$ evaluates to k iff $VW_1 \ldots W_r$ does.

The idea that the behaviour of PCF terms is completely determined by the functions they induce is expressed by saying that PCF is a *purely functional* programming language.

We may now see how to define a standalone type structure of PCF-definable functionals from the definition of PCF itself. This is analogous to the construction of $\mathsf{P}^{\mathit{eff}}$ from PCF^{++} mentioned at the end of Section 4.2.5.

DEFINITION 4.22 (PCF type structure). *Let B be the type structure consisting of PCF terms, and let $\mathsf{Q}^{\mathit{eff}}$ be the extensional collapse of B with respect to $\approx_0$.*

It follows easily from the context lemma that $\mathsf{Q}^{\mathit{eff}}$ is actually the *quotient* of B modulo observational equivalence. We may regard the type structure $\mathsf{Q}^{\mathit{eff}}$ as embodying the notion of "sequentially computable functional" represented by PCF.

It is natural to ask if one can give a more mathematically illuminating description of $\mathsf{Q}^{\mathit{eff}}$ than the syntactic definition given above. In particular, can we find a mathematically natural type structure Q within which $\mathsf{Q}^{\mathit{eff}}$ sits as an "effective substructure", in the way in which $\mathsf{P}^{\mathit{eff}}$ is a substructure of P? More or less equivalently, can we give a denotational interpretation of PCF such that $[\![U]\!] = [\![V]\!]$ iff $U \approx V$? In computer science terminology, such an interpretation would be called *fully abstract*.

Note that the interpretation of PCF in P does not fulfil this requirement. It follows from the adequacy theorem that if $[\![U]\!] = [\![V]\!]$ in P then $U \approx V$, but the converse fails. This is because one can find terms $U, V : \sigma \to \tau$ such that $[\![U]\!](x) = [\![V]\!](x)$ for all PCF-definable elements $x \in \mathsf{P}_\sigma$, but $[\![U]\!](x) \neq [\![V]\!](x)$ for some non-definable elements such as *por* (see Plotkin [1977]).

The problem of trying to give a good mathematical characterization of the "sequential" functionals at higher types, and in particular of finding a suitable type structure Q with the above properties, became known as the *full abstraction problem* for PCF, and was to receive much attention in theoretical computer science. The early investigators had in mind a denotational model consisting of CPOs of some kind, though chain completeness is not an essential requirement. Plotkin and Milner (Milner [1977]) showed for a large class of potential CPO models of PCF that full abstraction holds iff all the finite elements are PCF-definable. As shown by Milner, it follows that there is up to isomorphism just one (order-extensional) fully abstract CPO model. However, Milner's construction of this model was still syntactic in flavour and was not felt to yield a good mathematical characterization of sequentiality.

Characterizations of sequentiality were obtained easily enough for first order functions from $\mathbb{N}_\perp^r$ to $\mathbb{N}_\perp$. (By a mild abuse of notation we will write $\mathbb{N}_\perp^0 \to \mathbb{N}_\perp$ to mean $\mathbb{N}_\perp$, and $\mathbb{N}_\perp^{r+1} \to \mathbb{N}_\perp$ to mean $\mathbb{N}_\perp \to (\mathbb{N}_\perp^r \to \mathbb{N}_\perp)$.) The following definition (by induction on r) was given by Milner:

DEFINITION 4.23. *A monotone function $f : \mathbb{N}_\perp^r \to \mathbb{N}_\perp$ is* sequential *if either it is constant, or there is some i such that for all $x_i \in \mathbb{N}_\perp$ the function*

$$\Lambda x_1 \dots x_{i-1} x_{i+1} \dots x_r .\ f(x_1)\dots(x_i)\dots(x_r) \colon \mathbb{N}_\perp^{r-1} \to \mathbb{N}_\perp$$

is sequential.

Intuitively, f is sequential if at each stage in the computation of $f x_1 \ldots x_r$, either f can return an answer, or else it cannot because it needs to know the value of some argument x_i. (If there is only one such i, this is the argument that needs to be evaluated next in the computation; if there is more than one, we have some choice in which argument to evaluate next.) Similar (though not identical) ideas were present in Kleene's definition of unimonotonicity for type 2 oracles (Section 4.1.2). A different but equivalent characterization was also obtained independently by Vuillemin (Vuillemin [1973]).

It is easily shown that a *finite* element of $\mathsf{P}_{\overline{0}^r \to \overline{0}}$ is sequential iff it is PCF-definable, so that a model for PCF that included only sequential functions at first order types would be fully abstract for types of level 2. However, a counterexample due to Trakhtenbrot (Trakhtenbrot [1975]; see also Sazonov [1976a]) shows that not every *effective* sequential element of $\mathsf{P}_{\overline{0}^r \to \overline{0}}$ is PCF-definable — it can happen that at some stage some appropriate choice of index i exists, but there is no effective way to compute it.

The problem of characterizing sequentiality at higher types turned out to be much more difficult, and was the focus of a great deal of research effort in theoretical computer science. Milner (Milner [1977]) and later Mulmuley (Mulmuley [1987]) gave constructions of the fully abstract CPO model, but both of these referred in some way to the operational semantics of PCF and so did not qualify as an independent mathematical characterization. Several other models for PCF were constructed: Berry's *stable* and *bistable* models (Berry [1978]); the Berry-Curien *sequential algorithms* model (Berry and Curien [1982], Curien [1993]) based on Kahn and Plotkin's notion of *concrete data structure* (Kahn and Plotkin [1993]); and the Bucciarelli-Ehrhard *strongly stable* model (Bucciarelli and Ehrhard [1991b], Ehrhard [1993]). For a detailed survey of this material and further references, we recommend Ong [1995]. This line of research gave much insight into the difficulty of the full abstraction problem, and generated many counterexamples illustrating the subtlety of the notion of observational equivalence. We will not describe the above models in detail here, since as far as PCF is concerned the main point about them is that they are *not* fully abstract. However, both the sequential algorithms model and the strongly stable model turned out to be important in connection with other notions of computability, and we will return to them below.

Given that the only known constructions of a fully abstract model were felt to be unsatisfactory, it was natural to ask what exactly were the criteria for a "good" solution to the full abstraction problem. One possible criterion was proposed in Jung and Stoughton [1993]: when restricted to *finitary* PCF (that is, the fragment of PCF generated by the ground type of booleans), the model construction should be *effective*. In other words, given a simple type σ over the booleans it should be possible to effectively compute a complete description of the (finite) semantic object representing σ. This criterion rules out syntactic constructions such as Milner's; it admits the other models mentioned above, though they are not fully abstract.

If an extensional fully abstract model satisfying the Jung-Stoughton criterion existed, it would follow that observational equivalence for finitary PCF was decidable. However, around 1996 Loader showed:

THEOREM 4.24 (Loader [2001]). *Observational equivalence in finitary PCF is undecidable.*

The proof involves a tricky encoding of semi-Thue problems. Loader's result represents one of the most important advances in our understanding of PCF: it closes the door on a large class of attempts at the full abstraction problem, and indeed shows that it was not possible to give any good finitary analysis of PCF-sequentiality considered purely as a property of functions. (Remarkably, however, the fully abstract model for unary PCF — that is, PCF over a ground type with a single terminating value $\top$ — *is* effectively presentable; see Loader [1998].)

4.3.3. *Intensional semantics for PCF.* Another possible approach is to seek a good mathematical description of the *algorithms* embodied by PCF terms rather than of the functions they compute. This approach does in fact lead to good models of sequential computation, albeit of an "intensional" nature. One can regard such models as occupying a kind of middle ground between operational and denotational semantics as traditionally conceived: elements of the model are typically computation strategies with a dynamic operational behaviour, but many of the inessential details present in an operational definition of a language are abstracted out.

An approach of this kind was first successfully carried through by Sazonov (Sazonov [1975], [1976b], [1976c]), whose work was explicitly formulated in terms of Scott's LCF but was independent of the work of Milner and Plotkin, and indeed remained little known in the West until a brief description of it appeared in Hyland and Ong [2000]. In effect, Sazonov gave a model of PCF-style sequential computation in terms of Turing machines with oracles, much in the same spirit as Kleene's characterization of S1–S9 computations in Kleene [1962c], [1962d]. A Turing machine for a function of type $\overline{n}$ communicates with its argument (an oracle of type $\overline{n-1}$) by feeding it with a description of a Turing machine for type $\overline{n-2}$. As in Kleene [1962c], computations have the character that functionals of type $\overline{n-k}$ are represented by (codes for) Turing machines when k is even, and by pure oracles when k is odd. At the heart of Sazonov's work is his notion of the strategy followed by a Turing machine: sequentiality is enforced by a requirement that when a machine invokes an oracle, it must receive an answer before it can continue. Although Sazonov's formalization is somewhat complicated and rather dependent on the use of explicit codings for Turing machines, this approach succeeds in capturing the notion of PCF computability at all finite types and very closely anticipates many features of later models. (In fact, it could be argued that Sazonov had already accomplished what Kleene was trying to achieve in the papers from Kleene [1978] onwards!) An account of Sazonov's work framed in more modern computer science terms is given in Sazonov [1998].

A somewhat similar approach was pursued for many years by Gandy and his student Pani, who however concentrated more on the problem of characterizing the PCF-definable elements of P. This approach was apparently influenced by Kleene [1978], and emphasized the idea of computations as dialogues between two participants. The information available to the participants at each stage is represented by finite elements of P (or similar entities). Gandy's insights had a significant influence on the computer science community, though unfortunately Gandy never completed a written account of his ideas and their exact form remains unclear.

Within computer science itself, an abstract formulation of the essence of PCF sequentiality in terms of dialogue games was achieved around 1993, when three closely related models of PCF were obtained (simultaneously and more or less independently) in Abramsky, Jagadeesan and Malacaria (Abramsky, Jagadeesan, and Malacaria [2000]), Hyland and Ong (Hyland and Ong [2000]), and Nickau (Nickau [1994]). Again, the formal details can appear rather complicated, but the essential ideas can be easily grasped via an example. Using λ-calculus notation and some obvious abbreviations here for convenience, suppose we are given the PCF terms

$$\begin{aligned} F &\equiv \lambda gh.\, (g\ 3) + (g(\mathrm{Y}h)) && : \overline{1} \to \overline{1} \to \overline{0} \\ g &\equiv \lambda x.\, x + 2 && : \overline{1} \\ h &\equiv \lambda x.\, 4 && : \overline{1} \end{aligned}$$

and we wish to evaluate Fgh. The computation can be modelled as a dialogue between a Player P, who follows a strategy determined by F, and an Opponent O, whose strategy is determined by g and h. Moves in the game are either questions (written '?'), or answers (natural numbers) which are matched to previous questions. Furthermore, the moves are associated with occurrences of $\overline{0}$ in the type in question, according to their "meaning" in the context of the game. The game always starts with a question by Opponent, who asks for the final result of the computation. We show in Figure 2 the dialogue corresponding to the computation of Fgh, along with an informal paraphrase of the meaning of each move. (In fact, the order of moves here precisely matches the order of interactions between Turing machines and oracles in Sazonov's approach.)

In the game model of Hyland and Ong [2000], for instance, two main constraints on strategies are imposed: *well-bracketing* (every answer must match the most recent pending question) and *innocence* (roughly, the participants

FIGURE 2. A game play corresponding to a PCF computation.

	$(\overline{0} \to$	$\overline{0}) \to$	$(\overline{0} \to$	$\overline{0}) \to$	$\overline{0}$	
O:					?	What is $F(g)(h)$?
P:		?				What is $g(x)$?
O:	?					What is x here?
P:	3					$x = 3$.
O:		5				In that case, $g(x) = 5$.
P:		?				All right. Now what is $g(x')$?
O:	?					What is x' here?
P:				?		What is $h(y)$?
O:				4		$h(y) = 4$ (whatever y is).
P:	4					Then $x' = 4$.
O:		6				In that case, $g(x') = 6$.
P:					11	Well then, $F(g)(h) = 11$.

must decide on their moves purely on the basis of the answers received to previous questions, not on how these answers were obtained). Another feature of this model is that every move apart from the first one is explicitly *justified* by some previous move: answers are justified by the questions they answer, and questions are justified by earlier questions which open up the appropriate part of the game.

All of the game models mentioned yield definability results for PCF: every recursive strategy is the denotation of some PCF term. In fact, they all give rise to the same extensional type structure Q, which satisfies the criteria mentioned following Definition 4.22, and which seems to be the natural candidate for such a type structure. (It is, however, an open problem whether the Q_σ are all CPOs.) Note that the elements of Q are in general infinite equivalence classes of strategies, so these constructions do not have quite the finitary character that one might have liked (this is inevitable in view of Theorem 4.24). However, this does not mean that no useful analysis has been achieved. In our view, the main insight offered by the perspective of game semantics is the idea that a *parity* (O or P) may be consistently assigned to the implicit interactions between the subterms of a PCF term. This parity is extra structure present in the game models — it is not present in the raw operational definition of PCF. The game-theoretic analysis seems to us to be deep enough to resolve some interesting and purely syntactic questions about PCF, though significant results of this nature have yet to be worked out in detail.

The ideas of game semantics have also been successfully applied to yield characterizations of non-extensional notions of computability — see Section 6.2.

4.3.4. *Other work.* We now mention a few miscellaneous topics to conclude our survey of what is known about PCF computability.

A remarkable result of Sieber (Sieber [1992]) gives a complete characterization of the PCF-definable finite elements at types of level 2 as those that are invariant under a class of logical relations known as sequentiality relations. The idea of using invariance under logical relations to construct a model was carried much further by O'Hearn and Riecke (O'Hearn and Riecke [1995]) who obtained a fully abstract model in this way; however, their result depends only on general facts about λ-definability, and so it is unclear how much it reveals about sequentiality as such.

Some attention has also been given to other variants of PCF. The version considered above is the original call-by-name one, but one can equally well define a version of PCF with a call-by-value evaluation strategy, which can be naturally interpreted in CBV type structures (the idea essentially appears in Plotkin [1983, Chapter 3]). There is also a third possibility: the *lazy PCF* of Bloom and Riecke [1989]. The relationships between these languages were investigated in Sieber [1990], Riecke [1993], Longley [1995, Chapter 6]. The picture that has emerged is that these languages are sufficiently interencodable that, from a denotational perspective at least, they can all be regarded as embodying the same abstract "notion of computability". We will explain this point of view in more detail in Part II.

Another area of recent interest has been the relationship between PCF computability and the notions of *total* functional considered in Sections 3.3 and 3.4. The general idea here is to ask what total functionals can be computed in PCF, though this question can be made precise in several different ways. For example, we have already seen that C arises as an extensional collapse of P (Theorem 4.13); one can therefore ask which elements of C are represented by PCF-definable elements of P. It is not hard to see that all Kleene computable elements of C are PCF-definable in this way. More surprising is the fact that the fan functional Φ (see Section 3.3.1) is PCF-definable by means of a clever higher type recursion. This fact appeared in Berger's thesis (Berger [1990]), and was also independently known to Gandy. The following more general result was conjectured in Cook [1990] and Berger [1993], and proved by Normann in Normann [2000].

Theorem 4.25. *The PCF-definable elements of* C *are exactly those in* RC.

Equivalently (in the light of Theorems 4.13 and 4.17), the PCF-definable elements of C coincide exactly with the PCF^{++}-definable ones. Normann's proof is both ingenious and beautiful and is one of the highlights of the subject. Around the same time, Plotkin (Plotkin [1997]) considered other possible notions of totality in PCF (such as that given by an extensional collapse of the term model for PCF itself) and investigated the differences between the various notions.

Finally, a few papers have been devoted to the degree theory induced by the notion of relative PCF-definability. An element $x \in \mathsf{P}^{eff}$ (for example) is PCF-definable relative to $y \in \mathsf{P}^{eff}$ if there is a PCF-definable element f such that $f(y) = x$; the corresponding equivalence classes are known as *degrees of parallelism*. The lattice of degrees of parallelism was introduced by Sazonov (Sazonov [1976a]); like other lattices of degrees, its structure would appear to be extremely complicated. Sazonov mentioned several examples of distinct degrees and some relationships between them: for instance, the functions *por* and *exists* represent incomparable degrees. A few other results in a similar spirit appear in Trakhtenbrot [1975]. Bucciarelli (Bucciarelli [1995]) undertook a somewhat more systematic study of degrees of parallelism for first order functions; this line of investigation was pursued further by Lichtenthäler (Lichtenthäler [1996]). Degrees of parallelism can be seen as representing notions of computability intermediate between PCF and PCF^{++}; however, to date none of these intermediate notions have established themselves as being of independent mathematical interest.

4.4. The sequentially realizable functionals. Until the mid 1990s, it seemed reasonable to suppose that the only two respectable notions of hereditarily partial computable functional were those embodied by PCF and PCF^{++} — these were typically referred to as "sequential" and "parallel" computability. However, it emerged more recently that there is another good class of computable functionals which can reasonably be seen as embodying an alternative notion of "sequential" computability, more generous than the PCF one.

The basic idea can be given by a simple example. Let $\mathbb{M}_{rec}$ be the set of monotone computable functions from $\mathbb{N}_\perp$ to $\mathbb{N}_\perp$ (that is, $\mathbb{M}_{rec} = \mathsf{P}_1$), and consider the function $F : \mathbb{M}_{rec} \to \mathbb{N}_\perp$ defined by

$$F(g) = \begin{cases} 0 & \text{if } g(\perp) \in \mathbb{N} \ (\textit{i.e.}, \text{ if } g = \Lambda x.k \text{ for some } k \in \mathbb{N}), \\ 1 & \text{if } g(\perp) = \perp \text{ but } g(0) \in \mathbb{N}, \\ \perp & \text{otherwise } (\textit{i.e.}, \text{ if } g(0) = \perp). \end{cases}$$

Intuitively, the function F can be computed via the following strategy: given a function g, feed it the object $0 \in \mathbb{N}_\perp$ (that is, a program which when run will terminate giving the value 0), and then watch g closely to see whether it ever "looks at" its argument (that is, whether the above program is in fact ever run). If g returns a result without looking at its argument, we return 0; if g returns a result having looked at the argument, we return 1.

This is a perfectly effective and intuitively "sequential" way of computing F, as long as g is presented to us in some form of which it is sensible to ask whether it ever looks at its argument. The computation of F thus has to operate on some kind of algorithm or intensional representation for g rather than on its pure extension, although the *result* $F(g)$ is completely determined by the extension g. (As argued in Section 4.3.1, the same is true for PCF computations — the only difference is that here we allow some additional ways of manipulating the intensional representations.) It is easy to see that the function F cannot be defined in PCF: if $g_1 = \Lambda x.\mathit{if}\ (x)(0)(0)$ and $g_2 = \Lambda x.0$, then $g_1 \sqsubseteq g_2$ in the pointwise order but $F(g_1) \not\sqsubseteq F(g_2)$; thus F does not exist even in P since it is not (pointwise) monotone.

It may appear puzzling that a non-monotone function can be considered computable in some sense. The explanation hinges on the fact that computations here operate on intensional objects (as is the case with many of the definitions we have seen), and at this level, computable operations are indeed monotone. Thus, although $g_1 \sqsubseteq g_2$ extensionally, the *algorithm* that computes g_2 is not obtained by extending the algorithm that computes g_1. It is also worth noting that even at the extensional level, functions like F are monotone with respect to a different ordering known as the *stable* order. (Curiously, Kleene came across the possibility of algorithms such as the one described above for F — see Kleene [1985, §13.3] — but decided to rule them out by an extrinsic monotonicity requirement.)

It turns out that there is a mathematically natural type structure containing F and "all things like it", which we shall denote by R. This first appeared in the literature as the type structure arising from the *strongly stable* model of Bucciarelli and Ehrhard [1991b], a domain-theoretic model of PCF intended to capture certain aspects of sequential functionals, and conceived partly as a line of attack on the PCF full abstraction problem. The objects in this model are *dI-domains with coherence*, which are certain CPOs equipped with a class of finite subsets designated as *coherent sets*. The morphisms (equivalently the elements of function spaces) are the *strongly stable functions*, the continuous functions between such CPO which preserve coherent sets and least upper bounds of coherent sets. Though rather complicated to formulate, the construction of this model is as finitary and effective as could be desired, so that (for instance) equality of finite elements is decidable. Ehrhard later gave a simplified presentation in terms of *hypercoherences* (Ehrhard [1993]), and a slightly more abstract analysis of the relevant structure was given in Bucciarelli and Ehrhard [1993].

Another interesting characterization, given by Colson and Ehrhard in Colson and Ehrhard [1994], showed that R in some sense arises naturally from the class of (infinitary) first order sequential functions $\mathbb{N}_p^{\mathbb{N}} \to \mathbb{N}_\bot$. (For functions of this type, there is evidently only one reasonable notion of sequentiality. It can be characterized by a simple infinitary generalization of Definition 4.23, or equivalently in the style of Definition 4.27 below.) Note the analogy with Definition 4.15.

DEFINITION 4.26. *For each type σ we define a set R_σ, and a set R_σ^ω of functions from $\mathbb{N}_p^{\mathbb{N}}$ to R_σ, as follows:*

- $\mathsf{R}_0 = \mathbb{N}_\bot$.

- R_0^ω *is the set of sequential continuous functions* $\mathbb{N}_p^\mathbb{N} \to \mathbb{N}_\perp$.
- $\mathsf{R}_{\sigma\to\tau}$ *is the set of all functions* $f : \mathsf{R}_\sigma \to \mathsf{R}_\tau$ *such that for all* $g \in \mathsf{R}_\sigma^\omega$ *we have* $f \circ g \in \mathsf{R}_\tau^\omega$.
- $\mathsf{R}_{\sigma\to\tau}^\omega$ *is the set of all functions* $f : \mathbb{N}_p^\mathbb{N} \to \mathsf{R}_{\sigma\to\tau}$ *such that for all* $g \in \mathsf{R}_\sigma^\omega$ *the function* $\Lambda r : \mathbb{N}_p^\mathbb{N}.\, f(\mathit{fst}\ r)(g(\mathit{snd}\ r))$ *is in* R_τ^ω.

The above characterizations say nothing about whether the elements of R that are not definable in PCF are sequentially computable in any reasonable sense. However, Ehrhard showed in Ehrhard [1996] that there is indeed a computational aspect: every element of R can be in some sense computed by a Berry-Curien sequential algorithm. This line of investigation was continued in Ehrhard [1999], where it was shown that R is in fact the extensional collapse of the sequential algorithms model.

A somewhat simpler characterization in the same vein was discovered independently by van Oosten (Oosten [1999]) and Longley (Longley [2002]), who constructed R from a certain combinatory algebra $\mathcal{B}$. The definition of $\mathcal{B}$ hinges on the observation that a sequential algorithm for computing a function $F : \mathbb{N}_p^\mathbb{N} \to \mathbb{N}_\perp$ (at a given element $g \in \mathbb{N}_p^\mathbb{N}$) can be represented by an infinitely branching *decision tree*, in which each internal node is labelled with a "question" $?n$, (meaning "what is the value of $g(n)$?"), and each leaf either has an undefined label or is labelled with an "answer" $!n$ (meaning "the value of $F(g)$ is n"). Furthermore, such a decision tree can itself be easily coded by an element f of $\mathbb{N}_p^\mathbb{N}$. Likewise, a sequential algorithm of type $\mathbb{N}_p^\mathbb{N} \to \mathbb{N}_p^\mathbb{N}$ can be represented by an infinite forest of such trees, which can again be coded by an element of $\mathbb{N}_p^\mathbb{N}$. All this is very similar to the idea behind Kleene's *associates* (Section 3.3.1).

In the following definition we take $!n = 2n$ and $?n = 2n + 1$.

DEFINITION 4.27 (Van Oosten algebra).

(i) *Let play*: $\mathbb{N}_p^\mathbb{N} \times \mathbb{N}_p^\mathbb{N} \times \mathrm{Seq}(\mathbb{N}) \to \mathbb{N}_\perp$ *be the smallest partial function such that, for all* $f, g \in \mathbb{N}_p^\mathbb{N}$, $\alpha \in \mathrm{Seq}(\mathbb{N})$ *and* $n, m \in \mathbb{N}$,
  - *if* $f\langle\alpha\rangle = !n$ *then* $\mathit{play}(f, g, \alpha) = n$,
  - *if* $f\langle\alpha\rangle = ?n$ *and* $g(n) = m$ *then* $\mathit{play}(f, g, \alpha) = \mathit{play}(f, g, (\alpha; m))$.

(ii) *For* $f \in \mathbb{N}_p^\mathbb{N}$ *and* $n \in \mathbb{N}$, *write* f_n *for the least function such that* $f_n\langle\alpha\rangle = f\langle n; \alpha\rangle$ *for all* α. *Let* $|, \bullet$ *be the operations defined by*

$$f \mid g = \mathit{play}(f, g, [\,]), \qquad f \bullet g = \Lambda n.(f_n \mid g)$$

and let $\mathcal{B}$ *be the applicative structure* $(\mathbb{N}_p^\mathbb{N}, \bullet)$.

The construction of R from $\mathcal{B}$ is now a standard extensional collapse (see also Section 5 below). Define partial equivalence relations $\sim_\sigma$ on $\mathcal{B}$ by

$$f \sim_0 g \text{ iff } f(0) \simeq g(0) \quad \text{(for example)},$$
$$f \sim_{\sigma\to\tau} g \text{ iff } \forall x, y \in \mathcal{B}.\ x \sim_\sigma y \Rightarrow f \bullet x \sim_\tau g \bullet y$$

We may then define R by taking $\mathsf{R}_\sigma = \mathcal{B}/\sim_\sigma$, with application operations induced by $\bullet$.

This construction shows that all the elements of R can be computed or "realized" by sequential algorithms in some sense. In fact, the relationship between $\mathcal{B}$ and the Berry-Curien sequential algorithms model is very close: the relevant objects in the Berry-Curien category can all be obtained as retracts of $\mathcal{B}$ by standard categorical techniques. This and other results connecting up the known characterizations of R appeared in Longley [2002], where the elements of R were called the *sequentially realizable* functionals. Some further characterizations of R have recently been given in Hyland and Schalk [2002] and Laird [2002].

Longley also explicitly considered the effective analogue R^{eff} (which may be constructed either as a standalone type structure or as a substructure of R), and argued that this embodied a natural and compelling notion of sequential computability at higher types. One of the main results of Longley [2002] was that in both R and R^{eff} the type $\overline{2}$ is *universal* in the same sense in which $\mathbb{T}^\omega$ is universal in P and P^{eff} (see Theorem 4.14). A closely related fact is the existence of a type 3 functional $H \in \mathsf{R}^{eff}$ such that every element of R^{eff} is PCF-definable relative to H; indeed, one can define a programming language PCF +H with an effective operational semantics whose term model provides an alternative characterization of R^{eff}. (The operation H can in fact be implemented in existing higher order programming languages such as Standard ML; cf. Longley [1999c].)

As we have seen, the functional $F \in \mathsf{R}^{eff}$ mentioned above is not present in P^{eff}. On the other hand, the function $por \in \mathsf{P}^{eff}$ is not present in R^{eff}. We therefore have two incomparable notions of partial computable functional — this is somewhat analogous to the situation for the total type structures RC and HRC as described in Section 3.4. (See also the diagrams in Appendix A). Indeed, one can also make precise an "anti-Church's thesis" in the partial setting, to the effect that there is no possible type structure of computable functionals over $\mathbb{N}_\perp$ that subsumes both P^{eff} and R^{eff}. One version of such a result was given in Longley [2002, Section 11]; a slightly stronger version will be presented in Part II.

The type structures P^{eff} and R^{eff} both "contain" the type structure Q^{eff} of Definition 4.22 in some sense (see also Definition A.1). As a curiosity, however, it is worth noting that there are functionals present in both P^{eff} and R^{eff} that are not present in Q^{eff} (see Longley [2002, Section 11.2], or else Bucciarelli and Ehrhard [1994], [1991a] where a kind of "intersection" of P and R is constructed).

In conclusion we remark on a few points of comparison between the PCF-sequential and sequentially realizable functionals (see Longley [2002, Section 12] for a more detailed discussion). The evidence that R and R^{eff} are natural mathematical objects seems to us very compelling — indeed, at present we possess a much wider range of *prima facie* independent characterizations for R than for Q. Related to this is the fact that R appears to enjoy better structural properties than Q: for instance, the finitary analogue of R is effectively presentable (in contrast to Theorem 4.24 for Q); and R (unlike Q) possesses a universal type. However, one difficulty with $\mathsf{R}^{\mathit{eff}}$ from a practical point of view is that the universal functional H has a high inherent computational complexity (see Royer [2000]); this makes it unlikely that $\mathsf{R}^{\mathit{eff}}$ will ever become the staple notion of sequentially computable functional employed by higher order programming languages.

§5. **Realizability models.** We end our discussion of computable functionals by surveying some ideas from the study of *realizability models* that cross-cut several of the topics we have mentioned so far.

The concept of realizability was introduced by Kleene in Kleene [1945], who showed that the notion of recursive function could be used to give a constructive interpretation of arithmetic. Thereafter, many other kinds of realizability were introduced to give constructive interpretations of various logical systems, and hence establish metamathematical results (see Troelstra [1973]). Later, a more model-theoretic perspective emerged (Hyland [1982]), which made it clear that realizability could provide a common semantic setting for both logics and programming languages (such as typed λ-calculi). A survey of the history of realizability has recently been given in Oosten [2002].

The number and variety of notions of realizability that have been studied is very large (see *e.g.*, Oosten [1991] or Hyland [2002] for an overview). Here we shall give the definitions only for a class of models based on *standard* realizability, a simple generalization of Kleene's original notion — these are the models which have received the greatest attention so far. However, we will also allude briefly to realizability models of other kinds.

We start with the following definition, which was (essentially) introduced in Feferman [1975]:

DEFINITION 5.1 (PCAs). *A* partial combinatory algebra (PCA) *is a set A equipped with a partial "application" operation $\cdot : A \times A \rightharpoonup A$, in which there exist elements $k, s \in A$ such that for all $x, y, z \in A$ we have*

$$k \cdot x \cdot y = x, \quad s \cdot x \cdot y\downarrow, \quad s \cdot x \cdot y \cdot z \simeq (x \cdot z) \cdot (y \cdot z).$$

The definition of PCA leads to a rich computational structure: for instance, in any PCA A one can represent the natural numbers by means of an encoding due to Curry, and all recursive functions are then representable by elements of A. We may therefore think of a PCA as an untyped universe of computation in some abstract sense; this seems a reasonable point of view since many naturally arising PCAs are indeed "effective" in nature. Perhaps the leading example of a PCA is *Kleene's first model* K_1, consisting of the natural numbers with Kleene application: $m \cdot n \simeq \phi_m(n)$. Other interesting examples will be mentioned below.

The models we shall consider here are defined as follows:

DEFINITION 5.2 (PERs).

(i) *Given a PCA* $(A, \cdot)$, *a* partial equivalence relation (or PER) *on* A *is a symmetric, transitive binary relation* R *on* A, *i.e., an equivalence relation on a subset of* A. *We write* A/R *for the set of equivalence classes for* R.

(ii) *If* R, S *are PERs on* A, *the PER* S^R *is defined by*

$$S^R(a, a') \text{ iff } \forall b, b' \in A.\ R(b, b') \Rightarrow S(a \cdot b, a' \cdot b').$$

A morphism $f : R \to S$ *is an element of* $A/(S^R)$; *we say* a realizes f *if* $a \in f$. *We write* ***PER***(A) *for the category of PERs on* A *and morphisms between them.*

It is easy to show that ***PER***(A) is a cartesian closed category, with exponentials as suggested by the above definition. In fact, the categories ***PER***(A) turn out to have an extremely rich structure and to offer a common semantic framework for type theories, constructive logics and programming languages.

Girard originally introduced the category ***PER***(K_1) in order to give a semantics for his second order polymorphic λ-calculus, or "System F" (Girard [1972]). Later, ***PER***(K_1) was identified as an important subcategory of Hyland's *effective topos* (Hyland [1982]), a categorical model for higher order logic based on Kleene's realizability interpretation of arithmetic (Kleene [1945]). More generally, ***PER***(A) arises as a subcategory of the *standard realizability topos* ***RT***(A) (Hyland, Johnstone, and Pitts [1980]); we will refer loosely to the categories ***PER***(A), ***RT***(A) and their close relatives as *realizability models*. Here we shall concentrate on their connections with programming languages and computability.

One way of thinking about the above definitions (advocated by Mitchell and frequently stressed by the present author) is to regard a PCA A as a kind of abstract model of "machine level" computation, and to regard ***PER***(A) as a category of "datatypes", for a high level programming language implemented on this machine. In the case of K_1, for instance, this accords closely with what happens inside a computer: a high-level datatype consists of values which must ultimately be represented on the machine somehow by bit sequences (or let us say by natural numbers). Since two machine representations of some value might be indistinguishable from the point of view of the high-level language, we can think of the datatype as corresponding to a partial equivalence relation on $\mathbb{N}$. Abstracting from this situation, we can imagine any PCA A as providing a kind of primitive model of computation, and on this view all morphisms of ***PER***(A) are "computable" in the sense that they are realized by an element of A.

5.1. Type structures in realizability models. In any category of the categories $\boldsymbol{PER}(A)$, there is (up to isomorphism) a canonical "datatype of natural numbers" N, arising for instance from Curry's encoding of the natural numbers in the language of combinatory logic. (We will write N for the object of $\boldsymbol{PER}(A)$ representing the natural numbers, to distinguish it from the ordinary set $\mathbb{N}$ of natural numbers.) In addition, for most of the particular PCAs of interest, there is an object in $\boldsymbol{PER}(A)$ that stands out as being the obvious choice for $N_\perp$. We can therefore obtain type structures over $\mathbb{N}$ and $\mathbb{N}_\perp$ by repeated exponentiation in $\boldsymbol{PER}(A)$. From our point of view, these can be seen as type structures of "computable functionals" naturally arising from the notion of computability embodied by A. In the cases where A is a genuinely "effective" PCA, this will yield type structures of effectively computable functionals in some sense. We therefore have a rich supply of interesting constructions of type structures to consider.

In fact, many of the characterizations of type structures that we have already considered are easily seen to be of precisely this kind. As regards total type structures over $\mathbb{N}$, for instance, Kreisel's definition of HEO (Definition 3.16) is nothing other than the definition of the type structure over N in $\boldsymbol{PER}(K_1)$. Kleene's definition of C via associates (Definition 3.13) essentially arises in the same way from a PCA known as *Kleene's second model* K_2. This PCA was introduced in Kleene and Vesley [1965] — here the underlying set is $\mathbb{N}^\mathbb{N}$, and application is given by a minor modification of the operation $(- \mid -)$ mentioned in Section 3.3.1. (In fact, an associate for an element $x \in \mathsf{C}$ is essentially nothing other than a realizer for x in K_2.) The definition of HRC (Definition 3.18) arises similarly from the recursive submodel K_{2rec}. Scott's characterization of C via algebraic lattices is very close to the definition of the type structure over N in the Scott *graph model* $\mathcal{P}\omega$.

Other constructions can naturally be viewed as type structures in other kinds of realizability models. For instance, Bezem's construction of HEO as the extensional collapse of HRO (Theorem 3.20) corresponds to the type structure over N in the category $\boldsymbol{MPER}(A)$ of *modified PERs* on A, a natural subcategory of the *modified realizability topos* on A (see Oosten [1997]). In addition, several possible definitions of RC correspond to type structures in *relative realizability* models of the kind considered in Awodey, Birkedal, and Scott [2000].

The general programme of trying to identify the type structures over $\mathbb{N}$ arising from various PCAs was explicitly articulated in Beeson [1985, Chapter VI], where the cases of $\mathcal{P}\omega$ and $\mathcal{P}\omega_{re}$ (giving rise to C and HRC respectively) were considered. Similar results for other graph models were obtained by Bethke (Bethke [1988]). As one might expect, it would seem that all natural PCAs based solely on a notion of continuous function application give rise to the type structure C, and their recursive analogues give rise to HRC.

One can also view many definitions of type structures over $\mathbb{N}_\perp$ as arising from realizability models, but for this we need some additional ideas. A suitable way of talking about "computable partial functions" in categories such as $\boldsymbol{PER}(A)$ was provided by Rosolini's theory of *dominances* (Rosolini [1986]), inspired by ideas of Mulry. A dominance is a small piece of extra structure on a category which determines an abstract notion of "semidecidable predicate". Under certain conditions, a dominance gives rise to a *lifting* operation $X \mapsto X_\perp$ on objects; one may then identify computable partial functions $X \rightharpoonup Y$ with morphisms $X \to Y_\perp$. For instance, the natural choice of dominance on $\boldsymbol{PER}(K_1)$ gives rise to an object $N_\perp$ which may be defined (as a PER) by

$$N_\perp(m, n) \Longleftrightarrow \phi_m(0) \simeq \phi_n(0).$$

The morphisms $N \to N_\perp$ then correspond precisely to the partial recursive functions $\mathbb{N} \rightharpoonup \mathbb{N}$, as we might have hoped.

In Longley [1995], Longley developed explicitly the idea that different PCAs embody different notions of computability, and considered the problem of identifying the type structures over $N_\perp$ in particular models. The following result appeared in Longley [1995, Chapter 7]:

THEOREM 5.3. $\mathsf{T}(\boldsymbol{PER}(K_1), N_\perp) \cong \mathsf{P}^{\mathit{eff}}$.

This is very close in content as Ershov's generalized Myhill-Shepherdson theorem (Theorem 4.12); however, the formulation in terms of $\boldsymbol{PER}(K_1)$ is simpler insofar as here it is immediate that the required exponentials exist.

Longley drew particular attention to the coincidence between the type structure in $\boldsymbol{PER}(K_1)$ and the extensional term model for PCF^{++} (Section 4.2.5). Both of these constructions can be seen as very "pure" attempts at defining a class of hereditarily computable partial functionals (they do not require auxiliary concepts such as continuity), but they are totally different in spirit. In $\boldsymbol{PER}(K_1)$ we think of computations as taking place at the level of recursive indices, and an index realizes a functional whenever its action on indices for objects of lower type just "happens" to be extensional. In PCF^{++}, by contrast, computation takes place at a higher "symbolic" level, where extensionality is enforced by the design of the programming language, and the ways in which a function may interact with its argument seem *prima facie* to be much more restricted. That these two definitions yield the same structure at all finite types still appears to the present author to be a wonderful and surprising fact.

It was also shown in Longley [1995] that the type structure over (an obvious object) $N_\perp$ in $\boldsymbol{PER}(\mathcal{P}\omega)$ [resp. in $\boldsymbol{PER}(\mathcal{P}\omega_{re})$] coincided with P [resp. $\mathsf{P}^{\mathit{eff}}$]. This is not too surprising in view of the close connections between Scott domains and algebraic lattices.

Another interesting family of PCAs are the term models for untyped λ-calculi. For instance, let Λ^0/T be the set of pure closed untyped λ-terms modulo some reasonable theory T, regarded as a combinatory algebra. The following was conjectured implicitly (for a particular T) by Phoa (Phoa [1991]), and more explicitly (for a large class of theories T) by Longley (Longley [1995, Section 7.4]), who obtained some partial results and outlined a possible proof strategy:

CONJECTURE 5.4 (Longley-Phoa). *In the type structure over (an obvious choice of) $N_\perp$ in $\boldsymbol{PER}(\Lambda^0/T)$, every element is* PCF-*definable. In other words,*

$$\mathsf{T}(\boldsymbol{PER}(\Lambda^0/T), N_\perp) \cong \mathsf{Q}^{\mathit{eff}}.$$

There is fairly good evidence for this conjecture, but it has resisted extensive attempts at proof by the author and others. Moreover, even if proved, it is doubtful whether this result would give any useful information about the type structure $\mathsf{Q}^{\mathit{eff}}$, since the relevant objects in $\boldsymbol{PER}(\Lambda^0/T)$ appear to be quite intractable. The main interest in the conjecture is perhaps conceptual: it would provide an example of a highly non-trivial equivalence between two characterizations of $\mathsf{Q}^{\mathit{eff}}$, which in turn would provide evidence for the mathematical naturalness of this type structure. An interesting (and perhaps more useful) variant of the Longley-Phoa conjecture, involving a λ-calculus with certain constants, has been established by Streicher *et al* (Marz, Rohr, and Streicher [1999], Rohr [2002]), though this is a much easier result.

A more semantic example of a PCA which does give rise to $\mathsf{Q}^{\mathit{eff}}$ was constructed by Abramsky, based on the ideas of games and well-bracketed strategies. The proof that it does so is non-trivial and to some extent furnishes the same kind of evidence for the status of Q and $\mathsf{Q}^{\mathit{eff}}$ as would be provided by the Longley-Phoa conjecture; see Abramsky and Longley [2000].

Meanwhile, van Oosten and Longley constructed the PCA $\mathcal{B}$ described in Section 4.4; the type structure defined there is exactly the type structure over the obvious object $N_\perp$ in $\boldsymbol{PER}(\mathcal{B})$. It was this characterization that led to the systematic study of the sequentially realizable functionals in Longley [2002].

Various attempts have been made to generalize the construction of realizability models from untyped to typed structures. One way to do this was proposed by the present author in Longley [1999b]; we will make considerable use of this perspective in Parts II and III as a means of unifying much of the material described in this survey. This generalization also leads to many new ways of constructing type structures, though in most cases they turn out to be isomorphic to one of the type structures already known.

In conclusion, it appears that in almost all known "natural" examples of realizability models, the type structure over N is isomorphic to one of C, RC and HRC; while the type structure over (a natural choice of) $N_\perp$ is isomorphic to one of P, Q, R or their recursive analogues. This is consistent with the overall impression gained from the material in Sections 3 and 4 — namely, that a wide range of approaches to defining plausible notions of higher type computable functional actually leads to a relatively small and manageable handful of type structures. It would be interesting to know whether the Kleene computable functionals (*e.g.*, $\mathsf{KC}(A)$ for some A, or the type structure K mentioned in Section 4.3.1) can be obtained via a realizability construction in a natural way.

5.2. Domains in realizability models. Much of the work mentioned in the preceding section took place within the context of a general programme known as *synthetic domain theory*, which sought to identify good categories of "computational domains" (*e.g.*, for denotational semantics) within realizability models and similar categories. We include a short account of this programme here in order to give an impression of the background to the above results.

Synthetic domain theory was initiated by Scott in the early 1980s. The idea of this enterprise was to look for models of constructive set theory containing objects which, *by themselves* and without any additional structure, could serve as domains for denotational semantics. Domains would then simply be "sets" in some constructive universe, and the hope was that this might lead to a simpler, cleaner version of domain theory than the classical one.

Later work by Rosolini (Rosolini [1986]), Hyland (Hyland [1990]) and many other researchers sought to give axiomatic versions of synthetic domain theory: that is, a set of conditions on a model of constructive set theory (usually a topos) which suffice to ensure that it contains a good category of computational domains. Here realizability models provided the main source of motivating examples. Longley and Simpson (Longley and Simpson [1997]) developed a version of synthetic domain theory that applied uniformly to a wide range of realizability models, showing that very many PCAs gave rise to models of PCF at least. We refer to Rosolini's contribution in Fiore, Jung, Moggi, O'Hearn, Riecke, Rosolini, and Stark [1996] for a brief survey of synthetic domain theory and further references.

Related to this enterprise was the observation that many known categories of computational domains could be seen as full subcategories of particular realizability models. For instance, there is an obvious embedding of $\boldsymbol{EN}$ in $\boldsymbol{PER}(K_1)$ which preserves the exponentials that exist in $\boldsymbol{EN}$. Another example was provided by the work of McCarty (McCarty [1984]), who showed that the category of effectively given information systems (essentially equivalent to effective Scott domains) embeds fully in a model of intuitionistic ZF set theory based on Kleene realizability; this was recast in terms of $\boldsymbol{PER}(K_1)$ in Longley [1995].

Rather more easily, the category of Scott domains and its effective analogue can respectively be embedded in $\boldsymbol{PER}(\mathcal{P}\omega), \boldsymbol{PER}(\mathcal{P}\omega_{re})$. These categories play a central role in the work of Scott and his colleagues (Bauer, Birkedal, and Scott [2001]), who exploit the observation that PERs on $\mathcal{P}\omega$ are equivalent to countably based T_0 spaces equipped with an equivalence relation (such objects are termed *equilogical spaces*). The work of van Oosten and Longley on sequential realizability (Section 4.4) has shown that certain categories of sequential algorithms and hypercoherences arise as subcategories of $\boldsymbol{PER}(\mathcal{B})$.

Finally, Bauer has recently shown (Bauer [2001]) that much of the work of Weihrauch *et al* on representations of spaces via type two effectivity (Section 3.3.5) can be naturally understood in terms of the categories $\boldsymbol{PER}(K_2)$, $\boldsymbol{PER}(K_{2rec})$. All these results suggest that realizability models can provide an attractive setting for describing and relating many other kinds of models.

§6. Non-functional notions of computability. Thus far we have concentrated almost entirely on extensional notions of computability — that is, on notions of computable *functional.* One can also ask whether there are reasonable non-extensional notions of "computable operation" at higher types. Such notions have received relatively little attention by comparison with the extensional notions — perhaps because the very idea of an "intensional operation" seems rather hazy, and it is unclear *a priori* whether it is amenable to a precise mathematical formulation. We here briefly survey some known ideas that relate to this problem.

We have seen how notions of computable functional may be naturally embodied by extensional type structures (or substructures thereof). As a first attempt, therefore, we might propose that more general notions of computable operation could be identified simply with type structures without the extensionality requirement. A typical example would be the structure HRO of Definition 3.17. Many other examples arise from (non-well-pointed) cartesian closed categories: given any object X corresponding to N or $N_\perp$, interpret the simple types by repeated exponentiation and then apply the global elements functor $\mathrm{Hom}(1, -)$. This view seems somewhat unsatisfactory in that it is too concrete: for instance, different Gödel-numbering schemes can give rise to non-isomorphic variants of HRO, whereas we would presumably wish to consider all these variants as embodying essentially the same "notion of computability". This problem may be addressed by adopting the more refined point of view outlined in Section 6.3 below; in the meantime, however, we may at least collect examples of non-extensional type structures which might embody plausible notions of computable operation.

6.1. Structures over $\mathbb{N}$**.** Non-extensional type structures over $\mathbb{N}$ were systematically studied by Troelstra (Troelstra [1973]), who exploited them for metamathematical purposes: any such type structure containing suitable basic operations (essentially those of System T) can serve as a model for higher order arithmetic without the extensionality axiom. Two of the type structures considered by Troelstra are of particular interest from our point of view: the structure HRO of hereditarily recursive operations, and a type structure ICF of *intensional continuous functionals.* Both structures were first introduced by Kreisel (Kreisel [1958], [1962]).

Many of the relevant facts about HRO have already been described in Section 3.4. Let us call an element $F \in \mathrm{HRO}_2$ an *extensional operation* if

$F \cdot f = F \cdot f'$ whenever $f \cdot n = f' \cdot n$ for all n. A trivial example of a non-extensional operation is the element $G \in \mathsf{HRO}_2$ which given an element $x \in \mathsf{HRO}_1$ simply returns $x \in \mathsf{HRO}_0$. As a less trivial example of a non-extensional phenomenon at higher types, there is a *local modulus of continuity* operation $\Psi \in \mathsf{HRO}_{\bar{2}\to\bar{1}\to\bar{0}}$ with the following property: if $F \in \mathsf{HRO}_2$ is an extensional operation and $g \cdot n = g' \cdot n$ for all $n < \Psi \cdot F \cdot g$, then $F \cdot g = F \cdot g'$. (The existence of such a recursive operation is implicit in the original proof of the Kreisel-Lacombe-Shoenfield theorem; see Kreisel, Lacombe, and Shoenfield [1959].) By contrast, it is easily shown that no *extensional* local modulus of continuity operation is computable: that is, there is no element $\Psi \in \mathsf{HEO}$ with the above property. Observations such as these can be used to obtain a variety of consistency and independence results for theories of higher order arithmetic, and also to give us a feel for what operations are and are not computable in various settings.

The type structure ICF is the intensional counterpart of C (defined as in Definition 3.13) in the way that HRO is the intensional counterpart of HEO. That is, it is essentially obtained from Kleene's second model K_2 (see Section 5.1) in the same way in which HRO is obtained from K_1. For pure types, ICF may be defined as follows:

- $\mathsf{ICF}_0 = \mathbb{N}$, $\mathsf{ICF}_1 = \mathbb{N}^{\mathbb{N}}$;
- $\mathsf{ICF}_{n+1} = \{f \in \mathbb{N}^{\mathbb{N}} \mid \forall g \in \mathsf{ICF}_n.\, (f \mid g)\downarrow\}$

where $(f \mid g)$ is as defined in Section 3.3.1. As with the extensional type structures, one may also consider the recursive analogue of ICF, or its recursive substructure. It can be shown that ICF contains a local modulus of continuity operation as above, while C does not, and similarly for the recursive variants (see Troelstra [1973, §2.6]).

Troelstra also considered other non-extensional type structures, such as certain term models for System T, though these seem less appealing as candidates for notions of computability.

6.2. Structures over $\mathbb{N}_\perp$. Other non-extensional notions of computability have more recently been considered in computer science, where they arise naturally in connection with typed programming languages containing non-functional features such as *exceptions* or *state*. The term models for such programming languages frequently give rise to non-well-pointed cartesian closed categories, and hence to non-extensional type structures. One can regard such term models as defining notions of computability by themselves, though as in the extensional case, much of the interest lies in trying to provide other, more semantic characterizations of these type structures. Since the number of programming languages that have been considered in the computer science literature is very large, and for most of them little of interest is known from the point of view of computability, we will confine our attention here to languages for which some alternative characterization of the implicit notion of computability has been obtained.

One intensional notion of computability that has emerged as having good credentials is embodied by the language PCF+`catch` studied in Cartwright and Felleisen [1992], Cartwright, Curien, and Felleisen [1994], as well as by PCF with (first order) `callcc` (Kanneganti, Cartwright, and Felleisen [1993]) and by μ PCF (Ong and Stewart [1997]). All these languages are equivalent in the sense that their fully abstract term models (consisting of closed terms modulo observational equivalence) are isomorphic at the finite types. Here we will give a definition of PCF +`catch` as a representative of these languages:

DEFINITION 6.1 (PCF +`catch`).

(i) *Define the syntax of* PCF +`catch` *by augmenting the definition of* PCF (*Definition* 4.16) *with a constant* $\texttt{catch}_r : (\overline{0}^r \to \overline{0}) \to \overline{0}$ *for each* $r > 0$.

(ii) *Define a* context $E[-]$ *to be a* PCF +`catch` *term with a single occurrence of a hole '$-$' (in an obvious sense). The* evaluation contexts *are generated inductively as follows: $(-)$ is an evaluation context; and if $E[-]$ is an evaluation context then so are*

$$\texttt{succ}\,E[-],\ \ \texttt{pred}\,E[-],\ \ \texttt{if}\,E[-],\ \ E[-]V,\ \ \texttt{catch}_r(\lambda x_1 \cdots x_r.E[-]).$$

(*Intuitively, if $E[-]$ is an evaluation context, then in order to evaluate a term $E[U]$ the subterm U needs to be evaluated "next".*)

(iii) *Let $\rightsquigarrow$ be the smallest transitive relation on terms of* PCF +`catch` *satisfying the clauses for PCF (see Definition* 4.18) *together with the following two clauses*:

- *If $Ux_0 \dots x_{r-1} \rightsquigarrow E[x_i]$ where the $x_j : \overline{0}$ are fresh variables and $E[-]$ is an evaluation context (intuitively, if we cannot proceed further with the reduction without knowing x_i), then $\texttt{catch}_r\ U \rightsquigarrow \widehat{i}$.*
- *If $Ux_0 \dots x_{r-1} \rightsquigarrow \widehat{k}$, where the $x_j : \overline{0}$ are fresh variables (intuitively, if we can complete the computation without knowing any of the x_i), then $\texttt{catch}_r\ U \rightsquigarrow \widehat{r+k}$.*

Informally, $\texttt{catch}_r\ U$ evaluates $U\,x_0 \dots x_{r-1}$ and watches to see if U ever has to look at one of the x_i; if so, the computation is aborted and the index for the argument looked at is returned. It is easy to see how the functional F of Section 4.4 can be encoded in PCF+`catch`. Unlike F, however, the catch operators give rise to non-functional behaviour: for instance, we have

$$\texttt{catch}_2(\lambda xy.\, x + y) \rightsquigarrow 0, \quad \texttt{catch}_2(\lambda xy.\, y + x) \rightsquigarrow 1.$$

The fact that several proposed languages turn out to have the same expressivity as PCF+`catch` is already encouraging, but more significant is the following semantic characterization due to Cartwright, Curien and Felleisen (Cartwright, Curien, and Felleisen [1994]):

THEOREM 6.2. *The fully abstract term model for* PCF +`catch` *is isomorphic to the type structure over* $\mathbb{N}_\perp$ *arising from the category of effective sequential algorithms* (*that is, the effective analogue of the Berry-Curien model*; *see Berry and Curien* [1982]).

The original definition of the sequential algorithms model is rather heavy and we will not give it here. However, Longley showed that the relevant part of this model can be easily reconstructed from the van Oosten algebra $\mathcal{B}$ (Longley [2002, §4]); this gives a simpler alternative description of the type structure of Theorem 6.2.

Since then, the ideas of game semantics as developed by Abramsky *et al* have been successful in providing a semantic account of the expressivity of a number of programming languages. In Abramsky and McCusker [1999], for instance, it is shown that by imposing or not imposing the well-bracketing and innocence conditions on strategies (see Section 4.3.3) one can obtain a square of four categories of games corresponding to different computational paradigms, of which only one corresponds to a functional notion of computation (namely the model with both well-bracketing and innocence constraints, which corresponds to PCF). The model with innocence but not well-bracketing (studied in detail in Laird [1998], [1997]) turns out to correspond to PCF+`catch`: more precisely, the fully abstract term model for PCF+`catch` is a quotient of the type structure arising from this model. The models without innocence correspond to notions of computation involving memory or state; a full abstraction result for Idealized Algol (with respect to the non-innocent, well-bracketed game model) is obtained in Abramsky, Honda, and McCusker [1998].

In general, it would appear that categories of games and their correlations with programming languages provide a fruitful source of candidates for mathematically natural notions of non-extensional computability, and we expect more progress in this vein in the near future.

6.3. A realizability perspective. We have concentrated here on non-extensional type structures as a way to capture notions of computable operation. A somewhat more subtle perspective, making use of ideas from realizability, was outlined in Longley [1999a], [1999b]. The idea here is that given a realizability interpretation for a logic (say predicate logic for the simple types over $\mathbb{N}$ or $\mathbb{N}_\perp$) in which the realizers are drawn from some computational universe A (such as a PCA or a type structure of some kind), the set of realizable sentences gives information about what kinds of operation are computable in A. For example, whether the sentence

$$\forall F:\overline{2}.\ \forall g:\overline{1}.\ \exists n:\overline{0}.\ \forall g':\overline{1}.\ (\forall m<n.\ g(m)=g'(m)) \Rightarrow F(g)=F(g')$$

is realizable in A corresponds to whether local moduli of continuity are computable in A (see Section 6.1). We might propose, therefore, to *identify* notions of higher type computability with notions of realizability for such a logic. In some sense, this allows us to say what intensional operations are computable without having to commit ourselves to any concrete definition of "computable operation". This perspective will be further advocated and developed in detail in Part III.

This point of view is very general and leads to a large class of potential notions of computability: for instance, any untyped PCA implicitly embodies a notion of computable operation at higher types. Fortunately, however, the theory also allows us to say when two structures are equivalent from the point of view of computability, and this significantly cuts down the number of distinct notions to be considered. Even so, there are many notions of computability here competing for our attention, and much territory remains to be explored. It seems unclear as yet whether it is reasonable to hope ultimately for a small collection of genuinely natural notions, such as we have in the extensional setting, or whether there is in practice an unlimited range of plausible notions.

§7. **Conclusion and prospectus.** In this paper we have tried to trace the various lines of research to date that are relevant to the study of computability at higher types. Although these strands of research have been rather widely scattered across different areas of mathematical logic and computer science, it is our contention that when viewed together, the outline of a coherent subject area may be discerned.

The central conceptual question we are concerned with is "What are the good notions of computability at higher types, if there are any?" Our approach in this paper has been rather empirical: to collect a variety of different attempts at defining such a notion, and see what natural notions emerge. Our main concern has been to chart the history of the ideas that bear on the problem. The development of these ideas has (naturally enough) been rather haphazard, and we have made little attempt here to organize the material beyond what has been necessary to tell a coherent story.

Having collected and reviewed all this material, we are in a position to undertake a more systematic treatment. We will attempt this in the remaining two papers of the series, where we will try to show that the situation is less chaotic than it may at times have seemed during the course of this survey.

In Part II we will survey the material on notions of computable *functional* (that is, extensional notions of computability) within a uniform framework, presenting the important notions of computability, their various characterizations, their intrinsic properties, the relationships between different notions, and some discussion of their conceptual status. We will also include some results relating to the impossibility of a "Church's thesis" for higher type functionals.

We will argue that almost everything of interest that is known in the subject can be understood in terms of six basic notions of computability, which are represented in the world of "effective" type structures by RC, HRC, $\mathsf{P}^{\mathit{eff}}$, $\mathsf{Q}^{\mathit{eff}}$, $\mathsf{R}^{\mathit{eff}}$ and $\mathsf{K}^{\mathit{eff}}$ (the last of these being the type structure mentioned in Section 4.3.1, which will suffice to account for Kleene computability over all structures of interest). It seems to us probable that the picture here is by now reasonably complete: the territory has been fairly thoroughly explored, and all reasonable definitions of a natural class of effectively computable functionals seem to lead to one of the above notions.

Regarding more general notions of computable operation, the overall picture seems much less complete at present, but we are at least able to bring together a range of possible notions within a unified framework. In Part III we will develop in more detail the realizability perspective mentioned in Section 6.3, and will collect and organize much of what is known within this framework. We will see that a wide variety of results from recursion theory and computer science can be conveniently encapsulated in this setting. The framework used in Part III will in one sense subsume that of Part II, but the flavour of the two treatments will be rather different and we believe both to be valuable.

7.1. Acknowledgements. I am very grateful to the organizers of the Logic Colloquium for giving me the opportunity to lecture on this material, for providing me with the stimulus to write this article, for willingly tolerating its inordinate length, and for waiting so long for me to finish it. Many people have helped me to fill gaps in my knowledge of the subject, including Samson Abramsky, Ulrich Berger, Yuri Ershov, Solomon Feferman, Martin Hyland, Dag Normann, Gordon Plotkin, Helmut Schwichtenberg, Dana Scott, Alex Simpson and Stan Wainer. Thomas Streicher provided detailed comments on an earlier version of the paper, as did the anonymous referee who made many helpful suggestions for improving its clarity. Helpful feedback on later versions was provided by Samson Abramsky, Yiannis Moschovakis, Dag Normann, Jim Royer and Alex Simpson. I am also grateful to Florrie Kemp for her friendship.

During the writing of this paper I was supported by EPSRC Research Grants GR/L89532 "Notions of computability for general datatypes" and GR/N64571 "A proof system for correct program development".

Appendix A. Summary of type structures. For reference, we include here two diagrams summarizing the main type structures we have considered in this article, and the principal relationships between them. For the sake of clarity we show only the extensional type structures. We give separate diagrams for the type structures over $\mathbb{N}$ (Figure 3) and the type structures over $\mathbb{N}_\perp$ (Figure 4). For the purpose of the diagrams, we identify a language such as

FIGURE 3. Extensional type structures over $\mathbb{N}$.

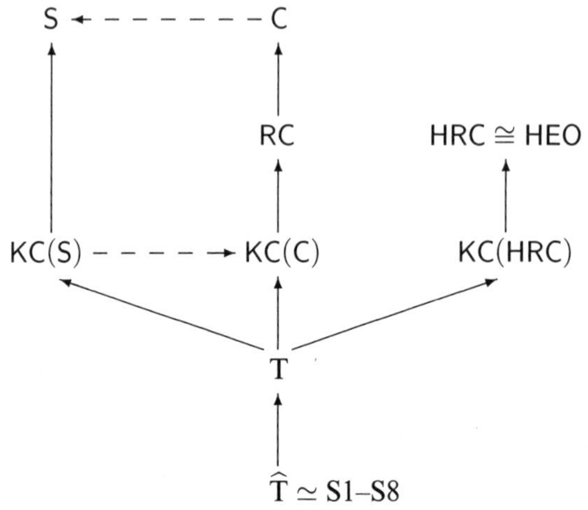

System T or PCF with its extensional term model (all the languages concerned have a unique non-trivial such model). Definitions of all these structures and languages in the main text may be located using the index of symbols.

The arrows in the diagrams correspond to morphisms of the following kind:

DEFINITION A.1. *Suppose A, B are type structures over X. A* morphism *$R\colon A \to B$ is a family of total relations R_σ from A_σ to B_σ such that $R_0 = \mathrm{id}_X$ and*

$$R_{\sigma\to\tau}(f, f') \wedge R_\sigma(x, x') \Longrightarrow R_\tau(f \cdot x, f' \cdot x').$$

These morphisms will be studied at length in Part II. For extensional type structures they may be thought of loosely as "embeddings", though not all of them are straightforward substructure inclusions.

The broken arrows represent morphisms that exist but do not seem to be particularly significant. A few other morphisms which we consider unimportant have been omitted.

Appendix B. Remarks on bibliography. In the following list of references, we have attempted to provide reasonably comprehensive bibliography for the field of higher type computability as delineated by this article. One of our aims in this survey has been to provide an accessible guide to the literature of the subject, and since almost all the works listed below are cited somewhere in the text, it is possible to view the entire article as an extended commentary on the bibliography.

FIGURE 4. Extensional type structures over $\mathbb{N}_\perp$.

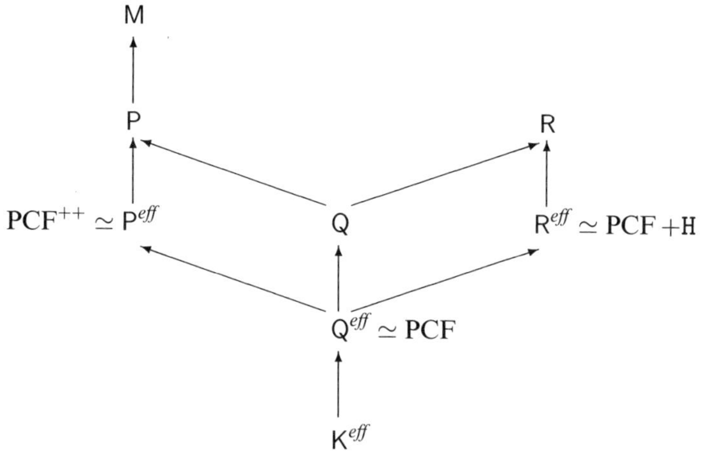

We have tried to include as many relevant published works as possible, along with any Ph.D. theses and other unpublished works that made important contributions to the subject. (The author would appreciate being informed of any significant omissions.) Unpublished technical reports whose contributions appeared soon afterwards in the published literature are usually not listed.

We have not attempted complete coverage of related areas in which ideas from higher type computability are applied (see Section 1.1), including only a selection of the more important works on these topics. For instance, there is a large literature on real number computability and computable analysis, of which only a few works have been included. We have given priority here to very early papers, and to works concerned explicitly with higher types over the reals.

We rather regret that we have not had the time for a fuller coverage of the literature on complexity at higher types. Fortunately, a thorough survey of this area has recently appeared in Irwin, Kapron, and Royer [2001a], [2001b].

Many helpful survey articles on closely related topics, and further references, can be found in Griffor [1999]. For an extensive bibliography on realizability, see Birkedal [1999].

REFERENCES

O. ABERTH [1980], ***Computable analysis***, McGraw-Hill, New York.

S. ABRAMSKY, K. HONDA, AND G. MCCUSKER [1998], *A fully abstract game semantics for general references*, ***Proceedings of 13th Annual Symposium on Logic in Computer Science***, IEEE, pp. 334–344.

S. Abramsky, R. Jagadeesan, and P. Malacaria [2000], *Full abstraction for PCF*, ***Information and Computation***, vol. 163, pp. 409–470.

S. Abramsky and J. R. Longley [2000], *Some combinatory algebras for sequential computation*, in preparation.

S. Abramsky and G. McCusker [1999], *Game semantics*, ***Computational Logic: Proceedings of the 1997 Marktoberdorf summer school*** (H. Schwichtenberg and U. Berger, editors), Springer-Verlag, pp. 1–56.

P. Aczel [1977], *An introduction to inductive definitions*, ***Handbook of mathematical logic*** (J. Barwise, editor), North-Holland, pp. 739–782.

P. Aczel and P. G. Hinman [1974], *Recursion in the superjump*, ***Generalized Recursion Theory*** (J. E. Fenstad and P. G. Hinman, editors), North-Holland, pp. 3–41.

J. Avigad and S. Feferman [1998], *Gödel's functional ("dialectica") interpretation*, ***Handbook of proof theory*** (S. R. Buss, editor), North-Holland, pp. 337–405.

S. Awodey, L. Birkedal, and D. S. Scott [2000], *Local realizability toposes and a modal logic for computability*, to appear in ***Mathematical Structures in Computer Science***.

S. Banach and S. Mazur [1937], *Sur les fonctions calculables*, ***Annales de la Société Polonaise de Mathématique***, vol. 16, p. 223.

H. P. Barendregt [1984], ***The lambda calculus: Its syntax and semantics***, revised ed., Studies in Logic and the Foundations of Mathematics, vol. 103, North-Holland.

H. P. Barendregt [1992], *Lambda calculi with types*, ***Handbook of logic in computer science*** (S. Abramsky, D. M. Gabbay, and T. S. E. Maibaum, editors), vol. 2, Oxford University Press, pp. 117–309.

A. Bauer [2000], ***The realizability approach to computable analysis and topology***, Ph.D. thesis, School of Computer Science, Carnegie Mellon University.

A. Bauer [2001], *A relationship between equilogical spaces and Type Two Effectivity*, ***Electronic Notes in Theoretical Computer Science*** (S. Brooks and M. Mislove, editors), vol. 45, Elsevier Science Publishers.

A. Bauer and L. Birkedal [2000], *Continuous functionals of dependent types and equilogical spaces*, ***Proceedings of Computer Science Logic 2000*** (P. G. Clote and H. Schwichtenberg, editors), Lecture Notes in Computer Science, vol. 1862, Springer-Verlag, pp. 202–216.

A. Bauer, L. Birkedal, and D. S. Scott [2001], *Equilogical spaces*, to appear in ***Theoretical Computer Science***.

A. Bauer, M. H. Escardó, and A. Simpson [2002], *Comparing functional paradigms for exact real-number computation*, ***ICALP***, Lecture Notes in Computer Science, vol. 2380, Springer, pp. 488–500.

M. Beeson [1985], ***Foundations of constructive mathematics***, Springer-Verlag.

S. Bellantoni, K.-H. Niggl, and H. Schwichtenberg [2000], *Higher-type recursion, ramification and polynomial time*, ***Annals of Pure and Applied Logic***, vol. 104, pp. 17–30.

U. Berger [1990], ***Totale Objekte und Mengen in der Bereichstheorie***, Ph.D. thesis, Munich.

U. Berger [1993], *Total sets and objects in domain theory*, ***Annals of Pure and Applied Logic***, vol. 60, pp. 91–117.

U. Berger [1997], ***Continuous functionals of dependent and transfinite types***, Habilitationsschrift, Ludwig-Maximilians-Universität München.

U. Berger [2000], *Minimisation vs. recursion on the partial continuous functionals*, draft paper.

U. Berger and H. Schwichtenberg [1991], *An inverse of the evaluation functional for typed λ-calculus*, ***Proceedings of 6th Annual Symposium on Logic in Computer Science***, IEEE, pp. 203–211.

J. A. Bergstra [1976], ***Computability and continuity in finite types***, Ph.D. thesis, University of Utrecht.

J. A. Bergstra [1978], *The continuous functionals and ^2E*, ***Generalized Recursion Theory II*** (J. E. Fenstad, R. O. Gandy, and G. E. Sacks, editors), North-Holland, pp. 39–53.

J. A. Bergstra and S. S. Wainer [1977], *The "real" ordinal of the 1-section of a continuous functional (abstract)*, ***The Journal of Symbolic Logic***, vol. 42, p. 440.

G. Berry [1978], *Stable models of typed lambda-calculi*, ***Proceedings of 5th International Colloquium on Automata, Languages and Programming***, Lecture Notes in Computer Science, vol. 62, Springer-Verlag, pp. 72–89.

G. Berry and P.-L. Curien [1982], *Sequential algorithms on concrete data structures*, ***Theoretical Computer Science***, vol. 20, no. 3, pp. 265–321.

I. Bethke [1988], ***Notes on partial combinatory algebras***, Ph.D. thesis, University of Amsterdam.

M. Bezem [1985a], *Isomorphisms between HEO and HRO^{E}, ECF and ICF^{E}*, ***The Journal of Symbolic Logic***, vol. 50, pp. 359–371.

M. Bezem [1985b], *Strongly majorizable functionals of finite type: a model for bar recursion containing discontinuous functionals*, ***The Journal of Symbolic Logic***, vol. 50, pp. 652–660.

M. Bezem [1988], *Equivalence of bar recursors in the theory of functionals of finite type*, ***Archive of Mathematical Logic***, vol. 27, pp. 149–160.

M. Bezem [1989], *Compact and majorizable functionals of finite type*, ***The Journal of Symbolic Logic***, vol. 54, pp. 271–280.

L. Birkedal [1999], *Bibliography on realizability*, ***Proceedings of Workshop on Realizability, Trento*** (L. Birkedal, J. van Oosten, G. Rosolini, and D. S. Scott, editors), published as Electronic Notes in Theoretical Computer Science 23 No. 1, Elsevier. Available via `http://www.elsevier.nl/locate/entcs/volume23.html`.

B. Bloom and J. G. Riecke [1989], *LCF should be lifted*, ***Proceedings of Conference on Algebraic Methodology and Software Technology***, Department of Computer Science, University of Iowa.

A. Bucciarelli [1993a], *Another approach to sequentiality: Kleene's unimonotone functions*, ***Proceedings of 9th Symposium on Mathematical Foundations of Programming Semantics, New Orleans***, Lecture Notes in Computer Science, vol. 802, Springer-Verlag, pp. 333–358.

A. Bucciarelli [1993b], ***Sequential models of PCF: some contributions to the domain-theoretic approach to full abstraction***, Ph.D. thesis, Dipartimento di Informatica, Università di Pisa.

A. Bucciarelli [1995], *Degrees of parallelism in the continuous type hierarchy*, ***Proceedings of 9th International Conference on Mathematical Foundations of Programming Semantics***.

A. Bucciarelli and T. Ehrhard [1991a], *Extensional embedding of a strongly stable model of PCF*, ***Proceedings of 18th International Conference on Automata, Languages and Programming, Madrid***, Lecture Notes in Computer Science, vol. 510, Springer-Verlag, pp. 35–46.

A. Bucciarelli and T. Ehrhard [1991b], *Sequentiality and strong stability*, ***Proceedings of 6th Annual Symposium on Logic in Computer Science***, IEEE, pp. 138–145.

A. Bucciarelli and T. Ehrhard [1993], *A theory of sequentiality*, ***Theoretical Computer Science***, vol. 113, pp. 273–291.

A. Bucciarelli and T. Ehrhard [1994], *Sequentiality in an extensional framework*, ***Information and Computation***, vol. 110, pp. 265–296.

R. Cartwright, P.-L. Curien, and M. Felleisen [1994], *Fully abstract semantics for observably sequential languages*, ***Information and Computation***, vol. 111, no. 2, pp. 297–401.

R. Cartwright and M. Felleisen [1992], *Observable sequentiality and full abstraction*, ***Proceedings of 19th Symposium on Principles of Programming Languages***, ACM Press, pp. 328–342.

A. Church [1936], *An unsolvable problem of elementary number theory*, ***American Journal of Mathematics***, vol. 58, pp. 345–363.

A. Church [1940], *A formulation of the simple theory of types*, ***The Journal of Symbolic Logic***, vol. 5, pp. 56–68.

D. A. Clarke [1964], ***Hierarchies of predicates of finite types***, Memoirs of the American Mathematical Society, vol. 51, American Mathematical Society.

L. COLSON AND T. EHRHARD [1994], *On strong stability and higher-order sequentiality*, ***Proceedings of 9th Annual Symposium on Logic in Computer Science***, IEEE, pp. 103–108.

S. COOK [1990], *Computability and complexity of higher type functions*, ***Proceedings of MSRI workshop on Logic from Computer Science*** (Y. Moschovakis, editor), Springer-Verlag, pp. 51–72.

S. B. COOPER [1999], *Clockwork or Turing U/universe? — Remarks on causal determinism and computability*, ***Models and Computability*** (S. B. Cooper and J. K. Truss, editors), Cambridge University Press, pp. 63–116.

P.-L. CURIEN [1993], ***Categorical combinators, sequential algorithms and functional programming***, second ed., Birkhäuser.

N. J. CUTLAND [1980], ***Computability***, Cambridge University Press.

M. DAVIS [1958], ***Computability and unsolvability***, McGraw-Hill.

M. DAVIS [1959], *Computable functionals of arbitrary finite type*, ***Constructivity in Mathematics: Proceedings of the colloquium held at Amsterdam, 1957*** (A. Heyting, editor), North-Holland, pp. 281–284.

J.-P. VAN DRAANEN [1995], ***Models for simply typed lambda-calculi with fixed point combinators and enumerators***, Ph.D. thesis, Catholic University of Nijmegen.

A. G. DRAGALIN [1968], *The computation of primitive recursive terms of finite type, and primitive recursive realization*, ***Zapiski Nauchnykh Seminarov Leningradskogo otdeleniia Matematicheskogo Instituta imeni V. A. Steklova***, vol. 8, pp. 32–45, translation in ***Seminars in Mathematics, V. A. Steklov Mathematical Institute, Leningrad, vol. 8 (1970), pp. 13–18***.

T. EHRHARD [1993], *Hypercoherences: a strongly stable model of linear logic*, ***Mathematical Structures in Computer Science***, vol. 3, pp. 365–385.

T. EHRHARD [1996], *Projecting sequential algorithms on strongly stable functions*, ***Annals of Pure and Applied Logic***, vol. 77, pp. 201–244.

T. EHRHARD [1999], *A relative PCF-definability result for strongly stable functions and some corollaries*, ***Information and Computation***, vol. 152, no. 1, pp. 111–137.

YU.L. ERSHOV [1971a], *Computable numerations of morphisms*, ***Algebra i Logika***, vol. 10, no. 3, pp. 247–308.

YU.L. ERSHOV [1971b], *La théorie des énumérations*, ***Actes du Congrès International des Mathématiciens, Nice 1970, Tome 1***, Gauthier-Villars, Paris, pp. 223–227.

YU.L. ERSHOV [1972], *Computable functionals of finite type*, ***Algebra i Logika***, vol. 11, no. 4, pp. 203–277, English translation in ***Algebra and Logic***, vol. 11 (1972), 203–242, AMS.

YU.L. ERSHOV [1973a], *Theorie der Numerierungen I*, ***Zeitschrift für mathematische Logik***, vol. 19, no. 4, pp. 289–388.

YU.L. ERSHOV [1973b], *The theory of A-spaces*, ***Algebra i Logika***, vol. 12, no. 4, pp. 369–416, English translation in ***Algebra and Logic***, vol. 12 (1973), 209–232, AMS.

YU.L. ERSHOV [1974a], *Maximal and everywhere defined functionals*, ***Algebra i Logika***, vol. 13, no. 4, pp. 210–255, English translation in ***Algebra and Logic***, vol. 13 (1974), 210–225, AMS.

YU.L. ERSHOV [1974b], *On the model G of the theory BR*, ***Soviet Mathematics, Doklady***, vol. 15, no. 4, pp. 1158–1160.

YU.L. ERSHOV [1975], *Theorie der Numerierungen II*, ***Zeitschrift für mathematische Logik***, vol. 21, no. 6, pp. 473–584.

YU.L. ERSHOV [1976a], *Constructions 'by finite'*, ***Proceedings of 5th International Congress on Logic, Methodology and Philosophy of Science***, London, Ontario, pp. 3–9.

YU.L. ERSHOV [1976b], *Hereditarily effective operations*, ***Algebra i Logika***, vol. 15, no. 6, pp. 642–654, English translation in ***Algebra and Logic***, AMS.

YU.L. ERSHOV [1977a], *Model C of the partial continuous functionals*, ***Logic Colloquium 1976***, North-Holland, pp. 455–467.

YU.L. ERSHOV [1977b], *Theorie der Numerierungen III*, ***Zeitschrift für mathematische Logik***, vol. 23, no. 4, pp. 289–371.

YU.L. ERSHOV [1977c], ***The theory of enumerations***, Monographs in Mathematical Logic and Foundations of Mathematics, Nauka, Moscow.

YU.L. ERSHOV [1996], ***Definability and computability***, Siberian School of Algebra and Logic, Plenum Publishing Corporation.

YU.L. ERSHOV [1999], *Theory of numberings*, ***Handbook of computability theory*** (E. R. Griffor, editor), North-Holland, pp. 473–503.

M. H. ESCARDÓ [1996], *PCF extended with real numbers*, ***Theoretical Computer Science***, vol. 162, pp. 79–115.

S. FEFERMAN [1975], *A language and axioms for explicit mathematics*, ***Algebra and logic*** (J. N. Crossley, editor), Springer-Verlag, pp. 87–139.

S. FEFERMAN [1977a], *Inductive schemata and recursively continuous functionals*, ***Logic Colloquium 1976***, North-Holland, pp. 373–392.

S. FEFERMAN [1977b], *Theories of finite type related to mathematical practice*, ***Handbook of mathematical logic*** (J. Barwise, editor), North-Holland, pp. 913–971.

S. FEFERMAN [1996], *Computation on abstract datatypes. The extensional approach, with an application to streams*, ***Annals of Pure and Applied Logic***, vol. 81, pp. 75–113.

J. E. FENSTAD [1978], *On the foundation of general recursion theory: Computations versus inductive definability*, ***Generalized Recursion Theory II*** (J. E. Fenstad, R. O. Gandy, and G. E. Sacks, editors), North-Holland, pp. 99–110.

J. E. FENSTAD [1980], ***General recursion theory***, Perspectives in Mathematical Logic, Springer.

M. P. FIORE, A. JUNG, MOGGI, O'HEARN, RIECKE, ROSOLINI, AND STARK [1996], *Domains and denotational semantics: History, accomplishments and open problems*, ***Bulletin of the European Association for Theoretical Computer Science***, vol. 59, pp. 227–256.

M. C. FITTING [1981], ***Fundamentals of generalized recursion theory***, Studies in Logic and the Foundations of Mathematics, vol. 105, North-Holland.

R. V. FREIVALDS [1978], *Effective operations and functionals computable in the limit*, ***Zeitschrift für mathematische Logik und Grundlagen der Mathematik***, vol. 24, pp. 193–206.

R. M. FRIEDBERG [1958a], *Four quantifier completeness: a Banach-Mazur functional not uniformly partial recursive*, ***Bulletin de l'Académie Polonaise des Sciences, Série des sciences mathématiques, astronomiques et physiques***, vol. 6, pp. 1–5.

R. M. FRIEDBERG [1958b], *Un contre-exemple relatif aux fonctionnelles récursives*, ***Comptes rendus hebdomadaires des séances de l'Académie des Sciences (Paris)***, vol. 247, pp. 852–854.

H. M. FRIEDMAN [1971], *Algorithmic procedures, generalized Turing algorithms, and elementary recursion theory*, ***Logic Colloquium 1969*** (R. O. Gandy and C. E. M. Yates, editors), North-Holland, pp. 113–137.

R. O. GANDY [1962], *Effective operations and recursive functionals (abstract)*, ***The Journal of Symbolic Logic***, vol. 27, pp. 378–379.

R. O. GANDY [1967a], *Computable functionals of finite type I*, ***Sets, Models and Recursion Theory*** (J. N. Crossley, editor), North-Holland, part II never appeared, pp. 202–242.

R. O. GANDY [1967b], *General recursive functionals of finite type and hierarchies of functionals*, ***Annales de la Faculté des Sciences, Université de Clermont-Ferrand***, vol. 35, pp. 5–24.

R. O. GANDY [1980], *Proofs of strong normalization*, ***To H. B. Curry: Essays on combinatory Logic, lambda calculus and formalism*** (J. R. Hindley and J. P. Seldin, editors), Academic Press.

R. O. GANDY AND J. M. E. HYLAND [1977], *Computable and recursively countable functions of higher type*, ***Logic Colloquium 1976***, North-Holland, pp. 407–438.

P. GIANNINI AND G. LONGO [1984], *Effectively given domains and lambda-calculus models*, ***Information and Control***, vol. 62, pp. 36–63.

J.-Y. GIRARD [1972], ***Interprétation functionelle et élimination des coupures de l'arithmétique d'ordre supérieur***, Ph.D. thesis, Paris.

J.-Y. GIRARD [1987], ***Proof theory and logical complexity***, vol. I, Bibliopolis, volume II has not appeared.

J.-Y. Girard [1988], *Normal functors, power series and λ-calculus*, ***Annals of Pure and Applied Logic***, vol. 37, no. 2, pp. 129–177.

K. Gödel [1931], *Über formal unentscheidbare Sätze der Principia Mathematica und verwandter Systeme I*, ***Monatshefte für Mathematik und Physik***, vol. 38, pp. 173–198, English translation in J. van Heijenoort, ed., ***From Frege to Gödel: A Source Book in Mathematical Logic, 1879–1931***, Harvard University Press, 1967, pp. 596–616.

K. Gödel [1958], *Über eine bisher noch nicht Erweiterung des finiten Standpunktes*, ***Dialectica***, vol. 12, pp. 280–287, English translation in Gödel [1990].

K. Gödel [1972], *On an extension of finitary mathematics which has not yet been used*, First published in Gödel [1990].

K. Gödel [1990], ***Collected works***, vol. II, Oxford University Press, edited by S. Feferman *et al.*

E. R. Griffor [1980], ***E-recursively enumerable degrees***, Ph.D. thesis, MIT, Cambridge, Massachusetts.

E. R. Griffor (editor) [1999], ***Handbook of computability theory***, Studies in Logic and the Foundations of Mathematics, vol. 140, North-Holland.

T. J. Grilliot [1967], ***Recursive functions of finite higher types***, Ph.D. thesis, Duke University.

T. J. Grilliot [1969a], *Hierarchies based on objects of finite type*, ***The Journal of Symbolic Logic***, vol. 34, pp. 177–182.

T. J. Grilliot [1969b], *Selection functions for recursive functionals*, ***Notre Dame Journal of Formal Logic***, vol. X, pp. 225–234.

T. J. Grilliot [1971], *On effectively discontinuous type-2 objects*, ***The Journal of Symbolic Logic***, vol. 36, pp. 245–248.

A. Grzegorczyk [1955a], *Computable functionals*, ***Fundamenta Mathematicae***, vol. 42, pp. 168–202.

A. Grzegorczyk [1955b], *On the definition of computable functionals*, ***Fundamenta Mathematicae***, vol. 42, pp. 232–239.

A. Grzegorczyk [1957], *On the definitions of computable real continuous functions*, ***Fundamenta Mathematicae***, vol. 44, pp. 61–71.

A. Grzegorczyk [1964], *Recursive objects in all finite types*, ***Fundamenta Mathematicae***, vol. 54, pp. 73–93.

L. A. Harrington [1973], ***Contributions to recursion theory in higher types***, Ph.D. thesis, MIT, Cambridge, Massachusetts.

L. A. Harrington [1974], *The superjump and the first recursively Mahlo ordinal*, ***Generalized Recursion Theory*** (J. E. Fenstad and P. G. Hinman, editors), North-Holland, pp. 43–52.

L. A. Harrington and A. S. Kechris [1975], *On characterizing Spector classes*, ***The Journal of Symbolic Logic***, vol. 40, pp. 19–24.

L. A. Harrington and D. MacQueen [1976], *Selection in abstract recursion theory*, ***The Journal of Symbolic Logic***, vol. 41, pp. 153–158.

J. P. Helm [1971], *On effectively computable operators*, ***Zeitschrift für mathematische Logik und Grundlagen der Mathematik***, vol. 17, pp. 231–244.

L. Henkin [1950], *Completeness in the theory of types*, ***The Journal of Symbolic Logic***, vol. 15, no. 2, pp. 81–91.

D. Hilbert [1925], *Über das Unendliche*, ***Mathematische Annalen***, vol. 95, pp. 161–190, English translation in J. van Heijenoort, ed., ***From Frege to Gödel: A Source Book in Mathematical Logic, 1879–1931***, Harvard University Press, 1967, pp. 367–392.

S. Hinata [1967], *Calculability of primitive recursive functionals of finite type*, ***Science Reports of the Tokyo Kyoiku Daigaku, A***, vol. 9, pp. 218–235.

S. Hinata and S. Tugué [1969], *A note on continuous functionals*, ***Annals of the Japan Association for Philosophy of Science***, vol. 3, pp. 138–145.

Y. HINATANI [1966], *Calculabilité des fonctionnels recursives primitives de type fini sur les nombres naturels*, ***Annals of the Japan Association for Philosophy of Science***, vol. 3, pp. 19–30.

P. G. HINMAN [1966], ***Ad astra per aspera: hierarchy schemata in recursive function theory***, Ph.D. thesis, University of California, Berkeley.

P. G. HINMAN [1969], *Hierarchies of effective descriptive set theory*, ***Transactions of the American Mathematical Society***, vol. 142, pp. 111–140.

P. G. HINMAN [1973], *Degrees of continuous functionals*, ***The Journal of Symbolic Logic***, vol. 38, pp. 393–395.

P. G. HINMAN [1978], ***Recursion-theoretic hierarchies***, Perspectives in Mathematical Logic, Springer-Verlag.

P. G. HINMAN [1999], *Recursion on abstract structures*, ***Handbook of computability theory*** (E. R. Griffor, editor), North-Holland, pp. 315–359.

W. A. HOWARD [1973], *Hereditarily majorizable functionals of finite type*, ***Metamathematical investigation of intuitionistic arithmetic and analysis*** (A. S. Troelstra, editor), Lecture Notes in Mathematics, vol. 344, Springer-Verlag, pp. 454–461.

J. M. E. HYLAND [1975], ***Recursion theory on the countable functionals***, Ph.D. thesis, University of Oxford.

J. M. E. HYLAND [1978], *The intrinsic recursion theory on the countable or continuous functionals*, ***Generalized Recursion Theory II*** (J. E. Fenstad, R. O. Gandy, and G. E. Sacks, editors), North-Holland, pp. 135–145.

J. M. E. HYLAND [1979], *Filter spaces and continuous functionals*, ***Ann. Math. Logic***, vol. 16, pp. 101–143.

J. M. E. HYLAND [1982], *The effective topos*, ***The L. E. J. Brouwer Centenary Symposium*** (A. S. Troelstra and D. van Dalen, editors), North-Holland, pp. 165–216.

J. M. E. HYLAND [1990], *First steps in synthetic domain theory*, ***Category Theory, Proceedings, Como*** (A. Carboni, M. C. Pedicchio, and G. Rosolini, editors), Lecture Notes in Mathematics, vol. 1488, Springer-Verlag, pp. 131–156.

J. M. E. HYLAND [2002], *Variations on realizability: realizing the propositional axiom of choice*, ***Mathematical Structures in Computer Science***, vol. 12, no. 3, pp. 295–318.

J. M. E. HYLAND, P. T. JOHNSTONE, AND A. M. PITTS [1980], *Tripos theory*, ***Mathematical Proceedings of the Cambridge Philosophical Society***, vol. 88, pp. 205–232.

J. M. E. HYLAND AND C.-H. L. ONG [2000], *On full abstraction for PCF: I, II and III*, ***Information and Computation***, vol. 163, pp. 285–408.

J. M. E. HYLAND AND A. SCHALK [2002], *Games on graphs and sequentially realizable functionals*, ***Proceedings of 17th Annual Symposium on Logic in Computer Science***, IEEE, pp. 257–264.

R. IRWIN, B. KAPRON, AND J. ROYER [2001a], *On characterizations of the basic feasible functionals, Part I*, ***Journal of Functional Programming***, vol. 11, pp. 117–153.

R. IRWIN, B. KAPRON, AND J. ROYER [2001b], *On characterizations of the basic feasible functionals, Part II*, to appear.

A. JUNG AND A. STOUGHTON [1993], *Studying the fully abstract model of PCF within its continuous function model*, ***Proceedings of International Conference on Typed Lambda Calculi and Applications, Utrecht***, Lecture Notes in Computer Science, no. 664, Springer-Verlag, pp. 230–245.

G. KAHN AND G. D. PLOTKIN [1993], *Concrete domains*, ***Theoretical Computer Science***, vol. 121, pp. 187–277, first appeared in French as INRIA-LABORIA ***Technical report***, 1978.

R. KANNEGANTI, R. CARTWRIGHT, AND M. FELLEISEN [1993], *SPCF: its model, calculus, and computational power*, ***Proceedings of REX Workshop on Semantics and Concurrency***, Lecture Notes in Computer Science, vol. 666, Springer-Verlag, pp. 318–347.

A. S. KECHRIS [1973], *The structure of envelopes: A survey of recursion theory in higher types*, MIT Logic Seminar notes.

A. S. Kechris and Y. N. Moschovakis [1977], *Recursion in higher types*, ***Handbook of mathematical logic*** (J. Barwise, editor), North-Holland, pp. 681–737.

D. P. Kierstead [1980], *A semantics for Kleene's j-expressions*, ***The Kleene Symposium*** (J. Barwise, H. J. Keisler, and K. Kunen, editors), North-Holland, pp. 353–366.

D. P. Kierstead [1983], *Syntax and semantics in higher-type recursion theory*, ***Transactions of the American Mathematical Society***, vol. 276, pp. 67–105.

S. C. Kleene [1936a], *General recursive functions of natural numbers*, ***Mathematische Annalen***, vol. 112, pp. 727–742.

S. C. Kleene [1936b], *λ-definability and recursiveness*, ***Duke Mathematical Journal***, vol. 2, pp. 340–353.

S. C. Kleene [1945], *On the interpretation of intuitionistic number theory*, ***The Journal of Symbolic Logic***, vol. 10, pp. 109–124.

S. C. Kleene [1952], ***Introduction to metamathematics***, Wolter-Noordhoff and North-Holland.

S. C. Kleene [1955a], *Arithmetical predicates and function quantifiers*, ***Transactions of the American Mathematical Society***, vol. 79, pp. 312–340.

S. C. Kleene [1955b], *Hierarchies of number-theoretic predicates*, ***Bulletin of the American Mathematical Society***, vol. 61, pp. 193–213.

S. C. Kleene [1959a], *Countable functionals*, ***Constructivity in Mathematics: Proceedings of the colloquium held at Amsterdam, 1957*** (A. Heyting, editor), North-Holland, pp. 81–100.

S. C. Kleene [1959b], *Recursive functionals and quantifiers of finite types I*, ***Transactions of the American Mathematical Society***, vol. 91, pp. 1–52.

S. C. Kleene [1962a], *Herbrand-Gödel-style recursive functionals of finite types*, ***Recursive function theory: Proceedings of Symposia on Pure Mathematics***, vol. 5, AMS, pp. 49–75.

S. C. Kleene [1962b], *Lambda-definable functionals of finite types*, ***Fundamenta Mathematicae***, vol. 50, pp. 281–303.

S. C. Kleene [1962c], *Turing-machine computable functionals of finite types I*, ***Logic, methodology and philosophy of science, Stanford***, pp. 38–45.

S. C. Kleene [1962d], *Turing-machine computable functionals of finite types II*, ***Proceedings of the London Mathematical Society***, vol. 12, pp. 245–258.

S. C. Kleene [1963], *Recursive functionals and quantifiers of finite types II*, ***Transactions of the American Mathematical Society***, vol. 108, pp. 106–142.

S. C. Kleene [1969], ***Formalized recursive functionals and formalized realizability***, Memoirs of the American Mathematical Society, vol. 89, American Mathematical Society.

S. C. Kleene [1978], *Recursive functionals and quantifiers of finite types revisited I*, ***Generalized Recursion Theory II*** (J. E. Fenstad, R. O. Gandy, and G. E. Sacks, editors), North-Holland, pp. 185–222.

S. C. Kleene [1980], *Recursive functionals and quantifiers of finite types revisited II*, ***The Kleene Symposium*** (J. Barwise, H. J. Keisler, and K. Kunen, editors), North-Holland, pp. 1–29.

S. C. Kleene [1982], *Recursive functionals and quantifiers of finite types revisited III*, ***Patras Logic Symposion*** (G. Metakides, editor), North-Holland, pp. 1–40.

S. C. Kleene [1985], *Unimonotone functions of finite types* (*Recursive functionals and quantifiers of finite types revisited IV*), ***Recursion Theory*** (A. Nerode and R. A. Shore, editors), Proceedings of Symposia in Pure Mathematics, vol. 42, pp. 119–138.

S. C. Kleene [1991], *Recursive functionals and quantifiers of finite types revisited V*, ***Transactions of the American Mathematical Society***, vol. 325, pp. 593–630.

S. C. Kleene and R. E. Vesley [1965], ***The foundations of intuitionistic mathematics***, North-Holland.

U. Kohlenbach [2002], *Foundational and mathematical uses of higher types*, ***Reflections on the foundations of mathematics*** (W. Sieg, R. Sommer, and C. Talcott, editors), Lecture Notes in Logic, vol. 15, A K Peters, pp. 92–116.

P. G. Kolaitis [1978], *On recursion in E and semi-Spector classes*, ***Cabal seminar 1976–77*** (A. S. Kechris and Y. N. Moschovakis, editors), Lecture Notes in Mathematics, vol. 689, Springer-Verlag, pp. 209–243.

P. G. Kolaitis [1979], *Recursion in a quantifier vs. elementary induction*, ***The Journal of Symbolic Logic***, vol. 44, pp. 235–259.

P. G. Kolaitis [1980], *Recursion and nonmonotone induction in a quantifier*, ***The Kleene Symposium*** (J. Barwise, H. J. Keisler, and K. Kunen, editors), North-Holland, pp. 367–389.

P. G. Kolaitis [1985], *Canonical forms and hierarchies in generalized recursion theory*, ***Proceedings of Symposia on Pure Mathematics***, vol. 42, AMS, pp. 139–170.

M. V. Korovina and O. V. Kudinov [2001], *Semantic characterisations of second-order computability over the real numbers*, ***Computer Science Logic***, Lecture Notes in Computer Science, vol. 2142, Springer-Verlag, pp. 160–173.

G. Kreisel [1958], *Constructive mathematics*, notes of a course given at Stanford University.

G. Kreisel [1959], *Interpretation of analysis by means of functionals of finite type*, ***Constructivity in Mathematics: Proceedings of the colloquium held at Amsterdam, 1957*** (A. Heyting, editor), North-Holland, pp. 101–128.

G. Kreisel [1962], *On weak completeness of intuitionistic predicate logic*, ***The Journal of Symbolic Logic***, vol. 27, pp. 139–158.

G. Kreisel, D. Lacombe, and J. R. Shoenfield [1957], *Fonctionnelles récursivement définissable et fonctionnelles récursives*, ***Comptes Rendus de l'Académie des Sciences, Paris***, vol. 245, pp. 399–402.

G. Kreisel, D. Lacombe, and J. R. Shoenfield [1959], *Partial recursive functionals and effective operations*, ***Constructivity in Mathematics: Proceedings of the colloquium held at Amsterdam, 1957*** (A. Heyting, editor), North-Holland, pp. 101–128.

L. Kristiansen and D. Normann [1997], *Total objects in inductively defined types*, ***Archive of Mathematical Logic***, vol. 36, pp. 405–436.

C. Kuratowski [1952], ***Topologie***, vol. I, Warsaw.

A. H. Lachlan [1964], *Effective operations in a general setting*, ***The Journal of Symbolic Logic***, vol. 29, pp. 163–178.

D. Lacombe [1955a], *Extension de la notion de fonction récursive aux fonctions d'une ou plusieurs variables réelles I*, ***Comptes Rendus de l'Académie des Sciences, Paris***, vol. 240, pp. 2478–2480.

D. Lacombe [1955b], *Extension de la notion de fonction récursive aux fonctions d'une ou plusieurs variables réelles II*, ***Comptes Rendus de l'Académie des Sciences, Paris***, vol. 241, pp. 13–14.

D. Lacombe [1955c], *Extension de la notion de fonction récursive aux fonctions d'une ou plusieurs variables réelles III*, ***Comptes Rendus de l'Académie des Sciences, Paris***, vol. 241, pp. 151–153.

D. Lacombe [1955d], *Remarques sur les opérateurs récursifs et sur les fonctions récursives d'une variable réelle*, ***Comptes Rendus de l'Académie des Sciences, Paris***, vol. 241, pp. 1250–1252.

J. Laird [1997], *Full abstraction for functional languages with control*, ***Proceedings of 12th Annual Symposium on Logic in Computer Science***, IEEE, pp. 58–67.

J. Laird [1998], ***A semantic analysis of control***, Ph.D. thesis, University of Edinburgh, examined March 1999.

J. Laird [2002], *Games and sequential algorithms*, submitted.

J. Lambek and P. J. Scott [1986], ***Introduction to higher-order categorical logic***, Cambridge Studies in Advanced Mathematics, vol. 7, Cambridge University Press.

B. Lichtenthäler [1996], *Degrees of parallelism*, ***Technical Report 96-01***, Fachgruppe Informatik, Siegen.

R. Loader [1997], *Equational theories for inductive types*, ***Annals of Pure and Applied Logic***, vol. 84, pp. 175–217.

R. LOADER [1998], *Unary PCF is decidable*, ***Theoretical Computer Science***, vol. 206, pp. 317–329.

R. LOADER [2001], *Finitary PCF is not decidable*, ***Theoretical Computer Science***, vol. 266, pp. 341–364.

M. H. LÖB [1970], *A model-theoretic characterization of effective operations*, ***The Journal of Symbolic Logic***, vol. 35, pp. 217–222, *a correction*, ibid., vol. 39, p. 225, 1974.

J. R. LONGLEY [1995], ***Realizability toposes and language semantics***, Ph.D. thesis, University of Edinburgh, available as ECS-LFCS-95-332.

J. R. LONGLEY [1999a], *Matching typed and untyped realizability*, ***Proceedings of Workshop on Realizability, Trento*** (L. Birkedal, J. van Oosten, G. Rosolini, and D. S. Scott, editors), published as Electronic Notes in Theoretical Computer Science 23 No. 1, Elsevier. Available via `http://www.elsevier.nl/locate/entcs/volume23.html`.

J. R. LONGLEY [1999b], *Unifying typed and untyped realizability*, unpublished note, available at `http://www.dcs.ed.ac.uk/home/jrl/unifying.txt`.

J. R. LONGLEY [1999c], *When is a functional program not a functional program?*, ***Proceedings of 4th International Conference on Functional Programming, Paris***, ACM Press, pp. 1–7.

J. R. LONGLEY [2002], *The sequentially realizable functionals*, ***Annals of Pure and Applied Logic***, vol. 117, no. 1, pp. 1–93.

J. R. LONGLEY [2004], *Universal types and what they are good for*, ***Domain theory, logic and computation: Proceedings of 2nd international symposium on domain theory, Chengdu*** (Guo-Qiang Zhang et al., editors), Kluwer.

J. R. LONGLEY AND G. D. PLOTKIN [1997], *Logical full abstraction and PCF*, ***Tbilisi Symposium on Language, Logic and Computation*** (J. Ginzburg et al., editor), SiLLI/CSLI, pp. 333–352.

J. R. LONGLEY AND A. K. SIMPSON [1997], *A uniform approach to domain theory in realizability models*, ***Mathematical Structures in Computer Science***, vol. 7, pp. 469–505.

G. LONGO AND E. MOGGI [1984a], *Cartesian closed categories of enumerations for effective type structures, Parts I and II*, ***Semantics of Data Types*** (G. Kahn, D. MacQueen, and G. Plotkin, editors), Springer-Verlag, pp. 235–255.

G. LONGO AND E. MOGGI [1984b], *The hereditary partial functionals and recursion theory in higher types*, ***The Journal of Symbolic Logic***, vol. 49, pp. 1319–1332.

F. LOWENTHAL [1976], *Equivalence of some definitions of recursion in a higher type object*, ***The Journal of Symbolic Logic***, vol. 41, pp. 427–435.

D. MACQUEEN [1972], ***Post's problem for recursion in higher types***, Ph.D. thesis, MIT, Cambridge, Massachusetts.

M. MARZ, A. ROHR, AND T. STREICHER [1999], *Full abstraction and universality via realisability*, ***Proceedings of 14th Annual Symposium on Logic in Computer Science***, IEEE, pp. 174–182.

S. MAZUR [1963], ***Computable analysis***, Rozprawy Matematyczne, vol. 33, Warsaw.

D. C. MCCARTY [1984], *Information systems, continuity and realizability*, ***Logics of Programs*** (E. Clarke and D. Cozen, editors), Lecture Notes in Computer Science, no. 164, Springer-Verlag, pp. 341–359.

R. MILNER [1977], *Fully abstract models of typed λ-calculi*, ***Theoretical Computer Science***, vol. 4, pp. 1–22.

R. MILNER, M. TOFTE, R. HARPER, AND D. MACQUEEN [1997], ***The definition of Standard ML (revised)***, MIT Press.

E. MOGGI [1988], *Partial morphisms in categories of effective objects*, ***Information and Computation***, vol. 73, pp. 250–277.

J. MOLDESTAD [1977], ***Computations in higher types***, Lecture Notes in Mathematics, vol. 574, Springer-Verlag.

Y. N. MOSCHOVAKIS [1967], *Hyperanalytic predicates*, ***Transactions of the American Mathematical Society***, vol. 129, pp. 249–282.

Y. N. Moschovakis [1969], *Abstract first order computability I, II*, ***Transactions of the American Mathematical Society***, vol. 138, pp. 427–464, 465–504.

Y. N. Moschovakis [1974a], ***Elementary induction on abstract structures***, North-Holland.

Y. N. Moschovakis [1974b], *On non-monotone inductive definability*, ***Fundamenta Mathematicae***, vol. 82, pp. 39–83.

Y. N. Moschovakis [1974c], *Structural characterizations of classes of relations*, ***Generalized Recursion Theory*** (J. E. Fenstad and P. G. Hinman, editors), North-Holland, pp. 53–79.

Y. N. Moschovakis [1976], *On the basic notions in the theory of induction*, ***Proceedings of 5th International Congress in Logic, Methodology and Philosophy of Science***, London, Ontario, pp. 207–236.

Y. N. Moschovakis [1981], *On the Grilliot-Harrington-MacQueen theorem*, ***Logic Year 1979–80***, Lecture Notes in Mathematics, vol. 859, Springer-Verlag, pp. 246–267.

Y. N. Moschovakis [1983], *Abstract recursion as a foundation for the theory of algorithms*, ***Computation and Proof Theory: Proceedings of the Logic Colloquium, vol. 2*** (E. Boerger, W. Oberschelp, M. M. Richter, B. Schinzel, and W.Thomas, editors), Lecture Notes in Mathematics, vol. 1104, Springer-Verlag, pp. 289–364.

Y. N. Moschovakis [1989], *The formal language of recursion*, ***The Journal of Symbolic Logic***, vol. 54, pp. 1216–1252.

K. Mulmuley [1987], ***Full abstraction and semantic equivalence***, MIT Press.

P. S. Mulry [1982], *Generalized Banach-Mazur functionals in the topos of recursive sets*, ***Journal of Pure and Applied Algebra***, vol. 26, pp. 71–83.

J. Myhill and J. C. Shepherdson [1955], *Effective operations on partial recursive functions*, ***Zeitschrift für mathematische Logik und Grundlagen der Mathematik***, vol. 1, pp. 310–317.

A. Nerode [1957], *General topology and partial recursive functionals*, ***Cornell Summer Institute of Symbolic Logic***, Cornell, pp. 247–251.

A. Nerode [1959], *Some Stone spaces and recursion theory*, ***Duke Mathematical Journal***, vol. 26, pp. 397–406.

H. Nickau [1994], *Hereditarily sequential functionals*, ***Proceedings of 3rd Symposium on Logical Foundations of Computer Science***, Lecture Notes in Computer Science, vol. 813, Springer-Verlag, pp. 253–264.

K.-H. Niggl [1993], *Subrecursive hierarchies on Scott domains*, ***Archive of Mathematical Logic***, vol. 32, pp. 239–257.

K.-H. Niggl [1999], *M^ω considered as a programming language*, ***Annals of Pure and Applied Logic***, vol. 99, pp. 73–92.

D. Normann [1978a], *A continuous functional with noncollapsing hierarchy*, ***The Journal of Symbolic Logic***, vol. 43, pp. 487–491.

D. Normann [1978b], *Set recursion*, ***Generalized Recursion Theory II*** (J. E. Fenstad, R. O. Gandy, and G. E. Sacks, editors), North-Holland, pp. 303–320.

D. Normann [1979a], *A classification of higher type functionals*, ***Proceedings of 5th Scandinavian Logic Symposium*** (F. V. Jensen, B. H. Mayoh, and K. K.Møller, editors), Aalborg University Press, pp. 301–308.

D. Normann [1979b], *Nonobtainable continuous functionals*, ***Proceedings of 6th International Congress on Logic, Methodology and Philosophy of Science***, Hanover, pp. 241–249.

D. Normann [1980], ***Recursion on the countable functionals***, Lecture Notes in Mathematics, vol. 811, Springer-Verlag.

D. Normann [1981a], *The continuous functionals: computations, recursions and degrees*, ***Annals of Mathematical Logic***, vol. 21, pp. 1–26.

D. Normann [1981b], *Countable functionals and the projective hierarchy*, ***The Journal of Symbolic Logic***, vol. 46, pp. 209–215.

D. Normann [1982], *External and internal algorithms on the continuous functionals*, ***Patras Logic Symposion*** (G. Metakides, editor), North-Holland, pp. 137–144.

D. Normann [1983], *Characterising the continuous functionals*, ***The Journal of Symbolic Logic***, vol. 48, pp. 965–969.

D. Normann [1989], *Kleene–spaces*, ***Logic colloquium 1988*** (Ferro, Bonotto, Valentini, and Zanardo, editors), Elsevier, pp. 91–109.

D. Normann [1997], *Closing the gap between the continuous functionals and recursion in* ^3E, ***Archive of Mathematical Logic***, vol. 36, pp. 269–287.

D. Normann [1999a], *The continuous functionals*, ***Handbook of computability theory*** (E. R. Griffor, editor), North-Holland, pp. 251–275.

D. Normann [1999b], *A Mahlo-universe of effective domains with totality*, ***Models and Computability*** (S. B. Cooper and J. K. Truss, editors), Cambridge University Press.

D. Normann [2000], *Computability over the partial continuous functionals*, ***The Journal of Symbolic Logic***, vol. 65, pp. 1133–1142.

D. Normann [2001], *The continuous functionals of finite types over the reals*, ***Domains and processes: Proceedings of 1st international symposium on domain theory, Shanghai*** (K. Keimel, G. Q. Zhang, Y. Liu, and Y. Chen, editors), Kluwer, pp. 103–124.

D. Normann [2002], *Exact real number computations relative to hereditarily total functionals*, ***Theoretical Computer Science***, vol. 284, no. 2, pp. 437–453.

D. Normann and C. Rørdam [2002], *The computational power of* M^ω, ***Mathematical Logic Quarterly***, vol. 48, no. 1, pp. 117–124.

D. Normann and S. S. Wainer [1980], *The 1-section of a countable functional*, ***The Journal of Symbolic Logic***, vol. 45, pp. 549–562.

P. G. Odifreddi [1989], ***Classical recursion theory, volume I***, Studies in Logic and the Foundations of Mathematics, vol. 125, Elsevier, second edition 1999.

P. W. O'Hearn and J. G. Riecke [1995], *Kripke logical relations and PCF*, ***Information and Computation***, vol. 120, no. 1, pp. 107–116.

C.-H. L. Ong [1995], *Correspondence between operational and denotational semantics*, ***Handbook of logic in computer science*** (S. Abramsky, D. Gabbay, and T. S. E. Maibaum, editors), vol. 4, Oxford University Press, pp. 269–356.

C.-H. L. Ong and C. A. Stewart [1997], *A Curry-Howard foundation for functional computation with control*, ***Proceedings of 24th Symposium on Principles of Programming Languages***, ACM Press, pp. 215–227.

J. van Oosten [1991], ***Exercises in realizability***, Ph.D. thesis, University of Amsterdam.

J. van Oosten [1997], *The modified realizability topos*, ***Journal of Pure and Applied Algebra***, vol. 116, pp. 273–289.

J. van Oosten [1999], *A combinatory algebra for sequential functionals of finite type*, ***Models and Computability*** (S. B. Cooper and J. K. Truss, editors), Cambridge University Press, pp. 389–406.

J. van Oosten [2002], *Realizability: a historical essay*, ***Mathematical Structures in Computer Science***, vol. 12, no. 3, pp. 239–264.

P. Päppinghaus [1985], ***Ptykes in Gödel's T und verallgemeinerte Rekursion über Mengen und Ordinalzahlen***, Habilitationsschrift, Hannover.

R. Péter [1951a], *Probleme der Hilbertschen Theorie der höheren Stufen von rekursiven Funktionen*, ***Acta mathematica Academiae Scientarum Hungaricae***, vol. 2, pp. 247–274.

R. Péter [1951b], ***Rekursive Funktionen***, Akademischer Verlag, Budapest, English translation published as ***Recursive Functions***, Academic Press, 1967.

W. K.-S. Phoa [1991], *From term models to domains*, ***Proceedings of Theoretical Aspects of Computer Software, Sendai***, Lecture Notes in Computer Science, vol. 526, Springer-Verlag.

R. Platek [1966], ***Foundations of recursion theory***, Ph.D. thesis, Stanford University.

R. Platek [1971], *A countable hierarchy for the superjump*, ***Logic Colloquium 1969*** (R. O. Gandy and C. E. M. Yates, editors), North-Holland, pp. 257–271.

G. D. PLOTKIN [1977], *LCF considered as a programming language*, ***Theoretical Computer Science***, vol. 5, pp. 223–255.

G. D. PLOTKIN [1978], $\mathbb{T}^{\omega}$ *as a universal domain*, ***Journal of Computer and System Sciences***, vol. 17, pp. 209–236.

G. D. PLOTKIN [1983], *Domains*, ***Technical report***, Department of Computer Science, University of Edinburgh.

G. D. PLOTKIN [1997], *Full abstraction, totality and PCF*, ***Mathematical Structures in Computer Science***, vol. 9, no. 1, pp. 1–20.

E. L. POST [1936], *Finite combinatory processes—formulation 1*, ***The Journal of Symbolic Logic***, vol. 1, pp. 103–105.

M. B. POUR-EL [1960], *A comparison of five "computable" operators*, ***Zeitschrift für mathematische Logik und Grundlagen der Mathematik***, vol. 6, pp. 325–340.

M. B. POUR-EL [1999], *The structure of computability*, ***Handbook of computability theory*** (E. R. Griffor, editor), North-Holland, pp. 315–359.

M. B. POUR-EL AND J. I. RICHARDS [1989], ***Computability in analysis and physics***, Springer-Verlag.

H. G. RICE [1953], *Classes of recursively enumerable sets and their decision problems*, ***Transactions of the American Mathematical Society***, vol. 74, pp. 358–366.

H. G. RICE [1956], *On completely recursively enumerable classes and their key arrays*, ***The Journal of Symbolic Logic***, vol. 21, pp. 304–308.

W. RICHTER [1967], *Constructive transfinite number classes*, ***Bulletin of the American Mathematical Society***, vol. 73, pp. 261–265.

J. G. RIECKE [1993], *Fully abstract translations between functional languages*, ***Mathematical Structures in Computer Science***, vol. 3, pp. 387–415.

H. ROGERS [1967], ***Theory of recursive functions and effective computability***, McGraw-Hill.

A. ROHR [2002], ***A universal realizability model of sequential functional computation***, Ph.D. thesis, Technical University of Darmstadt.

G. ROSOLINI [1986], ***Continuity and effectiveness in topoi***, Ph.D. thesis, Oxford; Carnegie-Mellon.

J. S. ROYER [2000], *On the computational complexity of Longley's H functional*, presented at Second International Workshop on Implicit Computational Complexity, UC/Santa Barbara.

G. E. SACKS [1971], *Recursion in objects of finite type*, ***Proceedings of International Congress of Mathematicians***, Gauthiers-Villars, Paris, pp. 251–254.

G. E. SACKS [1974], *The 1-section of a type n object*, ***Generalized Recursion Theory*** (J. E. Fenstad and P. G. Hinman, editors), North-Holland, pp. 81–96.

G. E. SACKS [1977], *The k-section of a type n object*, ***American Journal of Mathematics***, vol. 99, pp. 901–917

G. E. SACKS [1980], *Post's problem, absoluteness and recursion in finite types*, ***The Kleene Symposium*** (J. Barwise, H. J. Keisler, and K. Kunen, editors), North-Holland, pp. 201–222.

G. E. SACKS [1985], *Post's problem in E-recursion*, ***Proceedings of Symposia on Pure Mathematics***, vol. 42, AMS, pp. 177–193.

G. E. SACKS [1986], *On the limits of E-recursive enumerability*, ***Annals of Pure and Applied Logic***, vol. 31, pp. 87–120.

G. E. SACKS [1990], ***Higher recursion theory***, Springer-Verlag.

G. E. SACKS [1999], *E-recursion*, ***Handbook of computability theory*** (E. R. Griffor, editor), Elsevier, pp. 301–314.

L. E. SANCHIS [1967], *Functionals defined by recursion*, ***Notre Dame Journal of Formal Logic***, vol. 8, pp. 161–174.

L. E. SANCHIS [1992], ***Recursive functionals***, Studies in Logic and the Foundations of Mathematics, vol. 131, North-Holland.

L. P. SASSO [1971], ***Degrees of unsolvability of partial functions***, Ph.D. thesis, University of California, Berkeley.

V. YU. SAZONOV [1975], *Sequentially and parallelly computable functionals*, ***λ-calculus and Computer Science Theory: Proceedings of the symposium held in Rome***, Lecture Notes in Computer Science, vol. 37, Springer-Verlag, pp. 312–318.

V. YU. SAZONOV [1976a], *Degrees of parallelism in computations*, ***Proceedings of Symposium on Mathematical Foundations of Computer Science***, Lecture Notes in Computer Science, vol. 45, Springer-Verlag, pp. 517–523.

V. YU. SAZONOV [1976b], *Expressibility of functions in Scott's LCF language*, ***Algebra i Logika***, vol. 15, pp. 308–330.

V. YU. SAZONOV [1976c], *Functionals computable in series and in parallel*, ***Matematicheskii Zhurnal***, vol. 17, pp. 648–672.

V. YU. SAZONOV [1998], *An inductive definition of full abstract model for LCF (preliminary version)*, available from `http://csc.liv.ac.uk`.

B. SCARPELLINI [1971], *A model for barrecursion of higher types*, ***Compositio Mathematica***, vol. 23, pp. 123–153.

H. SCHWICHTENBERG [1975], *Elimination of higher type levels in definitions of primitive recursive functionals by means of transfinite recursion*, ***Logic Colloquium 1973*** (H. Rose and T. Shepherdson, editors), North-Holland, pp. 279–303.

H. SCHWICHTENBERG [1991], *Primitive recursion on the partial continuous functionals*, ***Informatik und Mathematik*** (M. Broy, editor), Springer-Verlag, pp. 251–269.

H. SCHWICHTENBERG [1996], *Density and choice for total continuous functionals*, ***Kreiseliana. About and around Georg Kreisel*** (P. Odifreddi, editor), A. K. Peters, Wellesley, Massachusetts, pp. 335–362.

H. SCHWICHTENBERG [1999], *Classifying recursive functions*, ***Handbook of computability theory*** (E. R. Griffor, editor), North-Holland, pp. 533–586.

D. S. SCOTT [1969], *A theory of computable functions of higher type*, unpublished seminar notes, University of Oxford. 7 pages.

D. S. SCOTT [1970], *Outline of a mathematical theory of computation*, ***Proceedings of 4th Annual Princeton Conference on Information Science and Systems***, pp. 165–176.

D. S. SCOTT [1972], *Continuous lattices*, ***Toposes, Algebraic Geometry and Logic*** (F. W. Lawvere, editor), Springer-Verlag.

D. S. SCOTT [1976], *Data types as lattices*, ***SIAM Journal of Computing***, vol. 5, no. 3, pp. 522–587.

D. S. SCOTT [1982], *Domains for denotational semantics*, ***Proceedings of 9th International Colloquium on Automata, Languages and Programming, Aarhus, Denmark*** (M. Nielsen and E. M. Schmidt, editors), Lecture Notes in Computer Science, no. 140, Springer-Verlag, pp. 577–610.

D. S. SCOTT [1993], *A type-theoretical alternative to ISWIM, CUCH, OWHY*, ***Theoretical Computer Science***, vol. 121, pp. 411–440, first written in 1969 and widely circulated in unpublished form since then.

J. R. SHOENFIELD [1962], *The form of the negation of a predicate*, ***Recursive function theory: Proceedings of Symposia on Pure Mathematics*** (J. C. E. Dekker, editor), vol. 5, AMS, pp. 131–134.

J. R. SHOENFIELD [1967], ***Mathematical logic***, Addison-Wesley.

J. R. SHOENFIELD [1968], *A hierarchy based on a type two object*, ***Transactions of the American Mathematical Society***, vol. 134, pp. 103–108.

K. SIEBER [1990], *Relating full abstraction results for different programming languages*, ***Proceedings of 10th Conference on Foundations of Software Technology and Theoretical Computer Science, Bangalore***, Lecture Notes in Computer Science, vol. 472, Springer-Verlag.

K. SIEBER [1992], *Reasoning about sequential functions*, ***Applications of Categories in Computer Science, Durham 1991*** (M. P. Fourman, P. T. Johnstone, and A. M. Pitts, editors), London

Mathematical Society Lecture Note Series, vol. 177, Cambridge University Press, pp. 258–269.

A. SIMPSON [1998], *Lazy functional algorithms for exact real functionals*, ***Mathematical Foundations of Computer Science, Proceedings***, Lecture Notes in Computer Science, vol. 1450, Springer-Verlag, pp. 456–464.

T. A. SLAMAN [1981], ***Aspects of E-recursion***, Ph.D. thesis, Harvard University.

T. A. SLAMAN [1985], *The E-recursively enumerable degrees are dense*, ***Proceedings of Symposia on Pure Mathematics***, vol. 42, AMS, pp. 195–213.

C. SPECTOR [1962], *Provably recursive functionals of analysis: a consistency proof of analysis by means of principles formulated in current intuitionistic mathematics*, ***Recursive function theory: Proceedings of Symposia on Pure Mathematics***, vol. 5, AMS, pp. 1–27.

D. SPREEN [1992], *Effective operators and continuity revisited*, ***Logical Foundations of Computer Science, Second International Symposium, Tver*** (A. Nerode and M. A. Taitslin, editors), Lecture Notes in Computer Science, vol. 620, Springer-Verlag, pp. 459–469.

D. SPREEN AND P. YOUNG [1983], *Effective operators in a topological setting*, ***Computation and Proof Theory: Proceedings of the Logic Colloquium, vol. 2*** (E. Boerger, W. Oberschelp, M. M. Richter, B. Schinzel, and W.Thomas, editors), Lecture Notes in Mathematics, vol. 1104, Springer-Verlag, pp. 437–451.

V. STOLTENBERG-HANSEN, I. LINDSTRÖM, AND E. R. GRIFFOR [1994], ***Mathematical theory of domains***, Cambridge Tracts in Theoretical Computer Science, no. 22, Cambridge University Press.

A. STOUGHTON [1991a], *Interdefinability of parallel operations in PCF*, ***Theoretical Computer Science***, vol. 79, pp. 357–358.

A. STOUGHTON [1991b], *Parallel PCF has a unique extensional model*, ***Proceedings of 6th Annual Symposium on Logic in Computer Science***, IEEE, pp. 146–151.

W. W. TAIT [1962], *A second order theory of functionals of higher type*, in Stanford report on the foundations of analysis, Stanford University.

W. W. TAIT [1967], *Intensional interpretations of functionals of finite type I*, ***The Journal of Symbolic Logic***, vol. 32, pp. 198–212.

M. B. TRAKHTENBROT [1975], *On representation of sequential and parallel functions*, ***Proceedings of 4th Symposium on Mathematical Foundations of Computer Science***, Lecture Notes in Computer Science, vol. 32, Springer-Verlag, pp. 411–417.

A. S. TROELSTRA [1973], ***Metamathematical investigation of intuitionistic arithmetic and analysis***, Lecture Notes in Mathematics, vol. 344, Springer-Verlag, second edition (with corrections): ILLC Prepublication Series X-93-05, University of Amsterdam, 1993.

G. S. TSEITIN [1959], *Algorithmic operators in constructive complete separable metric spaces*, ***Doklady Akademii Nauk***, vol. 128, pp. 49–52.

G. S. TSEITIN [1962], *Algorithmic operators in constructive metric spaces*, ***Trudy Matematicheskogo Instituta imeni V. A. Steklova***, vol. 67, pp. 295–361, translated in AMS Translations (2), vol. 64 (1966), 1–80.

J. V. TUCKER [1980], *Computing in algebraic systems*, ***Recursion theory: its generalizations and applications*** (F. R. Drake and S. S. Wainer, editors), London Mathematical Society Lecture Notes, vol. 45, Cambridge University Press, pp. 215–235.

T. TUGUÉ [1960], *Predicates recursive in a type-2 object and Kleene hierarchies*, ***Commentarii Mathematici Universitatis Sancti Pauli, Tokyo***, vol. 8, pp. 97–117.

A. M. TURING [1937a], *Computability and λ-definability*, ***The Journal of Symbolic Logic***, vol. 2, pp. 153–163.

A. M. TURING [1937b], *On computable numbers, with an application to the Entscheidungsproblem*, ***Proceedings of the London Mathematical Society***, vol. 42, pp. 230–265, *a correction*, ibid., vol. 42, pp. 455–546, 1937.

A. M. TURING [1939], *Systems of logic based on ordinals*, ***Proceedings of the London Mathematical Society***, vol. 45, pp. 161–228.

V. A. USPENSKII [1955], *On enumeration operators*, ***Doklady Akademii Nauk***, vol. 103, pp. 773–776.

J. VUILLEMIN [1973], *Correct and optimal implementations of recursion in a simple programming language*, ***Proceedings of 5th ACM Symposium on Theory of Computing***, pp. 224–239.

S. S. WAINER [1974], *A hierarchy for the 1-section of any type two object*, ***The Journal of Symbolic Logic***, vol. 39, pp. 88–94.

S. S. WAINER [1975], *Some hierarchies based on higher type quantification*, ***Logic Colloquium 1973*** (H. Rose and T. Shepherdson, editors), North-Holland, pp. 305–316.

S. S. WAINER [1978], *The 1-section of a non-normal type-2 object*, ***Generalized Recursion Theory II*** (J. E. Fenstad, R. O. Gandy, and G. E. Sacks, editors), North-Holland, pp. 407–417.

K. WEIHRAUCH [1985], *Type 2 recursion theory*, ***Theoretical Computer Science***, vol. 38, pp. 17–33.

K. WEIHRAUCH [2000a], ***Computability***, EATCS Monographs on Theoretical Computer Science, vol. 9, Springer-Verlag.

K. WEIHRAUCH [2000b], ***Computable analysis: an introduction***, Texts in Theoretical Computer Science, Springer-Verlag.

P. R. YOUNG [1968], *An effective operator, continuous but not partial recursive*, ***Proceedings of the American Mathematical Society***, vol. 19, pp. 103–108.

P. R. YOUNG [1969], *Toward a theory of enumerations*, ***Journal for the Association of Computing Machinery***, vol. 16, pp. 328–348.

I. D. ZASLAVSKII [1955], *Refutation of some theorems of classical analysis in constructive analysis*, ***Uspekhi Matematichekikh Nauk***, vol. 10, pp. 209–210.

I. D. ZASLAVSKII AND G. S. TSEITIN [1962], *On singular coverings and properties of constructive functions connected with them*, ***Trudy Matematicheskogo Instituta imeni V. A. Steklova***, vol. 67, pp. 458–502, translated in AMS Translations (2), vol. 98 (1971), 41–89.

J. I. ZUCKER [1971], ***Proof theoretic studies of systems of iterated inductive definitions and subsystems of analysis***, Ph.D. thesis, Stanford University.

INDEX

DIVISION OF INFORMATICS
UNIVERSITY OF EDINBURGH
EDINBURGH, EH9 3JZ, UK
E-mail: jrl@dcs.ed.ac.uk

THE CONTINUUM HYPOTHESIS

W. HUGH WOODIN

§1. Introduction. The now nearly 40 years since Cohen's discovery of the technique of forcing has seen the proliferation of independence results within set theory and beyond. A more recent trend though has been the discovery that some of these otherwise unsolvable problems do arguably have solutions. While these solutions are perhaps not yet universally accepted among set theorists, the claim is a serious one and these solutions have arisen as the result of rather elaborate mathematical investigations. Curiously the understanding of various independence results itself has played an important role.

So perhaps the view is now evolving that independence is not an insurmountable problem in set theory; perhaps there is after all a viable notion of (mathematical) truth for the transfinite, the vision is simply obscured by independence but not destroyed. I would go further and conjecture that fundamental questions such as that of Cantor's Continuum Hypothesis are solvable, or at the very least, in the case of the Continuum Hypothesis, claim that the theorem asserting otherwise has yet to be proved.

This paper is closely based on a series of three lectures given on the Continuum Hypothesis at the Logic Colloquium 2000 held at the Sorbonne in July of 2000. This account is also quite closely related to the two part series on the Continuum Hypothesis which appeared in the *Notices of the AMS*, in the summer of 2001,[1] though here there is some new material as well as additional details. Finally, in the time since this paper was first written there have been some important changes to several of the key definitions. Details are given in the section, *Added in proof*, at the end of this paper.

§2. The Continuum Hypothesis. As is well known, Cantor's Continuum Hypothesis (CH) asserts:

Suppose that $X \subseteq \mathbb{R}$ is an infinite set. Then either

$$\text{cardinality}(X) = \text{cardinality}(\mathbb{N})$$

or

$$\text{cardinality}(X) = \text{cardinality}(\mathbb{R}).$$

[1] [32] and [33].

Logic Colloquium 2000
Edited by R. Cori, A. Razborov, S. Todorčević, and C. Wood
Lecture Notes in Logic, 19

Of course the are many equivalent formulations. The following is an equivalent formulation, but equivalence requires the *Axiom of Choice*.

> *Suppose that* $\pi : \mathbb{R} \to \delta$ *is a surjection of* $\mathbb{R}$ *onto the cardinal* δ. *Then* $\delta \leq \omega_1$.

In certain natural settings (for example in the inner model $L(\mathbb{R})$) these two formulations can differ dramatically with first being true and the second being false.

But it is important to emphasize though that the problem of the Continuum Hypothesis is in the context of the *all* of the *Axioms of Set Theory*: These are the Zermelo–Fraenkel axioms together with the Axiom of Choice (ZFC). Nevertheless the possible splitting of the two formulations in natural inner models plays an important role in understanding the problem of the Continuum Hypothesis in V.

The first result concerning CH was obtained by Gödel, [10].

THEOREM 1 (Gödel, 1938). *Assume* ZFC *is consistent. Then so is* ZFC+CH. □

The modern era of set theory began with Cohen's discovery of the method of *forcing* and his application of this new method to show the following theorem, [3].

THEOREM 2 (Cohen, 1963). *Assume* ZFC *is consistent. Then so is*

$$\text{ZFC} + \text{"CH is false"}.$$ □

The current situation is the following. We can transform *models* of set theory with significant control over what is true in the resulting model. During the 40 years since Cohen's work a great number of set theoretical propositions have been shown to be independent. This currently cannot be accomplished, as successfully, for models of number theory. Arguably the intuition of a *true* model of number theory remains unchallenged, which is not to say that the *Peano Axioms* exhaust our intuitions for this model, but rather to say that the (fundamental) questions known to be formally independent of (say) the *Peano Axioms* can nevertheless be resolved.

The objects studied by mathematicians can be formalized as sets. But, as is frequently noted, a small fraction of the set theoretical universe given by the ZFC axioms is actually necessary for most of mathematics. One must then question whether the phenomena of independence, and our pondering of CH in particular, simply shows that (mathematically) "our reach exceeds our grasp".

While progress toward resolving CH, evidenced by a series of credible or at least interesting preliminary solutions, supports the claim that CH *can* be resolved; a *lack* of continuing progress naturally tends to support the claim that CH is meaningless. There certainly is at least one interesting preliminary solution: $V = L$; this is Gödel's *Axiom of Constructibility*. I will present another solution, yielding CH is false. Of course this solution could ultimately be as *wrong* as "$V = L$". At the very least, I hope the reader finds the story both interesting and surprising. Central to this "solution" is a sequence of theorems. It is precisely these mathematical developments that I shall be reviewing, for it is precisely these theorems which collectively constitute the justification for the solution I shall present. I begin with two simple definitions from set theory.

DEFINITION 3. (1) A set M is *transitive* if each element of M is a subset of M.

(2) The *transitive closure* of a set, M, is the set

$$\bigcap\{X \mid X \text{ is transitive and } M \subseteq X\}. \qquad \square$$

There is a natural stratification of the universe, V, of sets into an increasing hierarchy of transitive sets, $H(\kappa)$, indexed by the cardinals, κ, of V. Assuming the *Axiom of Choice*, this every set is a member of $H(\kappa)$ for some cardinal κ.

DEFINITION 4. Suppose κ is an infinite cardinal.

(1) A set M is of *hereditary cardinality* less than κ if $|M| < \kappa$, and if $|a| < \kappa$ for each a in the transitive closure of M.

(2) $H(\kappa)$ denotes the set of all sets M of hereditary cardinality less than κ. $\square$

There is an obvious incremental approach to understanding the *Continuum Hypothesis*. First understand $H(\omega)$, then understand $H(\omega_1)$, and then understand $H(\omega_2)$.

The *Continuum Hypothesis* concerns the structure of $H(\omega_2)$, so any reasonably complete understanding of $H(\omega_2)$ should resolve whether or not the *Continuum Hypothesis* is true.

The first of these structures, $H(\omega)$, is in essence simply the structure,

$$\langle \mathbb{N}, +, \cdot \rangle;$$

I leave this to the Number Theorists.

The second of these structures, $H(\omega_1)$, is in essence the structure,

$$\langle \mathcal{P}(\mathbb{N}), \mathbb{N}, +, \cdot, \in \rangle.$$

It is with the analysis of this structure that our story begins.

§3. The first step, $H(\omega_1)$. Two very simple operations on subsets of the Euclidean spaces, $\mathbb{R}^n$.

- (Projection) Suppose $X \subseteq \mathbb{R}^{n+1}$. The *projection* of X to $\mathbb{R}^n$ is the set

$$X^* = \{(a_1, \ldots, a_n) \mid \text{There exists } b \in \mathbb{R} \text{ such that } (a_1, \ldots, a_n, b) \in X.\}$$

- (Complements) Suppose $X \subseteq \mathbb{R}^n$. The *complement* of X is the set

$$X^* = \{(a_1, \ldots, a_n) \mid (a_1, \ldots, a_n) \notin X\}.$$

Closing the collection of closed subsets of $\mathbb{R}^n$ under these operations yields the collection of *projective* subsets of $\mathbb{R}^n$, more precisely:

DEFINITION 5 (Luzin, 1925). A set $X \subseteq \mathbb{R}^n$ is a *projective set* if it is generated from a closed subset of $\mathbb{R}^{n+k}$ in finitely many steps, taking projections and complements. □

Why consider the projective sets? The reason is quite simply because the structure of $H(\omega_1)$ can be reinterpreted as the structure of the projective sets. More formally, the projective sets correspond to the subsets of $H(\omega_1)$ which are *definable* (from parameters) in the structure $\langle H(\omega_1), \in\rangle$. This is a common method of logic, study a structure by studying the relations which can be defined in the structure.

The Banach–Tarski Paradox [30] shows that assuming the Axiom of Choice there is a partition of the unit ball of $\mathbb{R}^3$ into finitely many pieces from which two unit balls can be generated using rotations. This is a paradoxical decomposition.

Consider the following question:

> *Can there be a paradoxical decomposition of the unit ball of* $\mathbb{R}^3$ *into pieces each of which is a projective set*?

This question is raised merely to guide the subsequent discussion. In fact the question was never actually considered until after its answer was known.

A projective set $X \subseteq \mathbb{R}^n$ is *analytic* if it can be represented as the projection of the complement of the projection of a closed subset of $\mathbb{R}^{n+2}$ (so generated in 2 steps).

THEOREM 6 (Luzin, 1917). *Every analytic set is Lebesgue measurable.* □

As an immediate corollary,

COROLLARY 7. *There can be no paradoxical decomposition of the unit ball of* $\mathbb{R}^3$ *into pieces each of which is in the* σ*-algebra generated by analytic sets.* □

Of course the question on paradoxical projective decompositions suggests the more fundamental and natural question:

> *Are the projective sets Lebesgue measurable*?

This is the measure problem for the projective sets. I refer the reader to [13] and [19] for historical remarks on the measure problem and for a discussion of the history of the development of the subject of the projective sets in general.

By the 1920's the difficulty of the measure problem was apparent:

> (Luzin, 1925) "one does not know *and one will never know* [of the projective sets whether or not they are each Lebesgue measurable.]"

Part of the explanation of the difficulty of the measure problem was provided almost fifteen years later by Gödel [10].

THEOREM 8 (Gödel, 1938). *Assume* ZFC *is consistent. Then*

ZFC + "There is a non-measurable projective set"

is consistent. □

In fact an immediate corollary of the proof of this theorem is the (relative) consistency with the axioms of set theory that:

> *There exists a paradoxical decomposition of the unit ball of* $\mathbb{R}^3$ *into pieces each of which is the projection of the complement of an analytic set.*

Thus it is consistent that "pathological" sets be generated in one step from the analytics set; so in three steps from the closed sets.

An immediate corollary of Solovay's results [26] on the measure problem for the projective sets is the following theorem.

THEOREM 9 (Solovay, 1965). *Assume* ZFC *is consistent. Then so is* ZFC *together with the statement*:

> "There is no paradoxical decomposition of the unit ball of $\mathbb{R}^3$ into pieces each of which is a projective set." □

Thus already at $H(\omega_1)$ there are natural structural questions which are formally unsolvable. Any resolution of these problems requires new axioms.

> Gödel (1947): There might exist axioms so abundant in their verifiable consequences, shedding so much light upon a whole discipline, and furnishing such powerful methods for solving given problems (and even solving them, as far as possible, in a constructivistic way) that quite irrespective of their intrinsic necessity they would have to be assumed at least in the same sense as any established physical theory.

Axioms which fulfill this prophecy, at least for the projective sets, were proposed almost 40 years ago, but the realization that these axioms were the *correct* axioms took nearly 30 years. The axioms involve the notion of the determinacy of certain infinite games.

Fix $A \subseteq \mathbb{R}$. The game G_A involves two players: Player I and Player II alternate choosing

$$ _i \in \{0, 1\}$$

so that Player I chooses $_i$ for i odd, and Player II chooses $_i$ for i even. Player I begins by choosing $_1$, Player II then picks $_2$ and so forth.

Player I wins if

$$\sum_{i=1}^{\infty} {}_i 2^{-i} \in A$$

otherwise Player II wins (so this depends only on $A \cap [0, 1]$).

Let Seq be the set of all finite binary sequences. Strategies are simply functions

$$\tau \colon \mathrm{Seq} \to \{0,1\}.$$

A strategy, τ, is a *winning strategy* for Player I if following the strategy guarantees (no matter how Player II plays) victory for Player I. Similarly for Player II. The game G_A is *determined* if there is a winning strategy for one of the players. Obviously with the definition here, G_A is determined if and only if G_B is determined where $B = A \cap [0,1]$.

Assuming the *Axiom of Choice* it is not difficult to prove that there exists a set A such that G_A is *not* determined. But it is certainly not at all obvious that one can produce such a set A with the additional property that A be a projective set. Therefore (essentially: [21]) the following variation of the axiom that all sets are determined is not obviously false.

DEFINITION 10. *Projective Determinacy*: G_A is determined whenever A is a projective set. □

The axiom of *Projective Determinacy* solves the measure problem, [22].

THEOREM 11 (Mycielski–Swierczkowski, 1964). *Assume* Projective Determinacy. *Then every projective set is Lebesgue measurable.* □

So:

COROLLARY 12. *Assume* Projective Determinacy. *Then there is no paradoxical decomposition of the unit ball of* $\mathbb{R}^3$ *into projective pieces.* □

Morton Davis, [19], proved that, assuming *Projective Determinacy*, every uncountable projective set has cardinality $2^{\aleph_0}$. This established that, at least in one formulation, CH *holds* for the projective sets.

Surprisingly, the actual relationship between CH and the projective sets is far more subtle *even* assuming *Projective Determinacy*. The reason is simply that assuming *Projective Determinacy* there can be no projective wellordering of the reals and so restricting to the projective sets, otherwise equivalent versions of CH may (and do) fail to be equivalent.

One can also naturally analyze versions of the *Axiom of Choice* for the projective sets. As just noted there can be no projective wellordering of the reals assuming *Projective Determinacy*, nevertheless one version of the Axiom of Choice is not obviously excluded for the projective sets. This is the uniformization problem, [15].

Suppose that $A \subseteq \mathbb{R}^2$. A function, $f \colon \mathbb{R} \to \mathbb{R}$, *uniformizes* A if:

(1) $\mathrm{dom}(f) = \{x \in \mathbb{R} \mid A_x \neq \emptyset\}$;
(2) for all $x \in \mathrm{dom}(f)$, $(x, f(x)) \in A$;

where for each $x \in \mathbb{R}$, $A_x = \{y \in \mathbb{R} \mid (x,y) \in A\}$.

A partial function, $f \colon \mathbb{R} \to \mathbb{R}$, is *projective* if its graph is a projective subset of $\mathbb{R}^2$.

(Luzin, 1925) *Can every projective subset of the plane be uniformized by a projective function?*

An early and significant success in the study of the projective sets assuming determinacy was Moschovakis' solution of the uniformization problem, [19].

THEOREM 13 (Moschovakis, 1971). *Assume* Projective Determinacy. *Then every projective subset of the plane can be uniformized by a projective function.* □

Thus assuming *Projective Determinacy* one has a complete analysis of the *Axiom of Choice* for the projective sets. First, the *Axiom of Choice* fails projectively in that there is no projective wellordering of the reals. Second, the *Axiom of Choice* holds projectively in that if

$$F : \mathbb{R} \to V$$

is generalized projective then there is a choice function for F which is also generalized projective. Here a set M is generalized projective if there exist projective subsets of $\mathbb{R}^2$, E and $\sim$, and a projective set $U \subseteq \mathbb{R}$ such that

$$\langle N, M, \in\rangle \cong \langle \mathbb{R}, U, E\rangle / \sim$$

where N is the transitive closure of $\{M\}$. It is easily verified that a set $X \subseteq \mathbb{R}^n$ is generalized projective if and only if it is a projective set.

I also note for future reference that assuming *Projective Determinacy*, there are no essential uses of the *Axiom of Choice* in the analysis of the structure

$$\langle H(\omega_1), \in\rangle.$$

At this stage three questions come to mind.

(1) Is *Projective Determinacy* consistent?
(2) Does *Projective Determinacy* follow from combinatorial hypotheses?
(3) Can the consequences of *Projective Determinacy* be obtained directly from other axioms?

There is widespread agreement, now, that the axiom of *Projective Determinacy* is relatively consistent with the axioms of Set Theory. But this is a rather recent development and the axiom is certainly not a priori consistent.

Some of the consequences of *Projective Determinacy* for the projective sets *can* be directly obtained by combinatorial methods.

THEOREM 14 (Shelah–Woodin, 1984). *Assume there exist infinitely many Woodin cardinals. Then every projective set is Lebesgue measurable.* □

The potential mystery is resolved in the seminal theorem of Martin and Steel, [17].

THEOREM 15 (Martin–Steel, 1986). *Assume there exist infinitely many Woodin cardinals. Then every projective set is determined.* □

The situation was further clarified by:

THEOREM 16 (Woodin, 1987). *The following are equivalent:*

(1) Projective Determinacy
(2) *For each $k \in \mathbb{N}$, there exists a countable, transitive, model, $\mathcal{M}$, such that:*

$$\mathcal{M} \models \text{ZFC} + \text{“There exist } k \text{ Woodin cardinals.”}$$

and such that $\mathcal{M}$ is countably iterable. □

Here the notion that a transitive model, $\mathcal{M}$, is countably iterable is a fundamental notion from *Inner Model Theory*.

With the development of the *Core Model Theory* for Woodin cardinals it has become clear that *Projective Determinacy* is actually implied by a vast number of seemingly unrelated combinatorial propositions. The first results of this kind were obtained by Jensen through his *Covering Lemma* which enables one to prove the determinacy of all analytic sets from a wide variety of combinatorial statements. Current core model theory originates in seminal work of Dodd and Jensen. The extension of this theory to the realm of Woodin cardinals is due primarily to Steel, [28]. It is an accurate assessment that *Projective Determinacy* is ubiquitous in set theory.

A recent example involves axioms which attempt to solve the continuum problem by explicitly making the *Continuum Hypothesis* false. The axioms are *Forcing Axioms* which can be viewed as axioms which assert that certain generalizations of the *Baire Category Theorem* hold.

A complete Boolean algebra, $\mathbb{B}$, is *ccc* if every antichain in $\mathbb{B}$ is countable. The first forcing axiom proposed was *Martin's Axiom*:

> *Suppose that $\mathbb{B}$ is a complete Boolean algebra which is ccc. Let X be the Stone space of $\mathbb{B}$. Then X is not the union of $\aleph_1$ many meager subsets of X.*

One often cited intuition for such an axiom is that if CH is to fail then sets of cardinality ω_1 should behave, as far as possible, like countable sets.

Stronger forcing axioms are obtained by allowing a larger class of Boolean algebras. But, one *cannot* allow arbitrary Boolean algebras. Foreman, Magidor and Shelah identified the maximum possible class, [8]. The definition involves the notion of a closed unbounded subset of ω_1.

DEFINITION 17. A cofinal set $C \subseteq \omega_1$ is *closed and unbounded* if it is closed in the order topology of ω_1. □

We denote by $\mathcal{I}_{\text{NS}}$ the ideal

$$\mathcal{I}_{\text{NS}} \subset \mathcal{P}(\omega_1)$$

of sets $A \subset \omega_1$ whose complements contain a closed unbounded set.

DEFINITION 18 (Foreman–Magidor–Shelah, 1984). Suppose that $\mathbb{B}$ is a complete Boolean algebra. The Boolean algebra $\mathbb{B}$ is *stationary set preserving* if:

$$(\mathcal{I}_{\mathrm{NS}})^{V^{\mathbb{B}}} \cap V = (\mathcal{I}_{\mathrm{NS}})^{V}. \qquad \square$$

An equivalent definition is possible without using notions from the technique of forcing. For example $\mathbb{B}$ is stationary set preserving if the following holds. Suppose that $b \in \mathbb{B}$,

$$\langle b_\alpha : \alpha < \omega_1 \rangle$$

is a sequence of elements of $\mathbb{B}$, and $b \neq 0$.

Then there exists $0 < c \leq b$ such that either

(1) $c \wedge b_\alpha = 0$ for sufficiently large α, or
(2) for a closed unbounded set $C \subseteq \omega_1$, if $\gamma \in C$ then

$$c \wedge \Big(\bigwedge_{\alpha<\gamma} \Big(\bigvee_{\alpha<\eta<\gamma} b_\eta \Big) \Big) \neq 0.$$

THEOREM 19 (Foreman–Magidor–Shelah). *Suppose that $\mathbb{B}$ is a complete Boolean algebra and let X be the Stone space of $\mathbb{B}$. Suppose that for each open, nonempty, set,*

$$O \subseteq X,$$

the set O is not the union of $\aleph_1$ many meager subsets of X. Then $\mathbb{B}$ is stationary set preserving. □

Therefore the class of complete Boolean algebras which are stationary set preserving is the maximum possible class to which one might hope to generalize Martin's Axiom.

DEFINITION 20 (Foreman–Magidor–Shelah). *Martin's Maximum*: Suppose that $\mathbb{B}$ is a complete Boolean algebra and that $\mathbb{B}$ is stationary set preserving and let X be the Stone space of $\mathbb{B}$. Then X is not the union of $\aleph_1$ many meager subsets of X. □

For many applications a weaker version suffices.

DEFINITION 21 (Foreman–Magidor–Shelah). *Martin's Maximum*(c): Suppose that $\mathbb{B}$ is a complete Boolean algebra and that $\mathbb{B}$ is stationary set preserving. Suppose that $\mathbb{B} \setminus \{0\}$ has a dense set of cardinality at most $c = 2^{\aleph_0}$. Let X be the Stone space of $\mathbb{B}$. Then X is not the union of $\aleph_1$ many meager subsets of X.

One striking application is given in the following theorem which is closely related to results of Todorcevic.

THEOREM 22 (Foreman–Magidor–Shelah). *Assume* Martin's Maximum(c). *Then*

$$2^{\aleph_0} = \aleph_2. \qquad \square$$

So, curiously, requiring sets of cardinality ω_1 behave, as far as possible, like countable sets necessitates that the continuum be ω_2. Actually there are now a number of different proofs of this theorem exploiting various aspects of *Martin's Maximum*, or in some cases of weaker forcing axioms, but a common thread is *reflection* at ω_2. In fact there is an abstract theorem which shows that all forcing axioms are instances of asserting *generic supercompactness* for ω_2.

One application of *Core Model Theory* is the following theorem.

THEOREM 23. *Assume* Martin's Maximum(c). *Then* Projective Determinacy. $\square$

There is no known direct proof of this theorem. The method of proof is to use core model methods to show that for each k there exists a countable, iterable, model, $\mathcal{M}$, such that

$$\mathcal{M} \vDash \text{ZFC} + \text{“There exist } k \text{ Woodin cardinals”}.$$

Projective Determinacy follows by the theorem previously cited.

This is not an isolated occurrence. The methodology has been used to show that a wide variety of combinatorial statements actually imply *Projective Determinacy*. The natural occurrence of statements which imply the number theoretic statement that

$$\text{ZFC} + \text{“Projective Determinacy”}$$

is *consistent* but which do not imply *Projective Determinacy* is quite rare if at all.

Rephrased the only known examples of unsolvable problems about the projective sets, in the context of *Projective Determinacy*, are analogous to the known examples of unsolvable problems in *Number Theory*: these are Gödel sentences, and consistency statements. Thus I make the claim:

Projective Determinacy *is the* ***correct*** *axiom for the projective sets*;

so the ZFC axioms are *obviously* incomplete and moreover incomplete in a fundamental way. I would in fact go further and claim the following similarity of structures and axioms:[2]

$$\frac{\textit{Peano Axioms}}{\langle \mathbb{N}, +, \cdot \rangle} \sim \frac{\textit{Projective Determinacy} + \textit{Peano Axioms}}{\langle \mathcal{P}(\mathbb{N}), \mathbb{N}, +, \cdot, \in \rangle}$$

[2]Here the precise formulation of *Projective Determinacy* incorporates Σ_0-*Comprehension*.

§4. The second step, $H(\omega_2)$. Is there an analog of *Projective Determinacy* for the structure:

$$\langle H(\omega_2), \in \rangle?$$

Because of Cohen's method of forcing, there are necessary limitations, subject to very modest restrictions, [4] and [14].

(Levy–Solovay, 1964; Cohen, 1965) No large cardinal hypothesis can settle the Continuum Hypothesis.

Thus the resolution of the theory of the structure,

$$\langle H(\omega_2), \in \rangle,$$

if possible at all, necessitates a far deeper understanding of sets.

One example of the subtle aspects of this structure is given in the following theorem which is really the mathematical starting point for the "solution" to CH that I shall describe, [31].

THEOREM 24 (Woodin, 1991). *Suppose that* Martin's Maximum *holds. Then there exists a surjection,* $\rho \colon \mathbb{R} \to \omega_2$, *such that*

$$\{(x, y) \mid \rho(x) < \rho(y)\}$$

is a projective set. □

Thus assuming *Martin's Maximum* the following two claims hold. The first claim is that CH *holds projectively* in that if

$$X \subseteq \mathbb{R}$$

is an uncountable projective set, then $|X| = |\mathbb{R}|$. This is because *Projective Determinacy* must hold. The second claim is that CH *fails projectively* in that there exists a surjection, $\rho \colon \mathbb{R} \to \omega_2$, such that

$$\{(x, y) \mid \rho(x) < \rho(y)\}$$

is a projective set.

But there is a curious *asymmetry* which follows from the proofs of these claims. Assume there exist infinitely many Woodin cardinals. Then while there can be *no* projective "proof" of CH, there *can* be a projective "proof" of ¬CH. Martin, 25 years ago, implicitly cited the possibility that such an asymmetry holds, as weak evidence that CH may be false, [16]. The solution to CH that I shall be discussing is in essence based on a variation of this asymmetry.

Finally this phenomenon is quite possibly unique to CH. For example one can probably always force to achieve GCH is such a way that at every (uncountable) cardinal δ which is either of uncountable cofinality, or "small" and of cofinality ω, there is a surjection;

$$\pi \colon \delta^+ \to \mathcal{P}(\delta)$$

which is definable in the structure

$$\langle H(\delta^+), \in\rangle$$

from parameters.

On the other hand, if δ is a singular strong limit cardinal of uncountable cofinality then from *pcf* theory [25], there is a wellfounded relation

$$R \subset H(\delta^+)$$

which has rank α where $\alpha \geq 2^\delta$ and which is definable in the structure, $\langle H(\delta^+), \in\rangle$.

However the following theorems show that some of the features of $H(\omega_1)$ can reappear at $H(\lambda^+)$ for suitably "large" cardinals, λ, of cofinality ω.

The first theorem is a "perfect set" theorem. More precisely it is the version for λ of the theorem that assuming *Projective Determinacy*, uncountable projective sets contain uncountable closed subsets. The hypothesis of the theorem is a likely candidate for an axiom at λ which plays the role that the axiom of *Projective Determinacy* does at ω. But there is no claim that this hypothesis has anything to do with the determinacy of games.

THEOREM 25. *Suppose that there exists an elementary embedding*

$$j\colon L(V_{\lambda+1}) \to L(V_{\lambda+1})$$

with critical point below λ. Suppose that $X \subseteq \mathcal{P}(\lambda)$ is definable from parameters in the structure, $\langle H(\lambda^+), \in\rangle$. Then either

(1) $|X| \leq \lambda$ *or*

(2) $|X| = 2^\lambda$ *and X contains a "perfect set".* □

The second theorem is a version of the following fact. Suppose that CH holds and that $V[G]$ is a generic extension of V for adding δ many Cohen reals.

(1) Assume *Projective Determinacy* holds in V.
Then $(H(\omega_1))^V \prec (H(\omega_1))^{V[G]}$.

(2) Suppose that

$$\rho\colon \mathbb{R}^{V[G]} \to \alpha$$

is a surjection in $V[G]$ such that in $V[G]$ the set

$$\{(x, y) \mid \rho(x) < \rho(y)\}$$

is projective. Then $\alpha < \omega_2^V$.

THEOREM 26. *Suppose that there exists an elementary embedding*

$$j\colon L(V_{\lambda+1}) \to L(V_{\lambda+1})$$

with critical point below λ.

Then there exist a partial order $\mathbb{P}$ and a cardinal $\lambda^* < \lambda$ such that

$$V_{\lambda^*+1} \cong X \prec V_{\lambda+1}$$

and such that if $G \subseteq \mathbb{P}$ is V-generic then in $V[G]$:

(1) $(V_{\lambda^*+1})^V \prec (V_{\lambda^*+1})^{V[G]}$,
(2) $2^{\lambda^*} > (\lambda^*)^+$,
(3) *If R is a wellfounded relation definable from parameters in $\langle V_{\lambda^*+1}, \in\rangle^{V[G]}$, then*

$$\operatorname{rank}(R) < ((\lambda^*)^{++})^V. \qquad \square$$

§5. **Strong Logics.** Thus we are led to the following problem:

Even if there is an analog of *Projective Determinacy* for $H(\omega_2)$, how can we find it or even recognize it if we do find it?

Our solution is to take an abstract approach:

Can the theory of the structure

$$\langle H(\omega_2), \in\rangle$$

be finitely axiomatized (over ZFC) in a (reasonable) logic which extends first order logic?

Logics which extend first order logic are *strong* logics. We shall seek candidates for strong logics for which any possible axiomatization of the theory of the structure,

$$\langle H(\omega_2), \in\rangle$$

is immune to the difficulties posed by Cohen's method of forcing. For this we shall impose a requirement of *Generic Soundness* on the strong logic.

A strong logic, $\vdash_0$, is defined by:

(1) Specifying a collection of *test* structures, these are structures of the form

$$\mathcal{M} = (M, E)$$

where $E \subseteq M \times M$;

(2) Defining

$$\text{ZFC} \vdash_0 \phi$$

if for every test structure, $\mathcal{M}$, if

$$\mathcal{M} \models \text{ZFC}$$

then $\mathcal{M} \models \phi$.

A theory T is "$\vdash_0$-*consistent*" if there exists test structure, $\mathcal{M}$, with the property that $\mathcal{M} \models T$. Of course we shall only be interested in the case that the theory ZFC itself is "$\vdash_0$-*consistent*"; i.e., in the case that there exists a test structure, $\mathcal{M}$, such that $\mathcal{M} \models \text{ZFC}$.

The *smaller* the collection of test structures, the *stronger* the logic. Classical logic is the weakest logic and for example, β-logic is obtained by simply restricting to *transitive sets*,

$$\mathcal{M} = (M, \in).$$

The *Generic Soundness* requirement for a strong logic, $\vdash_0$ is formally defined below.

DEFINITION 27 (*Generic Soundness*). Suppose that $\mathbb{P}$ is a partial order, α is an ordinal and that

$$V_\alpha^{\mathbb{P}} \models \text{ZFC}.$$

Suppose that ZFC $\vdash_0 \phi$. Then $V_\alpha^{\mathbb{P}} \models \phi$. □

Our context for considering strong logics will require at the very least that there exists a proper class of Woodin cardinals, and so the requirement of *Generic Soundness* is nontrivial.

For a given strong logic, $\vdash_0$, the theory of

$$\langle H(\omega_2), \in \rangle$$

is "*finitely axiomatized over* ZFC" if there exists a sentence Ψ such that:

(1) For some α, $V_\alpha \models \text{ZFC} + \Psi$;
(2) For each sentence, ϕ, $\langle H(\omega_2), \in \rangle \models \phi$ if and only if

$$\text{ZFC} + \Psi \vdash_0 \text{“}H(\omega_2) \models \phi\text{”}.$$

For the problem of when the theory of the structure, $\langle H(\omega_2), \in \rangle$, *can* be finitely axiomatized over ZFC, the *weaker* the logic, the more interesting the problem; for the problem of when the theory of the structure, $\langle H(\omega_2), \in \rangle$, *cannot* be finitely axiomatized over ZFC, the *stronger* the logic, the more interesting the problem. For each of these problems we shall see that if there exists a proper class of Woodin cardinals then one can without loss of generality reduce to the case that the strong logic considered is both *definable* and *generically invariant*.

We shall define a specific strong logic, "Ω-logic", and analyze the following variation of the abstract problem: Can there exist a sentence Ψ such that for all ϕ either

(1) $\text{ZFC} + \Psi \vdash_\Omega \text{“}H(\omega_2) \models \phi\text{”}$, or
(2) $\text{ZFC} + \Psi \vdash_\Omega \text{“}H(\omega_2) \models \neg\phi\text{”}$;

and such that

$$\text{ZFC} + \Psi$$

is Ω-consistent?

The definition of Ω-logic involves the notion that a set $A \subseteq \mathbb{R}^n$ be *universally Baire*.

§6. Universally Baire sets. There is a *transfinite* hierarchy which extends the hierarchy of the projective sets, this is the hierarchy of the *universally Baire sets* [7].

DEFINITION 28 (Feng–Magidor–Woodin). A set $A \subseteq \mathbb{R}^n$ is *universally Baire* if for any continuous function, $F : \Omega \to \mathbb{R}^n$, where Ω is a compact Hausdorff space, the preimage of A,

$$\{p \in X \mid F(p) \in A\},$$

has the property of Baire in Ω; i.e., is open in Ω modulo a meager set. □

Every borel set $A \subseteq \mathbb{R}^n$ is universally Baire, moreover the universally Baire sets form a σ-algebra closed under preimages by borel functions

$$f : \mathbb{R}^n \to \mathbb{R}^m.$$

The universally Baire sets are Lebesgue measurable etc.

The fact that the universally Baire sets are determined follows from large cardinal assumptions. The following theorem of [7] is a corollary of the Martin–Steel Theorem. Neeman has improved the theorem reducing the hypothesis to just the assumption that there is one Woodin cardinal, [23].

THEOREM 29. *Suppose that there are two Woodin cardinals. Then every universally Baire set is determined.* □

If there exists a proper class of Woodin cardinals then the universally Baire sets are closed under continuous images. Therefore:

THEOREM 30. *Suppose that there are arbitrarily large Woodin cardinals. Then every projective set is universally Baire.* □

For purely expository reasons I shall follow the presentation of [33] in giving a brief summary of the Wadge hierarchy. This involves restricting attention to subsets of the Cantor set, so let $\mathbb{K} \subseteq [0, 1]$ be the Cantor set. Suppose that $A \subseteq \mathbb{K}$ and that $B \subseteq \mathbb{K}$. The set A is *reducible* to B if there exists a continuous function, $f : \mathbb{K} \to \mathbb{K}$, such that $A = f^{-1}(B)$. The set A is *strongly reducible* to B if the function f can be chosen such that for all $x, y \in \mathbb{K}$, $|f(x) - f(y)| \leq (1/2)|x - y|$.

There is a remarkably useful lemma of Wadge, [29]. The version of Wadge's lemma for projective subsets of $\mathbb{K}$ is:

LEMMA 31 (Wadge). *Suppose that the axiom* Projective Determinacy *holds and that A_0 and A_1 are projective subsets of $\mathbb{K}$. Then either A_0 is reducible to A_1 or A_1 is strongly reducible to $\mathbb{K} \backslash A_0$.* □

The proof of Wadge's lemma simply requires the determinacy of a set $B \subseteq [0, 1]$ which is the preimage of

$$(A_0 \times (\mathbb{K} \backslash A_1)) \cup ((\mathbb{K} \backslash A_0) \times A_1)$$

by a Borel function $F : [0, 1] \to \mathbb{R}^2$. Such sets, B, are necessarily universally Baire if both of the sets A_0 and A_1 are universally Baire. Therefore

THEOREM 32. *Suppose that there is a proper class of Woodin cardinals. Suppose that A_0 and A_1 are universally Baire subsets of $\mathbb{K}$. Then either A_0 is reducible to A_1 or A_1 is strongly reducible to $\mathbb{K}\backslash A_0$.* □

No set $A \subseteq \mathbb{K}$ can be strongly reducible to its complement, $\mathbb{K}\backslash A$. Therefore given two universally Baire subsets of $\mathbb{K}$ exactly one of the following must hold.

(1) Both A_0 and $\mathbb{K}\backslash A_0$ are strongly reducible to A_1, and A_1 is not reducible to A_0 (or to $\mathbb{K}\backslash A_0$).
(2) Both A_1 and $\mathbb{K}\backslash A_1$ are strongly reducible to A_0, and A_0 is not reducible to A_1 (or to $\mathbb{K}\backslash A_1$).
(3) A_0 and A_1 are reducible to each other; or $\mathbb{K}\backslash A_0$ and A_1 are reducible to each other.

Thus one can define an equivalence relation on the universally Baire subsets of the Cantor set by $A_0 \sim_{\mathrm{w}} A_1$ if (3) holds, and one can totally order the induced equivalence classes by defining for universally Baire sets, A_0 and A_1, $A_0 <_{\mathrm{w}} A_1$ if (1) holds.

Of course implicit in (1) is a partial order defined on all subsets of $\mathbb{K}$: $A < B$ if both A and $\mathbb{K}\backslash A$ are strongly reducible to B. In the context of determinacy assumptions, Martin proved that this partial order is wellfounded, [19]. In the absence of any determinacy assumptions, Martin's theorem can be formulated as follows.

THEOREM 33 (Martin). *Suppose that $\langle A_k : k \in \mathbb{N}\rangle$ is a sequence of subsets of $\mathbb{K}$ such that for all $k \in \mathbb{N}$, both A_{k+1} and $\mathbb{K}\backslash A_{k+1}$ are strongly reducible to A_k.*

Then there exists a continuous function, $g : \mathbb{K} \to \mathbb{K}$, such that $g^{-1}(A_1)$ does not have the property of Baire. □

An immediate corollary is that the partial order $<_{\mathrm{w}}$ is wellfounded (without any large cardinal assumptions). This is simply because the continuous preimages of a universally Baire set must have the property of Baire.

To summarize, assuming there is a proper class of Woodin cardinals, the universally Baire sets form a *wellordered* hierarchy under a suitable notion of complexity; and the projective sets define an initial segment. It follows from a theorem of Wadge that the sets $A \subseteq \mathbb{K}$ such that both A and $\mathbb{K}\backslash A$ are F_σ sets (countable unions of closed sets) define the initial segment of length ω_1, so the hierarchy is quite finely calibrated even when compared to the usual stratification of the Borel sets as Σ^0_α sets.

I shall argue later that this hierarchy is an abstract manifestation of the hierarchy of those *large cardinal axioms* which are suitably realized in V (and which admit a "weak inner model theory").

The structure of the universally Baire sets is a transfinite extension of $H(\omega_1)$; i.e., the universally Baire sets are transfinite extension of the projective sets. The only known examples of existential statements about the universally Baire sets which are independent correspond to generalizations of Gödel sentences, and consistency statements.

There is a natural generalization of classical first order logic which is defined from the universally Baire sets. This is Ω-logic; "proofs" in Ω-logic are witnessed by universally Baire sets which can without loss of generality be assumed to be subsets of $\mathbb{K}$. The rank of the witness in the Wadge order, $<_w$, yields a reasonable notion of the length of an Ω-proof. The universally Baire sets have a very complicated definition (from a set theoretical perspective) so the limitations of Ω-logic are not obvious: perhaps axioms can be found which settle, in Ω-logic, all questions about $H(\omega_2)$ and more.

The definition of Ω-logic involves the notion that a transitive set M is A-closed where A is universally Baire.

§7. *A*-closed sets. Suppose that $A \subseteq \mathbb{R}$ is universally Baire and suppose that $V[G]$ is a set generic extension of V. Then the set A has canonical interpretation as a set

$$A_G \subseteq \mathbb{R}^{V[G]}.$$

Suppose that $\lambda \in \mathrm{Ord}$, $\pi\colon \lambda^\omega \to \mathbb{R}$, and π satisfies the condition that for all $f, g \in \lambda^\omega$, if $f \neq g$ then

$$|\pi(f) - \pi(g)| < 1/(n+1)$$

where n is least such that $f(n) \neq g(n)$. If $\pi[\lambda^\omega] = A$ then π represents A as the projection of a tree on $\omega \times \lambda$. The function π naturally defines a function

$$\pi_G\colon (\lambda^\omega)^{V[G]} \to \mathbb{R}^{V[G]}.$$

If $A \neq \emptyset$ then the set A_G can be defined as follows,

$$A_G = \cup\left\{\pi_G\left[\lambda^\omega\right]^{V[G]} \mid \pi \in V, \lambda \in \mathrm{Ord}, \pi\left[\lambda^\omega\right] = A\right\}.$$

If there exists a proper class of Woodin cardinals then

$$\langle H(\omega_1), A, \in\rangle \prec \langle H(\omega_1)^{V[G]}, A_G, \in\rangle.$$

DEFINITION 34. Suppose that $A \subseteq \mathbb{R}$ is universally Baire and that M is a transitive set such that $M \vDash \mathrm{ZFC}$.

Then M is *A-closed* if for each partial order $\mathbb{P} \in M$, if $G \subseteq \mathbb{P}$ is V-generic then in $V[G]$:

$$A_G \cap M[G] \in M[G]. \qquad \square$$

The definition that M is A-closed actually makes sense if M is simply an ω-model.

LEMMA 35. *Suppose that (M, E) is an ω-model with*

$$(M, E) \vDash \mathrm{ZFC}.$$

Then the following are equivalent.

(1) *(M, E) is wellfounded.*
(2) *(M, E) is A-closed for each Π^1_1 set.* □

Therefore A-closure is a natural generalization of wellfoundedness.

§8. Ω-logic. Though the formal definition given below of Ω-logic is in the context that there exists a proper class of Woodin cardinals, it is possible to define Ω-logic just in the context of ZFC. But the definition would then be somewhat more technical because of the need to restrict the collection of universally Baire sets.

DEFINITION 36. Suppose that there exists a proper class of Woodin cardinals and ϕ is a sentence. Then

$$\text{ZFC} \vdash_\Omega \phi$$

if there exists a universally Baire set $A \subseteq \mathbb{R}$ such that $M \models \phi$ for any countable transitive set M such that $M \models \text{ZFC}$ and such that M is A-closed. □

Suppose that $A \subseteq \mathbb{R}$ is universally Baire. Then there exists a set $A^* \subseteq \mathbb{K}$ such that A^* is universally Baire and such that for all transitive sets M, if $M \models \text{ZFC}$ then M is A-closed if and only if M is A^*-closed. Thus in the definition of Ω logic, one can restrict to universally Baire subsets of $\mathbb{K}$ and then one can quite naturally define the length of an Ω proof as the rank of the witness in the Wadge ordering, $<_w$.

Let Γ^∞ be the set of all sets $A \subseteq \mathbb{R}$ such that A is universally Baire. For each set $\Gamma \subseteq \Gamma^\infty$, let

$$\text{“}\Omega_\Gamma\text{-logic”}$$

be the logic obtained by requiring that the witness $A \in \Gamma^\infty$ in the definition of Ω-logic be a finite subset $X \subseteq \Gamma$ and requiring that the countable transitive sets, M, be A-closed for each $A \in X$.

The indexed family of logics,

$$\langle \Omega_\Gamma\text{-logic} : \Gamma \subseteq \Gamma^\infty \rangle,$$

is a directed hierarchy of logics, with natural limit, Ω-logic. $\Omega_{\{\emptyset\}}$-logic is β-logic, but Ω_Γ-logic is strictly stronger where for example Γ is simply the set of all finite subsets of $\mathbb{R}$.

THEOREM 37 (Generic Invariance). *Suppose that there exists a proper class of Woodin cardinals and that ϕ is a sentence. Then for each partial order $\mathbb{P}$,*

$$(\text{ZFC} \vdash_\Omega \phi)^V$$

if and only if

$$(\text{ZFC} \vdash_\Omega \phi)^{V^{\mathbb{P}}}.$$ □

THEOREM 38 (Generic Soundness). *Suppose that there exists a proper class of Woodin cardinals. Suppose that $V_\alpha^{\mathbb{P}} \models \mathrm{ZFC}$ and that $\mathrm{ZFC} \vdash_\Omega \phi$. Then*

$$V_\alpha^{\mathbb{P}} \models \phi. \qquad \square$$

By the fundamental uniformization (scale) results of Moschovakis,

THEOREM 39. *Suppose that there exists a proper class of Woodin cardinals. Then for each sentence ϕ, $\mathrm{ZFC} \vdash_\Omega$ "$H(\omega_1) \models \phi$", if and only if $H(\omega_1) \models \phi$.* □

This theorem which essentially asserts projective absoluteness in Ω-logic is a priori a stronger theorem than the well known theorems on generic projective absoluteness. So something like the scale theorems are needed.

There are two immediate corollaries.

THEOREM 40. *Suppose that there exists a proper class of Woodin cardinals. Then for each sentence ϕ, either*

(1) $\mathrm{ZFC} \vdash_\Omega$ "$H(\omega_1) \models \phi$", *or*
(2) $\mathrm{ZFC} \vdash_\Omega$ "$H(\omega_1) \models \neg\phi$". □

THEOREM 41. *Suppose that there exists a proper class of Woodin cardinals. Then*

$$\mathrm{ZFC} \vdash_\Omega \text{Projective Determinacy} \qquad \square$$

The latter theorem is a special case of a far more general theorem. The essence of this more general theorem is that the Π_2 validities (of ZFC) in Ω-logic correspond *exactly* with the Π_2 consequences of those *large cardinal axioms* which are suitably realized in V (and which admit a "weak inner model theory"). This claim will be made precise later.

The question of whether there can exist analogs of determinacy for the structure

$$\langle H(\omega_2), \in \rangle$$

can now be given a precise formulation.

Can there exist a sentence Ψ such that for all ϕ either

- $\mathrm{ZFC} + \Psi \vdash_\Omega$ "$H(\omega_2) \models \phi$", *or*
- $\mathrm{ZFC} + \Psi \vdash_\Omega$ "$H(\omega_2) \models \neg\phi$";

and such that

$$\mathrm{ZFC} + \Psi$$

is Ω-consistent?

Why seek such such sentences? The reason is simply this. Given such a sentence Ψ, for each sentence, ϕ, the independence (in *classical* logic) of the proposition

$$\text{"}\langle H(\omega_2), \in \rangle \models \phi\text{"}$$

from Ψ must yield, in essence, a proposition about Ω-logic which is independent; i.e., it must yield a "bounded" statement about universally Baire sets which is independent, which *cannot be accomplished by forcing*.

Thus by adopting axioms which "settle" the theory of (for example) the structure, $\langle H(\omega_2), \in\rangle$, in Ω-logic, one recovers for the theory of this structure the *empirical* completeness currently enjoyed by the theory of the structure $\langle H(\omega), \in\rangle$ which is of course Number Theory. More speculatively, by adopting such axioms one might allow for the development of a truly rich theory for this structure, unencumbered by independence, Perhaps independence really has been a major obstacle. For evidence of this claim simply compare the theory of the projective sets as developed under the assumption of *Projective Determinacy* with collection of independence results about the projective sets over the base theory ZFC.

DEFINITION 42. (1) A cofinal set $C \subseteq \omega_1$ is *closed and unbounded* if it is closed in the order topology of ω_1.
(2) $\mathcal{I}_{\text{NS}}$ is the σ-ideal of all sets $A \subseteq \omega_1$ such that $\omega_1 \backslash A$ contains a closed unbounded set.
(3) A set $S \subseteq \omega_1$ is *stationary* if for each closed, unbounded, set $C \subseteq \omega_1$,

$$S \cap C \neq \emptyset.$$

(4) A set $S \subseteq \omega_1$ is *co-stationary* if the complement of S is stationary. □

It is the appearance of $\mathcal{I}_{\text{NS}}$, a *definable* countably additive uniform ideal, which is the key new feature, distinguishing $H(\omega_2)$ from $H(\omega_1)$. The countable additivity and the non-maximality of the ideal $\mathcal{I}_{\text{NS}}$ are consequences of the *Axiom of Choice*. The ideal $\mathcal{I}_{\text{NS}}$ plays an essential and fundamental role in the formulation of *Martin's Maximum*. Arguably, the stationary, co-stationary, subsets of ω_1 constitute the simplest true manifestation of the *Axiom of Choice*. Recall that assuming *Projective Determinacy*, there is really no manifestation within $H(\omega_1)$ of the *Axiom of Choice*.

DEFINITION 43. Axiom $(*)_\text{o}$:
(1) There is a proper class of Woodin cardinals.
(2) For each projective set, $X \subseteq \mathbb{R}$ and for each Π_2 sentence, ϕ, in the language for the structure $\langle H(\omega_2), \mathcal{I}_{\text{NS}}, X, \in\rangle$; If

$$\text{ZFC} + \text{“}\langle H(\omega_2), \mathcal{I}_{\text{NS}}, X, \in\rangle \vDash \phi\text{”}$$

is Ω-consistent, then

$$\langle H(\omega_2), \mathcal{I}_{\text{NS}}, X, \in\rangle \vDash \phi.$$ □

THEOREM 44. *Suppose that there exists a proper class of Woodin cardinals. Then every set*

$$X \in \mathcal{P}(\mathbb{R}) \cap L(\mathbb{R})$$

is universally Baire. □

Therefore it is natural to consider the expanded structures:

$$\langle H(\omega_2), \mathcal{I}_{\text{NS}}, X, \in\rangle;$$

where $X \subseteq \mathbb{R}$ and $X \in L(\mathbb{R})$; though for our purposes, one could restrict to only those sets X which are projective; i.e., to consideration of the axiom $(*)_{\text{o}}$.

Definition 45. Axiom $(*)$:

(1) There is a proper class of Woodin cardinals.
(2) For each set $X \subseteq \mathbb{R}$, with $X \in L(\mathbb{R})$, and for each Π_2 sentence in the language for the structure $\langle H(\omega_2), \mathcal{I}_{\text{NS}}, X, \in\rangle$; If

$$\text{ZFC} + \text{“}\langle H(\omega_2), \mathcal{I}_{\text{NS}}, X, \in\rangle \models \phi\text{”}$$

is Ω-consistent, then

$$\langle H(\omega_2), \mathcal{I}_{\text{NS}}, X, \in\rangle \models \phi. \qquad \square$$

Clearly $(*)$ implies $(*)_{\text{o}}$ and the latter axiom is a maximality principle for the structure

$$\langle H(\omega_2), \mathcal{I}_{\text{NS}}, \in\rangle$$

somewhat analogous to asserting algebraic closure for a field.

Theorem 46. *Suppose that there exists a proper class of Woodin cardinals and that there is an inaccessible cardinal which is a limit of Woodin cardinals. Then the theory*

$$\text{ZFC} + (*)$$

is Ω-consistent. $\square$

The axiom $(*)$ actually settles the theory of $L(H(\omega_2))$.

Theorem 47. *Suppose that there exists a proper class of Woodin cardinals. Then for each sentence ϕ, either*

(1) $\text{ZFC} + (*) \vdash_\Omega$ *“$L(H(\omega_2)) \models \phi$”, or*
(2) $\text{ZFC} + (*) \vdash_\Omega$ *“$L(H(\omega_2)) \models \neg\phi$”.* $\square$

The version of this theorem for the axiom $(*)_{\text{o}}$ is simply:

Theorem 48. *Suppose that there exists a proper class of Woodin cardinals. Then for each sentence ϕ, either*

(1) $\text{ZFC} + (*)_{\text{o}} \vdash_\Omega$ *“$H(\omega_2) \models \phi$”, or*
(2) $\text{ZFC} + (*)_{\text{o}} \vdash_\Omega$ *“$H(\omega_2) \models \neg\phi$”.* $\square$

Therefore the axioms $(*)_{\text{o}}$ and $(*)$ are natural examples of axioms which settle, in Ω-logic, the theory of

$$\langle H(\omega_2), \in\rangle,$$

but of course the axiom $(*)$ does more. The reason to even define $(*)_{\text{o}}$ is that a finer version of this theorem is true:

THEOREM 49. *Suppose there exists a proper class of Woodin cardinals and that ϕ is a Σ_k sentence. Then*

(1) $\mathrm{ZFC} + (*)_{\circ} \vdash_{\Omega_{\Sigma^1_{k+3}}}$ "$H(\omega_2) \models \phi$", *or*

(2) $\mathrm{ZFC} + (*)_{\circ} \vdash_{\Omega_{\Sigma^1_{k+3}}}$ "$H(\omega_2) \models \neg\phi$". □

Further:

THEOREM 50. *Suppose there exists a proper class of Woodin cardinals. The following are equivalent*:

(1) Axiom $(*)_{\circ}$.

(2) *For each $n < \omega$, for each Σ^1_n set,*[3] *$X \subseteq \mathbb{R}$, for each Π_2 sentence, ϕ; If*

$$\mathrm{ZFC} + \text{“}\langle H(\omega_2), \mathcal{I}_{\mathrm{NS}}, X, \in\rangle \models \phi\text{”}$$

is $\Omega_{\Sigma^1_{n+5}}$-consistent, then $\langle H(\omega_2), \mathcal{I}_{\mathrm{NS}}, X, \in\rangle \models \phi$. □

Thus the axiom $(*)_{\circ}$ settles the theory of the structure, $\langle H(\omega_2), \in\rangle$, in the weakest possible logic, this is Ω_Γ-logic, where

$$\Gamma = \left\{A \subseteq \mathbb{R} \mid A \text{ is } \Sigma^1_n \text{ for some n}\right\},^4$$

and further there is a recursive set of axioms, T, such that assuming there is a proper class of Woodin cardinals, $(*)_{\circ}$ holds if and only if,

$$\langle H(\omega_2), \in\rangle \models T.$$

This reformulation which requires only *Projective Determinacy* shows that the relevant aspect of $(*)_{\circ}$ is first order over $\langle H(\omega_2), \in\rangle$.

The axiom $(*)$ offers the opportunity to develop the theory of $\langle H(\omega_2), \in\rangle$ in an environment where independence is unlikely to be a factor. Will something interesting emerge?

§9. The axiom $(*)$ and $2^{\aleph_0}$. It is not difficult to show that if for some Π_2 sentence ϕ, the assertion

$$\text{“}\langle H(\omega_2), \in\rangle \models \phi\text{”}$$

settles the size of $2^{\aleph_0}$, (and is Ω-consistent), then the implied value *must* be $\aleph_2$. In other words if the axiom $(*)$ settles the size of the continuum it must imply that $2^{\aleph_0} = \aleph_2$.

In fact there is a Π_2 sentence, ψ_{AC}, which if true in the structure, $\langle H(\omega_2), \in\rangle$, implies that $2^{\aleph_0} = \aleph_2$. The assertion

$$\text{“}\langle H(\omega_2), \in\rangle \models \psi_{\mathrm{AC}}\text{”}$$

is Ω consistent and so the axiom $(*)$ implies that $2^{\aleph_0} = \aleph_2$.

[3] lightface!

[4] again, lightface.

Assuming *Martin's Maximum*, one can also show

$$\langle H(\omega_2), \in\rangle \models \psi_{\mathrm{AC}}.$$

This offers a different view of why *Martin's Maximum* implies that $c = \aleph_2$.

ψ_{AC} asserts:

Suppose S, T are stationary, co-stationary, subsets of ω_1. Then there exist:

(1) *a closed unbounded set $C \subseteq \omega_1$,*
(2) *a wellordering $\langle L, <\rangle$ of cardinality ω_1,*
(3) *a bijection $\pi\colon \omega_1 \to L$;*

such that for all $\alpha \in C$, $\alpha \in S$ if and only if

$$\text{ordertype}\left(\{\pi(\beta) \mid \beta < \alpha\}\right) \in T.$$

By standard methods this can be shown to be expressible in the required syntactical form, more precisely there is a Π_2 sentence ψ_{AC} such that

$$\langle H(\omega_2), \in\rangle \models \psi_{\mathrm{AC}}$$

if and only if the above statement holds.

THEOREM 51. *Assume ψ_{AC} holds. Then $2^{\aleph_0} = \aleph_2$.* □

There is a subtle aspect to this theorem. Suppose that CH holds and that

$$\langle x_\alpha\colon \alpha < \omega_1\rangle$$

is an enumeration of $\mathbb{R}$. Thus ψ_{AC} must fail. However: it is possible that there be *no* counterexample to ψ_{AC} which is *definable* from the given enumeration, $\langle x_\alpha\colon \alpha < \omega_1\rangle$.

§10. Connections with *Martin's Maximum*. An enhanced version of *Martin's Maximum*.

DEFINITION 52 (Foreman–Magidor–Shelah). *Martin's Maximum*$^{++}$: Suppose that $\mathbb{B}$ is a complete Boolean algebra and that $\mathbb{B}$ is stationary set preserving.

Suppose that $\langle D_\alpha\colon \alpha < \omega_1\rangle$ is a sequence of dense subsets of $\mathbb{B}\setminus\{0\}$ and that $\langle \tau_\alpha\colon \alpha < \omega_1\rangle$ is a sequence of terms in $V^{\mathbb{B}}$ for stationary subsets of ω_1.

Then there exists a filter $G \subset \mathbb{B}$ such that for all $\alpha < \omega_1$;

(1) $G \cap D_\alpha \neq \emptyset$,
(2) $\{\beta < \omega_1 \mid \text{For some } p \in G, p \Vdash \beta \in \tau_\alpha\}$ is stationary in ω_1. □

The next lemma is an easy consequence of the definitions.

LEMMA 53. *Assume* Martin's Maximum^{++}. *Suppose that $\mathbb{B}$ is a Boolean algebra, $\mathbb{B}$ is stationary set preserving and that $X \subseteq \mathbb{R}$ is universally Baire. Then*

$$\langle H(\omega_2), \mathcal{I}_{\mathrm{NS}}, X, \in\rangle \prec_{\Sigma_1} \langle H(\omega_2), \mathcal{I}_{\mathrm{NS}}, X, \in\rangle^{V^{\mathbb{B}}}. \qquad \square$$

This suggests the following weakening of *Martin's Maximum*$^{++}$ which generalizes a definition of Goldstern–Shelah, (as reformulated by Bagaria) [11] and [2].

DEFINITION 54. MM($*$): Suppose that $\mathbb{B}$ is a Boolean algebra, $\mathbb{B}$ is stationary set preserving, that $X \subseteq \mathbb{R}$ and $X \in L(\mathbb{R})$.

Then

$$\langle H(\omega_2), \mathcal{I}_{\mathrm{NS}}, X, \in \rangle \prec_{\Sigma_1} \langle H(\omega_2), \mathcal{I}_{\mathrm{NS}}, X, \in \rangle^{V^{\mathbb{B}}}. \qquad \square$$

A natural question is that of the relationship of the axioms ($*$) and MM($*$).

THEOREM 55. *Suppose that there exists a proper class of Woodin cardinals and there exists a strong cardinal.*

Then the following are equivalent.

(1) Axiom ($*$).
(2) $M \models$ MM($*$) *for **every** transitive inner model, M, such that $M \models$* ZFC,

$$H(\omega_2) \subset M$$

and such that the Woodin cardinals of M are unbounded in $M \cap$ Ord. $\square$

This theorem arguably explains the initial success of *Martin's Maximum* in resolving the combinatorics of subsets of ω_1.

§11. The general case for $H(\omega_2)$. There remains the basic problem: Is there an analog of the axiom ($*$) in the context of CH? For example can one complete the similarity:

$$\frac{? + \mathrm{CH}}{(*)} \sim \frac{\text{real closed} + \text{ordered}}{\text{algebraically closed}},$$

i.e., is there a "conditional" completion for the theory of

$$\langle H(\omega_2), \in \rangle$$

in the context that CH holds?

There are a number of generic absoluteness results that suggest some possibility of the existence of such a completion.

THEOREM 56. *Suppose that there exists a proper class of measurable Woodin cardinals.*

Suppose that CH *holds and that $\mathbb{B}$ is a complete Boolean algebra such that*

$$V^{\mathbb{B}} \models \mathrm{CH}.$$

Then Σ^2_1 sentences are absolute between V and $V^{\mathbb{B}}$. $\square$

Thus from suitable large cardinal assumptions: Σ^2_1-absoluteness holds modulo CH.

Further the following seems a quite plausible conjecture where $\diamond_G$ denotes the assertion:

$$\langle H(\omega_2), \in\rangle^V \equiv_{\Sigma_2} \langle H(\omega_2), \in\rangle^{V^{\mathrm{Coll}(\omega_1,\mathbb{R})}}.$$

The principle, $\diamond_G$, is a *generic* form of $\diamond$.

Suppose that there exists a proper class of supercompact cardinals.
Then for all Σ_2 sentences, ϕ, either
(1) $\mathrm{ZFC} + \diamond_G \vdash_\Omega$ "$H(\omega_2) \models \phi$", *or*
(2) $\mathrm{ZFC} + \diamond_G \vdash_\Omega$ "$H(\omega_2) \models \neg\phi$".

In other words the conjecture is that from suitable large cardinal assumptions, Σ_2^2-absoluteness holds modulo $\diamond_G$.

In summary, assuming suitable large cardinals axioms:

(1) Σ_1^2-absoluteness holds modulo CH;
(2) (conjecture) Σ_2^2-absoluteness holds modulo $\diamond_G$.

This certainly suggests the possibility of obtaining for each n, Σ_n^2-absoluteness modulo *some* combinatorial principle generalizing $\diamond$.

However we shall see that from the point of view of Ω-logic, Σ_2^2-absoluteness is the *strongest* absoluteness possible; again stationary, co-stationary, subsets of ω_1 play the pivotal role.

More precisely:

There cannot *exist a sentence Ψ such that for all Σ_2 sentences, ϕ, either*
(1) $\mathrm{ZFC} + \mathrm{CH} + \Psi \vdash_\Omega$ "$\langle H(\omega_2), \mathcal{I}_{\mathrm{NS}}, \in\rangle \models \phi$", *or*
(2) $\mathrm{ZFC} + \mathrm{CH} + \Psi \vdash_\Omega$ "$\langle H(\omega_2), \mathcal{I}_{\mathrm{NS}}, \in\rangle \models \neg\phi$";
and such that $\mathrm{ZFC} + \mathrm{CH} + \Psi$ *is Ω-consistent.*

In brief, $\Sigma_2^2(\mathcal{I}_{\mathrm{NS}})$-absoluteness cannot hold modulo *any* (Ω-consistent) axiom.

The details require yet more definitions. Recall that a set

$$X \subseteq \mathbb{N}$$

is a recursive set if it is Δ_1 definable in the structure $\langle H(\omega), \in\rangle$.

Definition 57. Suppose that there exists a proper class of Woodin cardinals. A set

$$X \subseteq \mathbb{N}$$

is Ω-*recursive* if there exists a universally Baire set, $A \subseteq \mathbb{R}$, such that X is Δ_1 definable in $L(A, \mathbb{R})$ from $\{\mathbb{R}\}$. □

The Ω-recursive sets form a transfinite generalization (in terms of complexity) of the recursive (i.e., Turing computable) sets. If the theory ZFC is consistent then a set $X \subseteq \mathbb{N}$ is recursive if and only if there exists a formula $\phi(x)$ such that

$$X = \{k \in \mathbb{N} \mid \mathrm{ZFC} \vdash \phi[k]\},$$

and such that

$$\mathbb{N}\backslash X = \{k \in \mathbb{N} \mid \mathrm{ZFC} \vdash (\neg\phi)[k]\}.$$

LEMMA 58. *Suppose that there exists a proper class of Woodin cardinals. A set*

$$X \subseteq \mathbb{N}$$

is Ω-recursive *if and only if there exists a formula* $\phi(x)$ *such that*:

(1) $X = \{k \in \mathbb{N} \mid \mathrm{ZFC} \vdash_\Omega \phi[k]\}$, *and*
(2) $\mathbb{N}\backslash X = \{k \in \mathbb{N} \mid \mathrm{ZFC} \vdash_\Omega (\neg\phi)[k]\}$. □

As an immediate corollary one obtains the following lemma.

LEMMA 59. *Suppose that* Ψ *is a sentence such that for all* ϕ *either*

(i) $\mathrm{ZFC} + \Psi \vdash_\Omega$ "$H(\omega_2) \models \phi$", *or*
(ii) $\mathrm{ZFC} + \Psi \vdash_\Omega$ "$H(\omega_2) \models \neg\phi$".

Let T *be the set of all sentences* ϕ *such that*

$$\mathrm{ZFC} + \Psi \vdash_\Omega \text{“}H(\omega_2) \models \phi\text{”}.$$

Then T *is* Ω*-recursive.* □

Thus assuming that the axiom $(*)$ holds, the theory of the structure, $\langle H(\omega_2), \in\rangle$, is Ω-recursive.

THEOREM 60. *Suppose that there exists a proper class of Woodin cardinals. Let* $T = \mathrm{Th}(H(\omega_2))$. *The following are equivalent.*

(1) T *is* Ω*-recursive.*
(2) *There exist a sentence* Ψ *and a cardinal* κ *such that*

$$V_\kappa \models \mathrm{ZFC} + \Psi$$

and such that for each sentence ϕ, *either*
(a) $\mathrm{ZFC} + \Psi \vdash_\Omega$ "$H(\omega_2) \models \phi$", *or*
(b) $\mathrm{ZFC} + \Psi \vdash_\Omega$ "$H(\omega_2) \models \neg\phi$". □

Thus the basic problem:

Under what circumstances can the theory of the structure

$$\langle H(\omega_2), \in\rangle$$

be finitely axiomatized, over ZFC, *in* Ω*-logic*?

naturally leads to the problem:

How complicated are the Ω*-recursive subsets of* $\mathbb{N}$?

The Ω-recursive such sets look potentially extremely complicated (because the definition involves universally Baire sets). The combinatorial analysis of the Ω-recursive sets involves combining elements of *Descriptive Set Theory* with elements of *Fine Structure Theory*. The elements from *Descriptive Set Theory* involve the axiom AD^+.

§12. AD$^+$. AD is the axiom that every set $A \subseteq \mathbb{R}$ is determined (following our definitions for the determinacy of a set of reals). A variation, AD$^+$, has emerged as the correct form of this axiom at least for the development of the associated structure theory. I very briefly review some of the basic definitions and theorems from the theory of AD$^+$. This is simply to provide some background for the subsequent discussion.

DEFINITION 61. Θ is the supremum of the ordinals α such that there exists a surjection

$$\pi : \mathbb{R} \to \alpha. \qquad \square$$

So if the reals can be wellordered, $\Theta = c^+$.

DEFINITION 62. Suppose $A \subseteq \mathbb{R}$. The set A is *$^\infty$-borel* if there exist a set $S \subset \mathrm{Ord}$, an ordinal α, and a formula $\phi(x_0, x_1)$ such that

$$A = \{y \in \mathbb{R} \mid L_\alpha[S, y] \vDash \phi[S, y]\}. \qquad \square$$

$A \subseteq \mathbb{R}$ is $^\infty$-borel if A has a *transfinite borel* code. A key feature is that the code be effective; i.e., that it be a set of ordinals. Assuming the *Axiom of Choice*, every set $A \subseteq \mathbb{R}$ is $^\infty$-borel.

A strong form of the requirement that a set $A \subseteq \mathbb{R}$ be $^\infty$-borel is that it be *Suslin*.

DEFINITION 63. Suppose $A \subseteq \mathbb{R}$ and $\lambda \in \mathrm{Ord}$. The set A is *λ-Suslin* if $A = \emptyset$ or if there exists a function, $\pi : \lambda^\omega \to \mathbb{R}$, such that:

(1) for all $f, g \in \lambda^\omega$, if $f \neq g$ then $|\pi(f) - \pi(g)| < 1/(n+1)$ where n is least such that $f(n) \neq g(n)$;
(2) $\pi[\lambda^\omega] = A$. $\square$

Notice that if A is λ-Suslin then A is naturally represented as the projection of a tree on $\omega \times \lambda$. This of course is the standard definition, but for subsets of the Euclidean space, $\mathbb{R}$, the definition given here is the more natural one. The set A is Suslin if it is λ-Suslin for some λ.

A fundamental notion is that of a Suslin cardinal first isolated by A. Kechris.

DEFINITION 64 (AD). An infinite cardinal κ is a *Suslin cardinal* if there exists a set $A \subseteq \mathbb{R}$ such that

(1) A is κ-Suslin,
(2) A is not δ-Suslin for all $\delta < \kappa$. $\square$

It is not difficult to show that assuming the *Axiom of Choice* every infinite cardinal $\kappa \leq c$ is a Suslin cardinal where $c = 2^{\aleph_0}$. But assuming AD the question of which cardinals are Suslin cardinals is far more subtle, indeed the study of the Suslin cardinals has been a major theme in the development of the structure theory of the projective sets and more generally in the structure theory of AD$^+$. Assuming AD, the first two uncountable Suslin cardinals are $\aleph_1$ and $\aleph_\omega$, these correspond with $\underset{\sim}{\Pi}^1_1$ sets and $\underset{\sim}{\Pi}^1_2$ sets respectively. In a technical tour de force Jackson [12] has shown (assuming AD) that the supremum of the Suslin cardinals given by the projective sets is $\aleph_{\varepsilon_0}$.

The definition of AD^+ involves the notion that a set be $^\infty$-borel together with the determinacy of certain ordinal games.

Definition 65 ($\text{ZF} + \text{DC}_\mathbb{R}$). AD^+:

(1) Suppose $A \subseteq \mathbb{R}$. Then A is $^\infty$-borel.
(2) Suppose $\lambda < \Theta$ and that $\pi : \lambda^\omega \to \mathbb{R}$ is a continuous function. Then for each $A \subseteq \mathbb{R}$ the set $\pi^{-1}[A]$ is determined. □

Taking $\lambda = 2$ and π to be the function:

$$\pi(f) = \sum_{i=1}^{\infty} f(i)2^{-(i+1)}$$

one easily obtains that AD^+ implies AD. In the cases that λ is uncountable, it is essential that the strategy in (2) be a strategy on λ; i.e., it must be a function

$$\tau : \lambda^{<\omega} \to \lambda.$$

I note that (assuming only ZF), there exists $X \subset \omega_1^\omega$ which is *not* determined.

The following lemma fails for the property of being Suslin, for example if AD holds in $L(\mathbb{R}^\#)$ then every set $A \in \mathcal{P}(\mathbb{R}) \cap L(\mathbb{R})$, is Suslin in $L(\mathbb{R}^\#)$, but this fails of course in $L(\mathbb{R})$. This is the reason that AD^+ is defined using the property that A is $^\infty$-borel.

Lemma 66 ($\text{ZF} + \text{DC}_\mathbb{R} + \text{AD}$). *Suppose that $A \subseteq \mathbb{R}$ is $^\infty$-borel and that*

$$L_\alpha(A, \mathbb{R}) \vDash \Sigma_1\text{-Replacement}.$$

Then $L_\alpha(A, \mathbb{R}) \vDash$ "A is $^\infty$-borel". □

The special case of ordinal determinacy that is the second clause in the definition of AD^+, follows from just AD if the specified set $A \subseteq \mathbb{R}$ is both Suslin and co-Suslin. This fact combined with the following lemma is the second half of the motivation of the definition of AD^+.

Lemma 67 (Moschovakis). ($\text{ZF} + \text{DC}_\mathbb{R} + \text{AD}$). *Suppose $A \subseteq \mathbb{R}$ and that*

$$\delta < (\Theta)^{L_\alpha(A,\mathbb{R})}.$$

Then $\mathcal{P}(\delta) \subset L_{\alpha+1}(A, \mathbb{R})$. □

As an immediate corollary of the previous two lemmas one obtains the theorem that AD^+ is absolute to inner models containing the reals.

Theorem 68 (AD^+). *Assume that M is a transitive inner model of* ZF *such that $\mathbb{R} \subset M$. Then $M \vDash \text{AD}^+$.* □

A fundamental open problem is the relationship of AD and AD^+. There are several partial results.

THEOREM 69 (Steel, Woodin) ($\mathrm{ZF} + \mathrm{DC}_{\mathbb{R}} + \mathrm{AD}$). *Let*

$$\gamma = \sup\{\kappa \mid \kappa \text{ is a Suslin cardinal}\}.$$

Then the set of Suslin cardinals is a closed subset of γ. □

Assuming AD^+ one can prove what seems a minor strengthening of this theorem.

THEOREM 70 (AD^+). *The set of Suslin cardinals is a closed subset of* Θ. □

However this improvement is exactly the potential difference between AD and AD^+. Note that the Suslin cardinals can fail to be closed in Θ only if they are bounded below Θ.

THEOREM 71. *Assume* $\mathrm{ZF} + \mathrm{DC}_{\mathbb{R}} + \mathrm{AD}$. *Then the following are equivalent.*

(1) AD^+.

(2) *The Suslin cardinals are closed in* Θ. □

An important aspect of the structure theory associated to AD^+ is the natural correspondence between transitive models, M, satisfying,

$$M \vDash \mathrm{ZF} + \mathrm{DC}_{\mathbb{R}} + \mathrm{AD}^+ + \text{“}V = L(\mathcal{P}(\mathbb{R}))\text{”}$$

and transitive models, N, satisfying

$$N \vDash \mathrm{ZFC} + \text{“There are infinitely many Woodin cardinals”}.$$

This correspondence is also quite useful in obtaining further partial results on the problem of whether AD implies AD^+.

The details of this correspondence are given in the next two theorems.

THEOREM 72 (Derived Model Theorem). *Suppose that* δ *is a limit of Woodin cardinals. Suppose that*

$$G \subset \mathrm{Coll}(\omega, <\delta)$$

is V*-generic and let*

$$\mathbb{R}_G = \cup\{\mathbb{R}^{V[G|\alpha]} \mid \alpha < \delta\}.$$

Let Γ_G *be the set of* $A \subseteq \mathbb{R}_G$ *such that*

(i) $A \in V(\mathbb{R}_G)$,

(ii) $L(\mathbb{R}_G, A) \vDash \mathrm{AD}^+$.

Then $L(\mathbb{R}_G, \Gamma_G) \vDash \mathrm{AD}^+$. □

Models which arise in this fashion are *derived models.*

THEOREM 73 (AD^+). *Assume that* $V = L(\mathcal{P}(\mathbb{R}))$. *There is a partial order* $\mathbb{P}$ *such that if* $G \subset \mathbb{P}$ *is* V*-generic then in* $V[G]$ *there exists an inner model*

$$N \subset V[G]$$

such that in $V[G]$:

(1) $N \vDash \mathrm{ZFC}$.

(2) ω_1^V *is a limit of Woodin cardinals in* N.

(3) *There exists an N-generic filter $g \subset \mathrm{Coll}(\omega, < \omega_1^V)$ such that $\mathbb{R}_g = \mathbb{R}_V$, and such that $\Gamma_g = (\mathcal{P}(\mathbb{R}))^V$.* □

So given a determinacy hypothesis: i.e., some strengthening of

$$\mathrm{ZF} + \text{“}V = L(\mathcal{P}(\mathbb{R}))\text{”} + \mathrm{AD}^+$$

one can find a large cardinal *companion*.

Corollary 74. *The following are equiconsistent*:

(1) ZF + AD,

(2) ZFC + *“There exist infinitely many Woodin cardinals”*. □

The following theorem shows that the consistency strength of

$$\mathrm{ZF} + \mathrm{DC} + \mathrm{AD} + \neg\mathrm{AD}^+$$

is fairly strong. The proof is an application of the basic machinery of derived models.

Theorem 75 ($\mathrm{ZF} + \mathrm{AD} + \mathrm{DC}_{\mathbb{R}}$). *Define*

$$\Gamma = \{A \subseteq \mathbb{R} \mid L(A, \mathbb{R}) \models \mathrm{AD}^+\}.$$

Then:

(1) $L(\Gamma, \mathbb{R}) \models \mathrm{AD}^+$;

(2) *Suppose that $\Gamma \neq \mathcal{P}(\mathbb{R})$ (i.e., that AD^+ fails) then*

$$L(\Gamma, \mathbb{R}) \models \mathrm{ZF} + \mathrm{DC} + \mathrm{AD}_{\mathbb{R}}.$$ □

Corollary 76. *Assume* ZF + AD + *“$V = L(\mathbb{R})$”. Then* AD^+. □

The next two theorems show that AD^+ is actually implied by strong versions of AD; e.g., $\mathrm{AD}_{\mathbb{R}}$ which is the axiom that if $A \subseteq \mathbb{R}^\omega$ then the *real game* given by the set A is determined. So in this case strategies are functions $\pi\colon \mathbb{R}^{<\omega} \to \mathbb{R}$, etc.

Theorem 77 (ZF + DC). *The following are equivalent.*

(1) AD + Uniformization.

(2) $\mathrm{AD}_{\mathbb{R}}$.

(3) AD + *Every set of reals is Suslin.* □

Assuming AD, if every set $A \subseteq \mathbb{R}$ is Suslin then necessarily the Suslin cardinals are cofinal in Θ. Therefore one obtains the following corollary from the previous theorems.

Corollary 78. *Assume* $\mathrm{ZF} + \mathrm{DC} + \mathrm{AD}_{\mathbb{R}}$. *Then* AD^+. □

In light of this corollary, it is natural to conjecture that AD is equivalent to AD^+ (over $\mathrm{ZF} + \mathrm{DC}_{\mathbb{R}}$).

The proofs of many of the deepest consequences of

$$\mathrm{ZF} + \mathrm{AD} + \text{“}V = L(\mathbb{R})\text{”}$$

are based on the presentation of $L(\mathbb{R})$ as an *inner model*; these proofs exploit the “smallness” of $L(\mathbb{R})$ in various essential ways.

Perhaps surprising then is that these consequences generalize, abstractly, to the theory:

$$\mathrm{ZF} + \mathrm{AD}^+ + \text{“}V = L(\mathcal{P}(\mathbb{R}))\text{”}.$$

The next theorem gives a typical example.

THEOREM 79 (AD^+). *Assume $V = L(\mathcal{P}(\mathbb{R}))$.*

(1) *The pointclass Σ_1^2 has the scale property.*
(2) *Suppose $\phi(x, y)$ is a Σ_1-formula and there exists a set X such that*

$$V \vDash \phi[X, \mathbb{R}].$$

There there exists X^* such that $V \vDash \phi[X^*, \mathbb{R}]$, and such that the set X^* is coded by a set $Z \subset \mathbb{R}$ with the property Z is Δ_1^2. □

The connection between the universally Baire sets and AD^+:

THEOREM 80. *Assume there exists a proper class of Woodin cardinals and let Γ^∞ be the set of all $A \subseteq \mathbb{R}$ such that A is universally Baire.*

(1) *Γ^∞ is a σ-algebra.*
(2) *For each $A \in \Gamma^\infty$, $\mathcal{P}(\mathbb{R}) \cap L(A, \mathbb{R}) \subset \Gamma^\infty$.*
(3) *For each $A \in \Gamma^\infty$, $L(A, \mathbb{R}) \vDash \mathrm{AD}^+$.* □

Combining the two previous theorems yields some information on the complexity of the Ω-recursive subsets of $\mathbb{N}$.

THEOREM 81. *Assume there exists a proper class of Woodin cardinals. Suppose that $T \subseteq \mathbb{N}$ is Ω-recursive, γ is a cardinal and that there is a Woodin cardinal below γ.*

Then T is definable in the structure $\langle H(\gamma), \in\rangle$ □

Improving this calculation requires a detour through *Fine Structure Theory*.

§13. Fine structure and inner models. L is Gödel's constructible universe:

$$L = \cup\{L_\eta \mid \eta \in \mathrm{Ord}\}$$

where;

(1) $L_0 = \emptyset$,
(2) $L_{\eta+1} = \{a \subseteq L_\eta \mid a \text{ is definable from parameters in } \langle L_\eta, \in\rangle\}$,
(3) If η is a limit ordinal then $L_\eta = \cup\{L_\alpha \mid \alpha < \eta\}$.

The detailed analysis of L is the *fine structure* of L; this was initiated, and mostly developed, by Jensen.

The generalization of L to *inner models* in which various large cardinal axioms hold, is the *Inner Model Program*. The goal is to understand these inner models and their fine structure.

For measurable cardinals the inner model was defined by Solovay. The fine structure was developed by Solovay, building on work of Kunen and Silver.

For cardinals below the level of superstrong cardinals, the inner models and their fine structure were defined and analyzed by Mitchell and Steel, [18]. However presently existence can be proved only at the level of Woodin cardinals; e.g., assume there exists a Woodin cardinal, then a Mitchell–Steel inner model for a Woodin cardinal exists. The best current result is due to Neeman; if there exists a Woodin cardinal which is a limit of strong cardinals then there exists a Mitchell–Steel inner model in which there is a Woodin cardinal which is a limit of strong cardinals. Note that if a Woodin cardinal is a limit of strong cardinals then necessarily it is a limit of Woodin cardinals.

Suppose that M, N are transitive sets,

$$M, N \models \mathrm{ZFC}\backslash\mathrm{Powerset},$$

and that, $j\colon M \to N$, is elementary embedding with critical point $\kappa \in M \cap \mathrm{Ord}$.

Suppose that $\kappa < \eta \leq j(\kappa)$. For each finite set $s \subset \eta$ let

$$E_s = \left\{A \subseteq [\kappa]^{|s|} \mid A \in M \text{ and } s \in j(A)\right\}.$$

Thus E_s is an M-*ultrafilter*. The set

$$E = \{(s, A) \mid s \in [\eta]^{<\omega} \text{ and } A \in E_s\}$$

is a (κ, η)-M-extender. If $V_{\kappa+1} \subset M$ then E is a (κ, η)-extender.

The Mitchell–Steel models are of the form $L[\tilde{E}]$ where

$$\tilde{E} \subset \mathrm{Ord} \times V$$

is a predicate defining a sequence of (partial) extenders; more precisely if $\eta \in \mathrm{dom}(\tilde{E})$ then

$$L_\eta[\tilde{E}] \models \mathrm{ZFC}\backslash\mathrm{Powerset}$$

and $(\tilde{E})_\eta$ is a (κ, η)-$L_\eta[\tilde{E}]$-extender (for some $\kappa < \eta$).

Thus if

$$\mathcal{P}(\kappa) \cap L_\eta[\tilde{E}] \neq \mathcal{P}(\kappa) \cap L[\tilde{E}],$$

then $(\tilde{E})_\eta$ is *not* a (κ, η)-$L[\tilde{E}]$-extender. It is because the extenders are partial that the development of the fine structure of the Mitchell–Steel models is necessary in order to even define the models.

Another inner model can *always* be defined; it is HOD. A set a belongs to HOD if there exist an ordinal α and a set, $A \subseteq \alpha$, such that $a \in L[A]$, and such that the set A is definable in the structure, $\langle V_\alpha, \in\rangle$, from ordinal parameters. The *Axiom of Choice* holds in HOD; *even* if the *Axiom of Choice* fails in the universe where HOD is computed. In general HOD is *not* absolute, it can change in passing from V to a generic extension of V.

By Vopenka's Theorem, V is a (class) generic extension of HOD. If the *Axiom of Choice* fails, then V is a (class) symmetric generic extension of HOD.

I recall that Θ denotes the supremum of the ordinals α such that there exists a surjection $\pi: \mathbb{R} \to \alpha$.

THEOREM 82. *Assume* $\mathrm{AD}^{L(\mathbb{R})}$. *Let* $\delta = (\Theta)^{L(\mathbb{R})}$. *Then in* $(\mathrm{HOD})^{L(\mathbb{R})}$, δ *is a Woodin cardinal.* □

A truly remarkable theorem of Steel:

THEOREM 83 (Steel). *Assume* $\mathrm{AD}^{L(\mathbb{R})}$. *Let* $\delta = (\Theta)^{L(\mathbb{R})}$. *Then*

$$(\mathrm{HOD})^{L(\mathbb{R})} \cap V_\delta$$

is a Mitchell–Steel model. □

Corollaries include:

THEOREM 84 (Steel). *Assume* $\mathrm{AD}^{L(\mathbb{R})}$. *Then the* Generalized Continuum Hypothesis *holds in* $(\mathrm{HOD})^{L(\mathbb{R})}$. □

But, what is $(\mathrm{HOD})^{L(\mathbb{R})}$? Curiously one has the following theorem.

THEOREM 85. $(\mathrm{HOD})^{L(\mathbb{R})}$ *is not a Mitchell–Steel model.* □

Nevertheless $(\mathrm{HOD})^{L(\mathbb{R})}$ is a model with fine structural properties. It belongs to a new, quite different, hierarchy of models. Further a fairly detailed analysis of this inner model is possible. One can calculate within $(\mathrm{HOD})^{L(\mathbb{R})}$ which cardinals are Suslin cardinals in $L(\mathbb{R})$, for example

$$(\aleph_\omega)^{L(\mathbb{R})}$$

is the least ordinal δ such that

$$(\mathrm{HOD})^{L(\mathbb{R})} \models \text{“}\delta \text{ is a Woodin cardinal in } L(V_\delta)\text{”}$$

and one can even show for example:

$$(\mathrm{HOD})^{L(\mathbb{R})} \models \text{“}V = \mathrm{HOD}\text{”}.$$

§14. Beyond $L(\mathbb{R})$.

DEFINITION 86. Assume AD^+ and that $V = L(\mathcal{P}(\mathbb{R}))$. Define

$$\langle \Theta_\alpha : \alpha \leq \Upsilon \rangle$$

by induction as follows.

(1) Θ_0 is the supremum of the ordinals η such that there exists a surjection

$$\pi: \mathbb{R} \to \eta$$

with π ordinal definable.

(2) $\Theta_\alpha = \sup \{\Theta_\beta \mid \beta < \alpha\}$ if α is a nonzero limit ordinal.

(3) If $\Theta_\alpha < \Theta$ then $\Theta_{\alpha+1}$ is the supremum of the ordinals η such that there exists a surjection, $\pi: \mathcal{P}(\Theta_\alpha) \to \eta$, with π ordinal definable. □

The "Θ" sequence, $\langle \Theta_\alpha : \alpha \leq \Upsilon \rangle$, refines Θ, yielding an effective sequence of approximations. The definition, in a slightly different form and in the more general context of AD, is due to Solovay.

Note that $\Theta_\Upsilon = \Theta$. Assuming AD^+ then $\mathrm{AD}_\mathbb{R}$ holds if and only if Υ is a (nonzero) limit ordinal; i.e., if and only if

$$\Theta = \sup \{\Theta_\alpha \mid \alpha < \Upsilon\}.$$

If $V = L(\mathbb{R})$ then assuming AD, Θ_0 is a Woodin cardinal in HOD and it is the only Woodin cardinal in HOD. If $V = L(\mathcal{P}(\mathbb{R}))$ then assuming AD^+, Θ_0 is the least Woodin cardinal in HOD, and for each $\alpha < \Upsilon$, Θ_α is a Woodin cardinal in HOD if and only if $\alpha = 0$ or α is not a limit ordinal.

The theorem that $\mathrm{HOD}^{L(\mathbb{R})}$ is not a Mitchell–Steel model generalizes to the following theorem which identifies an upper bound where HOD must diverge from the Mitchell–Steel models (and from essentially any hierarchy of inner models satisfying fairly modest definability conditions).

THEOREM 87 ($\mathrm{AD}^+ +$ "$V = L(\mathcal{P}(\mathbb{R}))$"). *HOD $\cap V_{\Theta_0+1}$ is not a Mitchell–Steel model.* □

Nevertheless as is the case for $\mathrm{HOD}^{L(\mathbb{R})}$, the HOD of a general AD^+ model which satisfies "$V = L(\mathcal{P}(\mathbb{R}))$" is a model in hierarchy of models with a fine structure. The fine structure associated to these models yields an *iteration theory*. Countable *iterable* models in this hierarchy can be compared and further this allows a comparison of the "parent" models of AD^+, *even* though these parent models models may not have the same reals. One corollary is the following improvement on the Derived Model Theorem.

THEOREM 88. *Suppose that δ is a limit cardinal and $\delta > \omega$. Suppose that*

$$G \subset \mathrm{Coll}(\omega, <\delta)$$

is V-generic and let

$$\mathbb{R}_G = \cup\{\mathbb{R}^{V[G|\alpha]} \mid \alpha < \delta\}.$$

Let Γ_G be the set of $X \subseteq \mathbb{R}_G$ such that there exists $Y \subseteq \mathbb{R}_G$ with

(1) $Y \in V(\mathbb{R}_G)$,
(2) $L(\mathbb{R}_G, Y) \models \mathrm{AD}^+$,
(3) *X is Δ_1-definable in $L(Y, \mathbb{R}_G)$ from $\{\mathbb{R}_G\}$.*

Suppose that $A \in \Gamma_G$ and that $B \in \Gamma_G$.

Then either $A \in L(\mathbb{R}_G, B)$ or $B \in L(\mathbb{R}_G, A)$. □

Thus Ω-logic can naturally be defined in ZFC. Further the A-closed transitive sets M, for the relevant universally Baire sets, $A \subseteq \mathbb{R}$, can be defined purely in *fine structural* terms.

Another corollary is that if $T \subseteq \mathbb{N}$ is Ω-recursive then T is definable in the structure

$$\langle H(\gamma), \in \rangle$$

for any *strong* limit cardinal $\gamma > \omega$. This is immediate for the structure $\langle H(\gamma^+), \in\rangle$.

This latter calculation can be still further refined yielding the following theorem.

THEOREM 89 (CH). *Suppose that there exists a proper class of Woodin cardinals and suppose that $T \subseteq \mathbb{N}$ is Ω-recursive. Then T is Δ_2 definable in the structure,*

$$\langle H(\omega_2), \mathcal{I}_{\mathrm{NS}}, \in\rangle. \qquad \square$$

Thus by Tarski's theorem on the undefinability of truth:

THEOREM 90. *Suppose that there exist a proper class of Woodin cardinals,*

$$V_\kappa \vDash \mathrm{ZFC} + \Psi$$

and for each sentence ϕ, either

(i) $\mathrm{ZFC} + \Psi \vdash_\Omega$ "$H(\omega_2) \vDash \phi$", *or*
(ii) $\mathrm{ZFC} + \Psi \vdash_\Omega$ "$H(\omega_2) \vDash \neg\phi$".

Then CH *is false.* □

We generalize the definition of the Ω-recursive subsets of $\mathbb{N}$ to define the Ω-recursive subsets of $\mathbb{R} \times \mathbb{R}$.

DEFINITION 91. Suppose that there exists a proper class of Woodin cardinals. A set, $X \subseteq \mathbb{R} \times \mathbb{R}$, is *$\Omega$-recursive* if there exists a universally Baire set, $A \subseteq \mathbb{R}$, such that X is Δ_1 definable in $L(A, \mathbb{R})$ from $\{\mathbb{R}\}$. □

The fundamental issue for the complexity of the Ω-recursive subsets of $\mathbb{N}$ is not whether CH holds or not but whether an effective version of CH holds. The analysis, under CH, of the complexity of the Ω-recursive subsets of $\mathbb{N}$ generalizes to yield the following dichotomy theorem.

THEOREM 92. *Assume there exists a proper class of Woodin cardinals and that $T \subseteq \mathbb{N}$ is Ω-recursive. Then either:*

(1) *T is Δ_2 definable in the structure, $\langle H(\omega_2), \mathcal{I}_{\mathrm{NS}}, \in\rangle$; or*
(2) *there exists a surjection, $\rho\colon \mathbb{R} \to \omega_2$, such that ρ is Δ_2-definable in the structure*

$$\langle H(\omega_2), \mathcal{I}_{\mathrm{NS}}, \in\rangle,$$

(without parameters) and such that the relation

$$R = \{(x, y) \mid \rho(x) < \rho(y)\}$$

is Ω-recursive. □

This suggests the followng definition.

Definition 93. $\Phi_{\neg\text{CH}}$: There exists a proper class of Woodin cardinals and there exists a surjection, $\rho\colon \mathbb{R} \to \omega_2$, such that ρ is Δ_2-definable in the structure

$$\langle H(\omega_2), \mathcal{I}_{\text{NS}}, \in \rangle,$$

(without parameters) and such that the relation

$$R = \{(x, y) \mid \rho(x) < \rho(y)\}$$

is Ω-recursive. □

For example, if $\underset{\sim}{\delta}^1_2 = \omega_2$ (and there is a proper class of Woodin cardinals) then $\Phi_{\neg\text{CH}}$ holds.

Therefore:

Theorem 94. *Suppose that there exist a proper class of Woodin cardinals,*

$$V_\kappa \models \text{ZFC} + \Psi$$

and for each Σ_2 sentence ϕ, either

(i) ZFC + $\Psi \vdash_\Omega$ "$\langle H(\omega_2), \mathcal{I}_{\text{NS}}, \in \rangle \models \phi$", *or*
(ii) ZFC + $\Psi \vdash_\Omega$ "$\langle H(\omega_2), \mathcal{I}_{\text{NS}}, \in \rangle \models \neg\phi$".

Then $\Phi_{\neg\text{CH}}$ holds and

$$\text{ZFC} + \Psi + \text{"There is a proper class of Woodin cardinals"} \vdash_\Omega \Phi_{\neg\text{CH}}.$$ □

(Note that if κ is a limit of Woodin cardinals then

$$\text{ZFC} + \Psi + \text{"There is a proper class of Woodin cardinals"}$$

is Ω-consistent.)

By a recent result, Theorem 94 cannot be improved; i.e., one cannot strengthen $\Phi_{\neg\text{CH}}$, (by attempting to reduce the complexity of the prewellordering which witnesses $\Phi_{\neg\text{CH}}$ within the (Wadge) hierarchy of the sets $A \subseteq \mathbb{R} \times \mathbb{R}$ which are Ω-recursive).

§15. The Ω Conjecture.

Is Ω-logic the *strongest* logic which satisfies the requirement of *Generic Soundness*?

Definition 95 (Ω^*-logic). Suppose that there exists a proper class of Woodin cardinals and that ϕ is a sentence. Then

$$\text{ZFC} \vdash_{\Omega^*} \phi$$

if for all ordinals α and for all partial orders $\mathbb{P}$ if

$$V_\alpha^{\mathbb{P}} \models \text{ZFC},$$

then $V_\alpha^{\mathbb{P}} \models \phi$. □

Generic Soundness is immediate for Ω^*-logic and trivially, Ω^*-logic is the strongest possible logic satisfying this requirement. It is perhaps surprising that the property of generic invariance also holds for Ω^*-logic.

THEOREM 96 (Generic Invariance). *Suppose that there exists a proper class of Woodin cardinals. Suppose that ϕ is a sentence. Then for each partial order $\mathbb{P}$,*

$$(\text{ZFC} \vdash_{\Omega^*} \phi)^V$$

if and only if

$$(\text{ZFC} \vdash_{\Omega^*} \phi)^{V^{\mathbb{P}}}. \qquad \square$$

The following conjecture is the **Ω Conjecture**.

> *Suppose that there exists a proper class of Woodin cardinals. Then for each Π_2 sentence ϕ,*
>
> $$\text{ZFC} \vdash_{\Omega^*} \phi$$
>
> *if and only if* $\text{ZFC} \vdash_{\Omega} \phi$.

In some sense the Ω Conjecture is simply the conjecture that Ω-logic is the strongest logic which satisfies *Generic Soundness*; i.e., that Ω-logic is precisely Ω^*-logic. While Steel has noted that the latter equivalence is strictly speaking false, the following lemma recovers the spirit of this equivalence.

LEMMA 97. *Suppose that there exists a proper class of Woodin cardinals. Then the following are equivalent*:

(1) *The Ω Conjecture*;
(2) *For each sentence ϕ, if there exists a finite set $T \subset$ ZFC such that*

$$T \vdash_{\Omega^*} \phi,$$

then there exists a finite set $S \subset$ ZFC such that $S \vdash_{\Omega} \phi$. $\square$

We define the notion of an Ω^*-recursive subset of $\mathbb{R}$.

DEFINITION 98. Suppose that there exists a proper class of Woodin cardinals. A set $A \subseteq \mathbb{R}$ is *Ω^*-recursive* if there exists a formula $\phi(x)$ such that:

(1) $A = \{r \mid \text{ZFC} \vdash_{\Omega^*} \phi[r]\}$;
(2) For all partial orders, $\mathbb{P}$, if $G \subseteq \mathbb{P}$ is V-generic then for each $r \in \mathbb{R}^{V[G]}$, either

$$V[G] \models \text{“ZFC} \vdash_{\Omega^*} \phi[r]\text{”},$$

or $V[G] \models \text{“ZFC} \vdash_{\Omega^*} (\neg\phi)[r]\text{”}$. $\square$

If the Ω Conjecture holds then a set $A \subseteq \mathbb{R}$ is Ω^*-recursive if and only if it is Ω-recursive. So some evidence for the Ω Conjecture is provided by the following theorem.

THEOREM 99. *Suppose that there exists a proper class of Woodin cardinals. Suppose that $A \subseteq \mathbb{R}$ is Ω^*-recursive. Then A is universally Baire.* $\square$

An immediate corollary is that if $A \subseteq \mathbb{R}$ is Ω^*-recursive then the set A is determined. This is yet another version of the connection between generic absoluteness and determinacy.

§16. **Connections with the logic of large cardinal axioms.** The next definition is not intended to characterize "large cardinal axioms" but rather isolate a feature shared by essentially all those axioms currently viewed as large cardinal axioms. There are trivial exceptions: for example the large cardinal axiom, "κ is a supercompact cardinal", is an exception, but the large cardinal axiom, "There exists an inaccessible cardinal δ above κ such that κ is supercompact cardinal in V_δ" is not an exception.

DEFINITION 100. $(\exists x\phi)$ is a *large cardinal axiom* if

(1) $\phi(x)$ is a Σ_2-formula;

(2) (As a theorem of ZFC) if κ is a cardinal such that

$$V \models \phi[\kappa]$$

then κ is strongly inaccessible and for all partial orders $\mathbb{P} \in V_\kappa$,

$$V^{\mathbb{P}} \models \phi[\kappa]. \qquad \square$$

It seems a plausible conjecture that there are large cardinal axioms $(\exists x\phi_1)$ and $(\exists x\phi_2)$ such that one can always force a proper class of witnesses for either (assuming a proper class of inaccessible cardinals) but such that there cannot simultaneously exist a proper class of witnesses for $(\exists x\phi_1)$ and a proper class of witnesses for $(\exists x\phi_2)$.

This possibility is one reason for the following definition.

DEFINITION 101. Suppose that $(\exists x\phi)$ is a large cardinal axiom.

Then V is *ϕ-closed* if for every set, X, there exist a transitive set, M, and $\kappa \in M \cap \mathrm{Ord}$ such that

(1) $M \models \mathrm{ZFC}$,

(2) $X \in M_\kappa$ and $M \models \phi[\kappa]$. $\square$

Note that if $(\exists x\phi)$ is a large cardinal axiom and if there exists a proper class of inaccessible cardinals, κ, such that $\phi[\kappa]$ holds, then V is ϕ-closed. On the other hand if V is ϕ-closed then for each inaccessible cardinal κ,

$$V_\kappa \models \text{“}V \text{ is } \phi\text{-closed”},$$

and so the converse fails.

Assuming there is a proper class of Woodin cardinals then there do exist large cardinal axioms $(\exists x\phi_1)$ and $(\exists x\phi_2)$ such that

$$\mathrm{ZFC} \vdash_\Omega \text{“}V \text{ is both } \phi_1\text{-closed and } \phi_2\text{-closed”};$$

such that both

$$\mathrm{ZFC} + \text{“}\phi_1 \text{ holds for a proper class of cardinals”}$$

and

$$\text{ZFC} + \text{“}\phi_2 \text{ holds for a proper class of cardinals”}$$

are Ω-consistent; but such that

$$\text{ZFC} + \text{“}\phi_1 \text{ and } \phi_2 \text{ each hold for proper classes of cardinals”}$$

is not even consistent.

The following is an easy consequence of the definitions.

LEMMA 102. *Suppose there there exists a proper class of Woodin cardinals and that Ψ is a Π_2 sentence. The following are equivalent.*

1) $\text{ZFC} \vdash_\Omega \Psi$.
2) *There is a large cardinal axiom $(\exists x \phi)$ such that*
   (a) $\text{ZFC} \vdash_\Omega$ *“V is ϕ-closed”,*
   (b) $\text{ZFC} +$ “V is ϕ-closed” $\vdash \Psi$. □

An immediate corollary of this lemma is that the Ω Conjecture is equivalent to:

> *Suppose that there exists a proper class of Woodin cardinals and that $(\exists x \phi)$ is a large cardinal axiom. The following are equivalent.*
> (1) *V is ϕ-closed.*
> (2) $\text{ZFC} \vdash_\Omega$ *“V is ϕ-closed”.*

Thus the Ω Conjecture implies that Ω-logic is simply the natural logic associated to the set of large cardinal axioms $(\exists x \phi)$ for which V is ϕ-closed.

§17. The Ω Conjecture and iteration hypotheses. Suppose that $F : V \to V$ is a (class) function. The function F *satisfies condensation* if for all (limit) $\eta \in \text{Ord}$ such that

$$F[V_\eta] \subseteq V_\eta$$

and for all elementary substructures

$$X \prec \langle V_\eta, F \cap V_\eta, \in \rangle,$$

if F_X is the image of $F \cap X$ under the transitive collapse of X then $F_X \subset F$.

For example, define $F : V \to V$ by: $F(a) = L_\alpha$ if $a = \alpha$ for some $\alpha \in \text{Ord}$; and $F(a) = \emptyset$ otherwise. Then the function F satisfies condensation.

THEOREM 103. *Suppose that there exists a proper class of Woodin cardinals and that $(\exists x \phi)$ is a large cardinal axiom. The following are equivalent.*

1) $\text{ZFC} \vdash_\Omega$ “V is ϕ-closed”.
2) *There exists a countable structure, $\langle M, \tilde{E}, \delta \rangle$, such that*
   (a) *M is transitive and $M \models \text{ZFC}$,*
   (b) *$M \models$ “V is ϕ-closed”,*
   (c) *$\tilde{E} \in M$ and in M, $\tilde{E}$ is an extender sequence which witnesses that δ is a Woodin cardinal,*

(d) $\langle M, \tilde{E}\rangle$ *has a transfinite iteration strategy which satisfies condensation.* □

This theorem suggests formulating the **Ω Iteration Hypothesis**:

Suppose that there exists a proper class of Woodin cardinals. Then there exists $(\kappa, \tilde{E}, \delta)$ *such that*

(1) κ *is strongly inaccessible,*
(2) $\delta < \kappa$ *and* δ *is a Woodin cardinal,*
(3) $\tilde{E} \subset V_\delta$ *and* $\tilde{E}$ *is an extender sequence which witnesses* δ *is a Woodin cardinal;*

and such that for some countable elementary substructure,

$$\langle M_X, \tilde{E}_X, \delta_X\rangle \cong X \prec \langle V_\kappa, \tilde{E}, \delta\rangle,$$

$\langle M_X, \tilde{E}_X\rangle$ *has a transfinite iteration strategy which satisfies condensation.*

Clearly in light of the previous theorem, the Ω *Iteration Hypothesis* implies the Ω Conjecture. At present this seems to be the most promising approach to proving the Ω Conjecture. The Ω *Iteration Hypothesis* is not obviously equivalent to the Ω Conjecture but assuming there exists a proper class of measurable cardinals which are limits of Woodin cardinals,

$$\text{ZFC} + \text{“The } \Omega \text{ Iteration Hypothesis”}$$

is Ω-consistent.

§18. The Ω Conjecture and inner model theory. The Ω Conjecture cannot be refuted by any large cardinal axiom which admits an inner model theory based on any seemingly reasonable notion of *comparison*. This is a vague claim which I shall make more precise but first I define some abstract requirements for an inner model theory.

Definition 104. Suppose that $(\exists x\phi)$ is a large cardinal axiom. $(\exists x\phi)$ *admits a weak inner model theory* if there exists a formula $\Phi(x, y)$ such that the following three conditions hold where for each transitive set, M,

$$I_\Phi^M = \{(a, b) \mid M \models \Phi[a, b]\}.$$

Suppose that M is a transitive model of ZFC and that in M there is a proper class of Woodin cardinals and a proper class of cardinals for which ϕ holds. Then:

(1) I_Φ^M is a function,

$$I_\Phi^M : M \cap \mathcal{P}(M \cap \text{Ord}) \to M,$$

such that for all $a \in M \cap \mathcal{P}(M \cap \text{Ord})$,
(a) $|N|^M = |a \cup \omega|^M$,
(b) N is transitive, $a \in N_\delta$, and $N \models \phi[\delta]$,

(c) $N \models \mathrm{ZFC}$,
where $(\delta, N) = I^M_\Phi(a)$.

(2) If $\mathbb{P} \in M$ and $G \subseteq \mathbb{P}$ is M-generic, then $I^M_\Phi = I^{M[G]}_\Phi \cap M$.

(3) Suppose that κ is a measurable cardinal in M such that in M, κ is a limit of Woodin cardinals and a limit of cardinals for which ϕ holds in M_κ.
Then $I^M_\Phi \cap M_\kappa = I^{M_\kappa}_\Phi$. □

Here is an example. Let $(\exists x \phi_0)$ be the large cardinal axiom where $\phi_0(x)$ asserts: "x is a measurable cardinal". Let $\Phi_0(x, y)$ assert: "x is a set of ordinals and y is the the ω-model of $x^\dagger$". Then Φ_0 witnesses that the large cardinal axiom $(\exists x \phi_0)$ admits a weak inner model theory.

For the following theorem it is necessary only that the witness $\Phi(x, y)$, that $(\exists x \phi)$ admits a weak inner model theory, satisfy Definition 104 just in cases that M is a set generic extension of V. This is simply the assertion that there exists a function, $I\colon H(\omega_1) \cap \mathcal{P}(\omega_1) \to H(\omega_1)$, such that:

(1) for all $a \in \mathrm{dom}(I)$, if $N = I(a)$ then N is transitive, $N \models \mathrm{ZFC}$, and there exists $\delta \in N$ such that $a \in N_\delta$ and $N \models \phi[\delta]$;

(2) the set of $x \in \mathbb{R}$ such that x codes an element of the graph of I is Ω^*-recursive.

Condition (2) is in turn equivalent, in this case, to the requirement that specified set be Ω-recursive. This is by the basis theorem for Σ^2_1 and by the theorem that the Ω^*-recursive sets are universally Baire.

Theorem 105. *Suppose that there exists a proper class of Woodin cardinals and there exists a proper class of strong cardinals. Suppose that $(\exists x \phi)$ is a large cardinal axiom, there is a proper class of cardinals for which ϕ holds, and that $(\exists x \phi)$ admits a weak inner model theory. Then*

$$\mathrm{ZFC} \vdash_\Omega \text{"}V \text{ is } \phi\text{-closed."}$$ □

There is also an approximate converse.

Theorem 106. *Suppose that there exists a proper class of Woodin cardinals, $(\exists x \phi)$ is a large cardinal axiom and that*

$$\mathrm{ZFC} \vdash_\Omega \text{"}V \text{ is } \phi\text{-closed."}$$

Then there is a large cardinal axiom $(\exists x \psi)$ such that

(1) ZFC $\vdash$ "If V is ψ-closed then V is ϕ-closed."

(2) *V is ψ-closed.*

(3) *$(\exists x \psi)$ admits a weak inner model theory.* □

These theorems show in effect that the Ω Conjecture is essentially equivalent to the conjecture that the large cardinal axioms $(\exists x \phi)$ for which V is ϕ-closed and which admit a weak inner model theory, are cofinal in the hierarchy of large cardinal axioms; more precisely to the conjecture:

Suppose that there exists a proper class of Woodin cardinals, $(\exists x\phi)$ is a large cardinal axiom and that V is ϕ-closed. Then there is a large cardinal axiom $(\exists x\psi)$ such that

(1) ZFC $\vdash_\Omega$ "If V is ψ-closed then V is ϕ-closed.".
(2) *V is ψ-closed.*
(3) *$(\exists x\psi)$ admits a weak inner model theory.*

THEOREM 107. *Suppose that there exists a proper class of Woodin cardinals and there exists a proper class of strong cardinals. Suppose that $(\exists x\phi)$ is a large cardinal axiom, there is a proper class of cardinals for which ϕ holds, and that $(\exists x\phi)$ admits a weak inner model theory.*

Then there exist a transitive set M and $\delta \in M$ such that

(1) *$M \models (\exists x\phi)$,*
(2) *$M \models$ "δ is a measurable limit of Woodin cardinals",*
(3) *$M_\delta \models$ "The Ω Conjecture holds".* □

Thus if some large cardinal hypothesis implies that the Ω Conjecture fails in all rank initial segments of V, then the hypothesis is beyond the reach of any type of inner model theory based on comparison.

§19. The Ω Conjecture and the consistency hierarchy. Suppose there exists a proper class of Woodin cardinals and let

$$\Gamma^\infty = \{A \subseteq \mathbb{R} \mid A \text{ is universally Baire}\}.$$

The large cardinal axioms $(\exists x\phi)$ such that

$$\text{ZFC} \vdash_\Omega \text{"}V \text{ is } \phi\text{-closed"}$$

naturally define a wellordered hierarchy: $\phi_1 \leq \phi_2$ if the shortest Ω-proof

$$\text{ZFC} \vdash_\Omega \text{"}V \text{ is } \phi_1\text{-closed"}$$

has length less than or equal to the shortest Ω-proof that ZFC $\vdash_\Omega$ "V is ϕ_2-closed". Equivalently, $\phi_1 \leq \phi_2$ if for each universally Baire set $A \subseteq \mathbb{R}$ such that

$$M \models \text{ZFC} + \text{"}V \text{ is } \phi_2\text{-closed"}$$

for all countable A-closed transitive sets, M, such that $M \models$ ZFC, there exists a real x such that

$$M \models \text{ZFC} + \text{"}V \text{ is } \phi_1\text{-closed"}$$

for all countable A-closed transitive sets, M, such that $M \models$ ZFC and $x \in M$.

If the Ω Conjecture *holds* in V then this hierarchy includes *all* large cardinal axioms $(\exists x\phi)$ such that V is ϕ-closed. Further if the Ω Conjecture is *provable*, then this hierarchy is in essence a (coarse) version of the consistency hierarchy. This, arguably, accounts for the *empirical* fact that all large cardinal axioms are comparable.

Under certain circumstances, one can define a (pre)wellordering on all large cardinal axioms $(\exists x\phi)$ such that V is ϕ-closed without assuming that the Ω Conjecture holds in V. The brief discussion of this naturally leads to the notion of when a large cardinal axiom $(\exists x\phi)$ admits a *strong inner model theory*.

LEMMA 108. *Assume that there exists a proper class inaccessible limits of Woodin cardinals and let Γ^∞ be the set of all $A \subseteq \mathbb{R}$ such that A is universally Baire. Suppose that*

$$L(\Gamma^\infty, \mathbb{R}) \nvDash \mathrm{AD}$$

and suppose that $(\exists x\phi)$ is a large cardinal axiom such that V is ϕ-closed. Then there exists $A \in \Gamma^\infty$ such that for all sets X there exists a transitive set M such that

(1) *$M \vDash \mathrm{ZFC}$ + "There is a proper class of Woodin cardinals",*
(2) *$X \in M$ and $M \vDash$ "V is ϕ-closed",*

and such that M is not A-closed. □

Now suppose that there exists a proper class of Woodin cardinals and that

$$L(\Gamma^\infty, \mathbb{R}) \nvDash \mathrm{AD}$$

where Γ^∞ is the set of all universally Baire subsets of $\mathbb{R}$. For each large cardinal axiom, $(\exists x\phi)$, such that V is ϕ-closed let $A_\phi \subseteq \mathbb{K}$ be a witness to the lemma of minimum rank in the Wadge order. Now define $\phi_1 \leq \phi_2$ by comparing the Wadge ranks of A_{ϕ_1} and A_{ϕ_2}. If the Ω Conjecture is provable then this order is simply a coarser version of the order defined above by comparing the minimum possible lengths of Ω-proofs that V is ϕ-closed.

It seems quite plausible that one might be able to prove the Ω Conjecture assuming that

$$L(\Gamma^\infty, \mathbb{R}) \nvDash \mathrm{AD},$$

but as the following theorem illustrates this may not be of much help in proving the full conjecture.

THEOREM 109. *Suppose that there exists a proper class of Woodin cardinals, κ is supercompact, $G \subset \mathrm{Coll}(\omega, 2^\kappa)$ is V-generic and that $V[G][H]$ is a set generic extension of $V[G]$. Then there is an elementary embedding*

$$j\colon (L(\Gamma^\infty, \mathbb{R}))^{V[G]} \to (L(\Gamma^\infty, \mathbb{R}))^{V[G][H]}. \qquad \square$$

Now if δ is a Woodin cardinal which is a limit of Woodin cardinals and if $H \subset \mathrm{Coll}(\omega, <\delta)$ is V-generic then in $V[H]$,

$$(L(\Gamma^\infty, \mathbb{R}))^{V[H]} \vDash \mathrm{AD}^+.$$

Thus it is a corollary of the theorem that if there is a proper class of Woodin cardinals, κ is supercompact, and if $V[G]$ is a set generic extension of V in which $(2^\kappa)^V$ is countable then,

$$\left(L\left(\Gamma^\infty, \mathbb{R}\right)\right)^{V[G]} \models \mathrm{AD}^+.$$

Nevertheless the current test question for inner model theory is the following. Suppose that $(\exists x \phi)$ is a large cardinal axiom.

Is there an inner model M such that the following hold?

(1) $M \models \mathrm{ZFC} + (\exists x \phi) +$ *"There is a proper class of Woodin cardinals";*

(2) *Suppose that $x \in \mathbb{R} \cap M$. Then* $(\mathrm{ZFC} \vdash_\Omega \text{"}x \in \mathrm{HOD}\text{"})^M$; *equivalently there exists $A \in (\Gamma^\infty)^M$ such that x is ordinal definable in* $(L(A, \mathbb{R}))^M$.

One can weaken this requiring (2) to hold only for $x \in \mathbb{R} \cap (\mathrm{HOD})^M$. This then becomes a reasonable question for a fairly large class of Σ_2 sentences not just those which are large cardinal axioms; e.g., for the sentence which asserts that there exists α such that $V_\alpha \models \mathrm{ZFC} +$ *Martin's Maximum.*

All of these considerations suggest defining the notion that a large cardinal axiom $(\exists x \phi)$ admits a *strong inner model theory*. The details require several preliminary definitions.

DEFINITION 110. Suppose there exists a proper class of Woodin cardinals.

(1) 0^Ω is the set of sentences ϕ such that $\mathrm{ZFC} \vdash_\Omega \phi$.

(2) $\underset{\sim}{0}^\Omega$ is the set of pairs $(\phi(x), r)$ such that $\phi(x)$ is a formula, $r \in \mathbb{R}$ and $\mathrm{ZFC} \vdash_\Omega \phi[r]$. □

As usual we regard 0^Ω as a subset of ω and $\underset{\sim}{0}^\Omega$ as a subset of $\mathbb{R}$. If ZFC is consistent then $0'$ is recursively equivalent to the set of all sentences ϕ such that $\mathrm{ZFC} \vdash \phi$. Thus 0^Ω is the generalization of $0'$ to Ω-logic. The set $\underset{\sim}{0}^\Omega$ is (borel) equivalent to the set of all pairs $(\phi(x, y), r)$ such that $\phi(x, y)$ is a formula, $r \in \mathbb{R}$ and there exists a universally Baire set A with the property that $L(A, \mathbb{R}) \models \phi[r, A]$.

If κ is a strong cardinal (and there exists a proper class of Woodin cardinals) then $\underset{\sim}{0}^\Omega$ is universally Baire in $V[G]$ where $G \subset \mathrm{Coll}(\omega, 2^{2^\kappa})$. So granting fairly modest large cardinal axioms, one can always force to obtain that $\underset{\sim}{0}^\Omega$ is universally Baire. Further if $\underset{\sim}{0}^\Omega$ is universally Baire in V then this must hold in all generic extensions of V. The central question for inner model theory is whether a given large cardinal axiom actually proves that $\underset{\sim}{0}^\Omega$ is universally Baire.

DEFINITION 111. Suppose that $(\exists x \phi)$ is a large cardinal axiom. $(\exists x \phi)$ *admits a strong inner model theory* if there exists a formula $\Phi(x, y)$ such that the following three conditions hold where for each transitive set, M,

$$I_\Phi^M = \{(a, b) \mid M \models \Phi[a, b]\}.$$

Suppose that M is a transitive model of ZFC and that in M there is a proper class of Woodin cardinals, a proper class of strong cardinals and a proper class of cardinals for which ϕ holds.

Suppose that $M \models$ "$\underset{\sim}{0}^{\Omega}$ is universally Baire".

(1) I_{Φ}^{M} is a function,

$$I_{\Phi}^{M} : M \cap \mathcal{P}(M \cap \mathrm{Ord}) \to M,$$

such that for all $a \in M \cap \mathcal{P}(M \cap \mathrm{Ord})$,

(a) $|N|^{M} = |a \cup \omega|^{M}$,

(b) N is transitive, $a \in N_{\delta}$, and $N \models \phi[\delta]$,

(c) $N \models$ ZFC + "There is a proper class of Woodin cardinals",

(d) $M \models$ "N is $\underset{\sim}{0}^{\Omega}$-closed",

where $(\delta, N) = I_{\Phi}^{M}(a)$.

(2) If $\mathbb{P} \in M$ and $G \subseteq \mathbb{P}$ is M-generic, then $I_{\Phi}^{M} = I_{\Phi}^{M[G]} \cap M$.

(3) Suppose that in M, κ is a measurable cardinal which is a limit of Woodin cardinals. Suppose further that $(\underset{\sim}{0}^{\Omega})^{M} = (\underset{\sim}{0}^{\Omega})^{M_{\kappa}}$. Then $I_{\Phi}^{M} \cap M_{\kappa} = I_{\Phi}^{M_{\kappa}}$. □

Assume that the Ω Conjecture holds. Then the large cardinal axioms $(\exists x \phi)$ which admit a strong inner model theory are cofinal in the hierarchy of large cardinal axioms (among those large cardinal axioms $(\exists x \psi)$ such that V is ψ-closed).

We require one last definition. Suppose that $(\exists x \phi)$ is a large cardinal axiom. Let $\phi'(x)$ be a formula expressing

(1) x is a Woodin cardinal which is a limit of Woodin cardinals,

(2) x is a limit of cardinals for which ϕ holds.

THEOREM 112. *Suppose there exists a proper class of Woodin cardinals which are limits of strong cardinals, $(\exists x \phi)$ is a large cardinal axiom and that there exists a proper class of cardinals for which ϕ holds.*

Suppose that both $(\exists x \phi)$ and $(\exists x \phi')$ admit a strong inner model theory.

Then there exists a transitive set M such that:

(1) *$M \models$ ZFC + $(\exists x \phi)$ + "There is a proper class of Woodin cardinals";*

(2) *$M \models$ "The Ω Conjecture holds";*

(3) *Suppose $x \in \mathbb{R} \cap M$. Then there exists $A \in (\Gamma^{\infty})^{M}$ such that x is ordinal definable in $(L(A, \mathbb{R}))^{M}$.* □

As I have noted, it seems plausible that one might be able to prove the Ω Conjecture from the assumption that $\underset{\sim}{0}^{\Omega}$ is not universally Baire. The following theorem shows that this hypothesis does essentially suffice to establish the result on generic absoluteness and CH. The only change is the introduction of real parameters.

THEOREM 113. *Suppose that there exists a proper class of Woodin cardinals and that $\underset{\sim}{0}^{\Omega}$ is not universally Baire. Suppose that Ψ is a sentence such that for* *all partial orders $\mathbb{P}$, for all Σ_2 formulas $\phi(x)$, and for all $r \in (\mathbb{R})^{V^{\mathbb{P}}}$, either*

$$\left(\mathrm{ZFC} + \Psi \vdash_{\Omega^*} \text{“}\langle H(\omega_2), \mathcal{I}_{\mathrm{NS}}, \in\rangle \models \phi[r]\text{”}\right)^{V^{\mathbb{P}}}$$

or

$$\left(\mathrm{ZFC} + \Psi \vdash_{\Omega^*} \text{“}\langle H(\omega_2), \mathcal{I}_{\mathrm{NS}}, \in\rangle \models (\neg\phi)[r]\text{”}\right)^{V^{\mathbb{P}}}$$

Then $\mathrm{ZFC} + \Psi \vdash_{\Omega^*} \neg\mathrm{CH}$. □

Another variation of the previous theorem shows that, assuming there is a proper class of Woodin cardinals and a proper class of strong cardinals, if generic absoluteness holds for the theory of $\langle H(\omega_2), \in\rangle$ in real parameters conditioned on a Σ_2 sentence which implies CH (and which holds in some generic extension of V), then the set $\underset{\sim}{0}^{\Omega}$ is Δ_2-definable.

As I have noted, the conclusion of Lemma 108 suffices to define a prewellordering of those large cardinal axioms $(\exists x \phi)$ for which V is ϕ-closed. One can further show that the definition of this prewellordering is absolute to set generic extensions of V. If the conclusion of Lemma 108 *fails* to hold in V then in fact the notion that a set A be universally Baire is itself Δ_2-definable (which implies that $\underset{\sim}{0}^{\Omega}$ is Δ_2-definable).

THEOREM 114. *Suppose that there exists a proper class of inaccessible limits of Woodin cardinals, $(\exists x \phi)$ is a large cardinal axiom and V is ϕ-closed. Let Γ^∞ be the set of all $A \subseteq \mathbb{R}$ such that A is universally Baire. Then either*:

(1) *there exists $A \in \Gamma^\infty$ such that for all sets X there exists a transitive set M such that*
  (a) $M \models \mathrm{ZFC}$ + *“There is a proper class of Woodin cardinals”*,
  (b) *$X \in M$ and $M \models$ “V is ϕ-closed”*,
  and such that M is not A-closed; *or*
(2) *Γ^∞ is Σ_2-definable.* □

§20. Concluding remarks. Even if the Ω Conjecture is true so that the criterion of (conditional) generic absoluteness for the theory of $H(\omega_2)$ necessitates that CH be false, one still seems forced to acknowledge that the argument against CH is fundamentally different than the argument for *Projective Determinacy*. This is the enduring legacy of the nature of the independence of CH as demonstrated through forcing. For me this is a novel feature of the problem and it reaffirms the claim that the resolution of CH would be an instance of a genuinely new metamathematical phenomenon; unlike anything that has happened before. The attempts to justify large cardinal axioms by demonstrating their necessary influence on “small” sets, despite having yielded remarkable theorems, cannot produce similar instances and cannot by themselves really succeed to justify any large cardinal axioms.

The body of results on the projective sets, including the relationships between *Projective Determinacy* and Woodin cardinals, is not (for me) a convincing argument *at all* for the axiom that Woodin cardinals exist. Indeed if one enlarges the scope of ones considerations to include inner models of AD^+ then it becomes apparent that while the role of transitive models with Woodin cardinals in the structure theory of AD^+ has become a central one (e.g., the role of HOD etc.), this has little if anything to do with the position of Woodin cardinals within the hierarchy of large cardinals. This is simply because the transitive models used in the analysis all satisfy the "smallness" condition, relative to their specifed extender sequences, that there exist no pair of cardinals $\kappa < \delta$ such that κ is δ-strong and δ is a Woodin cardinal.

The structure theory of the projective sets stands on its own; so must, I would claim, any (new) structure theory in the realm of the finite sets based on or motivated by considerations of large cardinals axioms.

On the other hand, if in the course of our mathematical investigations we come to acknowledge that there really is some unambiguous conception of the transfinite realm, beyond merely a catalog of formal consequences of axioms, then to me something truly remarkable will have been discovered. While such a discovery need not entail a resolution of the Continuum Hypothesis, resolving the Continuum Hypothesis would almost certainly provide a powerful argument for such an eventuality. Still, perhaps in reaction to this, there is a tendency to claim that the Continuum Hypothesis is inherently vague and that this is simply the end of the story. But any legitimate claim that CH is inherently vague must have a mathematical basis, at the very least a theorem or a collection of theorems.[5]

My own view is that the independence of CH from ZFC and from ZFC together with large cardinal axioms, does not provide this basis. I would hope this is the minimum metamathematical assessment of the solution to CH that I have presented. Instead, for me, the independence results for CH simply show that CH is a difficult problem.

One could[6,7] adopt the position that the entire menagerie of independence results is the "solution" to CH (and to the other unsolvable problems of set theory as well). I would call this position, in its most platonistic form, Ω^*-formalism. For me the viability of Ω^*-formalism hinges on the Ω Conjecture. For example suppose that the Ω Conjecture fails badly in that the set, $\{\phi \mid \text{ZFC} \vdash_{\Omega^*} \phi\}$, which is evidently Π_2 definable in V, turns out, under suitably plausible assumptions, to be recursively equivalent to the set of *all* Π_2 sentences ϕ such that $V \models \phi$. Then no essential complexity is lost in adopting Ω^*-formalism and so perhaps in the end it becomes a reasonable position in

[5]For a collection of differing views on CH and the issue of new axioms see [6].

[6]Cohen in [1].

[7]Perhaps also, Shelah in [24].

complete harmony some unambiguous conception of the transfinite.[8] On the other hand if the Ω Conjecture is true then Ω^*-formalism is in essence just Ω-formalism and so this position seems a rejection of the transfinite beyond $H(c^+)$, where $c = 2^{\aleph_0}$, and ultimately merely a variant of the position that the true transfinite extension of the integers is *not* set theory but simply the consistency hierarchy. In an extreme form this position rejects the possibility of an unambiguous conception of uncountable sets even while acknowledging the subject has been a useful source of new insights into the (hereditarily) finite sets.

The difficulty here is that the hierarchy of large cardinal axioms as given by Ω logic seems clearly an unambiguous conception, moreover this is independent of whether the Ω Conjecture is actually true or false. No known techniques or theorems can be used to seriously challenge this view. Further the definition of this hierarchy necessarily involves *uncountable sets*, no insight into the realm of (hereditarily) countable sets can possibly resolve this hierarchy. So for me there is *already* an instance of a clear glimpse into the realm of the uncountable. What remains to be seen is its extent and whether it will eventually encompass a resolution of CH, or for that matter of any other question which is both Δ_2 (provably equivalent to both a Σ_2 sentence and to a Π_2 sentence) and independent of ZFC in Ω^* logic.

Several axiomatic approaches designed to yield positive solutions to CH have been proposed; for example, [9] and [20]. Both these approaches seem still somewhat premature, indeed in each approach some of the candidates discussed are not even known to be consistent.[9] So the underlying combinatorial principles are not well understood in either case and clearly more work is needed. At present there is no known axiomatic solution yielding CH which can really compete with $(*)$, but this of course does *not* imply that no such solution exists. It is interesting to note that none of the proposed approaches yields axioms which are first order over $H(\omega_2)$. But there is no a priori reason why the axioms which reveal the theory of $H(\omega_2)$ should be first order over this structure.

Finally, CH is a very special case of the GCH. There are several arguments for this claim. First CH concerns $H(\omega_2)$ and this seems clearly a special fragment of V. Second, transfinite combinatorics, particularly at the successors of (large) singular cardinals (for example those above a supercompact), may be fairly robust; i.e., the technique of forcing seems to be of very limited use. Moreover, as Solovay's early results on the *Singular Cardinals Hypothesis* [27], and the more recent results of Shelah's *pcf* theory [25] vividly illustrate, the phenomenon of independence may not be as widespread, or at the very least is far more subtle, in this realm. I mention several problems to illustrate this.

---

[8] which incidentally fails to resolve CH.

[9] for example: the axiom of [9] that ω_1 is generically ω_2-huge by an ω_1-dense ideal.

For each pair $\delta_0 < \delta_1$ of (infinite) regular cardinals let

$$S^{\delta_1}_{\delta_0} = \{\alpha < \delta_1 \mid \operatorname{cof}(\alpha) = \delta_0\},$$

and let $\mathcal{I}^{\delta_1}_{\mathrm{NS}} \subset \mathcal{P}(\delta_1)$ be the ideal of *nonstationary* subsets of δ_1. Clearly $S^{\delta_1}_{\delta_0}$ is a stationary subset of δ_1. If $\delta_1 > \omega_1$ then $S^{\delta_1}_{\delta_0}$ is also co-stationary in δ_1.

Can the following hold? (if there is a supercompact cardinal?)

(1) For each uncountable regular cardinal, δ_1,

$$(\mathcal{P}(\delta_1) \cap \mathrm{HOD})/\mathcal{I}^{\delta_1}_{\mathrm{NS}}$$

is atomic. (*Limited definable splitting of definable stationary sets into stationary subsets.*)

(2) For each pair $\delta_0 < \delta_1$ of infinite regular cardinals,

$$(\mathcal{P}(S^{\delta_1}_{\delta_0}) \cap \mathrm{HOD})/\mathcal{I}^{\delta_1}_{\mathrm{NS}}$$

is trivial. (*No nontrivial definable splitting of definable stationary sets into stationary subsets.*)

There are partial results, the consistency of (1) can be obtained (even together with a supercompact cardinal) but at present only from extremely strong assumptions:

THEOREM 115 (ZF). *Suppose there exists κ such that for all λ there exists an elementary embedding*

$$j : V \to V$$

with critical point κ and such that $j(\kappa) > \lambda$.

Then there exists a partial $\mathbb{P}$ and an ordinal α such that the following hold in $V^{\mathbb{P}}_{\alpha}$:

(1) ZFC + *"There is a supercompact cardinal."*

(2) *For each uncountable regular cardinal, δ_1,*

$$(\mathcal{P}(\delta_1) \cap \mathrm{IIOD})/\mathcal{I}^{\delta_1}_{\mathrm{NS}}$$

is atomic. □

There is no known consistency proof from a large cardinal hypothesis consistent with the *Axiom of Choice* and the consistency of (2) is completely open.

At this point one cannot rule out a resolution of CH (and if the Ω Conjecture fails, of all sentences which are provably Δ_2 in ZFC) based on some natural structural axioms of a nearly traditional large cardinal character. I note the following theorem.

THEOREM 116. *Suppose that there exists an elementary embedding*

$$j : L(V_{\lambda+1}) \to L(V_{\lambda+1})$$

with critical point below λ.

Then for each (infinite) regular cardinal $\delta < \lambda$,

$$\left|\left(\mathcal{P}(S_{\delta}^{\lambda^+}) \cap L(V_{\lambda+1})\right)/\mathcal{I}\right| < \lambda$$

where $\mathcal{I}$ *is the nonstationary ideal on* λ^+ *as computed in* $L(V_{\lambda+1})$. □

Now suppose there exists an elementary embedding

$$j\colon L(V_{\lambda+1}) \to L(V_{\lambda+1})$$

with critical point below λ. Which of the following can hold?

(1) For each (infinite) regular cardinal $\delta < \lambda^+$,

$$\left(\mathcal{P}(S_{\delta}^{\lambda^+})/\mathcal{I}_{\text{NS}}^{\lambda^+}\right)^{L(V_{\lambda+1})}$$

is trivial.

(2) (1) + CH (or (1) + "$\underset{\sim}{\delta}_2^1 < \omega_2$").

(3) (1) + $\neg$CH (or (1) + "$\underset{\sim}{\delta}_2^1 = \omega_2$").

The potentially subtle point is that if (1) holds then this is not obviously preserved under adding even just one Cohen real. The difficulty is that if c is Cohen generic over V then very likely,

$$\left(L(V_{\lambda+1})\right)^{V[c]} \neq L(V_{\lambda+1})[c].$$

On the other hand if CH fails and if $G \subseteq \text{Coll}(\omega_1, \mathbb{R})$ is V-generic then while it *is* true that

$$\left(L(V_{\lambda+1})\right)^{V[G]} = L(V_{\lambda+1})[G],$$

it is not difficult to show that (1) must *fail* in $V[G]$. It is not known if (1) can ever hold, but the theorem above certainly suggests it is plausible that (1) holds.

§21. Some open problems.

1. Assume $L(\mathbb{R}) \models$ AD. Must $\Theta^{L(\mathbb{R})} \leq \omega_3$?
2. Suppose that ϕ_1 and ϕ_2 are Σ_2 sentences such that both

$$\text{ZFC} + \phi_1$$

and

$$\text{ZFC} + \phi_2$$

are each Ω-consistent. Must

$$\text{ZFC} + \phi_1 + \text{"}V^{\mathbb{P}} \models \phi_2 \text{ for some semiproper } \mathbb{P}\text{"}$$

be Ω-consistent? *Conjecture*: "Yes" (from 2 supercompact cardinals.)

3. Suppose that ϕ_1 and ϕ_2 are Π_2 sentences such that both
$$\text{ZFC} + \text{CH} + \text{“}\langle H(\omega_2), \mathcal{I}_{\text{NS}}, \in\rangle \models \phi_1\text{”}$$
and
$$\text{ZFC} + \text{CH} + \text{“}\langle H(\omega_2), \mathcal{I}_{\text{NS}}, \in\rangle \models \phi_2\text{”}$$
are Ω-consistent. Let $\phi = (\phi_1 \wedge \phi_2)$.
Must
$$\text{ZFC} + \text{CH} + \text{“}\langle H(\omega_2), \mathcal{I}_{\text{NS}}, \in\rangle \models \phi\text{”}$$
be Ω-consistent? *Conjecture*: No.
4. Let $\diamond_{\text{G}}^{++}$ assert:
$$\langle H(\omega_2), \mathcal{I}_{\text{NS}}, \in\rangle^V \equiv_{\Sigma_2} \langle H(\omega_2), \mathcal{I}_{\text{NS}}, \in\rangle^{V^{\text{Coll}(\omega_1,\mathbb{R})}}.$$
Suppose that for each Π_2 sentence ϕ, if
$$\text{ZFC} + \diamond_{\text{G}}^{++} + \text{“}\langle H(\omega_2), \mathcal{I}_{\text{NS}}, \in\rangle \models \phi\text{”}$$
is Ω-consistent then $\langle H(\omega_2), \mathcal{I}_{\text{NS}}, \in\rangle \models \phi$. Let δ^∞ be the supremum of the ordinals α such that there exists a surjection, $\pi\colon \mathbb{R} \to \alpha$, such that the induced prewellordering, $\{(x, y) \mid \pi(x) < \pi(y)\}$, is ω_1-universally Baire. Must $\delta^\infty = \omega_2$?
5. Let X be the set of Π_2 sentences, ϕ, such that
$$\text{ZFC} + \text{CH} + \text{“}\langle H(\omega_2), \in\rangle \models \phi\text{”}$$
is Ω-consistent. Is X an Ω-recursive set?
6. Can there exist a sentence Ψ such that for all Σ_2 sentences, ϕ, either
   - $\text{ZFC} + \text{CH} + \Psi \vdash_\Omega \text{“}H(\omega_2) \models \phi\text{”}$, or
   - $\text{ZFC} + \text{CH} + \Psi \vdash_\Omega \text{“}H(\omega_2) \models \neg\phi\text{”}$;

   and such that $\text{ZFC} + \text{CH} + \Psi$ is Ω-consistent?

REMARK. Recent results based on the determinacy of some games defined by Neeman provide strong evidence that $\diamond_{\text{G}}$ works and good evidence that $\diamond$ works. Less clear is what large cardinal hypothesis might suffice to provide a positive answer. Modulo failure of inner model theory, the existence of superstrong cardinals cannot suffice.

7. (Ω Conjecture) Assume there exists a proper class of Woodin cardinals. Let ϕ be a Σ_2 sentence. Then the following are equivalent.
   (a) $\text{ZFC} + \phi$ is Ω-consistent.
   (b) There exists a partial order $\mathbb{P}$ such that $V^{\mathbb{P}} \models \phi$.
8. Assume there exists an elementary embedding
$$j\colon L(V_{\lambda+1}) \to L(V_{\lambda+1})$$
with critical point below λ. Can either, or both, of the following hold?
   (a) $(\text{HOD})^{L(V_{\lambda+1})} \models \text{GCH}$.

(b) Suppose that $x \in \mathbb{R}$ is OD in $L(V_{\lambda+1})$. Then there exists $A \in (\Gamma^\infty)^{V_\lambda}$ such that x is OD in $L(A, \mathbb{R})$; i.e., in V_λ, x is Ω-recursive in a countable ordinal.

9. Assume there exists an elementary embedding

$$j\colon L(V_{\lambda+1}) \to L(V_{\lambda+1})$$

with critical point below λ. Suppose that the following both hold.

(a) Every set $A \in \mathcal{P}(\lambda^+) \cap L(V_{\lambda+1})$ is definable from parameters in $\langle H(\lambda^+), \in \rangle$.

(b) In $L(V_{\lambda+1})$, the club filter on λ^+ is an ultrafilter on each (maximal) subset of λ^+ of constant cofinality.

Can CH hold? Can CH fail? Can $\underset{\sim}{\delta}^1_2 < \omega_2$? Can $\underset{\sim}{\delta}^1_2 = \omega_2$?

Remark. It is not known if (a) and/or (b) can even hold. But there are close approximations which must necessarily hold and it is not known if (a) *can* fail to hold. By analogy with $L(\mathbb{R})$ in the context of AD one would expect that it is possible for (a) and (b) to hold.

10. Assume there exists an elementary embedding

$$j\colon L(V_{\lambda+1}) \to L(V_{\lambda+1})$$

with critical point below λ.

Define $\underset{\sim}{\delta}^\lambda_\omega$ to be the supremum of the ordinals α such that there exists a surjection,

$$\rho\colon V_{\lambda+1} \to \alpha$$

such that the relation, $\{(x, y) \mid \rho(x) < \rho(y)\}$, is definable in $V_{\lambda+1}$ from parameters. Must $\underset{\sim}{\delta}^\lambda_\omega < \lambda^{++}$?

Remark. $\underset{\sim}{\delta}^\lambda_\omega$ is the analog for $V_{\lambda+1}$ of $\underset{\sim}{\delta}^1_\omega$. It is known that

$$(\aleph_{\lambda+\lambda})^{L(V_{\lambda+1})} < \underset{\sim}{\delta}^\lambda_\omega.$$

§22. Added in proof. Since this paper was written there have been some important changes in the principle definitions relating to Ω-logic. Specifically the definition of the *logical relation*, ZFC $\models_\Omega \phi$, has been introduced, replacing the definition of ZFC $\vdash_{\Omega^*} \phi$, and this in turn has led to a slight modification of the definition of ZFC $\vdash_\Omega \phi$. Further in the new approach one defines the relations, $T \models_\Omega \phi$, and, $T \vdash_\Omega \phi$, where T is an arbitrary (possibly empty) theory in the language of set theory.

Eliminating the definition of Ω^*-logic in favor of defining the logical relation $\models_\Omega$ was suggested by Patrick Dehornoy, [5].

Here are the new definitions (with T = ZFC).

Definition 1' (Ω-logic). Suppose that there is a proper class of Woodin cardinals.

Then ZFC $\models_\Omega \phi$ if for all partial orders, $\mathbb{P}$, for all ordinals, α, if

$$V_\alpha^{\mathbb{P}} \models \text{ZFC}$$

then $V_\alpha^{\mathbb{P}} \models \phi$. □

Definition 2' ($\vdash_\Omega$). Suppose that there exists a proper class of Woodin cardinals and ϕ is a sentence. Then

$$\text{ZFC} \vdash_\Omega \phi$$

if there exists a universally Baire set $A \subseteq \mathbb{R}$ such that for any countable transitive set M, if $M \models$ ZFC and if M is A-closed, then $M \models$ "ZFC $\models_\Omega \phi$". □

With these definitions the Ω Conjecture is reformulated as:

Suppose that there exists a proper class of Woodin cardinals. Then for each sentence ϕ,

$$\text{ZFC} \vdash_\Omega \phi$$

if and only if ZFC $\models_\Omega \phi$.

or equivalently;

Suppose that there exists a proper class of Woodin cardinals. Then for each sentence ϕ,

$$\emptyset \vdash_\Omega \phi$$

if and only if $\emptyset \models_\Omega \phi$.

In both approaches the content of Ω Conjecture is the same, its statement is simply more elegant in the new approach. It becomes simply the conjecture that the completeness theorem holds for Ω-logic.

Finally some very recent results on *suitable extender sequences* strongly suggest both that no known large cardinal hypothesis can refute the Ω Conjecture and also that Σ_2^2 conditional absoluteness is not possible. Thus the answer to both questions 5 and 6 in the problem list is very likely no.

The details of the results on suitable extender sequences will appear in a paper which is in preparation.

REFERENCES

[1] Donald J. Albers, Gerald L. Alexanderson, and Constance Reid (editors), ***More mathematical people***, Contemporary conversations, Harcourt Brace Jovanovich Publishers, Boston, MA, 1990.

[2] Joan Bagaria, *Bounded forcing axioms as principles of generic absoluteness*, ***Archives in Mathematical Logic***, vol. 39 (2000), no. 6, pp. 393–401.

[3] Paul Cohen, *The independence of the Continuum Hypothesis*, ***Proceedings of the National Academy of Sciences of the United States of America***, vol. 50, 1963, pp. 1143–1148.

[4] Paul J. Cohen, ***Set theory and the Continuum Hypothesis***, W. A. Benjamin, Inc., New York-Amsterdam, 1966.

[5] Patrick Dehornoy, *Progrès récents sur l'Hypothèse du Continu [d'après Woodin]*, ***Séminaire Bourbaki***, 55ème année, 2002–2003, #915.

[6] Solomon Feferman, Harvey M. Friedman, Penelope Maddy, and John R. Steel, *Does mathematics need new axioms?*, ***The Bulletin of Symbolic Logic***, vol. 6 (2000), no. 4, pp. 401–446.

[7] Qi Feng, Menachem Magidor, and Hugh Woodin, *Universally Baire sets of reals*, ***Set Theory of the Continuum (Berkeley, CA, 1989)***, Springer, New York, 1992, pp. 203–242.

[8] M. Foreman, M. Magidor, and S. Shelah, *Martin's Maximum, saturated ideals, and nonregular ultrafilters, I*, ***Annals of Mathematics (2)***, vol. 127 (1988), no. 1, pp. 1–47.

[9] Matthew Foreman, *Generic large cardinals: new axioms for mathematics?*, ***Proceedings of the International Congress of Mathematicians, vol. II (Berlin, 1998)***, 1998, (electronic), pp. 11–21.

[10] Kurt Gödel, ***The Consistency of the Continuum Hypothesis***, Princeton University Press, Princeton, N.J., 1940.

[11] Martin Goldstern and Saharon Shelah, *The bounded proper forcing axiom*, ***The Journal of Symbolic Logic***, vol. 60 (1995), no. 1, pp. 58–73.

[12] Steve Jackson, ad *and the projective ordinals*, ***Cabal seminar 81–85***, Springer, Berlin, 1988, pp. 117–220.

[13] Akihiro Kanamori, *The emergence of descriptive set theory*, ***From Dedekind to Gödel, Boston, MA, 1992***, Kluwer Academic Publishers, Dordrecht, 1995, pp. 241–262.

[14] A. Lévy and R. M. Solovay, *Measurable cardinals and the Continuum Hypothesis*, ***Israel Journal of Mathematics***, vol. 5 (1967), pp. 234–248.

[15] N. Luzin, *Sur les ensembles projectifs de M. Henri Lebesgue*, ***Comptes Rendus Hebdomadaires des Séances de l'Académie des Sciences, Paris***, vol. 180 (1925), pp. 1572–1574.

[16] Donald A. Martin, *Hilbert's first problem: the Continuum Hypothesis. Mathematical developments arising from Hilbert problems*, ***Proceedings of the Symposium on Pure Mathematics, Northern Illinois University, De Kalb, Illinois, 1974***, American Mathematical Society, Providence, R.I., 1976, pp. 81–92.

[17] Donald A. Martin and John R. Steel, *A proof of projective determinacy*, ***Journal of the American Mathematical Society***, vol. 2 (1989), no. 1, pp. 71–125.

[18] William J. Mitchell and John R. Steel, ***Fine structure and iteration trees***, Springer, Berlin, 1994.

[19] Yiannis N. Moschovakis, ***Descriptive set theory***, North-Holland, Amsterdam, 1980.

[20] Jan Mycielski, *Axioms which imply GCH*, ***Fundamenta Mathematicae***, (2002).

[21] Jan Mycielski and H. Steinhaus, *A mathematical axiom contradicting the axiom of choice*, ***Bull. Acad. Polon. Sci. Sér. Sci. Math. Astronom. Phys.***, vol. 10 (1962), pp. 1–3.

[22] Jan Mycielski and S. Świerczkowski, *On the Lebesgue measurability and the axiom of determinateness*, ***Fundamenta Mathematicae***, vol. 54 (1964), pp. 67–71.

[23] Itay Neeman, *Optimal proofs of determinacy*, ***The Bulletin of Symbolic Logic***, vol. 1 (1995), no. 3, pp. 327–339.

[24] Saharon Shelah, *The future of set theory*, ***Set Theory of the Reals (Ramat Gan, 1991)***, Bar-Ilan Univ., Ramat Gan, 1993, pp. 1–12.

[25] ———, ***Cardinal arithmetic***, Oxford Science Publications, The Clarendon Press Oxford University Press, New York, 1994.

[26] Robert M. Solovay, *A model of set-theory in which every set of reals is Lebesgue measurable*, ***Annals of Mathematics (2)***, vol. 92 (1970), pp. 1–56.

[27] ———, *Strongly compact cardinals and the GCH*, ***Proceedings of the Tarski Symposium, (Proceedings of Symposia in Pure Mathematics, vol. XXV, University of California, Berkeley, California, 1971)***, American Mathematical Society, Providence, R.I., 1974, pp. 365–372.

[28] John R. Steel, ***The core model iterability problem***, Springer, Berlin, 1996.

[29] W. WADGE, *Degrees of complexity of subsets of the Baire space*, ***Notices of the American Mathematical Society***, vol. 19 (1972), pp. A–714.

[30] STAN WAGON, ***The Banach-Tarski paradox***, Cambridge University Press, Cambridge, 1993, with a foreword by Jan Mycielski, corrected reprint of the 1985 original.

[31] W. HUGH WOODIN, ***The Axiom of Determinacy, forcing axioms, and the nonstationary ideal***, Walter de Gruyter & Co., Berlin, 1999.

[32] ———, *The Continuum Hypothesis. I*, ***Notices of the American Mathematical Society***, vol. 48 (2001), no. 6, pp. 567–576.

[33] ———, *The Continuum Hypothesis. II*, ***Notices of the American Mathematical Society***, vol. 48 (2001), no. 7, pp. 681–690.

DEPARTMENT OF MATHEMATICS
UNIVERSITY OF CALIFORNIA, BERKELEY
BERKELEY, CA 94720-3840, USA
E-mail: woodin@math.berkeley.edu

PARTICIPANT PHOTOGRAPHS

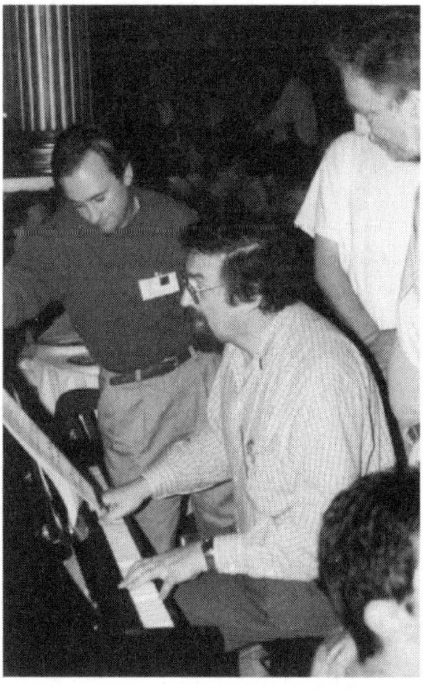

Maurice Boffa (1939–2001)

ARTICLES

BOUNDED FORCING AXIOMS AND THE SIZE OF THE CONTINUUM

DAVID ASPERÓ

Abstract. We prove that if BMM holds and $X^\sharp$ does not exist for some $X \in H(\omega_2)$, then $2^{\aleph_0} = \aleph_2$. Also, if a strong form of BMM holds and $X^\sharp$ does not exist for some set X, then also $2^{\aleph_0} = \aleph_2$. On the other hand, and building on previous work of Woodin, we show that, if BMM holds, $r^\sharp$ exists for every real r, the second uniform indiscernible is ω_2 and $NS^+_{\omega_1}$ forces that the generic ultrapower has an initial well-founded segment of order type $\check{\omega}_2 + 1$, then ψ_{AC} holds, and so $2^{\aleph_0} = \aleph_2$.

§1. Notation and preliminary facts. Let κ be a cardinal and X a set. $Coll(\kappa, X)$ denotes the partially ordered set (*poset*, for short) whose elements are all functions $p \subseteq \kappa \times X$ of size less than κ and such that $q \leq p$ (q is stronger than p) if and only if $p \subseteq q$.

$[X]^\kappa$ and $[X]^{<\kappa}$ denote $\{x \subseteq X : |x| = \kappa\}$ and $\{x \subseteq X : |x| < \kappa\}$, respectively.

$C \subseteq [X]^\kappa$ is a *club* (*of* $[X]^\kappa$) if and only if for every $x \in [X]^\kappa$ there is $y \supseteq x$ such that $y \in C$, and for every $\lambda < \kappa$ and every $\subseteq$-increasing sequence $\langle x_\alpha : \alpha \in \lambda\rangle$ of elements of C, $\bigcup_{\alpha<\lambda} x_\alpha \in C$.

$A \subseteq [X]^\kappa$ is a *stationary subset* (*of* $[X]^\kappa$) if and only if $A \cap C \neq \emptyset$ for every club $C \subseteq [X]^\kappa$.

If $f : [X]^{<\omega} \longrightarrow X$ and $Y \subseteq X$ is such that, for every $s \in [Y]^{<\omega}$, $f(s) \in Y$, then we say that Y *is closed under* f. Also, for every $Y \subseteq X$, let $cl_f(Y)$ denote the *closure of* Y *under* f, i.e., the $\subseteq$-minimal subset of X including Y and closed under f.

It is clear that, for every infinite cardinal κ and every $f : [X]^{<\omega} \longrightarrow X$, the set of all elements of $[X]^\kappa$ closed under f is a club of $[X]^\kappa$. The following fact, which will be frequently used, is well-known:

Lemma 1.1. [12] *For every club $C \subseteq [X]^{\aleph_0}$ there is a function $f : [X]^{<\omega} \longrightarrow X$ such that all $x \in [X]^{\aleph_0}$ which are closed under f are in C.*

I thank the organizing commitee of the LC2000 for giving me the opportunity of presenting some of the results in this paper (Section 2). I also wish to thank S. Todorčević for his useful comments, especially those concerning Theorems 2.1 and 3.11.

Logic Colloquium 2000
Edited by R. Cori, A. Razborov, S. Todorčević, and C. Wood
Lecture Notes in Logic, 19

If $\omega_1 \subseteq X$, $A \subseteq [X]^{\aleph_0}$ is a *projective stationary subset* (*of* $[X]^{\aleph_0}$) [7] if, for every stationary $T \subseteq \omega_1$, $\{x \in A\colon x \cap \omega_1 \in T\}$ is a stationary subset of $[X]^{\aleph_0}$.

Given a set X and $A \subseteq [X]^{\aleph_0}$, we define the following poset $\mathbb{P}_A$: $p \in \mathbb{P}_A$ if and only if p is a strictly $\subseteq$-increasing and $\subseteq$-continuous (i.e., if $\alpha \in dom(p)$ is a limit ordinal, then $p(\alpha) = \bigcup_{\beta<\alpha} p(\beta)$) $\nu + 1$-sequence of elements of A for some countable ordinal ν. $q \leq p$ if and only if $p \subseteq q$.

We will frequently use the following fact:

LEMMA 1.2. [7] *Let X be a set and let A be a stationary subset of $[X]^{\aleph_0}$. Then $\mathbb{P}_A$ forces the existence of a $\subseteq$-strictly increasing and $\subseteq$-continuous sequence $\langle x_\alpha\colon \alpha < \omega_1\rangle$ of members of A such that $X = \bigcup_{\alpha<\omega_1} x_\alpha$.*

Suppose, in addition, that $\omega_1 \subseteq X$. Then $\mathbb{P}_A$ is stationary-set-preserving (i.e., for some condition $p \in \mathbb{P}_A$, $p \Vdash_{\mathbb{P}_A}$ "$\check{S}$ is stationary" for every stationary $S \subseteq \omega_1$) if and only if A is a projective stationary subset of $[X]^{\aleph_0}$.

DEFINITION 1.1. *Let Γ be a class of posets. The Bounded Forcing Axiom for Γ $BFA(\Gamma)$ is the following statement:*

For every $\mathbb{P} \in \Gamma$ and every set $\mathcal{A}$ of size at most $\aleph_1$ consisting of maximal antichains of $\mathbb{P}$, all of size at most $\aleph_1$, there exists a filter $G \subseteq \mathbb{P}$ which is $\mathcal{A}$-generic (i.e., $G \cap A \neq \emptyset$ for all $A \in \mathcal{A}$).

If Γ is the class of all proper (semiproper) (stationary-set-preserving) posets, then we let *BPFA* (*BSPFA*) (*BMM*) denote $BFA(\Gamma)$.

The following result shows that bounded forcing axioms are natural absoluteness principles for $H(\omega_2)$.

FACT 1.3. [2] *Let Γ be a class of complete Boolean algebras. Then $BFA(\Gamma)$ holds if and only if for every $a \in H(\omega_2)$ and every Σ_1 formula $\varphi(x)$, if there is some $\mathbb{P} \in \Gamma$ such that $\Vdash_{\mathbb{P}} \varphi(\check{a})$, then $H(\omega_2) \models \varphi(a)$.*

DEFINITION 1.2. *Let T be a first order theory in the language of set theory and let Γ be a class of posets in T (i.e., $T \vdash \forall \mathbb{P} \in \Gamma(\mathbb{P}$ is a poset$)$).*

A formula $\varphi(x)$ is a T-provably Γ-persistent predicate if and only if

$$T \vdash \forall x \left[\varphi(x) \rightarrow (\forall \mathbb{P} \in \Gamma(\Vdash_{\mathbb{P}} \varphi(\check{x})))\right].$$

We will typically take $T = ZFC$ and suppress the '*ZFC*-' prefix.

My original inspiration for Definition 1.2 came from [14], where the notion of *provably ccc-persistent predicate* is defined.

DEFINITION 1.3. *Let Γ be a class of partially ordered sets and let Σ be a class of formulas of the language of set theory. Then $BFA(\Gamma, \Sigma)$ denotes the following schema:*

Let $\varphi(x)$ be a formula in Σ. Given $a \in H(\omega_2)$, if there is some $\mathbb{P}$ in Γ such that $\Vdash_{\mathbb{P}} \varphi(\check{a})$, then $\varphi(a)$.

Suppose Σ is the class of all provably Γ-persistent predicates. Then $BFA(\Gamma)'$ denotes $BFA(\Gamma, \Sigma)$ and, for every natural number $n \geq 1$, $BFA(\Gamma)'_n$ denotes $BFA(\Gamma, \Sigma \cap \Sigma_n)$, where Σ_n is the class of all Σ_n formulas of the language of set theory.

Note that, because of the undefinability of truth, one can only express $BFA(\Gamma, \Sigma)$ as an axiom schema in general. However, due to the definability of the satisfaction relation for Σ_n formulas, if there is some $n < \omega$ such that $\Sigma \subseteq \Sigma_n$, then $BFA(\Gamma, \Sigma)$ is first order expressible.

If Γ is the class of all proper (semiproper) (stationary-set-preserving) posets, we write $BPFA'$ ($BSPFA'$) (BMM') and $BPFA'_n$ ($BSPFA'_n$) (BMM'_n) for $BFA(\Gamma)'$ and $BFA(\Gamma)'_n$, respectively.

Note that, whenever $\Gamma_0 \subseteq \Gamma_1$, the class of provably Γ_1-persistent predicates is included in the class of provably Γ_0-persistent predicates, and therefore $BFA(\Gamma_1)'$ ($BFA(\Gamma_1)'_n$) does not necessarily imply $BBFA(\Gamma_0)$ ($BFA(\Gamma_0)'_n$).

Fact 1.3 makes it reasonable to call the statements defined in 1.3 *Generalized Bounded Forcing Axioms.*

We say that an ordinal α is *Σ_n-correct* if, for every Σ_n formula $\varphi(x)$ and every $a \in V_\alpha$, $V_\alpha \models \varphi(a)$ if and only if $V \models \varphi(a)$.

As to the consistency strength of generalized bounded forcing axioms we have:

FACT 1.4.

(a) [10] *BPFA, BSPFA and 'there is an inaccessible Σ_2-correct cardinal' are all equiconsistent (modulo ZFC).*

(b) [1] *If $n \geq 2$, $BPFA'_n$, $BSPFA'_n$ and 'there is an inaccessible Σ_n-correct cardinal' are all equiconsistent (modulo ZFC).*

(c) [18] *Suppose ZFC + 'there is a proper class of Woodin cardinals' is consistent. Then so is ZFC + BMM.*

(d) [1] *Suppose $n \geq 2$ and there is an inaccessible Σ_n-correct cardinal which is the limit of a sequence of strongly compact cardinals. Then one can force $BFA(\Gamma, \Sigma)$, where Γ is the class of all stationary-set-preserving posets and Σ is the class of all provably semiproper persistent Σ_n predicates.*

Suppose $BFA(\Gamma)$ is a bounded forcing axiom extending $MA_{\aleph_1}$. It is well-known that $BFA(\Gamma)$ then implies $2^{\aleph_0} = 2^{\aleph_1} > \aleph_1$ [13]. It is also known that already the unbounded forcing axiom for the class of all σ-closed $*$ ccc posets implies $2^{\aleph_0} = \aleph_2$ (see [3], [15] and [17]). Therefore, an important question is whether any natural generic absoluteness principle for $H(\omega_2)$—i.e., any (generalized) bounded forcing axiom—already implies this.

§2. When the universe is not too far from being L-like.

The following two-person game appears in [9] and in [17].

DEFINITION 2.1. *Let $\lambda \geq \omega_2$ be a regular cardinal, let $\nu < \omega_1$ and let $h\colon [\lambda]^{<\omega} \longrightarrow \lambda$. Then $\mathcal{G}_{\nu,h}$ is the following game of length ω with players I and II:*

At stage n, player I plays an interval I_n of ordinals in λ and an ordinal $\xi_n \in I_n$ and player II plays an ordinal $\mu_n < \lambda$.

If $n \geq 1$, I is required to play I_n so that $inf(I_n) > \mu_{n-1}$.

Player I wins if and only if, letting $X = cl_h(\{\xi_n : n < \omega\} \cup \nu)$, $X \subseteq \bigcup_{n<\omega} I_n$ and $X \cap \omega_1 = \nu$.

It turns out (see [17]) that, given $h : [\lambda]^{<\omega} \longrightarrow \lambda$, there are club many $\nu < \omega_1$ such that player I has a winning strategy for $\mathcal{G}_{\nu,h}$.

THEOREM 2.1. *Suppose the universe is a set generic extension of $L[X]$ for some $X \in H(\omega_2)$ (more generally, suppose that, for some $X \in H(\omega_2)$ and some regular cardinal $\lambda \geq \omega_2$, every club of λ includes a club of λ belonging to $L[X]$). If $BPFA'_2$ holds, then $2^{\aleph_0} = \aleph_2$.*

PROOF. Let $W = \{\alpha < \lambda : cf^{L[X]}(\alpha) = \omega\}$. Recall (see [5]) that in $L[X]$ there is a $\diamondsuit_\lambda(W)$-sequence $\langle Y_\alpha : \alpha \in W\rangle$, i.e., $Y_\alpha \subseteq \alpha$ for all $\alpha \in W$ and, for every $Y \in \mathcal{P}^{L[X]}(\lambda)$, $\{\alpha \in W : Y \cap \alpha = Y_\alpha\}$ is a stationary subset of λ in $L[X]$. Let $W' = \{\alpha \in W : sup(Y_\alpha) = \alpha\}$ and let $\overline{C} = \langle C_\alpha : \alpha \in W'\rangle$ be a sequence in $L[X]$ such that, for every $\alpha \in W'$, C_α is a subset of Y_α cofinal in α and of order type ω. Let C be a club of λ in $L[X]$. In $L[X]$ there are stationarily many $\alpha \in W$ such that $Y_\alpha = C \cap \alpha$. Also, since the set of all limit points of C is a club in $L[X]$, it follows that in $L[X]$ there are stationarily many α's in W' such that $C_\alpha \subseteq Y_\alpha = C \cap \alpha$.

For every $\alpha \in W'$, let $C_\alpha(n)$ denote the n-th element of C_α. Fix a real $r \subseteq \omega$ such that both r and $\omega \backslash r$ are infinite. Let us see that the set $\mathcal{S}_{r,\overline{C}}$ of all $x \in [\lambda]^{\aleph_0}$ such that

$$x \cap \left[C_{sup(x)}(n),\quad C_{sup(x)}(n+1)\right) \neq \emptyset \text{ if and only if } n \in r$$

is a projective stationary subset of $[\lambda]^{\aleph_0}$:

Fix a stationary $S \subseteq \omega_1$ and $h : [\lambda]^{<\omega} \longrightarrow \lambda$.

We know that, for some $\nu \in S$, player I has a winning strategy σ for the game $\mathcal{G}_{\nu,h}$ in Definition 2.1. Let θ be a large enough cardinal and let $\langle N_i : i < \lambda\rangle$ be a strictly $\subseteq$-increasing and $\subseteq$-continuous $\in$-chain of elementary substructures of $H(\theta)$ of size less than λ such that, for all $i < \lambda$, $\sigma \in N_i$ and $N_i \cap \lambda \in \lambda$. Then, $D = \{N_i \cap \lambda : i < \lambda\}$ is a club of λ, and so there is some club E of λ in $L[X]$ such that $E \subseteq D$. Then, for some $\alpha \in W'$, $C_\alpha \subseteq E \subseteq D$. For each n, let i_n be such that $N_{i_n} \cap \lambda = C_\alpha(n)$. Since $\sigma \in N_{i_n}$ for all n, it is easy to find a play of $\mathcal{G}_{\nu,h}$ won by player I in which player I plays $I_n \subseteq [N_{i_n} \cap \lambda,\ N_{i_{n+1}} \cap \lambda)$ exactly when $n \in r$. But then, letting $x = cl_h(\{\xi_n : n < \omega\} \cup \nu)$, it turns out that x is an element of $\mathcal{S}_{r,\overline{C}}$ closed under h and such that $x \cap \omega_1 = \nu \in S$. This shows that $\mathcal{S}_{r,\overline{C}}$ is a projective stationary subset of $[\lambda]^{\aleph_0}$.

Now, applying $BPFA'_2$ to $Coll(\omega_1, \lambda)$ and noting that $\mathcal{S}_{r,\overline{C}}$ remains a stationary subset of $[\lambda]^{\aleph_0}$ in any stationary-set-preserving forcing extension of $V^{Coll(\omega_1,\lambda)}$, we get $\delta_r < \omega_2$ and a sequence

$$\overline{C}_r = \left\langle C^r_\alpha : \alpha < \delta_r, cf^{L[X]}(\alpha) = \omega\right\rangle$$

in $L_{\omega_2}[X]$ such that the set $\mathcal{S}_{r,\overline{C}_r}$ of all $x \in [\delta_r]^{\aleph_0}$ such that

$$x \cap \left[C^r_{sup(x)}(n),\ C^r_{sup(x)}(n+1)\right) \neq \emptyset \text{ if and only if } n \in r$$

is a stationary subset of $[\delta_r]^{\aleph_0}$.

Now suppose $2^{\aleph_0} > \aleph_2$. Then, as $|L_{\omega_2}[X]| = \aleph_2$, there is $Y \subseteq \mathcal{P}(\omega)$ of size $\aleph_3$ such that $\delta_r = \delta_{r'} = \delta_0$ and $\overline{C}_r = \overline{C}_{r'} = \overline{C}_0$ for all $r, r' \in Y$. It follows that $\mathcal{S}_{r,\overline{C}_0} \cap \mathcal{S}_{r',\overline{C}_0} = \emptyset$ for all $r \neq r'$ in Y, but this is a contradiction, since $|\delta_0| = \aleph_1$ and so $[\delta_0]^{\aleph_0}$ cannot be partitioned into more than $\aleph_1$-many stationary sets. ⊣

Corollary 2.2. *Suppose the universe is $L[X]$ for some set X. More generally, suppose that, for some set X, there is a cardinal λ such that, for every regular cardinal $\mu \geq \lambda$, every club of μ includes a club in $L[X]$. If $BPFA'_3$ holds, then $2^{\aleph_0} = \aleph_2$.*

Proof. Notice that $Coll(\omega_1, X)$ forces the following statement:

$(*_1)$: There is $X \in H(\omega_2)$ and λ such that, for every regular cardinal $\mu \geq \lambda$, every club of μ includes a club in $L_{\mu^+}[X]$.

$(*_1)$ is clearly Σ_3 expressible with ω_1 as a parameter. Also, provided $(*_1)$ holds, it gets preserved under any subsequent set forcing extension. It follows that, by $BPFA'_3$, $(*_1)$ holds. Now, the result follows from Theorem 2.1. ⊣

The idea for the coding of a given real using a $\diamondsuit$-sequence $\overline{C}$ in the proof of Theorem 2.1 is due to Todorčević.

Theorem 2.3. *Suppose that, for some $X \in H(\omega_2)$, $X^\sharp$ does not exist. If BMM holds, then $2^{\aleph_0} = \aleph_2$.*

Proof. Let $X \in H(\omega_2)$. If $X^\sharp$ does not exist, then Jensen's Covering Lemma applies to $L[X]$ (see [5]), i.e., for every set of ordinals Y there is Z such that $Y \subseteq Z$, $Z \in L[X]$ and $|Z| = |Y| + \aleph_1$. Therefore, for every limit $\alpha < \omega_3$, if $cf(\alpha) \leq \omega_1$ and $\omega_2 < \alpha$, then $cf^{\,L[X]}(\alpha) < \omega_2^V < \alpha$. It follows that

$$E = \{\alpha < \omega_3 \colon cf(\alpha) = \omega, cf^{\,L[X]}(\alpha) < \alpha\}$$

is a stationary subset of ω_3.

As $\alpha \longrightarrow cf^{\,L[X]}(\alpha)$ is regressive on E, there is $\alpha_0 < \omega_3$ such that

$$F = \{\alpha \in E \colon cf^{\,L[X]}(\alpha) = \alpha_0\}$$

is stationary.

For every $\alpha \in F$, let C_α be the $<_{L[X]}$-first cofinal set $C \subseteq \alpha$ in order type α_0.

A standard argument shows that there is some $\xi_0 < \alpha_0$ such that, for every $\beta < \omega_3$,

$$F_\beta = \{\alpha \in F \colon C_\alpha(\xi_0) \geq \beta\}$$

is a stationary subset of ω_3:

Suppose otherwise. Then, for every $\xi < \alpha_0$ there is some $\beta_\xi < \omega_3$ and some club $D_\xi \subseteq \omega_3$ such that $C_\alpha(\xi) < \beta_\xi$ for every $\alpha \in D_\xi \cap F$.

Let $\beta = sup_{\xi<\alpha_0} \beta_\xi < \omega_3$ and $D = \bigcap_{\xi<\alpha_0} D_\xi$. D is a club of ω_3 and, if $\alpha \in D \cap F$, then $C_\alpha(\xi) < \beta$ for every $\xi < \alpha_0$. Hence, $D \cap F \subseteq \beta + 1$, but this is a contradiction.

Again, $\alpha \longrightarrow C_\alpha(\xi_0)$ is regressive on F_β for every $\beta < \omega_3$. Thus, there is some $\beta_0 < \omega_3$ such that

$$F_{\omega_3,\alpha_0,\xi_0,\beta_0} = \{\alpha \in F : C_\alpha(\xi_0) = \beta_0\}$$

is a stationary and co-stationary subset of ω_3.

Fix a partition $\langle S_n : n < \omega\rangle$ of ω_1 into stationary sets. Pick $r \subseteq \omega$ and let $S_r = \bigcup_{n\in r} S_n$. Let

$$\mathcal{S}_r^{\omega_3,\alpha_0,\xi_0,\beta_0} = \left\{X \in [\omega_3]^{\aleph_0} : X \cap \omega_1 \in S_r \text{ if and only if } sup(X) \in F_{\omega_3,\alpha_0,\xi_0,\beta_0}\right\}.$$

Next we check that $\mathcal{S}_r^{\omega_3,\alpha_0,\xi_0,\beta_0}$ is a projective stationary subset of $[\omega_3]^{\aleph_0}$:

Fix $H : [\omega_3]^{<\omega} \longrightarrow \omega_3$ and T a stationary subset of ω_1. Let $n \in \omega$ be such that $T \cap S_n$ is a stationary subset of ω_1.

Suppose $n \in r$. Take $\alpha < \omega_3$, $\omega_1 < \alpha$ such that α is closed under H and $\alpha \in F_{\omega_3,\alpha_0,\xi_0,\beta_0}$. Since $cf(\alpha) = \omega$, there is a countable $X \subseteq \alpha$ such that X is cofinal in α and closed under H and $X \cap \omega_1 \in T \cap S_n$.

Similarly if $n \notin r$ (take α as above, $\alpha \notin F_{\omega_3,\alpha_0,\xi_0,\beta_0}$).

Now apply BMM to the standard poset $\mathbb{Q}_r$ $(= \mathbb{P}_{\mathcal{S}_r^{\omega_3,\alpha_0,\xi_0,\beta_0}})$ that shoots an ω_1-club through $\mathcal{S}_r^{\omega_3,\alpha_0,\xi_0,\beta_0}$.

By Lemma 1.2, $\mathbb{Q}_r$ is stationary-set-preserving and forces the following statement:

$(*_2)$: There are $\delta_r < \omega_2$, $\alpha_r < \delta_r$, $\xi_r < \alpha_r$, $\beta_r < \delta_r$ and a $\subseteq$-continuous strictly $\subseteq$-increasing sequence $\langle X_\nu^r : \nu < \omega_1\rangle$ of countable subsets of δ_r such that $\delta_r = \bigcup_{\nu<\omega_1} X_\nu^r$ and, for every $\nu < \omega_1$, $X_\nu^r \cap \omega_1 \in S_r$ if and only if $sup(X_\nu^r) \in F_{\delta_r,\alpha_r,\xi_r,\beta_r}$.

Note that "$\gamma \in F_{\delta_r,\alpha_r,\xi_r,\beta_r}$" is expressed by "$\gamma < \delta_r$, $cf(\gamma) = \omega$, $cf^{L[X]}(\gamma) = \alpha_r$ and, if C is the $<_{L[X]}$-least cofinal $D \subseteq \gamma$ in order type α_r, then the ξ_r-th member of C is β_r".

It follows that $(*_2)$ can be expresssed by means of a Σ_1 sentence with r, ω_1, $\langle S_n : n < \omega_1\rangle$ and X as parameters and, by BMM, it is true.

Now suppose $2^{\aleph_0} > \aleph_2$. Then there are $r \neq r'$ such that $\langle \delta_r, \alpha_r, \xi_r, \beta_r\rangle = \langle \delta_{r'}, \alpha_{r'}, \xi_{r'}, \beta_{r'}\rangle$.

Suppose $n \in r\backslash r'$ (the argument when $n \in r'\backslash r$ is symmetrical to this). $C = \{X_\nu^r : \nu < \omega_1\} \cap \{X_\nu^{r'} : \nu < \omega_1\}$ is a club of $[\delta_r]^{\aleph_0}$, $C = \{Y_\nu : \nu < \omega_1\}$. Let $\nu < \omega_1$ be such that $Y_\nu \cap \omega_1 \in S_n$. Then $sup(Y_\nu) \in F_{\delta_r,\alpha_r,\xi_r,\beta_r} = F_{\delta_{r'},\alpha_{r'},\xi_{r'},\beta_{r'}}$, contrary to the fact that $n \notin r'$. ⊣

Corollary 2.4. *Suppose $X^\sharp$ does not exist for some set of ordinals X. If BMM_2' holds, then $2^{\aleph_0} = \aleph_2$.*[1]

---

[1]Added in proof: Building on the proof of Theorem 2.3 and on Jensen's techniques for coding the universe with a real, Schindler has proved that in this result one can in fact replace BMM_2

PROOF. Since no forcing notion can add any sharp, $Coll(\omega_1, X)$ forces that there is a set $X \in H(\omega_2)$ such that $X^\sharp$ does not exist. Moreover, this is a set forcing persistent statement. Since "there is $X \in H(\omega_2)$, $\kappa > \omega_1$ and H such that $H = H(\kappa)$ and $H \models$ '$X^\sharp$ does not exist' " is Σ_2 expressible with ω_1 as a parameter, by BMM'_2 it follows that there really is $X \in H(\omega_2)$ whose sharp does not exist. Now apply Theorem 2.3. ⊣

§3. When the universe is very far from being L-like. Recall the following statement ψ_{AC} considered by Woodin ([18], Definition 5.12):

DEFINITION 3.1.

ψ_{AC}: *If S and T are stationary and co-stationary subsets of ω_1, then there is an ordinal δ, a bijection $\pi\colon \omega_1 \longrightarrow \delta$ and a club $C \subseteq \omega_1$ such that*

$$\{\nu < \omega_1 : o.t.\,(\pi``\nu) \in T\} \cap C = S \cap C$$

The following simple application of ψ_{AC} is due to Woodin ([18], Lemma 5.13).

FACT 3.1. ψ_{AC} *implies* $2^{\aleph_1} = \aleph_2$.

PROOF. Fix a pairwise disjoint sequence $\langle S_\alpha : \alpha < \omega_1\rangle$ of stationary subset of ω_1 and a stationary and co-stationary subset T of ω_1. For every nonempty $X \subseteq \omega_1$, $X \neq \omega_1$, let $S_X = \bigcup_{\alpha \in X} S_\alpha$. Then, S_X is a stationary and co-stationary subset of ω_1 and so, applying ψ_{AC} to S_X and T, we get an ordinal $\delta_X < \omega_2$, a bijection $\pi_X\colon \omega_1 \longrightarrow \delta_X$ and a club $C_X \subseteq \omega_1$ such that $\{\nu < \omega_1 : o.t.(\pi_X``\nu) \in T\} \cap C_X = S_X \cap C_X$. Suppose $2^{\aleph_1} > \aleph_2$. Then there would be distinct $X, Y \subseteq \omega_1$ such that $\delta_X = \delta_Y$. Let $\alpha \in X \setminus Y$ (if $\alpha \in Y \setminus X$, we argue symmetrically). Let $D = \{\nu < \omega_1 : \pi_X``\nu = \pi_Y``\nu\}$. $D$ is a club of $\omega_1$. Pick $\nu \in D \cap C_X \cap C_Y \cap S_\alpha$. Then, $o.t.(\pi_Y``\nu) = o.t.(\pi_X``\nu) \in T$, which is a contradiction, since $\nu \notin S_Y$. ⊣

The following result is due to Woodin [18].

THEOREM 3.2. *Suppose BMM holds and there is a measurable cardinal. Then ψ_{AC} holds.*

PROOF. Let κ be a measurable cardinal. We want to see that, whenever S and T are stationary and co-stationary subsets of ω_1,

$$\mathcal{S}_{S,T} = \left\{X \in [\kappa]^{\aleph_0} : X \cap \omega_1 \in S \text{ if and only if } o.t.(X) \in T\right\}$$

is a projective stationary subset of $[\kappa]^{\aleph_0}$. In order to prove this it suffices to check that, whenever S and T are stationary subsets of ω_1,

$$\left\{X \in [\kappa]^{\aleph_0} : X \cap \omega_1 \in S \text{ and } o.t.(X) \in T\right\}$$

is a stationary subset of $[\kappa]^{\aleph_0}$.

---

with BMM; he has actually shown that if ω_1 is inaccessible to reals and BMM holds, then the sharp of every set of ordinals exists.

Suppose otherwise. Let θ be a large enough regular cardinal (e.g., $\theta = (2^\kappa)^+$). We will build a strictly $\subseteq$-increasing and $\subseteq$-continuous sequence $\langle N_i : i < \omega_1 \rangle$ of countable elementary substructures of $H(\theta)$ containing κ, S and T and such that

(i) $N_0 \cap \omega_1 \in S$, and
(ii) $N_i \cap \kappa$ is a proper initial segment of $N_{i+1} \cap \kappa$ for every $i < \omega_1$.

This is enough since then, for some i_0, $o.t.(N_{i_0} \cap \kappa) \in T$. But, as $N_{i_0} \preccurlyeq H(\theta)$ and $S, T \in N_{i_0}$, in N_{i_0} there is some $H: [\kappa]^{<\omega} \longrightarrow \kappa$ such that no $X \in [\kappa]^{\aleph_0}$ with $X \cap \omega_1 \in S$ and $o.t.(X) \in T$ is closed under H. However, $N_{i_0} \cap \kappa$ is a countable subset of κ closed under H, which is a contradiction. It follows that $\{X \in [\kappa]^{\aleph_0} : X \cap \omega_1 \in S \text{ and } o.t.(X) \in T\}$ is a stationary subset of $[\kappa]^{\aleph_0}$.

So suppose N_i has been found.

Let $U \in N_i$ be a normal κ-complete ultrafilter on κ.

Let η be the least ordinal in $\bigcap(U \cap N_i)$. Note that, since U is κ-complete, $\bigcap(U \cap N_i) \in U$ and so η exists, and also that $\kappa \backslash \alpha \in U \cap N_i$ for every $\alpha \in N_i \cap \kappa$ and so $N_i \cap \kappa \subseteq \eta$.

Let

$$N_{i+1} = \{f(\eta) : f \text{ is a function with domain } \kappa,\ f \in N_i\}.$$

As the constant function with value a is in N_i for every $a \in N_i$, $N_i \subseteq N_{i+1}$.

Also, as $id_\kappa(\eta) = \eta$, $\eta \in N_{i+1}$.

Let us check that $N_{i+1} \cap \eta = N_i \cap \kappa$: Suppose $\gamma \in N_{i+1} \cap \eta$. For some function f in N_i with domain κ, $f(\eta) = \gamma$. Then, for some $x \in U \cap N_i$, f is regressive on x: Otherwise there would be some $x \in U \cap N_i$ such that $f(\alpha) \notin \alpha$ for all $\alpha \in x$. But, as $\eta \in x$, $f(\eta) \notin \eta$.

Hence, as U is a normal filter, f is constant on a set y in $U \cap N_i$ with value $\alpha \in N_i$. Finally, $\gamma = f(\eta) = \alpha$, since $\eta \in y$.

There only remains to see that $N_{i+1} \preccurlyeq H(\theta)$: Let $a_0, \dots a_{n-1} \in N_{i+1}$, let $\exists x \varphi(x, x_0, \dots x_{n-1})$ be a formula and suppose $H(\theta) \models \exists x \varphi(x, a_0, \dots a_{n-1})$. Let $f_0, \dots f_{n-1}$ be functions in N_i with domain κ such that $f_0(\eta) = a_0, \dots f_{n-1}(\eta) = a_{n-1}$. There is a function $f \in H(\theta)$ such that $dom(f) = \kappa$ and, for every $\alpha \in \kappa$, $H(\theta) \models \varphi(f(\alpha), f_0(\alpha), \dots f_{n-1}(\alpha))$ in case there is some $b \in H(\theta)$ such that $H(\theta) \models \varphi(b, f_0(\alpha), \dots f_{n-1}(\alpha))$. By elementarity, there is also such a function $f_\star$ in N_i. But then, $H(\theta) \models \varphi(f_\star(\eta), a_0, \dots a_{n-1})$ and $f_\star(\eta) \in N_{i+1}$.

We have just seen that $\mathcal{S}_{S,T}$ is a projective stationary subset of $[\kappa]^{\aleph_0}$ whenever S and T are stationary and co-stationary subsets of ω_1. But then, by Lemma 1.2, the standard poset $\mathbb{Q}_{S,T}$ $(= \mathbb{P}_{\mathcal{S}_{S,T}})$ that shoots an ω_1-club through $\mathcal{S}_{S,T}$ with countable conditions is stationary-set-preserving.

$\mathbb{Q}_{S,T}$ forces the following statement:

$(*_3)$: There is an ordinal $\delta < \omega_2$ and a strictly $\subseteq$-increasing and $\subseteq$-continuous sequence $\langle X_\nu : \nu < \omega_1 \rangle$ of countable subsets of $[\delta]^{\aleph_0}$ such that $\delta =$

$\bigcup_{\nu<\omega_1} X_\nu$ and such that, for every $\nu < \omega_1$, $X_\nu \cap \omega_1 \in S$ if and only if $o.t.(X_\nu) \in T$.

Since $(*_3)$ is a Σ_1 statement with ω_1, S and T as parameters, by *BMM* there really is such a δ and such a decomposition $\langle X_\nu : \nu < \omega_1 \rangle$ of δ. Now let π be any bijection between ω_1 and δ. Since $\{X_\nu : \nu < \omega_1\}$ is a club of $[\delta]^{\aleph_0}$, there is a club $C \subseteq \omega_1$ such that, for every $\nu \in C$, $\pi``\nu = X_\nu$ and $X_\nu \cap \omega_1 = \nu$. Thus, for every $\nu \in C$, $\nu \in S$ if and only if $o.t.(\pi``\nu) \in T$. ⊣

The following are essentially Definition 3.5 and Lemmas 3.8, 3.10 and 4.36, respectively, in [18]. For the reader's convenience, we reproduce the proofs of the lemmas.

DEFINITION 3.2. *Let M be a transitive set and let I be a (perhaps proper) class of M which, in M, is a normal uniform ideal on ω_1^M. A sequence*

$$\langle \langle \mathcal{M}_\beta, I_\beta \rangle, G_\alpha, j_{\alpha,\beta} : \alpha < \beta < \gamma \rangle$$

is an iteration of $\langle M, I \rangle$ if the following holds:

(1) $\mathcal{M}_0 = \langle M, \in |M \rangle$ *and* $I_0 = I$.
(2) $j_{\alpha,\beta} : \mathcal{M}_\alpha \longrightarrow \mathcal{M}_\beta$ *is a commuting family of elementary embeddings.*
(3) *For each* $\beta < \gamma$, $I_\beta = j_{0,\beta}(I_0)$.
(4) *For each η such that $\eta + 1 < \gamma$, G_η is $(\mathcal{P}(\omega_1)\backslash I_\eta)^{\mathcal{M}_\eta}$-generic over $\mathcal{M}_\eta$, $\mathcal{M}_{\eta+1}$ is the $\mathcal{M}_\eta$-ultrafilter of $\mathcal{M}_\eta$ by G_η and $j_{\eta,\eta+1}$ is the induced elementary embedding.*
(5) *For each limit $\beta < \gamma$, $\mathcal{M}_\beta$ is the direct limit of $\{\mathcal{M}_\alpha : \alpha < \beta\}$ and for all $\alpha < \beta$, $j_{\alpha,\beta}$ is the induced elementary embedding.*

If γ is a limit ordinal, γ is the length of the iteration. Otherwise, the length of the iteration is δ such that $\delta + 1 = \gamma$.

A pair $\langle \mathcal{N}, J \rangle$ is an iterate of $\langle M, I \rangle$ if it occurs in an iteration of $\langle M, I \rangle$.

If an iterate $\mathcal{N}$ of M is well-founded, we may identify it with its transitive collapse.

$\langle M, I \rangle$ is iterable if and only if every iterate of $\langle M, I \rangle$ is well-founded.

$\langle M, I \rangle$ is weakly iterable if and only if $\omega_1^{\mathcal{N}}$ is well-founded for every iterate $\langle \mathcal{N}, J \rangle$ of $\langle M, I \rangle$.

If $I = (NS_{\omega_1})^M$, then we write

$$\langle \mathcal{M}_\beta, G_\alpha, j_{\alpha,\beta} : \alpha < \beta < \gamma \rangle$$

for $\langle \langle \mathcal{M}_\beta, I_\beta \rangle, G_\alpha, j_{\alpha,\beta} : \alpha < \beta < \gamma \rangle$.

We also say that a model $\mathcal{N}$ is an iterate of M if $\langle \mathcal{N}, (NS_{\omega_1})^{\mathcal{N}} \rangle$ is an iterate of $\langle M, (NS_{\omega_1})^M \rangle$ and that M is iterable (weakly iterable) if every iterate of M is well-founded (if, for every iterate $\mathcal{N}$ of M, $\omega_1^{\mathcal{N}}$ is well-founded).

From now on, let *ZFC** denote a suitably large fragment of *ZFC*. The following two facts are easily proved.

FACT 3.3. *Let x be a real coding a pair $\langle M, I\rangle$, where M is a countable transitive set and $I \subseteq M$ is, in M, a normal uniform ideal on ω_1^M. Then, 'the pair coded by x is iterable' is a Π^1_2 assertion about x.*

FACT 3.4. *Suppose that M and M^* are transitive models of ZFC* such that*

(i) $\omega_1^M = \omega_1^{M^*}$,
(ii) $\mathcal{P}(\omega_1)^M = \mathcal{P}(\omega_1)^{M^*}$.

Suppose that either

(iii) $\mathcal{P}(\mathcal{P}(\omega_1))^M = \mathcal{P}(\mathcal{P}(\omega_1))^{M^*}$, *or*
(iv) $M^* \models NS_{\omega_1}$ *is* $\aleph_2$*-saturated*.

Suppose $\langle \mathcal{M}_\beta, G_\alpha, j_{\alpha,\beta}\colon \alpha < \beta < \gamma\rangle$ is an iteration of M. Then there is a unique iteration $\langle \mathcal{M}^*_\beta, G^*_\alpha, j^*_{\alpha,\beta}\colon \alpha < \beta < \gamma\rangle$ of M^* such that, for all $\alpha < \beta < \gamma$,

(1) $\omega_1^{\mathcal{M}_\beta} = \omega_1^{\mathcal{M}^*_\beta}$,
(2) $\mathcal{P}(\omega_1)^{\mathcal{M}_\beta} = \mathcal{P}(\omega_1)^{\mathcal{M}^*_\beta}$, *and*
(3) $G_\alpha = G^*_\alpha$.

Suppose further that $M \in M^*$. Then, for all $\beta < \gamma$, $j^*_{0,\beta}(M) \in \mathcal{M}^*_\beta$ and there is an elementary embedding $k_\beta\colon \mathcal{M}_\beta \longrightarrow j^*_{0,\beta}(M)$ such that $j^*_{0,\beta} \restriction M = k_\beta \circ j_{0,\beta}$.

And similarly if $M^* \in M$.

LEMMA 3.5. *Suppose M is a transitive model of ZFC* in which NS_{ω_1} is precipitous. Suppose $\langle \mathcal{M}_\beta, G_\alpha, j_{\alpha,\beta}\colon \alpha < \beta < \gamma\rangle$ is an iteration of M such that $\gamma \leq M \cap Ord$. Then $\mathcal{M}_\beta$ is well-founded for all $\beta < \gamma$.*

PROOF. Suppose otherwise and let $\langle \gamma_0, \kappa_0, \eta_0\rangle$ be the least triple of ordinals in M, relative to the lexicographical order, such that

(1.1) $M \models cf(\kappa_0) > \omega_1$,
(1.2) $\eta_0 < \kappa_0$,
(1.3) there is an iteration $\langle \mathcal{N}_\beta, G_\alpha, j_{\alpha,\beta}\colon \alpha < \beta < \gamma_0 + 1\rangle$ of $V_{\kappa_0} \cap M$ such that $j_{0,\gamma_0}(\eta_0)$ is not well-founded.

Notice that η_0 is a limit ordinals and that, since NS_{ω_1} is precipitous, so is γ_0.

Fix an iteration $\langle \mathcal{N}_\beta, G_\alpha, j_{\alpha,\beta}\colon \alpha < \beta < \gamma_0 + 1\rangle$ witnessing that $\langle \gamma_0, \kappa_0, \eta_0\rangle$ is the above defined triple of ordinal.

Choose $\beta^* < \gamma_0$ and $\eta^* < j_{0,\beta^*}(\eta_0)$ such that $j_{\beta^*,\gamma_0}(\eta^*)$ is not well-founded.

Let $\langle \mathcal{M}_\beta, G_\alpha, j^*_{\alpha,\beta}\colon \alpha < \beta < \gamma_0 + 1\rangle$ be the induced iteration of M given by Fact 3.4 (κ_0 is a limit ordinal such that $M \models cf(\kappa_0) > \omega_1$). By the minimality of γ_0 it follows that $\mathcal{M}_\beta$ is well-founded for all $\beta < \gamma_0$.

Now, it turns out that $\langle \gamma_0, \kappa_0, \eta_0\rangle$ can be correctly defined in M: $\langle \gamma_0, \kappa_0, \eta_0\rangle$ is the least triple of ordinals in M, relative to the lexicographical order, such that

(2.1) $M \models cf(\kappa_0) > \omega_1$,
(2.2) $\eta_0 < \kappa_0$;

(2.3) there exist $X \in M$ and $p \in Coll(\omega, X)$ such that, in M, $p \Vdash_{Coll(\omega,X)}$ 'there is an iteration $\langle \mathcal{N}^*_\beta, G^*_\alpha, j^*_{\alpha,\beta} \colon \alpha < \beta < \check{\gamma}_0 + 1\rangle$ of $V_{\check{\kappa}_0} \cap M$ such that $j^*_{0,\gamma_0}(\check{\eta}_0)$ is not well-founded'.

Certainly, if $p \in Coll(\omega, X)$ is as in (2.3), then there really is such an ill-founded iteration. On the other hand, if there really is such an iteration, X is large enough, $G \subseteq Coll(\omega, X)$ is generic over M and x_0, x_1 and x_2 are reals coding γ_0, $V_{\kappa_0} \cap M$ and η_0 respectively in $M[G]$ then, as $M[G]$ is $\Sigma^1_1(x_0, x_1, x_2)$-correct, some $p \in G$ is as in (2.3).

Further, since $\mathcal{M}_{\beta^*}$ is well-founded, the same considerations apply to $\mathcal{M}_{\beta^*}$ and so $\langle j^*_{0,\beta^*}(\gamma_0), j^*_{0,\beta^*}(\kappa_0), j^*_{0,\beta^*}(\eta_0)\rangle$ must be the triple as defined in V for $\mathcal{M}_{\beta^*}$. However, the tail of the iteration $\langle \mathcal{N}_\beta, G_\alpha, j_{\alpha,\beta} \colon \alpha < \beta < \gamma_0+1\rangle$ starting at β^* is an iteration of $j^*_{0,\beta^*}(V_{\kappa_0} \cap M) = V_{j^*_{0,\beta^*}(\kappa_0)} \cap \mathcal{M}^*_\beta$ of length at most γ_0 and $\gamma_0 + 1 \leq j_{0,\beta^*}(\gamma_0) + 1$. Further, the image of η^* under this iteration is ill-founded. This is a contradiction since $\eta^* < j^*_{0,\beta^*}(\eta_0)$. ⊣

Lemma 3.6. *Suppose M is a countable transitive model of ZFC* and $I \subseteq M$ is, in M, a normal uniform ideal on ω_1^M such that $\langle M, I\rangle$ is a weakly iterable pair. Suppose J is a normal uniform ideal on ω_1. Then there exists an iteration $\langle \mathcal{M}^*, I^*\rangle$ such that*

(1) $j(\omega_1^M) = \omega_1$,
(2) $J \cap \mathcal{M}^* = I^*$.

Proof. Fix a sequence $\langle A_{k,\nu} \colon k < \omega, \nu < \omega_1\rangle$ of J-positive sets which are pairwise disjoint. J is normal and so each $A_{k,\nu}$ is a stationary subset of ω_1. We may also assume that $A_{k,\nu} \cap (\nu + 1) = \emptyset$ for all k and ν.

Fix a surjection

$$f \colon \omega \times \omega_1^M \longrightarrow H(\omega_2)^M \backslash I$$

such that

(1.1) for all $k < \omega$, $f \restriction (k \times \omega_1^M) \in M$,
(1.2) for all $A \in M$, if A has size $\aleph_1^M$ in M and if $A \subseteq \mathcal{P}(\omega_1^M)\backslash I$, then $A \subseteq range(f \restriction (k \times \omega_1^M))$ for some $k \in \omega$, and
(1.3) for every $k < \omega$ and every $A \in range(f \restriction (k \times \omega_1^M))$, if A is an ω_1^M-sequence of elements of $H(\omega_2)^M\backslash I$, then $range(A) \subseteq range(f \restriction (k \times \omega_1^M))$.

Suppose $j \colon \langle M, I\rangle \longrightarrow \langle \mathcal{M}^*, I^*\rangle$ is an iteration. Then we define

$$j(f) = \cup\{j\,(f \restriction (k \times \omega_1^M)) : k \in \omega\}$$

and it is easily verified that the range of $j(f)$ is $H(\omega_2)^{\mathcal{M}^*}\backslash I^*$.

We construct an iteration $\langle\langle \mathcal{M}_\beta, I_\beta\rangle, G_\alpha, j_{\alpha,\beta} \colon \alpha < \beta \leq \omega_1\rangle$ such that, for each $\alpha < \omega_1$, if $\omega_1^{\mathcal{M}_\alpha} \in A_{k,\nu}$, then $j_{0,\alpha}(f)(k, \nu) \in G_\alpha$ (since all $\omega_1^{\mathcal{M}_\alpha}$ are well-founded, we identify them with their transitive collapses).

The set $C = \{j_{0,\alpha}(\omega_1^M) : \alpha < \omega_1\}$ is a club of ω_1. Thus for each $B \subseteq \omega_1$ such that $B \in \mathcal{M}_{\omega_1}$ and $B \notin j_{0,\omega_1}(I)$ there exist $k < \omega$ and $\nu < \omega_1$ such that

$$(C \backslash \alpha) \cap A_{k,\nu} \subseteq B \cap A_{k,\nu}$$

for some $\alpha < \omega_1$. On the other hand, if $B \subseteq \omega_1$, $B \in \mathcal{M}_{\omega_1}$ and $B \in j_{0,\omega_1}(I)$, then $B \cap C = \emptyset$.

It follows that $J \cap \mathcal{M}_{\omega_1} = I_{\omega_1}$. ⊣

We also reproduce Woodin's proof of Theorem 3.7, as it will be used in Theorem 3.12.

THEOREM 3.7 (Woodin). *Suppose NS_{ω_1} is precipitous and BMM holds. Then ψ_{AC} holds.*

PROOF. This is Lemma 10.95 in [18]. Let $\gamma = |\mathcal{P}(\mathcal{P}(\omega_1))|^+$. By the same kind of argument as in the proof of Theorem 3.2, it is enough to prove that, whenever S and T are stationary subsets of ω_1,

$$\{X \in [\gamma]^{\aleph_0} : X \cap \omega_1 \in S \text{ and } o.t.(X) \in T\}$$

is a stationary subset of $[\gamma]^{\aleph_0}$. Fix $H : [\gamma]^{<\omega} \longrightarrow \gamma$. Let $\kappa = (2^{\aleph_1})^+$ and let $\theta > \gamma$ be large enough. Let $G \subseteq Coll(\omega, H(\kappa))$ be generic over V. Note that, since $|H(\kappa)| = |\mathcal{P}(\mathcal{P}(\omega_1))|$, $\gamma = \omega_1^{V[G]}$.

CLAIM 3.8. *In $V[G]$ there is an elementary embedding $j : H^V(\theta) \longrightarrow \mathcal{M}$ such that $\omega_1^V \in j(S)$ and $j(T) \subseteq \gamma$ is a stationary subset of $\gamma = \omega_1$.*

PROOF OF CLAIM. Work in $V[G]$: Note that, by Fact 3.4 and as $\mathcal{P}(\mathcal{P}(\omega_1))^{H^V(\kappa)} = \mathcal{P}(\mathcal{P}(\omega_1))^{H^V(\theta)}$, every iteration $\langle \mathcal{M}_\beta, G_\alpha, j_{\alpha,\beta} : \alpha < \beta \leq \gamma\rangle$ of $H^V(\kappa)$ induces a unique iteration $\langle \mathcal{N}_\beta, G_\alpha, j^*_{\alpha,\beta} : \alpha < \beta \leq \gamma\rangle$ of $H^V(\theta)$. Furthermore, since $\gamma < \theta$ and $H^V(\theta) \models NS_{\omega_1}$ is precipitous, by Lemma 3.5 this iteration must be well-founded and since, by Fact 3.4, $\mathcal{M}_\beta \subseteq \mathcal{N}_\beta$ for all $\beta < \gamma$, $\langle \mathcal{M}_\beta, G_\alpha, j_{\alpha,\beta} : \alpha < \beta \leq \gamma\rangle$ is also well-founded. Hence, as $H^V(\kappa)$ is countable, there is such an iteration in $V[G]$ with $\omega_1^V \in j_{0,\gamma}(S)$ (take $f(0,0) = A_{0,0} = S$ in the proof of Lemma 3.6) and $j_{0,\gamma}(T)$ a stationary subset of ω_1. Again, by Fact 3.4, this induces a unique iteration $\langle \mathcal{N}_\beta, G_\alpha, j^*_{\alpha,\beta} : \alpha < \beta \leq \gamma\rangle$ of $H^V(\theta)$ and we certainly have that $\omega_1^V \in j^*_{0,\gamma}(S)$ and that $j^*_{0,\gamma}(T)$ is a stationary subset of ω_1. Take $\mathcal{M} = \mathcal{N}_\gamma$ and $j = j^*_{0,\gamma}$. ⊣

Now let $D = \{\alpha \in \gamma : [\alpha]^{<\omega}$ is closed under $H\}$. D is a club of γ and so there is some $\alpha \in D \cap j(T)$. Let $X = j``\alpha$. Note that X is a countable subset of $j(\alpha)$ such that

(1) $X \cap j(\omega_1^V) = \omega_1^V \in j(S)$,
(2) X is closed under $j(H)$ and
(3) $o.t.(X) = \alpha \in j(T)$.

In $\mathcal{M}$, let $f : \omega \longrightarrow \omega_1^V$ and $g : \omega \longrightarrow \alpha$ be bijections. Let $\mathbb{P}$ the following poset of finite approximations to a set satisfying (1)–(3) above:

$t \in \mathbb{P}$ if and only if $t = \langle n_t, x_t, h_t\rangle$, where

(1.1) $n_t \in \omega$,
(1.2) $x_t \in [j(\alpha)]^{<\omega}$,
(1.3) $h_t \colon x_t \longrightarrow \alpha$ is order-preserving, and
(1.4) $g``n_t \subseteq range(h_t)$ and $f``n_t \subseteq x_t \cap j(\omega_1^V) \subseteq \omega_1^V$.

$t <_{\mathbb{P}} s$ if and only if

(2.1) $n_s < n_t$,
(2.2) $x_s \subseteq x_t$,
(2.3) $h_s \subseteq h_t$, and
(2.4) $j(H)``[x_s]^{<\omega} \subseteq x_t$.

Note that $j``\alpha$ gives place in $V[G]$ to an infinite $<_{\mathbb{P}}$-chain. Hence, $<_{\mathbb{P}}$ is not well-founded in $V[G]$ and, by absoluteness of well-foundedness, neither it is in $\mathcal{M}$ (here it is enough to use that $\omega_1^{\mathcal{M}}$ is well-founded, so that an order-preserving map $R \colon \mathbb{P} \longrightarrow \omega_1^{\mathcal{M}}$ in $\mathcal{M}$ would really witness well-foundedness of $<_{\mathbb{P}}$ in $V[G]$). But then, in M there is an infinite $<_{\mathbb{P}}$-chain $\langle t_n \colon n \in \omega \rangle$. By the definition of $\mathbb{P}$, $X = \bigcup_n x_{t_n}$ is a countable subset of $j(\alpha) \subseteq j(\omega_2)$ such that (1)–(3) hold for X. Now we are done by elementarity. ⊣

If κ is an uncountable regular cardinal, $S \in NS_\kappa^+$ and $f, g \colon \kappa \longrightarrow Ord$, let $f <_S^* g$ if and only if, for club-many $\nu \in S$, $f(\nu) < g(\nu)$. We write $<^*$ when $S = \kappa$. Since NS_κ is an ω_1-complete ideal, $<_S^*$ is a well-founded relation and we can define the *Galvin–Hajnal norm* of $g \colon \kappa \longrightarrow Ord$,

$$\|g\|_S = sup\,\{\|f\|_S + 1 \colon f \colon \kappa \longrightarrow Ord,\ f <_S^* g\}\,.$$

If $S = \kappa$, we write $\| \ \|$ for $\| \ \|_S$. Note that, if $T \subseteq S$ are stationary subsets of κ,

$$\|g\|_S \leq \|g\|_T.$$

If G is a NS_κ^+-generic filter over V, let $j \colon V \longrightarrow M = V^\kappa \cap V$ be the induced elementary embedding. The following fact is well-known:

FACT 3.9. *Given an ordinal $\gamma < \kappa^+$ and a one-to-one mapping $e_\gamma \colon \gamma \longrightarrow \kappa$, the function g_γ with domain κ such that*

$$g_\gamma(\nu) = o.t.\ (\{\xi < \gamma \colon e_\gamma(\xi) < \nu\})$$

for every $\nu < \kappa$ is such that

$$\Vdash_{NS_\kappa^+} \check{g}_\gamma \text{ represents } \check{\gamma} \text{ in } M,$$

meaning that NS_κ^+ forces that the set of M-ordinals below the ordinal represented by g_γ is well-ordered in order type γ. We say that g_γ is a canonical function for γ.

In [6], the following weak form of Chang's Conjecture is considered:

DEFINITION 3.3. *Let $\kappa = \lambda^+$ be a successor cardinal. The weak Chang's Conjecture for κ, $wCC(\kappa)$, is the following statement:*

For every first order structure $\mathcal{M}$ of a countable language such that $\kappa^+ \subseteq \mathcal{M}$ there is some $\alpha < \kappa$ such that, for every $\beta < \kappa$, there is some $X \preccurlyeq \mathcal{M}$ with $X \cap \kappa \subseteq \alpha$ and $o.t.(X \cap \kappa^+) > \beta$.

Consider the following strengthening of $(\lambda^{++}, \lambda^+) \Longrightarrow (\lambda^+, \lambda)$:

DEFINITION 3.4. *Let $\kappa = \lambda^+$ be a successor cardinal. Then, $CC^*(\kappa)$ is the following statement:*

For every club $C \subseteq [\kappa^+]^\lambda$ there is a club $D \subseteq [\kappa^+]^\lambda$, $D \subseteq C$, such that for every $M \in D$ there is some $N \in D$ with $M \subsetneq N$ and $N \cap \kappa = M \cap \kappa$.

Consider also the following strengthening of $wCC(\kappa)$:

DEFINITION 3.5. *Let $\kappa = \lambda^+$. Then $wCC(\kappa)^*$ is the following statement:*

For every first order structure $\mathcal{M}$ of a countable language such that $\kappa^+ \subseteq \mathcal{M}$ there are club-many ordinals α in κ such that for every $\beta < \kappa$ there is some $X \preccurlyeq \mathcal{M}$ with $sup(X \cap \kappa) = \alpha$ and $o.t.(X \cap \kappa^+) > \beta$.

FACT 3.10. *Let $\kappa = \lambda^+$.*

(a) *$CC^*(\kappa)$ implies $wCC(\kappa)^*$ and $(\kappa^+, \kappa) \Longrightarrow (\kappa, \lambda)$ implies $wCC(\kappa)$.*
(b) *$wCC(\kappa)^*$ holds if and only if $\Vdash_{NS^+_\kappa} j(\kappa) = \kappa^+$ if and only if for every $g \in \kappa^\kappa$ and every $S \in NS^+_\kappa$, $\|g\|_S < \kappa^+$.*
(c) *$\not\Vdash_{NS^+_\kappa} j(\kappa) > \kappa^+$ implies that, for every $g \in \kappa^\kappa$, $\|g\| < \kappa^+$.*
(d) *$wCC(\kappa)$ holds if and only if for every $g \in \kappa^\kappa$, $\|g\| < \kappa^+$.*

PROOF. (a), (c) and one of the equivalences of (b) are easy and the other equivalence in (b) follows from a trivial modification of the proof of (d) in [6]. ⊣

Theorem 3.11 is due jointly to Todorčević and to myself.

THEOREM 3.11. *Assume BMM and suppose S is a stationary subset of ω_1 such that*

$$S \Vdash_{NS^+_{\omega_1}} Ord^M \text{ has an initial well-founded segment of length } \check{\omega}_2 + 1$$

Then $S \Vdash_{NS^+_{\omega_1}} j(\check{\omega}_1) = \check{\omega}_2$. In fact, the following stronger statement holds:

For every $g \in \omega_1^{\omega_1}$ there is $\gamma < \omega_2$ and a canonical function g_γ for γ such that, for club-many $\nu \in S$, $g(\nu) < g_\gamma(\nu)$ (i.e., $S \Vdash_{NS^+_{\omega_1}} [g]_M < \gamma$).[2]

(j denotes the generic elementary embedding $j\colon V \longrightarrow M = V^{\omega_1} \cap V \subseteq V[G]$.)

PROOF. Start fixing a one-to-one $e_\gamma\colon \gamma \longrightarrow \omega_1$ for every $\gamma < \omega_2$ and let $g_\gamma(\nu) = o.t.(\{\xi < \gamma\colon e_\gamma(\xi) < \nu\})$ for all $\nu \in \omega_1$. Working towards a contradiction, suppose that $S \not\Vdash_{NS^+_{\omega_1}} j(\check{\omega}_1) = \check{\omega}_2$. By shrinking S if necessary, we

[2]Added in proof: Deiser and Donder [4] have proved that the statement that every function from ω_1 into ω_1 is bounded on a club by a canonical function of some ordinal less than ω_2 is equiconsistent with the existence of an inaccessible limit of measurable cardinals. Hence, this statement is much stronger than $wCC(\omega_1)$, which is equiconsistent with just an almost $< \omega_1$-Erdős cardinal, a large cardinal property weaker than being ω_1-Erdős [6].

may assume that there is a function $g \in \omega_1^{\omega_1}$ such that

$$S \Vdash_{NS^+_{\omega_1}} \check{g} \text{ represents } \check{\omega_2} \text{ in } M.$$

Note that, for every $h \in \omega_1^{\omega_1}$, if $T = \{\nu \in S : h(\nu) < g(\nu)\}$ is stationary, then there is some $\gamma < \omega_2$ and some stationary $T_\gamma \subseteq T$ such that $h(\nu) < g_\gamma(\nu)$ for all $\nu \in T_\gamma$. Hence, we may assume that, for every $\nu \in S$, $g(\nu)$ is a limit ordinal.

For every $n \in \omega$ fix $a_n \in \omega_1^{\omega_1}$ such that, for each $\nu \in S$, $\langle a_n(\nu) : n \in \omega\rangle$ is an increasing sequence converging to $g(\nu)$. For every $n \in \omega$ and every $\gamma < \omega_2$ let

$$B_{n,\gamma} = \{\nu \in S : a_n(\nu) < g_\gamma(\nu)\}.$$

Then, $\mathcal{B}_n = \{B_{n,\gamma} : \gamma < \omega_2\}$ is predense below S for every n.

For every n let $\mathbb{Q}(\mathcal{B}_n, \omega_1 \backslash S)$ be the poset for shooting, with countable conditions, an ω_1-club through a generic ω_1-enumeration of $\mathcal{B}_n \cup \{\omega_1 \backslash S\}$ (see [8]). $\mathbb{Q}(\mathcal{B}_n, \omega_1 \backslash S) = Coll(\omega_1, \omega_2) * \dot{\mathbb{Q}}_n$, where $\dot{\mathbb{Q}}_n$ is defined in $V^{Coll(\omega_1,\omega_2)}$ as follows:

Let $G : \omega_1 \longrightarrow \omega_2^V$ be the canonical generic bijection. Let

$$\nabla_G B_{n,\xi} = \left\{\nu < \omega_1 : \exists \nu' < \nu, \nu \in B_{n,G(\nu')}\right\}.$$

Then, $\dot{\mathbb{Q}}_n$ is the standard poset for shooting a club through $(\nabla_G B_{n,\xi}) \cup \omega_1 \backslash S$ with countable conditions.

Using the fact that $\mathcal{B}_n$ is predense below S, it is easily seen that $\mathbb{Q}(\mathcal{B}_n, \omega_1 \backslash S)$ is stationary-set-preserving and that it adds a bijection $G : \omega_1 \longrightarrow \omega_2^V$ and a strictly increasing, continuous and cofinal function $C : \omega_1 \longrightarrow \omega_1$ such that, for every $\nu \in \omega_1$, either $C(\nu) \notin S$ or else, for some $\nu' < \nu$, $C(\nu) \in B_{n,G(\nu')}$.

By BMM, for every n there is $\delta_n < \omega_2$, a sequence $\langle e_{n,\gamma} : \gamma \leq \delta_n\rangle$ where, for each $\gamma \leq \delta_n$, $e_{n,\gamma} : \gamma \longrightarrow \omega_1$ is one-to-one, and a club $C_n \subseteq \omega_1$ such that for every $\nu \in C_n \cap S$ there is some $\gamma(n,\nu) < \delta_n$ such that $e_{n,\delta_n}(\gamma(n,\nu)) < \nu$ and $a_n(\nu) < o.t.(\{\xi < \gamma(n,\nu) : e_{n,\gamma(n,\nu)}(\xi) < \nu\})$.

Let $C = \bigcap_{n\in\omega} C_n$ and let $\gamma = (sup_{n\in\omega,\nu\in C\cap S}\, \gamma(n,\nu)) + 1$. As $S \Vdash_{NS^+_{\omega_1}} \gamma = [g_\gamma]_M <_M [g]_M$, there is a club $D \subseteq \omega_1$ such that $g_\gamma(\nu) < g(\nu)$ for all $\nu \in D \cap C \cap S$. Hence, there is some stationary $T \subseteq D \cap C \cap S$ and some $n \in \omega$ such that $g_\gamma(\nu) < a_n(\nu)$ for all $\nu \in T$. But then, as $\nu \longrightarrow e_{n,\delta_n}(\gamma(n,\nu))$ is regressive on T, there is some stationary $T_0 \subseteq T$ and some $\gamma_0 < \delta_n$ such that $a_n(\nu) < o.t.(\{\xi < \gamma_0 : e_{n,\gamma_0}(\xi) < \nu\})$ for all $\nu \in T_0$. Finally, as $\nu \longrightarrow o.t.(\{\xi < \gamma_0 : e_{n,\gamma_0}(\xi) < \nu\})$ is a canonical function for $\gamma_0 < \gamma$, there is a club $E \subseteq \omega_1$ such that, for every $\nu \in E \cap T_0$,

$$g_\gamma(\nu) < a_n(\nu) < o.t.\left(\{\xi < \gamma_0 : e_{n,\gamma_0}(\xi) < \nu\}\right) < g_\gamma(\nu).$$

This contradiction shows that $S \Vdash_{NS^+_{\omega_1}} j(\check{\omega_1}) = \check{\omega_2}$.

Now fix $g \in \omega_1^{\omega_1}$. Then, if we let $A_\gamma = \{\nu \in S : g(\nu) < g_\gamma(\nu)\}$ for every $\gamma < \omega_2$, $\mathcal{A} = \{A_\gamma : \gamma < \omega_2\}$ is predense below S. Therefore, the poset

$\mathbb{Q}(\mathcal{A}, \omega_1 \backslash S)$ for shooting, with countable conditions, an ω_1-club through a generic ω_1-enumeration of $\mathcal{A} \cup \{\omega_1 \backslash S\}$, is stationary-set-preserving. An application of *BMM* to $\mathbb{Q}(\mathcal{A}, \omega_1 \backslash S)$ gives us an ordinal $\delta < \omega_2$, a sequence $\langle \overline{e_\gamma} : \gamma \leq \delta \rangle$ of one-to-one functions $\overline{e_\gamma} : \gamma \longrightarrow \omega_1$ and a club $C \subseteq \omega_1$ such that for all $\nu \in C$ there is $\gamma < \delta$ such that $\overline{e_\delta}(\gamma) < \nu$ and $g(\nu) < o.t.\{\xi < \gamma : \overline{e_\gamma}(\xi) < \nu\}$.

Consider the club $\overline{C}$ of all $\nu \in C$ with the following properties:

(a) $\{\gamma < \delta : e_\delta(\gamma) < \nu\} = \{\gamma < \delta : \overline{e_\delta}(\gamma) < \nu\}$,

(b) For every $\gamma < \delta$, if $e_\delta(\gamma) < \nu$, then $\{\xi < \gamma : \overline{e_\gamma}(\xi) < \nu\} \subseteq \{\xi < \gamma : e_\delta(\xi) < \nu\}$.

But now we are done, since $\overline{C} \cap S \subseteq A_\delta$. ⊣

THEOREM 3.12. *Suppose $r^\sharp$ exists for every real r, $u_2 = \omega_2$ and $\Vdash_{NS^+_{\omega_1}}$ "Ord^M has an initial well-founded segment of length $\check{\omega}_2 + 1$". Assume BMM. Then ψ_{AC} holds.*

PROOF. This follows closely the proof of Theorem 3.7, using the proof of Lemma 3.16 in [18].

Asssume *BMM* in V. Let $\gamma = |\mathcal{P}(\mathcal{P}(\omega_1))|^+$, let S and T be stationary subsets of ω_1 and let $H : [\gamma]^{<\omega} \longrightarrow \gamma$. We want to find some $X \in [\gamma]^{\aleph_0}$ closed under H such that $X \cap \omega_1 \in S$ and $o.t.(X) \in T$. Let $\kappa = (2^{\aleph_1})^+$ and let $\theta > \gamma$ be large enough. Let $G \subseteq Coll(\omega, H(\kappa))$ be generic over V.

CLAIM 3.13. *In $V[G]$ there is an elementary embedding $j : H(\theta)^V \longrightarrow \mathcal{M}$ such that $\omega_1^V \in j(S)$ and $j(T) \subseteq \gamma$ is a stationary subset of $\gamma = \omega_1$.*

PROOF OF CLAIM. Work in $V[G]$: Fix a sequence $\langle \gamma_n : n \in \omega \rangle$ cofinal in ω_2^V. Since $u_2 = \omega_2$, for every $n \in \omega$ there is some $z_n \in \mathbb{R} \cap H(\omega_2)^V$ such that $rank(\mathcal{M}((z_n)^\sharp, \omega_1^V + 1)) > \gamma_n$ where, given a real a and an ordinal α, $\mathcal{M}(a^\sharp, \alpha)$ is the Skolem closure inside $L[a]$ of the first α Silver indiscernibles for $L[a]$ (see Proposition 14.18(b) in [11]).

Now let $\langle \mathcal{M}_\beta, G_\alpha, j_{\alpha,\beta} : \alpha < \beta \leq \gamma \rangle$ be any iteration of $H(\kappa)^V$ of length γ. Since $H(\omega_2)^V \models NS_{\omega_1}$ is $\aleph_2$-saturated, by Fact 3.4, the above iteration induces a unique iteration $\langle \mathcal{M}^*_\beta, G_\alpha, j^*_{\alpha,\beta} : \alpha < \beta \leq \gamma \rangle$ of $H(\omega_2)^V$. Notice that, for every $\beta \leq \gamma$, $\{j^*_{0,\beta}(\gamma_n) : n \in \omega\}$ is cofinal in $Ord^{\mathcal{M}^*_\beta}$. Also, for each $n \in \omega$,

$$j^*_{0,\beta}(\mathcal{M}((z_n)^\sharp, \omega_1^V + 1)) = \mathcal{M}^{\mathcal{M}^*_\beta}((z_n)^\sharp, \omega_1^{\mathcal{M}^*_\beta} + 1),$$

which is isomorphic to $\mathcal{M}((z_n)^\sharp, \omega_1^{\mathcal{M}^*_\beta} + 1)$ (provided $\omega_1^{\mathcal{M}^*_\beta}$ is an ordinal). Thus, if $\omega_1^{\mathcal{M}^*_\beta}$ is well-founded, then so is $\mathcal{M}^*_\beta$. Also, by Theorem 3.11, $\Vdash_{NS^+_{\omega_1}}$ $j(\check{\omega}_1) = \check{\omega}_2$ in V. This is first order expressible in $H(\omega_2)^V$. Therefore, for every $\beta < \gamma$, if $\mathcal{M}^*_\beta$ is well-founded then, by elementarity between $H(\omega_2)^V$ and $\mathcal{M}^*_\beta$, $\omega_1^{\mathcal{M}^*_{\beta+1}} = Ord^{\mathcal{M}^*_\beta}$. It follows that, for every $\beta \leq \gamma$, $\omega_1^{\mathcal{M}^*_\beta}$ is well-founded and so is $\mathcal{M}^*_\beta$. Thus, $H(\omega_2)^V$ is iterable. In particular, $H(\kappa)^V$ is weakly iterable. Then, by Lemma 3.6, there is some iteration $\langle \mathcal{M}_\beta, G_\alpha, j_{\alpha,\beta} : \alpha < \beta \leq \gamma \rangle$ of $H(\kappa)^V$ such that $\omega_1 \in j_{0,\gamma}(S)$ and $j(T)$ is a stationary subset of γ. Finally, by Fact 3.4 and since $\mathcal{P}(\mathcal{P}(\omega_1))^{H(\kappa)^V} = \mathcal{P}(\mathcal{P}(\omega_1))^{H(\theta)^V}$, this iteration induces an iteration $\langle \mathcal{N}_\beta, G_\alpha, k_{\alpha,\beta} : \alpha < \beta \leq \gamma \rangle$ of $H(\theta)^V$ such that $\omega_1^V \in k_{0,\gamma}(S)$ and $k_{0,\gamma}(T)$ is a stationary subset of γ. Let $M = \mathcal{N}_\gamma$ and $j = k_{0,\gamma}$. ⊣

The rest of the proof is like in the proof of Theorem 3.7. ⊣

The following question still remains open:

QUESTION 3.1. *Does BMM imply* $2^{\aleph_0} = \aleph_2$?

REFERENCES

[1] D. ASPERÓ, ***Bounded forcing axioms and the continuum***, Ph.D. thesis, U. Barcelona, 2000.

[2] J. BAGARIA, *Bounded forcing axioms as principles of generic absoluteness*, ***Archive for Mathematical Logic***, vol. 39 (2000), pp. 393–401.

[3] M. BEKKALI, ***Topics in set theory***, Lecture Notes in Mathematics, vol. 1476, Springer–Verlag, Berlin, 1991.

[4] O. DEISER and H. D. DONDER, *Canonical functions, non-regular ultrafilters and Ulam's problem on* ω_1, ***The Journal of Symbolic Logic***, vol. 68 (2003), pp. 713–739.

[5] K. DEVLIN, ***Constructibility***, Springer–Verlag, Berlin, 1984.

[6] H. D. DONDER and P. KOEPKE, *On the consistency strength of 'accessible' Jonsson cardinals and the weak Chang's Conjecture*, ***Annals of Pure and Applied Logic***, vol. 25 (1983), pp. 233–261.

[7] Q. FENG and T. JECH, *Projective stationary sets and strong reflection principle*, ***Journal of the London Mathematical Society***, vol. 58 (1998), pp. 271–283.

[8] M. FOREMAN, M. MAGIDOR, and S. SHELAH, *Martin's Maximum, saturated ideals, and non-regular ultrafilters. Part I*, ***Annals of Mathematics***, vol. 127 (1988), pp. 1–47.

[9] M. GITIK, *Nonsplitting subsets of* $\mathcal{P}_\kappa(\kappa^+)$, ***The Journal of Symbolic Logic***, vol. 50 (1985), pp. 881–894.

[10] M. GOLDSTERN and S. SHELAH, *The Bounded Proper Forcing Axiom*, ***The Journal of Symbolic Logic***, vol. 60 (1995), pp. 58–73.

[11] A. KANAMORI, ***The higher infinite***, Springer–Verlag, Berlin, 1994.

[12] D. KUEKER, *Countable approximations and Löwenheim–Skolem theorems*, ***Annals of Mathematical Logic***, vol. 11 (1977), pp. 57–103.

[13] D. MARTIN and R. SOLOVAY, *Internal Cohen extensions*, ***Annals of Mathematical Logic***, vol. 2 (1970), pp. 143–178.

[14] J. STAVI and J. VÄÄNÄNEN, *Reflection principles for the continuum*, preprint.

[15] S. TODORČEVIĆ, ***Partitions problems in topology***, Contemporary Mathematics, vol. 84, American Mathematical Society, Providence, Rhode Island, 1989.

[16] ———, *Localized reflection and fragments of PFA*, to appear in ***DIMACS Series in Discrete Mathematics and Theoretical Computer Science***.

[17] B. VELIČKOVIĆ, *Forcing axioms and stationary sets*, ***Advances in Mathematics***, vol. 94 (1992), pp. 256–284.

[18] H. WOODIN, ***The axiom of determinacy, forcing axioms, and the nonstationary ideal***, Series in Logic and its Applications, no. 1, De Gruyter, Berlin, New York, 1999.

INSTITUT FÜR FORMALE LOGIK, UNIVERSITÄT WIEN,
WÄHRINGERSTRASSE 25, 1090 WIEN, AUSTRIA
E-mail: aspero@logic.univie.ac.at

HILBERT'S WIDE PROGRAM

WILLIAM EWALD

Otto Blumenthal's elegant biographical sketch of Hilbert, written in 1922 in celebration of Hilbert's sixtieth birthday, begins with a long discussion of the "Hilbert Problems" address, delivered here one hundred years ago next month. That address, says Blumenthal, epitomizes both Hilbert's manner of working and his personality — the selection of problems that are difficult without being inaccessible; their formulation in terms so clear as to render them intelligible to the average person in the street; and then their solution within an axiomatic system that combines simplicity with full rigor. "Hilbert," says Blumenthal, "is the man of problems. He collects and solves existing ones and points out new ones. His biography can be recounted in terms of problems. The birth of a man is chance, but his development is his own work."

Blumenthal then lists Hilbert's astonishing range of accomplishments, spanning the entire breadth of mathematics: his early work on invariant theory; his contributions to algebraic number theory; his book on the foundations of geometry; his solution of Waring's Problem and the rehabilitation of the Dirichlet Principle; his work in integral equations and mathematical physics; and one could add his work in logic and proof-theory, large parts of which still lay in the future.

This designation of Hilbert as "the man of problems" neatly catches a widely held view of his mathematical accomplishment. In particular, the standard view of his 1900 Paris lecture sees it as a *tour de force* of mathematical problem-posing — as it were, the supreme "Mathematical Games and Puzzles" column, but at the highest level of mathematical sophistication, displaying profound insight into the problems that were to guide the development of mathematics in the twentieth century.

This way of understanding the Hilbert Problems address combines readily with three widespread and interrelated assumptions about Hilbert that have collectively become a part of the Hilbert folklore. It would be too strong to call them myths, since each contains a considerable kernel of truth; but, taken together, and without qualification, they present a picture of Hilbert that

I should like to thank Michael Hallett, Ralf Haubrich, Albert Krayer, Ulrich Majer, Tilman Sauer, and Wilfried Sieg for helpful conversations and comments on earlier drafts.

Logic Colloquium 2000
Edited by R. Cori, A. Razborov, S. Todorčević, and C. Wood
Lecture Notes in Logic, 19

plays down the interconnections and the philosophical unity of his thought, and that is impossible to reconcile what we now know from his unpublished writings.

The first of these assumptions about Hilbert is most clearly stated by Hermann Weyl in his famous obituary notice:

> Hilbert helped the reviewer of his work greatly by seeing to it that it is rather neatly cut into different periods during each of which he was almost exclusively occupied with one particular set of problems. If he was engrossed in integral equations, integral equations seemed everything; dropping a subject, he dropped it for good and turned to something else. It was in this characteristic way that he achieved universality. ([25], p. 135)

Weyl's portrait undoubtedly captures a central aspect of Hilbert's creativity, and many writers have commented on his tendency to focus his research with single-minded intensity on a particular branch of mathematics, but then to abandon it and move on.

The second widespread assumption is related. It is that, in contrast to thinkers like Frege or Russell, whose work was embedded within broad philosophical programs and therefore retains an interest which goes beyond their purely technical accomplishments, Hilbert was above all a working mathematician, a sceptic about vaguely-formulated philosophical theses, a technical virtuoso whose primary business was the solving of concrete problems. On this view, Hilbert was not so much a synthesizing intellect, trying to obtain a comprehensive, philosophical view of his subject, but a mathematician's mathematician, a collector and solver of problems, moving from challenge to challenge like a mountain climber, but with no underlying theme beyond the desire to climb as many peaks as possible.

This picture derives some sustenance from the writings of Hilbert himself. His best-known utterances on the foundations of mathematics date from the 1920s and are filled with passionate argument against Weyl and Brouwer. He deplores the "havoc" caused by the paradoxes, blaming them for calling into question "cherished" methods of proof, and for threatening mathematicians with expulsion from "Cantor's Paradise." At one point at the very end of his career he refers to the critics of the *tertium non datur* as "philosophers disguised as mathematicians" — and he makes quite clear his opinion of philosophers and their "twaddle." ([17]. The editors of his collected papers [18] deleted these polemical remarks, presumably with his approval.) He goes on to declare proudly (twice) that, in his proof theory, "the world has been rid, once and for all, of the question of the foundations of mathematics as such."

It is easy to hear in these remarks the exasperation of a professional mathematician, impatient to get on with the business of proving theorems. Philosophy and the paradoxes, it seems, are for him a distraction and an irritation;

and the very point of proof theory is to banish such annoying issues from mathematics forever. In contrast to his more philosophical contemporaries, Hilbert appears to offer us only a technical response to a particular mathematical problem; his significance today is therefore merely historical, and strictly limited by the fate of his consistency program.

The third piece of folklore is that Hilbert's work specifically in foundations was inspired by the logical and set-theoretical antinomies, and by the so-called crisis in the foundations of mathematics. Hilbert was certainly aware of Cantor's paradox as early as 1897 (the documents are collected in [3], pp. 923–940) and in fact alludes to it briefly both in his statement of the Second Problem in 1900, and at the very end of his paper *On the Concept of Number* [9]. In consequence, it is often assumed that his Second Problem, and indeed his entire program in the foundations of mathematics, was a straightforward reaction to the difficulties posed by the paradoxes. (This view has even appeared in the popular press: a recent article in the German newspaper *Die Zeit* [20] treats Hilbert's call for a consistency proof as a reaction to Russell's Paradox, without noticing that Russell did not discover his Paradox until nearly two years *after* the Hilbert Problems Address.)

A highly influential version of this third thesis is to be found in Weyl's obituary. According to Weyl, Hilbert's work in foundations divides into two sharply distinct periods. The first is the period around 1900, when he was occupied with the axioms of geometry. Then came a long period when he devoted himself exclusively to other pursuits and put foundational research aside. "Meanwhile," says Weyl,

> the difficulties concerning the foundations of mathematics had reached a critical stage, and the situation cried for repair. Under the impact of undeniable antinomies in set theory, Dedekind and Frege had revoked their own work on the nature of numbers and arithmetical propositions, Bertrand Russell had pointed out the hierarchy of types which, unless one decides to 'reduce' them by sheer force, undermine the arithmetical theory of the continuum; and finally L.E.J. Brouwer by his intuitionism had opened our eyes and made us see how far generally accepted mathematics goes beyond such statements as can claim real meaning and truth founded on evidence. I regret that in his opposition to Brouwer, Hilbert never openly acknowledged the profound debt which he, as well as all other mathematicians, owes Brouwer for this revelation.
>
> Hilbert was not willing to make the heavy sacrifices which Brouwer's standpoint demanded, and he saw, at least in outline, a way by which the cruel mutilation could be avoided. At the same time he was alarmed by signs of wavering loyalty within the ranks of mathematicians, some of whom openly sided with Brouwer. My

> own article on the *Grundlagenkrise* in *Math. Zeit.* vol. 10 (1921), written in the excitement of the first postwar years in Europe, is indicative of the mood. Thus Hilbert returns to the problem of foundations in earnest. He is convinced that complete certainty can be restored 'without committing treason to our science.' There is anger and determination in his voice when he proposes 'die Grundlagenfragen einfürallemal aus der Welt zu schaffen.' ([25], p. 157)

The chronology here is subtle and significant. On Weyl's account, Hilbert for many years entirely abandoned foundational research, only to be drawn back into the fray by the paradoxes and by the writings of Brouwer and Weyl. And if one looks at the published record alone, this chronological development, seen as a progression from Brouwer to Hilbert, appears eminently plausible, since Hilbert's series of proof-theoretic papers start to appear in print with his paper [15], and that paper does, in fact, begin with an extensive critique of the views of Brouwer and Weyl.

It is important to observe the narrowing way in which these three theses support one another. The first thesis isolates the phases of Hilbert's career into several discrete areas of interest, each disconnected from the others. The second then says that, within each area of interest, Hilbert's work can be viewed as the effort to solve certain well-defined and narrow mathematical problems. And then, finally, the third thesis tells us that, specifically within the area of foundations, Hilbert's aim was to solve a particular problem, namely, the problem posed by the paradoxes, and to "rid the world of the question of the foundations of mathematics."

The standard folklore adds to these theses something that *does* deserve to be called a myth, namely, the assumption that Hilbert's philosophy of mathematics was that of a coarse "formalist." In Ramsey's words,

> the formalist school, of whom the most eminent representative is now Hilbert, have pronounced [the propositions of mathematics] to be meaningless formulae to be manipulated according to certain arbitrary rules, and they hold that mathematical knowledge consists in knowing what formulas can be derived from what others consistently with the rules. Such being the propositions of mathematics, their account of its concepts, for example the number 2, immediately follows: '2' is a meaningless mark occurring in these meaningless formulae. ([22], p. 338)

But this depiction (which is not confined to Ramsey) is scarcely accurate enough even to be considered a caricature. G. H. Hardy puts his finger on the problem:

> [I]f I thought that this really was the beginning and end of formalism, I should agree with Ramsey's rather contemptuous rejection of it. But is it really credible that this is a fair account of Hilbert's

> view, the view of a man who has probably added to the structure of significant mathematics a richer and more beautiful aggregate of theorems than any other mathematician of his time? I can believe that Hilbert's philosophy is as inadequate as you please, but not that an ambitious mathematical theory which he has elaborated is trivial or ridiculous. It is impossible to suppose that Hilbert denies the significance and reality of mathematical concepts, and we have the best of reasons for refusing to believe it: 'the axioms and demonstrable theorems,' he says himself, 'which arise in our formalistic game, are the images of the ideas which form the subject matter of ordinary mathematics.'

Hardy's analysis is amply confirmed by the lecture notes. When formal arithmetical calculi are first introduced (as they are in two sets of lectures from 1920) they are presented as a technical device for exploring the problem of consistency; but there is no suggestion that "meaningless marks" are to be identified with numbers, or that mathematical thought is to be abandoned in favor of the manipulation of formulas: still less that these calculi constitute a philosophy of mathematics. The point here goes well beyond the understanding of technical work in proof theory, and of the role played by contentual considerations in the metamathematics. In fact, Hilbert neither adopted nor accepted the label "formalist," and in his lectures and correspondence and conversation tended rather to disparage "mere calculation with formulas" (so much so that Blumenthal ([2], p. 394) felt compelled to point out that Hilbert, too, at times depended on formal calculations). Indeed, his first mathematical triumph, in the theory of invariants, consisted precisely in turning his back on the complicated formal calculations of Gordan, isolating the central ideas, and providing a direct, abstract proof of the existence of a basis. Likewise, in the Preface to the *Zahlbericht* (in a passage Hardy surely had seen) he wrote,

> I have attempted to avoid the great calculating apparatus of Kummer so that here, too, Riemann's principle will be satisfied, that one should execute proofs, not by calculations, but solely by thought.

This Riemannian ideal was to be a *leitmotif* throughout his career, and it pulled him in a very different direction than the philosophical views ascribed to him by Ramsey. Similarly, a careful examination even of his published writings alone is enough to cause discomfort to the other theses about Hilbert. He repeatedly (e.g. in [12], §32) declares the dilemma posed by the paradoxes to have been *solved* in 1908 by Zermelo's axiomatization of set theory; his important foundational paper, "Axiomatic Thought," was delivered in 1917, well before the *Grundlagenkrise* and the writings of Brouwer and Weyl (who are not mentioned) are said to have drawn him back into foundational research; the thesis that his interest lay in discrete, unrelated problems is hard to square with his repeated insistence in the Paris address on the unity of mathematics and on the fundamental importance of physics to pure mathematics. But only when one turns to his voluminous lecture notes on the foundations of mathematics and natural science (now in the process of being published by Springer, in five large volumes) does it become clear just how untenable are the three theses in their undiluted form.

To be sure, if one looks at the published record alone then Hilbert's career tidily divides into several distinct periods, exactly as the first thesis says. But the lecture notes tell a far more tangled story, and anyone who looks at even the titles of his lectures is much more likely to be struck by the continuity of his interests than by sudden leaps. Already by the first years of the century he had lectured on physics and geometry, algebra and logic, integral equations and invariants — and he was still holding lectures on precisely the same topics in the 1920s. The focus of his research may have shifted; but certain titles, like *Zahlentheorie* or *Grundlagen der Geometrie* or *Mechanik* or *Zahlbegriff und Quadratur des Kreises*, accompany him throughout his career, and indeed it is hard to find any subject, in any of his four decades of teaching, that he did not also lecture on during each of the other three decades.

But more importantly, when one turns to the notes themselves one finds constant cross-references from his work in one field to his work in another. He is constantly dropping hints to his students about how geometry is related to physics, and physics to set theory and logic; but most of these hints he excised from his published papers. A good example is provided by his important 1917–18 lecture notes on mathematical logic. As Wilfried Sieg [23] has carefully documented, Hilbert had already returned to the foundations of mathematics in earnest in 1917, and his lectures present the first fully modern treatment of mathematical logic, substantially identical to the monograph published in 1928 under the highly misleading rubric of "Hilbert and Ackermann." (Ackermann's contribution was far more that of an editorial assistant than a co-author, as a careful comparison of the lecture notes with the published monograph makes clear.)

But fully the first quarter of the lecture notes (representing at least a month in the classroom) are devoted to a detailed examination of the axiomatic method. This entire discussion, which clearly played a major motivating role in Hilbert's thinking about logic, was omitted from the published 1928 monograph, thereby obscuring the interconnections he emphasized in his lectures between geometry, logic, arithmetic, physics, and the axiomatic method. Moreover, there is no mention whatsoever in the 1917–18 lectures of any *Grundlagenkrise*, or indeed of Brouwer or Weyl: the logical work is motivated by systemic considerations that grow directly out of the discussion of the axiomatic method. The paradoxes are discussed only at the end of the course, where they are treated as an important lesson in why one needs to be careful in the formulation of higher-order logic. But they are not presented as a motivating factor in Hilbert's logical investigations, and still less as a shattering disaster for mathematics as a whole. The discussion is cool and a bit dry, without the slightest taint of hysteria.

The cumulative effect of the three theses is to split Hilbert's thought into fragments, and to lose sight of much of what is distinctive about his intellectual enterprise. In particular, there is an almost irresistible temptation to view the history of logic in the early years of the twentieth century through the lens of the paradoxes, and to assimilate Hilbert's foundational work, and especially his consistency program, to the technical logical work of Russell — or, in a related manner, to view the development of his proof-theoretic program in the twenties as commencing with a reaction against Brouwer. And this way of viewing Hilbert — of separating his accomplishments in foundations from his accomplishments in mainstream mathematics — has led to the sort of denigration of his logical work found in Hans Freudenthal, who, after praising his contributions to geometry and algebra, says that "according to the standards set by Hilbert himself, his ideas in foundations of mathematics look poor and shallow." Hilbert, he says, was driven by an "obsession" to prove the consistency of mathematics; and this obsession ("the tremendous problem of Hilbert's psychological makeup") led him into the "delusive profundity" of proof theory. His "naïve" views were decisively refuted by Gödel in 1931. "At closer look, 1931 is not the turning point but the starting point of foundations of mathematics as it has developed since. But then Hilbert can hardly be counted among the predecessors, as could Löwenhim and Skolem. This is a sad statement, but it would be a sadder thing if those who know nothing more about Hilbert than his work in foundations of mathematics judged his genius on this evidence." [4]

As an antidote, I wish here to focus on the 1900 Paris address, and in particular on the Second Problem, the problem of the consistency of arithmetic, and shall make two claims, one negative and one positive. The first is that it is a mistake to view Hilbert's call for a consistency proof as primarily a reaction to the paradoxes. To say this is not, of course, to deny that the paradoxes were important for Hilbert, for he tells us himself that they were. The task is rather to determine exactly *how* and *why* they were important; and my negative claim is that, in sharp contrast to Russell, the paradoxes did not inspire the consistency program, and that their importance for Hilbert was subordinate to other, broader and more systematic considerations. My second, positive claim is that those considerations were richer and more complex than is often recognized. Far from being a narrow, technical reaction to an isolated problem, the question of consistency for Hilbert has a programmatic significance and indeed might well have appeared on his list of problems even if the paradoxes had never been communicated to him by Cantor.

For reasons of space my focus will be limited to the early Hilbert, and I shall have little to say about the logical writings after 1917–18. But if the interpretation sketched here is correct then these laterdevelopments are fully harmonious with the Hilbert of the Hilbert Problems address, and that address is best viewed, not simply as an insightful list of twenty-three challenging problems (which of course it was) but also as the expression of a carefully-meditated and comprehensive view of the *unity* of mathematics that can broadly be called philosophical and that he pursued throughout his scientific career. This view of mathematics represents Hilbert's response to a complex set of developments that had swept through mathematics in the nineteenth century, and the crucial point to observe is that his Second Problem (and his later work in logic and proof theory) were not simply being pursued for their own sake or to clear up certain local difficulties posed by the paradoxes, but also for the light they could shed on these more comprehensive developments within mathematics as a whole.

§1. To bring out the distinctive traits of Hilbert's attitude to foundations and the paradoxes it will be helpful, by way of contrast, to recall some familiar facts about Russell. Russell tells us that the emotional drive which led him to philosophy was the desire to find a substitute for religious belief. He tells the story in his *Autobiography* and elsewhere of his loss of faith and of his first contact with Euclid. "I wished to believe that some knowledge is certain," he wrote, "and I thought the best hope of finding certain knowledge was in mathematics."

It is important to observe that there were two distinct components to Russell's search; both are clearly on display in the *Principles*. First, he wanted mathematics to be *certain*; second, he wanted it to be *objective*, that is, to be a set of truths *about* something. For if mathematics were to satisfy the craving for religious belief, it would have to reveal an impersonal, eternal realm; it must not be just an empty game played with words. This second element, the quest for objectivity, shows itself in his search for the "indefinables" of mathematics, and was ultimately called into question, not by the paradoxes, but by the *Tractatus*. "I thought of mathematics with reverence," he later recalled, "and suffered when Wittgenstein led me to regard it as nothing but tautologies."

With this motivation to philosophy the discovery of the paradoxes could only have struck Russell as a major blow, shaking his hope of finding certainty in mathematics. "Intellectual sorrow descended upon me in full measure," he wrote; the intellectual honeymoon he had been enjoying since he had first encountered the work of Peano was at an end. He later said that although he considered the paradoxes as being in a certain sense trivial and not worthy of the attention of a grown man, he nevertheless felt honor-bound to attempt a solution. The theory of types, his principal technical accomplishment, was the direct result; and it is important to observe that its *sole* reason for existence was to overcome the problem posed by the paradoxes.

§**2.** These various elements — the quest for certainty and objectivity, the pursuit of a logicist reduction, the discovery of paradox, and the technical response of the theory of types — make up the core of Russell's relation to the paradoxes. Before we turn to Hilbert it will be useful to observe that Russell's was not the only possible response. Indeed, as Bernays pointed out long ago, the apocalyptic character of the paradoxes in Russell is in part an artifact of his underlying reductionist ambitions. If, like Kronecker or Poincaré, for example, you believe that mathematicians have direct access to the natural numbers and that there is therefore no need to ground arithmetic in logic, then the discovery of the paradoxes will threaten, not *arithmetic*, but something else — logic, say, or Cantorian set theory. Poincaré's unruffled response is well known: "Logic is no longer sterile: it has given birth to contradiction."

There is an important general issue here about the way in which working mathematicians of the nineteenth century dealt with paradox, and it is useful to ask why nobody before Hilbert had attempted to provide explicit *proofs* of consistency. Certainly paradoxical problems were abundantly familiar to mathematicians of the nineteenth century, in geometry, in algebra, in the foundations of the calculus, and elsewhere. One need only think of Berkeley's criticisms of Newton in the *Analyst*, or of Bolzano's *Paradoxes of the Infinite*. Why did Newton or Gauss or Cauchy not propose dealing with such issues by means of syntactic consistency proofs? Why, in other words, did Hilbert's Second Problem not emerge a century earlier?

The answer cannot be that a syntactic formulation of the consistency problem was entirely beyond the intellectual resources of the age, for in fact Lambert, in an unpublished manuscript [19], *did* give a purely syntactic formulation of the consistency problem for the special case of the axiom of parallels. But there is no sign that Lambert attempted to exploit his own suggestion, and it would not naturally have occurred to mathematicians of the nineteenth century to pose the consistency question in this way.

The reason has to do with a profound change that took place in mathematics in the nineteenth century. The story is complex, and I have tried to document it at length in [3]. But very roughly, for a mathematician of the age of Gauss, mathematics divided into two parts: arithmetic and geometry. The truths of mathematics were just exactly that: *truths* about a particular *subject matter*, arithmetic being the *science of quantity*, just as geometry was the *science of space*. And this fact both underwrote the reliability of all genuine mathematics, and stood in the way of a purely syntactic conception of consistency. If a contradiction occurred the proper response was to re-trace your steps and find out where the falsehood had entered your reasoning; if a symbol like $\sqrt{-1}$ seemed to make no sense, the proper response was to search for the mathematical object it could be taken to refer to. There is a persistent confusion in the writings of mathematicians of the nineteenth century between the manipulation of a *symbol* and the manipulation of an *object*; and even at the end of the century, when mathematicians like Helmholtz or Klein tried to establish the legitimacy of non-euclidean geometry, they did not conceive their task in terms of axiomatically presented deductive theories, but instead tried to find geometrical *interpretations* under which the theorems of non-euclidean geometry would be *true*.

This digression should now make clear one response that was available to mathematicians confronted by Russell's or Cantor's paradox: be more careful, locate the point at which their reasoning about the objects in question led to falsehood, and move on. Of course this response would not answer the epistemological worries about the completed infinite that had worried Kronecker; but then, neither would Russell's theory of types. And in any case those worries had been around at least since Berkeley's *Analyst*, and mathematicians had largely ignored them, developing the mathematical theory of the continuum in spite of the objections.

As we now know, there was a hidden tension in this state of affairs. Once it became clear that there were several possible consistent geometries, once mathematicians had developed hundreds of new algebraic structures, and once they had begun to explore the complex relationships between geometry and arithmetic, it became much more difficult to think of mathematics as dividing into two distinct parts, a set of truths about number, and a set of truths about space. The history here is extremely complex, and the issues were far from clear to the participants, but in retrospect we can see that there were two broad responses to the increased abstractness of mathematics. One, the *objectivist* response, was to cling to the referential view of mathematics, but to attempt to find something more general than space or number for mathematics to be true of — sets, or logical objects, or, as in Brouwer, mental constructions. The other response — the *abstractionist* response — was to detach mathematics from particular referents, to take it altogether out of the realm of the actual and to view it as a postulational science about possible entities. (These two tendencies stand in obvious tension with one another; but it is important to observe that they could co-exist in the thought of somebody like Russell, who in the *Principles* both upheld a version of the postulational view and pursued the search for ultimate mathematical primitives.)

§3. Let us now turn our attention to Hilbert. It is not difficult to find numerous points of continuity with the views of Russell. Both agree that the paradoxes are an important problem. Both agree, contrary to Kronecker and Poincaré, that the concept of natural number cannot simply be taken for granted, but itself demands to be given a deeper foundation. Both wish to preserve as much of classical mathematics as possible. In one of Hilbert's published utterances [15], in reply to Weyl and Brouwer, he says that "the goal of finding a secure foundation for mathematics, which had been called into question by the paradoxes, is also my own." It is tempting, in the light of such statements, to view Hilbert as engaged upon the same sort of quest for epistemic certainty as drove Russell.

But if we turn to the early writings we find no mention of certainty as an important issue, still less of mathematics as a substitute for religious faith. The entire response to the paradoxes is more reserved. As we saw, Hilbert had been informed by Cantor of the greatest number paradox already in 1897, and he even alludes to Cantor's paradox in his statement of the Second Problem. He may well have known of Russell's Paradox by that time as well, for when Frege communicated it to him he replied (in a letter dated November 7, 1903) that that particular paradox, and others, had been known in Göttingen for four or five years. Yet in sharp contrast to Russell he devoted little comment to the issue, either in public or in private, until about 1904, and then without any of the sense of calamity we find in Russell.

But there is an even more striking fact about Hilbert's Second Problem. It is important to remember that the problem of the consistency of mathematical theories does not arise for the first time in the Hilbert Problems address. It is already a central theme in his monograph [8] on the foundations of geometry, and in the unpublished lectures he delivered in the preceding years. There he provides relative consistency proofs for various geometries, including euclidean plane geometry.

This fact should make us pause and wonder whether his primary motivation could have been the avoidance of paradox. Questions had indeed been raised about the validity of various branches of nineteenth-century mathematics, from the foundations of real analysis to non-euclidean geometry to set theory. But not even the hint of a paradox had arisen in elementary plane euclidean geometry — "school geometry" — the very paradigm of a theory whose consistency was established by direct appeal to the self-evident truth of Euclid's axioms. Why, then, did Hilbert find it necessary to provide a consistency proof for a theory whose consistency nobody had ever doubted?

Note that it is not enough to say that Hilbert was pursuing an ideal of absolute rigor; for from the traditional point of view geometry was *already* fully rigorous. Indeed, if anything the theory of the real numbers on which he proposed to ground the consistency of geometry was more problematical.

To get a sense of what animated him it is necessary to bear in mind his geometrical work as a whole. The *Grundlagen* had yielded a rich bounty of results in the ancient field of euclidean geometry, using model-theoretic techniques to develop, not only a relative consistency proof and independence results, but new non-Archimedean geometries, a topological characterization of the plane, insights into the nature of continuity, and deep new theorems about the relationship between geometry and arithmetic. These were striking accomplishments, but still more striking was the manner in which Hilbert had achieved his success. First and most obviously — and already in contrast to Klein or Helmholtz — his results depended upon a rigorous use of axiomatics, with all the axioms being set forth explicitly, and with deductions proceeding solely in accord with logic. This axiomatic approach was necessary if Hilbert was to accomplish his goal of determining exactly which bits of the edifice of geometry rested on which, a task which in turn divided into several subquestions. *First* there was the task of showing that the known theorems of geometry could all be derived from Hilbert's axiomatic base. *Second*, and even more central to Hilbert's project (as Michael Hallett has shown in his study of the geometrical lecture notes [6]) was the task of establishing that certain things could *not* be proved from a given set of axioms — most obviously in his proofs that the various axioms are independent of one another. But to this category of impossibility results, which Hilbert emphasizes in the concluding paragraphs of the *Grundlagen*, also belong Hilbert's concern with "purity of method" and with the attempt, for instance, to determine the exact deductive weight of a theorem like Desargues's Theorem within euclidean geometry as a whole.

To *pose* these questions one needed a strict axiomatic formulation of euclidean geometry; but to *solve* them one needed something else: the ability to ask meta-questions about the axiom system itself and to investigate questions of dependence and independence by exhibiting models of various sets of axioms and formulas. It is important to note that both these elements — the formulation of explicit axioms and the model-theoretic investigation of deductive relationships — were foreign to the geometric tradition of Riemann, Helmholtz, and Klein. Those geometers do not work in terms of axiom systems, and although they do, in a sense, show how to provide models for non-euclidean geometries, the discussion of the logical issues is left at an intuitive level. Certainly there is no attempt to pose the sort of questions Hilbert addresses. These investigations in turn depended upon yet a *third* feature of Hilbert's approach, namely the ability to generate a wide multiplicity of different models for various sets of formulas; and in fact a great deal of the *Grundlagen* consists precisely in the ingenious construction of models to establish the precise deductive power of various axioms and theorems.

It is at this point that Hilbert's affinity towards the second of the tendencies in nineteenth century mathematics mentioned above — the tendency to abstract away from any particular subject matter for mathematics — becomes evident. Hilbert was not just operating within an axiom system (as had done many mathematicians since the time of Euclid), but was operating with a radically new conception of what an axiom is. For earlier geometers axioms were understood as fundamental *truths* about some realm of *objects* from which other *truths* of mathematics could then be deduced. But Hilbert's investigations required him to sever the link between geometrical axioms and any particular domain of objects, and consequently to abandon the view that axioms state referential truths at all. And this fact separates him from Russell's quest for an eternal, changeless realm of mathematical objects, and produces a different reaction to the paradoxes.

There are thus three elements to Hilbert's new approach: axiomatization, abstraction, and the deliberate generation of a multiplicity of interpretations. It is not clear exactly when Hilbert arrived at his new conception of axiomatics, but there is good evidence that as early as 1891 he was remarking to people in conversation that "It must be possible to replace in all geometric statements the words *point*, *line*, *plane* by *table*, *chair*, *mug*," and in a sense all his subsequent geometrical investigations grow out of this insight. It is true that other mathematicians at about the same time, notably in Italy, were investigating abstract geometrical axiom systems, and that various elements of his conception of axiomatics can be found in earlier figures such as Pasch or Peano or Veronese. But four quarters are not the same thing as a silver dollar. Hilbert was the first to show how to combine these various elements into a powerful new tool, and then to *use* that tool to produce deep, novel results in the familiar field of elementary geometry.

So far I have been speaking only about the axioms of geometry. But of course the technique of giving a variable letter or operational sign multiple interpretations is above all an *algebraic* idea, and in the course of the nineteenth century algebraists like Hamilton, Cayley, Benjamin Peirce, Sylvester, Grassmann, Kummer, Kronecker, and Dedekind had discovered a new world of algebraic structures and had begun to lay the foundations for modern abstract algebra. Hilbert was of course intimately familiar with these new techniques and had exploited them both in his work on invariants and in the *Zahlbericht*. There is a great deal more that needs to be said on the way in which algebra and geometry interacted in the thought of the Hilbert of the 1890s;[1] for the present we merely note Blumenthal's comment that, "For Hilbert the charm of

[1]Above all, it should be stressed that Dedekind is a major precursor of Hilbert's foundational ideas. Wilfried Sieg ([23] and [*forthcoming*]) rightly observes that Dedekind in "*Was sind und was sollen die Zahlen*?" understood himself to be giving a consistency proof for the theory of natural numbers, and that the Hilbertian themes of axiomatization, abstraction, and multiple interpretations are all prefigured in his work; Michael Hallett, in [5], has also carefully explored the relationship of Dedekind to Hilbert.

It is important also to observe that there are significant differences that separate Hilbert from Dedekind; this is a large topic that can only be briefly sketched here.

1. Hilbert [9] distinguishes between two different ways of introducing the numbers (and, more generally, of presenting any mathematical field): the *genetic method* and the *axiomatic method*. The former, he says, is exemplified in the literature of arithmetic; the latter, in the literature of geometry. Although Hilbert does not here mention Dedekind, his description of the genetic method is a plausible description of at least one major strand of Dedekind's thought, and Hilbert's remarks can be taken as highlighting what is distinctive and novel about his own axiomatic approach. In contrast to Dedekind, he explicitly begins with a set of axioms for the real numbers, and investigates the resulting system; in so doing, he extends the axiomatic methods of his *Grundlagen der Geometrie* into the realm of arithmetic. (It is of course true, as Hilbert was surely aware, that one can easily extract a set of axioms for the natural numbers from *Was sind und was sollen die Zahlen?*, and that Peano in fact did so. But "could have" is not the same as

the field of elementary geometry was that here he saw the simplest example on which he could construct his ideal, abstracted from number theory, of a complete proof-architecture independent of the theory of the natural numbers." ([15], p. 68.)

§4. Let us now turn to Hilbert's Paris lecture. Hilbert was speaking as a mathematician to other mathematicians, and interested above all in the future progress of his science. He himself was already famous for three things:

"did," and the actual fact of this extraction makes a large difference to how the entire subject is conceived, and to what problems one chooses to investigate.)

2. This issue of axiomatics is related to the subtle issue of the extent to which Dedekind and the early Hilbert thought that the basic concepts of a mathematical theory could be disinterpreted, and viewed as having arbitrary reference. Certainly abstract structural concerns are very much to the fore in *Was sind*? and in the supplements to Dirichlet-Dedekind (collected in [3]); but there is also another powerful strand in Dedekind's thought, in which he views the numbers as "creations of the human mind," and takes himself to be explicating the structure of a realm of *objects*. So in his algebraic work he speaks, not of ideals or fields in the abstract, but of fields or groups *of numbers*. By contrast, Hilbert's comments on axiomatics, starting with his observation that "It must be possible to replace in all geometric statements the words *point*, *line*, *plane* with *table*, *chair*, *mug*" contain a stronger sense of the power of disinterpretation. (Some of his early comments on axiomatics, in [8], [9], and [10], already can be read to contain an implicitly purely syntactic, fully disinterpreted, conception of axiomatics; but Hilbert's comments waver, and as Sieg ([23] and *forthcoming*) stresses, there is an *existential* aspect to Hilbertian axiomatics, so that as late as the 1917–18 lecture notes he does not consider formal theories in isolation, but rather axioms together with the assumption of a structure satisfying the axioms.) The exegetical issues are delicate, and cannot be explored here; but it seems clear that the existential strand in Hilbert takes him closer to Dedekind, and the structuralist strand in Dedekind takes him closer to Hilbert, than a stark rendering of the contrast between "genetic" and "axiomatic" mathematics would imply.

3. Nevertheless, Hilbert, in explicitly extracting axioms for the real numbers, extended the axiomatic techniques of geometry into arithmetic in ways that Dedekind had never envisioned; and to this extent the contrast Hilbert draws in [9] is fully justified. In particular he had in the *Grundlagen der Geometrie* explicitly posed meta-systemic questions, not about systems of *objects*, but about systems of *axioms*, and he had put the idea of disinterpretation to powerful use by deliberately generating a variety of deviant models in order to investigate the relations of dependence and independence among various axiomatic systems. Hilbert differs from Dedekind, too, both in the explicitness and in the scope of his program. Dedekind does not mention consistency at all as a central goal in *Was sind*? itself, but only in his unpublished letter to Keferstein (reproduced in [24]); Hilbert, by contrast, presents consistency and independence as the core properties to be established for any axiomatic theory, and explicitly links consistency to a general criterion of mathematical existence. He also, in extending axiomatic techniques from geometry to arithmetic, was aware that he was proposing a new grounding for the whole of mathematics (including, as he says in [9], mathematical physics). Very broadly, Hilbert's turn-of-the-century accomplishment was to marry the abstractionist tendencies of modern algebra to the set-theory of Cantor and Dedekind and to the axiomatic techniques that had been familiar in geometry since Euclid. What is novel is not so much the individual strands as the way they have been combined into a mathematical tool of great power and originality; historians who seek to play down Hilbert's originality by observing that this idea was already to be found in Pasch, and that one in Peano, have missed the point.

the solution of Gordan's problem, the *Zahlbericht*, and the *Grundlagen der Geometrie*. These accomplishments, and especially the work in geometry, had depended for their success on a novel approach that carried mathematical abstraction to new heights, and that exploited the use of a plurality of models to investigate structural similarities and interdependencies. This new approach certainly recommended itself to Hilbert for a variety of reasons, and his remarks on it, both in the address and in his unpublished lectures, can be grouped into two categories.

One set of remarks deals with what might be called the *problem-solving* aspect of the new technique. Hilbert points out that the abstract axiomatic approach not only serves the cause of mathematical rigor, but serves the cause of clarity, making evident exactly what depends on what, and thus allowing problems to be stated and solved in full generality. If Russell's writings on the foundations of mathematics are stamped by the word "certainty," for Hilbert the equivalent word is "*Fruchtbarkeit*" — fecundity. He repeatedly tells his students that the techniques of axiomatization and abstraction, of seeking out multiple interpretations, of pursuing the "purity of method" and independence proofs are all tools that will promote the fertility of mathematical research; and he gives copious examples from the history of nineteenth century algebra and geometry of how such techniques had given birth to entire new domains of mathematical research.

But Hilbert was not simply recommending his new approach as a set of technologies for solving problems, and there is also a second, *programmatic* aspect to his remarks. I have already indicated how the mathematics of the nineteenth century had gradually emancipated itself from the view that mathematics is the science of quantity, on the one hand, and of space, on the other. Hilbert takes a radical position on this question. Mathematics is liberated, not only from the foreign notions of space, time, and motion, as mathematicians like Bolzano and Weierstrass had long ago urged, but from being the study of any actual objects whatever. Mathematics becomes a purely hypothetical science. We postulate some objects having certain properties, and then, purely on the basis of the given properties, deduce what further properties any such objects must have. As Hilbert said in some lectures in 1922, "To have carried out a separation between the things of thought of the axiomatic framework and the real things of the actual world is the chief contribution of axiomatics." ([16], p. 122.) As for the question, *which* hypothetical entities can be a legitimate object of mathematical study, Hilbert's answer is equally sweeping: *any* entities that can be specified without falling into contradiction. This is the celebrated criterion of mathematical existence which he presented in his formulation of the Second Problem: "if it can be shown that the attributes assigned to a concept can never lead to a contradiction ... I say that the concept (for example of a number or of a function satisfying certain conditions) is thereby proved." The proposal that lies tacitly embedded within the discussion of the Second Problem is at least as important as the problem itself: the abstract axiomatic method, which Hilbert had unveiled only the year before, and which had done such service in geometry, is now to be extended to all of mathematics. The entire subject, pure and applied, is to be given a new grounding.

Seen in this light, the problem of the consistency of arithmetic is not just another random problem. For if mathematics is no longer about a particular subject matter, and if, rather than relying on mathematical existence to secure consistency, one instead reverses matters and grounds existence on consistency, then the problem of establishing consistency becomes central to the entire enterprise.

How are consistency proofs to be provided? Hilbert had shown how to provide such a proof for geometry by reducing the problem to the consistency of arithmetic. It was of course a natural next step to inquire about the consistency of arithmetic, but that fact by itself would probably not have been enough to have earned the problem a spot on the list of Hilbert Problems. There was another factor as well. For Hilbert, as for Gauss, arithmetic was the fundamental mathematical science *überhaupt*, and although the exact content of this doctrine changed over the years, it was a constant in Hilbert's thought. So the problem of arithmetic had a certain conceptual priority, and this fact in turn had a couple of further consequences. First, it meant that, in contrast to geometry, the consistency of arithmetic would have to be secured by some means other than that of a relative consistency proof. That is, this problem would require the development of a novel *kind* of consistency proof; and there are signs that already by the time of the Paris address Hilbert may have been contemplating some sort of syntactic approach to the problem (although the details are far from clear).

Secondly, it is important not to lose sight of Hilbert's concern that each of his twenty-three problems be fertile of novel mathematical results. What sort of fertility could he have hoped for from his Second Problem? — His investigations into the consistency of geometry had provided a rich harvest of results, and in 1900 there was every reason to expect that an investigation of the reasons for the consistency of arithmetic, especially if the investigation were carried out by novel means, would prove similarly rewarding. It is only necessary to observe that if this were any part of Hilbert's motivation for the Second Problem, subsequent events confirmed his hopes in the most dramatic way. The point is not sufficiently appreciated, but in the early years of the twentieth century Hilbert's call for a consistency proof met with general incomprehension. Yet it is ultimately to his reflections on abstract axiomatics, and on the need for a consistency proof, that we owe, in time, the sharp distinction between syntax and semantics; the explicit formulation of questions of completeness, consistency, and decidability; the investigation of the logical scaffolding of axiomatic theories — in other words, the seeds of model theory, proof theory, and recursion theory, which make up the core of modern mathematical logic.

Hilbert famously did not include Fermat's Last Theorem on his list of problems because, in his opinion, it had already made its contribution by stimulating Kummer, Dedekind, and Kronecker to develop algebraic number theory. The point of a good problem, he said, was not to answer individual riddles or to add to a pile of theorems, but to build new mathematical structures, to create new theoretical understanding in what he called the "mathematische Gedankenwelt." And, in this sense, the point of his call for a consistency proof was not primarily to establish that arithmetic is consistent. Cartesian Demons were never his preoccupation.

§5. We are now in a position to see that Hilbert's foundational investigations had different sources than Russell's, and that he was pursuing different goals. Russell's deepest concerns in foundations revolve around the question of certainty. How can we know that mathematics is true? What justification do we have for believing in a timeless, Platonic heaven of mathematical objects? — And with this epistemic motivation, the paradoxes for Russell were a major intellectual calamity, striking to the heart of his entire enterprise.

But the picture with Hilbert is very different, and if one takes him to be offering answers to *Russell's* questions, the result can only be severe misunderstanding. For him, the deepest questions center, not on certainty, but on the axiomatic method, and on the new mathematics pioneered in the nineteenth century. There is no sign that his work in foundations was driven by a fear of paradox or that he was seeking to reassure himself about the truth of mathematics.

To say this is not, of course, to deny that the paradoxes were important for Hilbert, nor that he devoted a great deal of energy, both early and late, to a penetrating analysis of their origins and implications. But the point to be emphasized is that they were important to Hilbert in a different way, and for different reasons, than they were to Russell. For Russell, the paradoxes were the *sole* reason for much of his logical work, and without them he would have had no reason to develop the theory of types or the axiom of reducibility. But for Hilbert the paradoxes are a *subordinate* issue within a consistency program that has much deeper reasons for existing.[2]

---

[2]In particular, Hilbert rapidly realized that the paradoxes which Cantor communicated to him in 1897 and 1898 undermined the approach to the foundations of mathematics to be found in Dedekind's *Was sind und was sollen die Zahlen?* However it should also be observed, as noted above, that Dedekind, too, had pursued consistency proofs for the theory of natural number, but well before the discovery of the paradoxes. (I also note in passing the brute archival fact, pointed out to me by Michael Hallett, that the first mention in Hilbert's lectures of the need for

But there is a deeper contrast between Hilbert and Russell, and one that belies the familiar caricature of Hilbert with which we began. Hilbert is often portrayed as a master technician, interested in a vast array of isolated problems, but without the comprehensive vision of a Frege or a Russell. We can now see that this picture understands him almost precisely backwards.

When in the formulation of the Second Problem he introduces his new conception of mathematical existence his remarks are the fruit of a decade of sustained reflection on geometry and algebra, on the nature of mathematics as a whole and on the profound changes it had undergone at the hands of Riemann and Dedekind and Cantor. The traditional ideal of objectivity — that is, the ideal of finding some *objects* for mathematics to be *about* — is one that he here explicitly rejects, breaking decisively with the older view, to which Russell still adhered.

Moreover, as he repeatedly explains in the lecture notes (and indeed as he says both in the Paris address and in [15], §14) he saw in the new axiomatic method a highly fertile set of tools both for unifying mathematics and for the solution of mathematical problems — tools whose power he had already demonstrated in his work on geometry and algebraic number theory. Throughout the Hilbert problems themselves are scattered problems whose statement or whose resolution demand precisely the tools of axiomatics: not just the first two, but also, for example, the third (independence results in definition of Euclidean volume), the sixth (axioms of physics), and the tenth (algorithms for diophontine equations). Those tools were to become *the* distinguishing mark of twentieth-century mathematics, through the work, not only of Hilbert himself and of his students, but of his colleague Emmy Noether and her student van der Waerden, and later of Bourbaki, whose encyclopedic efforts put the axiomatic stamp on the whole of modern mathematics.

It is vital not to lose sight at this juncture of another crucial aspect of the Hilbert Problems address. Most of the problems do *not* directly involve abstract axiomatics. Some (like the Riemann hypothesis) are entirely classical; and many others come straight from mathematical physics. Hilbert, in celebrating his new, non-referential conception of axiomatics, was not for a moment recommending that mathematics cut itself off from the real world, and that mathematicians henceforth merely deduce abstract theorems from arbitrary axiomatic premises. On the contrary (and *pace* Ramsey), that is a view he roundly dismissed in some lectures from 1919:

> If the view [that mathematics is just tautologies grounded in definitions] were correct, then mathematics would have to be nothing but a heap of logical inferences piled up on top of each other. There would have to occur a choiceless concatenation of inferences, where

a consistency proof occurs in 1894 — that is, fully three years before Cantor informed him of the paradox of the greatest number; the motivation of course is his new conception of axiomatics.)

> logical inference alone was the driving force. But in reality there is no question of such arbitrariness; rather we see that the formation of concepts in mathematics is constantly accompanied by intuition and experience, so that mathematics as a whole is a non-arbitrary, closed structure.[3] ([14], p. 5)

The point about "intuition and experience" is echoed in §§13–15 of the Paris address, where he stresses the importance of problems drawn from physics. Already some years earlier he had distinguished between the *naïve*, the *formal*, and the *critical* stages in the development of a mathematical theory. Axiomatics for Hilbert belongs to the final, critical stage, after a field of knowledge has sufficiently ripened to the point where an axiomatic investigation is warranted. But as he says in §13, the *source* of most mathematics is observation of the external world: even elementary geometry and elementary arithmetic originally arose in this way. The same thing is true for more advanced mathematics. It is not by axiomatics alone, nor by the solution of concrete problems alone, that mathematics advances, but by what in §15 he calls "the constantly repeated, alternating interplay between thought and experience." It is for this reason that mathematical physics plays such a large role in his thinking about the foundations of mathematics, both in the Hilbert Problems address and throughout his lecture notes.

We have now located three differences between Russell and Hilbert which underlie and go beyond their differing reactions to the paradoxes. First, whereas Russell came to foundations from epistemology and occupied himself with only those parts of mathematics directly relevant to his philosophical concerns, Hilbert traveled the opposite route, coming to foundations with a comprehensive grasp of the entirety of mathematics, and searching for an underlying conceptual unity. Secondly, while Russell still clung to the referential, nineteenth-century ideal of mathematical objectivity, Hilbert was creating the conceptual framework for the axiomatic mathematics of the twentieth. Third, while Russell sought a Platonic realm of eternal mathematical objects, Hilbert's foundational work was deeply rooted in applied mathematics and the world of physics. — It is perhaps not surprising that, on the strictly mathematical side, Hilbert approached the philosophy of mathematics with a broader understanding of the whole of mathematics, and a far more prescient understanding of where the subject was headed than the other great logicians of the turn of the century. But there is a philosophical counterpoint to his

[3] "Wäre die dargelegte Ansicht zutreffend, so müsste die Mathematik nichts anderes als eine Anhäufung von übereinander getürmten logischen Schlüssen sein. Es müsste ein wahlloses Aneinanderreihen von Folgerungen stattfinden, bei welchem das logische Schliessen allein die treibende Kraft wäre. Von einer solchen Willkür ist aber tatsächlich keine Rede; vielmehr zeigt sich, dass die Begriffsbildungen in der Mathematik beständig durch Anschauung und Erfahrung geleitet werden, sodass im grossen und ganzen die Mathematik ein willkürfreies, geschlossenes Gebilde darstellt."

thought as well, one that contradicts the thesis that he was solely interested in solving a maximal number of unrelated problems. As Blumenthal shrewdly observed,

> Some see in Hilbert the paradigm of the pure logician, the son of the City of Pure Reason. I believe that Hilbert himself wishes to be judged differently. The longer I know him, the more I recognize in him the man of philosophy who, once he had become aware of his own power, always has his eye fixed on a supreme goal and pursues it along a carefully-considered path: the goal of a unified, closed world-picture, at least in the narrower domain of the exact sciences. It may be that this impulse dwells in the mathematician as such; the science of infinite connections encourages such a thing. But the impulse is of no help if it is not guided by a great, ordering point of view. This Hilbert found in axiomatics. It seems to me beyond doubt that for posterity he will always be the axiomatist, the man who gave logic a new form of expression. ([15], p. 71.)

Hilbert's progression led from mathematics to philosophy; and he was to reject Russell's technical work in logic ultimately for philosophical reasons. Although he appears to have been tempted for a time by logicism (and although he certainly admired Russell's achievements in the axiomatization of logic, as his laudatory remarks in [12] make clear), in his 1920 lectures and afterwards he subjected the theory of ramified types and the axiom of infinity to searching examination, and found them inadequate to his own purposes. The analysis is intricate, and cannot be explored here; but Hilbert saw the axiom of reducibility and the axiom of infinity as *ad hoc* solutions, implausible and awkward. More importantly he saw them as inadequate to the constructivist standpoint. In introducing explicit, infinitary existential assumptions, the Russellian axioms failed to address the *epistemic* worries about the completed infinite that lay at the heart of Kronecker's objections to the set theory of Dedekind and Cantor; nor did they point the way forward to a consistency proof, or to an analysis of the (finitary) mental capacities that, in Hilbert's view, lie at the root of all mathematical thought. On *these* matters it is ironically Russell whose theory of types and axiom of reducibility look today like a narrow technical response to the paradoxes, while Hilbert's sensitivity to the nuances of finitary reasoning displays a surer grasp of the deep philosophical issues, and a clearer understanding of the way in which logic must develop if it is to address the philosophical questions raised by the new mathematics.

§6. We began with a widespread view of Hilbert as a "man of problems," a narrowly technical virtuoso, driven by the paradoxes (and later by Brouwer) to propose his consistency program, and to adopt the philosophical theory that mathematics is nothing but "the manipulation of meaningless formulae according to certain arbitrary rules." We have seen that no part of this portrait is correct; and the cause of future clarity may be served if we now distinguish between two enterprises, both of which have an entitlement to be called "Hilbert's program."

The *narrow* Hilbert program is the program to prove the consistency ultimately of formalized second-order arithmetic by finitary means. (Actually, it is misleading to speak of "the" program, since Hilbert never formulated the program precisely, never specified exactly what would count as a solution, and repeatedly added technical refinements to his emerging proof theory.) But it is this narrow program that sought the solution of a relatively specific technical problem, that invoked (though only as a programmatic device) the study of "meaningless" formal systems, and that seemed to have been designed as a reply to Brouwer, Weyl, and the antinomies.[4] With some effort (it takes effort) one can formulate the narrow program so narrowly that it is squarely refuted by the Gödel incompleteness theorems. (Gödel himself, as is well known, was cautious about making any such claim.)

The *wide* Hilbert program, in contrast, consists of two parts, both extremely ambitious. First, he sought to extend the new techniques of set theory and abstract axiomatics, inaugurated by Dedekind and Cantor and elaborated by himself, to the whole of mathematics, both pure and applied. Secondly, it was his special insight that these new methods could themselves be made the subject of a precise, axiomatic analysis, and that the result was likely to be of fundamental importance to mathematics as a whole. As he said in a favorite metaphor, "Just as the physicist investigates his apparatus and the astronomer investigates his location; just as the philosopher practices the critique of reason; so, in my opinion, the mathematician must secure his theorems by a critique of his proofs, and for this he needs proof theory." ([15], §59)

That Hilbert saw the two parts as related in this way emerges perhaps most clearly in his 1917–18 lecture notes, the landmark text that launched modern mathematical logic. He devoted the first month of the lectures to explaining the axiomatic method and its importance for modern mathematics. (It is suggestive that, at the same time, he was giving a course on "mathematical principles" with Noether, of which no protocol exists.) Only after this preamble did he turn to the axiomatic development of logic, which he presents both as an extension of the axiomatic method into the virgin territory of logic, and also as the creation of a set of tools for the investigation of the foundations of axiomatic theories themselves.

---

[4]An anonymous referee has pointed out that the common misunderstanding of the programmatic character of Hilbert's formalism is encouraged by an unfortunate mistranslation of a phrase from *Über das Unendliche* in [Benacerraf and Putnam 1983, p. 197]: "it is consistent with our finitary viewpoint to deny any meaning to logical symbols . . . ." The rendering in van Heijenoort ([24], p. 381) is more accurate: "but we will be consistent in our course if we now divest the logical signs, too, of all meaning, just as we did the mathematical ones . . . ."

As we have already seen, both parts of the wide program have delivered a rich bounty. Abstract algebra and the axiomatic method have penetrated almost every branch of mainstream mathematics; and the foundational investigations have given us the core apparatus of modern logic. It should by now be clear that, for Hilbert, mathematical logic was situated squarely within mainstream mathematics, and was anything but a fringe exercise, separate from his other research.

Why has Hilbert's conception of foundations disappeared from view? One reason has been belabored throughout this paper: the ease with which he can be misunderstood as an unsubtle formalist, seeking to answer questions posed by others, and offering a narrow consistency program that was decisively refuted by Gödel. A second is the fact that much of his best writing on these matters occurs in the unpublished lecture notes, and that the published record gives a distorted picture of the development of his ideas (so that, to mention only two examples, the extent of his foundational involvement with theoretical physics drops from sight, or so that Hilbert-Ackermann appears near the end of his proof-theoretic investigations, instead of at the beginning). A third is the fact that Hilbert's foundational ideas (which of course evolved over the four decades of his professional career) must be pieced together from widely scattered texts, against a rapidly shifting background in mathematics and physics.[5]

But despite those serious limitations on our knowledge, it is manifest that when Hilbert spoke here a century ago he was not simply putting forward a list of twenty three challenging mathematical puzzles. The game was much larger. Only a year before he had shown what could be accomplished in geometry with abstract axiomatic reasoning. He was now proposing to extend those methods to arithmetic and analysis and physics, and to re-ground all of mathematics. In so doing, he was giving voice to a complicated mixture of insights and aspirations that had deep roots in the work of Riemann, Cantor, and Dedekind, and that were closely bound up with a comprehensive view of the unity of mathematics and of the possibilities for its future. Those aspirations revolved around the abstract axiomatic conception of mathematics; and that conception in turn drew attention to the question of consistency. But it also raised other questions as well. What are the limits of the new axiomatic techniques, and how are they to be justified? On what capacities does abstract mathematical reasoning rest? What is the relationship between a mathematical theory and its interpretations? What is the nature of mathematical proof,

[5]Even the imminent publication of his lecture notes on geometry, logic, and foundations of physics will not fully remove this last problem: the notes on mainstream mathematics (which make up about three quarters of the whole) are likely to prove necessary to a full reconstruction of the development of his thought, even on foundational questions. So we are still many years away from a satisfactory intellectual biography of Hilbert; and further still from a full understanding of the natural sciences in Göttingen prior to 1933.

and what is the relation of axiomatics to logic? — These questions, central to his wide program, are still only tacit in the Paris address, but they summon up a range of problems, and suggest technical solutions, that in 1900 as yet only Hilbert had glimpsed, and for now only imperfectly. They were come more clearly into focus and to occupy him increasingly in the coming decades; his work in foundations can be seen as a sustained attempt to grapple with them, and, in the characteristically Hilbertian touch, to show how to subject these philosophical problems to mathematical analysis.

REFERENCES

[1] OTTO BLUMENTHAL, *David Hilbert*, ***Die Naturwissenschaften***, vol. 4 (1922), pp. 67–72.

[2] ———, *Lebensgeschichte* (of David Hilbert), vol. 3 of D. HILBERT, ***Gesammelte Abhandlungen***, Springer, Berlin, 1935, pp. 388–429.

[3] WILLIAM EWALD, ***From Kant to Hilbert: A source book in the foundations of mathematics***, vol. 2, Clarendon Press, Oxford, 1996.

[4] HANS FREUDENTHAL, *Hilbert, David*, ***Dictionary of scientific biography*** (C. Gillespie et al., editors), vol. 6, 1976, pp. 388–395.

[5] MICHAEL HALLETT, *Hilbert and logic*, ***Quebec studies in the philosophy of science*** (M. Marion and R.S. Cohen, editors), vol. 1, 1995, pp. 135–187.

[6] ———, *Hilbert on Geometry, Number, and Continuity*, Unpublished manuscript, 1997.

[7] G. H. HARDY, *Mathematical proof*, ***Mind***, vol. 38 (1929), pp. 1–25, (Reprinted in [3]).

[8] DAVID HILBERT, ***Grundlagen der Geometrie***, Teubner, Leipzig, 1899.

[9] ———, *Über den Zahlbegriff*, ***Jahresbericht der Deutschen Mathematiker-Vereinigung***, vol. 8 (1899), pp. 180–194, (English translation in [3]).

[10] ———, *Mathematische Probleme*, ***Nachrichten der Königlichen Gesellschaft der Wissenschaften zu Göttingen***, (1900), pp. 253–297, (Partial English translation in [3]).

[11] ———, *Über die Grundlagen der Logik und der Arithmetik*, ***Verhandlungen des Dritten Internationalen Mathematiker-Kongresses in Heidelberg vom 8. bis 13. August 1904***, Teubner, Leipzig, 1905, English translation in [24].

[12] ———, *Axiomatisches Denken*, ***Mathematische Annalen***, vol. 78 (1917), pp. 405–415, (English translation in [3]).

[13] ———, *Prinzipien der Mathematik*, Lecture notes, written out by Paul Bernays, and on file in the Mathematisches Institut, Göttingen, 1917–1918, forthcoming in the Hilbert Edition, Springer Verlag.

[14] ———, *Natur und mathematisches Erkennen*, Lecture notes, written out by Paul Bernays, and on file in the Mathematisches Institut, Göttingen, 1919, forthcoming in the Hilbert Edition, Springer Verlag.

[15] ———, *Neubegründung der Mathematik. Erste Mitteilung*, ***Abhandlungen aus dem mathematischen Seminar der Hamburgischen Universität***, vol. 6 (1922), pp. 65–85, (English translation in [3]).

[16] ———, *Wissen und mathematisches Denken*, Lecture notes, written out by Wilhelm Ackermann, and on file in the Mathematisches Institut, Göttingen, 1922–1923, forthcoming in the Hilbert Edition, Springer Verlag.

[17] ———, *Die Grundlegung der elementaren Zahlentheorie*, ***Mathematische Annalen***, vol. 104 (1931), pp. 485–494, (English translation in [3]).

[18] ———, ***Gesammelte Abhandlungen***, vol. 3, Springer, Berlin, 1935.

[19] JOHANN HEINRICH LAMBERT, *Theorie der Parallellinien*, ***Magazin für reine und angewandte Mathematik für 1786***, pp. 137–164, 325–358, (Written in 1766; partial English translation in [3]).

[20] Annette Lessmöllmann, *Brillantes Versagen*, ***Die Zeit***, 21 June 2001, p. 29.

[21] Volcker Peckhaus, ***Hilbertprogramm und Kritische Philosophie***, Vandenhoeck und Ruprecht, Göttingen, 1990.

[22] Frank P. Ramsey, *The foundations of mathematics*, ***Proceedings of the London Mathematical Society***, vol. 25, part 5 (1925), pp. 338–384.

[23] Wilfried Sieg, *Hilbert's programs: 1917–1922*, ***The Bulletin of Symbolic Logic***, vol. 5 (1999), pp. 1–44.

[24] Jean van Heijenoort, ***From Frege to Gödel: A source book in mathematical logic***, Harvard, Cambridge, 1967.

[25] Hermann Weyl, *David Hilbert and his mathematical work*, ***Bulletin of the American Mathematical Society***, vol. 50 (1944), pp. 612–654, (Reprinted in H. Weyl, ***Gesammelte Abhandlungen***, vol. 4, pp. 130–172. Page references are to this reprinting.).

UNIVERSITY OF PENNSYLVANIA
3400 CHESTNUT ST.
PHILADELPHIA, PENNSYLVANIA 19104, USA
E-mail: wewald@law.upenn.edu

RIGIDITY CONJECTURES

ILIJAS FARAH

Abstract. If $\mathcal{B}$ is a Boolean algebra and $\mathcal{I}, \mathcal{J}$ are two of its ideals, when are their quotients, $\mathcal{B}/\mathcal{I}$ and $\mathcal{B}/\mathcal{J}$, isomorphic? A sufficient condition is the existence of an automorphism of $\mathcal{B}$ that sends $\mathcal{I}$ onto $\mathcal{J}$. We conjecture that in the case when $\mathcal{B}$ is the power-set of the natural numbers, $\mathcal{P}(\mathbb{N})$, and $\mathcal{I}, \mathcal{J}$ are its 'simply definable' ideals (see §1) other than the Fréchet ideal, Fin, this condition is also necessary. This conjecture is an attempt to subsume the results of [28, 40, 41, 16, 17, 15, 39, 7, 6, 20, 21], where some of its instances have already been proved. Similar phenomena occur in other categories, for example in quotient groups (see [8]).

§1. Introduction.

As a Boolean algebra, $\mathcal{P}(\mathbb{N})$ is well understood: it is completely generated by countably many atoms (the singletons). We will concentrate on quotients of this algebra. An *ideal* $\mathcal{I}$ on $\mathbb{N}$ is an ideal of the Boolean algebra $\mathcal{P}(\mathbb{N})$, i.e., it is a family of subsets of $\mathbb{N}$ closed under taking finite unions and subsets of its elements.

Consider $\mathcal{P}(\mathbb{N})$ with its Cantor-set topology, obtained by identifying subsets of $\mathbb{N}$ with their characteristic functions. This enables us to take the advantage of Borel structure and Haar measure in the study of subsets of $\mathcal{P}(\mathbb{N})$. We can therefore talk about F_σ, Borel, analytic, and so on, ideals on $\mathbb{N}$. (A set is *analytic* if it is a continuous image of a Borel set of the reals.)

Instead of $\mathbb{N}$, we may consider ideals on $\mathbb{Q}$, some countable ordinal, or any countable set, since certain ideals can be visualized more easily in this way. To simplify the notation, in our arguments we will pretend that all of our ideals live on $\mathbb{N}$. Some examples of ideals:

$$P_1 = \{A \subseteq \mathbb{N} : 1 \notin A\},$$
$$\text{Fin} = \{A \subseteq \mathbb{N} : A \text{ is finite}\},$$
$$\mathcal{I}_{1/n} = \Big\{A \subseteq \mathbb{N} : \sum_{n \in A} \frac{1}{n} < \infty\Big\},$$

The author acknowledges support received from the National Science Foundation via grant DMS-0070798, PSC-CUNY grant #62785-00-31, and the NSERC.

Logic Colloquium 2000
Edited by R. Cori, A. Razborov, S. Todorčević, and C. Wood
Lecture Notes in Logic, 19

$$\mathcal{Z}_0 = \left\{ A \subseteq \mathbb{N} : \limsup_{n\to\infty} \frac{|A \cap n|}{n} = 0 \right\},$$

$$\mathrm{NWD}(\mathbb{Q}) = \{A \subseteq \mathbb{Q} : A \text{ is nowhere dense}\},$$

$$\mathcal{I}_{\omega^2} = \{A \subseteq \omega^2 : \text{the order-type of } A \text{ is less than } \omega^2\}.$$

Definitions of some of these ideals depend on a parameter, changing of which leads to a whole class of ideals with similar properties (see §2). All of the above ideals have simple definitions. In the Cantor-set topology, the first one is both closed and open, the next two are F_σ, the next two are $F_{\sigma\delta}$, and the last one is $G_{\delta\sigma\delta}$.

To $\mathcal{I}$ we associate an equivalence relation: $AE_{\mathcal{I}}B$ if and only if $A\Delta B \in \mathcal{I}$. We will denote the $E_{\mathcal{I}}$-equivalence class of X by $[X]_{\mathcal{I}}$. This relation is a congruence, and $\mathcal{P}(\mathbb{N})/\mathcal{I}$ is the *quotient algebra over* $\mathcal{I}$.

BASIC QUESTION. *How does a change of the ideal* $\mathcal{I}$ *effect the change of its quotient algebra,* $\mathcal{P}(\mathbb{N})/\mathcal{I}$?

First of all, this question needs to be put in a proper context. It is clear what we mean by saying that two Boolean algebras are isomorphic.

DEFINITION 1.1. Ideals $\mathcal{I}$ and $\mathcal{J}$ are *Rudin–Keisler isomorphic*, $\mathcal{I} \approx_{\mathrm{RK}} \mathcal{J}$, if there are sets $A \in \mathcal{I}$ and $B \in \mathcal{J}$ and a bijection $h\colon (\mathbb{N} \setminus B) \to (\mathbb{N} \setminus A)$ such that for all $X \subseteq \mathbb{N}$ we have $X \in \mathcal{I}$ if and only if $h^{-1}(X) \in \mathcal{J}$.

If $\mathcal{I} \approx_{\mathrm{RK}} \mathcal{J}$ and both $\mathcal{I}$ and $\mathcal{J}$ contain an infinite set, then we may assume that $A = B = \emptyset$, and in this case h is a permutation of the natural numbers. Thus the map $A \mapsto h^{-1}(A)$, used in the lemma below, is an automorphism of $\mathcal{P}(\mathbb{N})$.

LEMMA 1.2. *For any two ideals* $\mathcal{I}, \mathcal{J}$, *if* $\mathcal{I} \approx_{\mathrm{RK}} \mathcal{J}$ *then* $\mathcal{P}(\mathbb{N})/\mathcal{I} \approx \mathcal{P}(\mathbb{N})/\mathcal{J}$.

PROOF. Define $\Phi\colon \mathcal{P}(\mathbb{N})/\mathcal{I} \to \mathcal{P}(\mathbb{N})/\mathcal{J}$ by

$$\Phi\left([X]_{\mathcal{I}}\right) = \left[h^{-1}(X)\right]_{\mathcal{J}}.$$

Since $h^{-1}(Y) \in \mathcal{J}$ if and only if $Y \in \mathcal{I}$, Φ is an isomorphic embedding. Since the restriction of h to $\mathbb{N} \setminus B$ is one-to-one, Φ is an isomorphism. ⊣

Does Lemma 1.2 have a converse? A positive answer, that $\mathcal{I} \approx_{\mathrm{RK}} \mathcal{J}$ if and only if $\mathcal{P}(\mathbb{N})/\mathcal{I} \approx \mathcal{P}(\mathbb{N})/\mathcal{J}$, would also be the strongest possible answer to Basic Question. It is therefore not surprising that the answer to an unrestricted version of this question is negative. If $\mathcal{I}$ and $\mathcal{J}$ are maximal proper ideals, then both $\mathcal{P}(\mathbb{N})/\mathcal{I}$ and $\mathcal{P}(\mathbb{N})/\mathcal{J}$ are two-element Boolean algebras, and therefore isomorphic. But if $\mathcal{J} \supseteq \mathrm{Fin}$ is a maximal proper ideal, it is not Rudin–Keisler isomorphic to P_1 (consider $h^{-1}(\{1\})$).

PROBLEM 1.3. *Isolate the optimal conditions under which Lemma* 1.2 *has a converse.*

By the above example, it is necessary to exclude the possibility of having a nonprincipal maximal ideal. Our approach is to restrict our attention to the ideals that are 'simply definable' in some way. This approach was taken in [7] where we have considered only the ideals that are analytic subsets of $\mathcal{P}(\mathbb{N})$. (Recall that a subset of $\mathcal{P}(\mathbb{N})$ is *analytic* if it is a continuous image of a Borel set of reals.) In §4.1 we will consider (arguably) the largest class of reasonably definable ideals.

§**2. Ideals on** $\mathbb{N}$. Before we proceed with the analysis of Problem 1.3, let us get acquainted with some examples of ideals. From now on we will consider only the ideals that contain all finite subsets of $\mathbb{N}$, since the principal ideals do not give rise to interesting quotients.

An $A \subseteq \mathbb{N}$ is $\mathcal{I}$*-positive* if $A \notin \mathcal{I}$. We write $\mathcal{I} \restriction A$ for $\mathcal{I} \cap \mathcal{P}(A)$ and consider this to be an ideal on A. An ideal $\mathcal{I}$ is *dense* if every $\mathcal{I}$-positive set A has an infinite subset belonging to $\mathcal{I}$.

2.1. Summable ideals. If $f : \mathbb{N} \to \mathbb{R}^+$ then

$$\mathcal{I}_f = \Big\{A \subseteq \mathbb{N} : \sum_{n \in A} f(n) < \infty\Big\}$$

is an example of a *summable* ideal. Note that this ideal is *proper* (i.e., $\mathbb{N} \notin \mathcal{I}_f$) if and only if $\sum_{n=1}^{\infty} f(n) = \infty$ and that it is dense if and only if $\lim_n f(n) = 0$. Since $\mathcal{I}_f = \bigcup_{m \in \mathbb{N}} \{A : \sum_{n \in A} f(n) \leq m\}$, each $\mathcal{I}_f$ is an F_σ ideal.

A simple (even F_σ) characterization of all pairs (f, g) such that $\mathcal{I}_f \approx_{\mathrm{RK}} \mathcal{I}_g$ can be obtained from [7, Lemma 1.12.5].

2.2. Density ideals. Let $\mathbb{N} = \bigcup_{n=0}^{\infty} I_n$ be a partition of $\mathbb{N}$ into pairwise disjoint finite sets, and let $\mu = \{\mu_n\}_{n=0}^{\infty}$ be a sequence of measures such that μ_n concentrates on I_n. Then

$$\mathcal{Z}_\mu = \{A \subseteq \mathbb{N} : \lim_{n \to \infty} \mu_n(A \cap I_n) = 0\}$$

is an $F_{\sigma\delta}$ ideal. Such ideals are called *density ideals* and they were introduced and studied in [7, §13]. Both Fin and $\mathcal{Z}_0$ are density ideals (see [7, Theorem 1.13.3] for the latter). Dense density ideals are very different from the summable ideals (see [7, Proposition 1.13.14]).

There are many pairs (μ, ν) such that $\mathcal{Z}_\mu \not\approx_{\mathrm{RK}} \mathcal{Z}_\nu$ [7, Theorem 1.13.12], but no simple characterization of such pairs is known.

2.3. Ideals induced by submeasures. A map $\phi \colon \mathcal{P}(\mathbb{N}) \to [0, \infty]$ is a *submeasure* if $\phi(\emptyset) = 0$, and ϕ is monotonic and subadditive. It is moreover *lower semicontinuous* if $\phi(A) = \lim_{n \to \infty} \phi(A \cap n)$. To a lower semicontinuous submeasure ϕ we associate an ideal

$$\mathrm{Exh}(\phi) = \{A \subseteq \mathbb{N} : \lim_{n \to \infty} \phi(A \setminus n) = 0\}.$$

Such ideals are always P-ideals. (An ideal $\mathcal{I}$ is a *P-ideal* if for every sequence A_n ($n \in \mathbb{N}$) of elements of $\mathcal{I}$ there is an $A_\infty \in \mathcal{I}$ such that $A_n \setminus A_\infty$ is finite, for all n.) For example, Fin, all summable ideals, and all density ideals are P-ideals. Each summable ideal can be written in the form $\mathrm{Exh}(\phi)$: let $\phi(A) = \sum_{n\in A} f(n)$. Similarly, $\mathcal{Z}_\mu = \mathrm{Exh}(\sup_n \mu_n)$. The ideal $\mathrm{Exh}(\phi)$ is dense if and only if $\lim_n \phi(\{n\}) = 0$. By a result of Solecki [32, 33], every analytic P-ideal is of the form $\mathrm{Exh}(\phi)$ for some lower semicontinuous submeasure ϕ. In particular, all analytic P-ideals are automatically $F_{\sigma\delta}$.

2.4. F_σ ideals. A G_δ ideal that contains Fin has to be equal to $\mathcal{P}(\mathbb{N})$, by the Baire Category Theorem. Hence F_σ ideals are the simplest nontrivial ideals on $\mathbb{N}$. By a result of K. Mazur [26, Lemma 1.2] an ideal $\mathcal{I}$ is F_σ if and only if

$$\mathcal{I} = \mathrm{Fin}(\phi) = \{A \subseteq \mathbb{N} : \phi(A) < \infty\}.$$

for some lower semicontinuous submeasure ϕ. In spite of their simplicity, F_σ-ideals form a very rich and well-studied structure (see [23]).

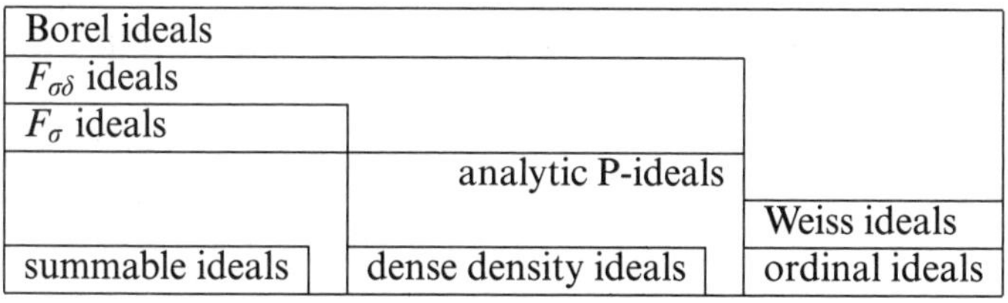

FIGURE 1. Classes of Borel ideals.

2.5. Nonpathological ideals. A submeasure ϕ is *nonpathological* if for all A in the domain of ϕ we have $\phi(A) = \sup_{\mu\leq\phi} \mu(A)$, where the supremum is taken over all measures μ pointwise dominated by ϕ. (This definition of a nonpathological submeasure was introduced in [7] and it differs somewhat from the standard one; what we call pathological is sometimes called *weakly pathological.*) An ideal is *nonpathological* if it is of the form $\mathrm{Exh}(\phi)$ or $\mathrm{Fin}(\phi)$ for some lower semicontinuous nonpathological submeasure ϕ. All summable and all density ideals are obviously nonpathological. Nonpathological ideals were introduced in [7, §1] in an attempt to describe a class of ideals for which Todorcevic's conjecture (Conjecture 4.1) is true. Many ideals occurring in the literature are nonpathological. In §4 we will see that the nonpathological ideals have a property related to Basic Question and Problem 1.3.

2.6. Other $F_{\sigma\delta}$ ideals. While the structure of F_σ ideals and analytic P-ideals is well-understood, some very natural questions on the structure of $F_{\sigma\delta}$ ideals are still open (see [34]). An interesting example of an $F_{\sigma\delta}$ ideal, suggested to us by Michal Hrusák, is $\mathrm{NWD}(\mathbb{Q})$. It is $F_{\sigma\delta}$: If $\mathcal{B}$ is a countable basis for $\mathbb{Q}$ consisting of nonempty sets, then $A \in \mathrm{NWD}(\mathbb{Q})$ if and only if

$(\forall U \in \mathcal{B})(\exists V \in \mathcal{B})(V \subseteq U \text{ and } V \cap A = \emptyset\}$. By [11], this ideal is moreover homogeneous (an ideal $\mathcal{I}$ is *homogeneous* if $\mathcal{I} \approx_{\text{RK}} \mathcal{I} \restriction A$ for every $\mathcal{I}$-positive set A). This answers [7, Question 3.7.6], where it was asked whether there are any homogeneous analytic ideals other than Fin and the ideals in §§2.7–2.8 below.

2.7. Ordinal ideals. For a countable ordinal α let $\mathcal{I}_\alpha$ be the family of all subsets of α of strictly smaller order type. This family is an ideal if and only if α is an *indecomposable ordinal*, i.e., if α is not equal to the sum of two strictly smaller ordinals; equivalently, if $\alpha = \omega^\beta$ for some ordinal β. These ideals can also be considered as iterated Fubini products of the ideal Fin (see [21]). All the ordinal ideals are clearly homogeneous, and $\mathcal{I}_\omega = \text{Fin}$ is the only P-ideal among them. All $\mathcal{I}_\alpha$ are Borel ideals, and each $\mathcal{I}_{\omega^\alpha}$ is by a result of Zafrany $\Sigma^0_{2\alpha}$-complete. In particular, these ideals have arbitrarily high Borel complexity, and they are pairwise Rudin–Keisler nonisomorphic.

2.8. Cantor–Bendixson ideals. Let X be a countable topological space, and let α be a countable ordinal. Let

$$\text{CB}_\alpha(X) = \{Y \subseteq X : \text{ the Cantor–Bendixson rank of } Y \text{ is } < \alpha\}.$$

Then $\text{CB}_\alpha(X)$ is an ideal if and only if α is additively indecomposable. A special case of Cantor–Bendixson ideals are the *Weiss ideals*, $\mathcal{W}_{\omega^\alpha} = \text{CB}_\alpha(\omega^\alpha)$. This is the ideal of all subsets of ω^α that do not contain a closed copy of ω^α, and it was suggested by W. Weiss [42]. The ideal $\mathcal{W}_\omega = \text{Fin}$ is the only P-ideal among these ideals. These ideals are also homogeneous, have arbitrarily high Borel complexities, and they are pairwise Rudin–Keisler nonisomorphic [21]. They are also Rudin–Keisler nonisomorphic to all ordinal ideals except Fin.

§3. Liftings. We begin our analysis of Problem 1.3 by looking at the connecting maps between quotients. Let $\Phi\colon \mathcal{P}(\mathbb{N})/\mathcal{I} \to \mathcal{P}(\mathbb{N})/\mathcal{J}$ be a homomorphism. A map $\Phi_*\colon \mathcal{P}(\mathbb{N}) \to \mathcal{P}(\mathbb{N})$ such that $[\Phi_*(X)]_\mathcal{J} = \Phi([X]_\mathcal{I})$ for all X, or equivalently, such that the diagram ($\pi_\mathcal{I}$ is the natural projection of $\mathcal{P}(\mathbb{N})$ to $\mathcal{P}(\mathbb{N})/\mathcal{I}$)

$$\begin{array}{ccc}
\mathcal{P}(\mathbb{N}) & \xrightarrow{\Phi_*} & \mathcal{P}(\mathbb{N}) \\
\big\downarrow{\scriptstyle \pi_\mathcal{I}} & & \big\downarrow{\scriptstyle \pi_\mathcal{J}} \\
\mathcal{P}(\mathbb{N})/\mathcal{I} & \xrightarrow{\Phi} & \mathcal{P}(\mathbb{N})/\mathcal{J}
\end{array}$$

commutes, is a *lifting* (or *representation*) of Φ. The reader should be warned that we **do not** require Φ_* to have any algebraic properties; in particular it need not be a homomorphism.

The simplest way to describe a homomorphism is via one of its liftings. The homomorphism defined in the proof of Lemma 1.2 has a particularly simple lifting; it is a map of the form

$$\Phi_h(A) = h^{-1}(A)$$

for some $h\colon \mathbb{N} \to \mathbb{N}$. We say that such a lifting is *completely additive*, since it preserves the infinitary Boolean operations, namely it satisfies the formulas

$$\Phi_*(A^{\complement}) = (\Phi_*(A))^{\complement},$$
$$\Phi_*(A) = \bigcup_{n\in A} \Phi_*(\{n\}).$$

If Φ is an isomorphism and it has a completely additive lifting Φ_h, then h is a Rudin–Keisler isomorphism between the underlying ideals. Thus Problem 1.3 is tightly associated with the question which homomorphisms have completely additive liftings. We also consider *additive liftings*, the liftings that preserve the finitary Boolean operations of $\mathcal{P}(\mathbb{N})$.

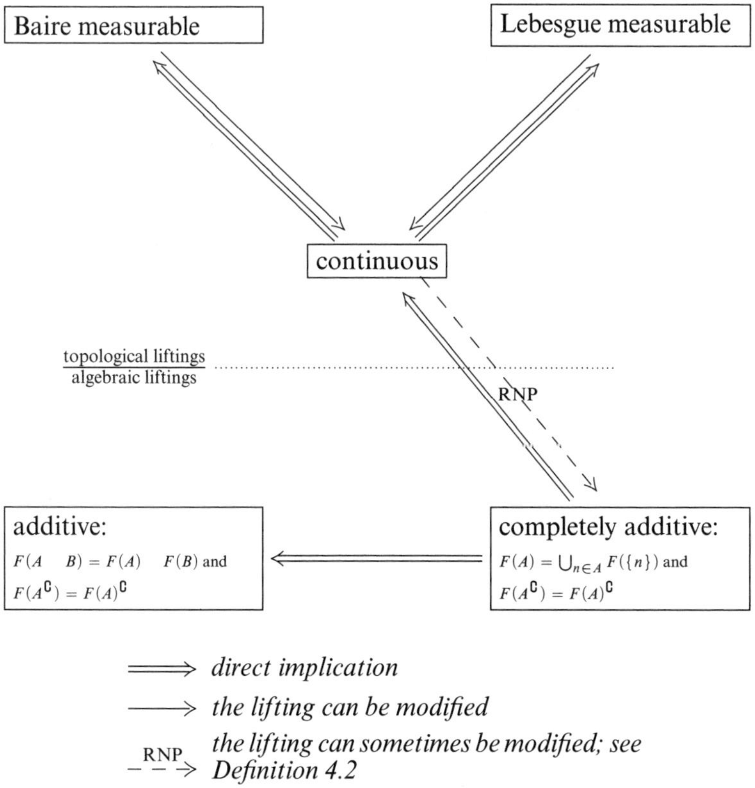

FIGURE 2. Types of liftings.

3.1. Topological liftings. Since we consider $\mathcal{P}(\mathbb{N})$ with its Cantor space topology, the methods of set theory can be applied to liftings that are simple in a topological way, for example continuous, Lebesgue-measurable or Baire-measurable.

A subset of $\mathcal{P}(\mathbb{N})$ is *meager* if it is a countable union of nowhere dense sets. A set $X \subseteq \mathcal{P}(\mathbb{N})$ has the *Property of Baire* if the symmetric difference $X \Delta U$ is meager for some open U. A function is *Baire-measurable* if the preimage of every open set has the property of Baire.

Relationship between simple liftings is given in Figure 2. The fact that every Baire-measurable lifting can be turned into a continuous one was proved in [40, p. 132] and [39, Theorem 3]. An analogous result for Lebesgue-measurable liftings was proved independently in [20] and [13, Proposition 1C]. See [8] for more on transformations between simple types of liftings.

Finitely additive liftings that are not completely additive did not attract much attention, but they may turn out to be interesting. Every known homomorphism from $\mathcal{P}(\mathbb{N})/\mathrm{Fin}$ into itself that can be constructed without using additional Set-theoretic axioms has an additive lifting (see the definition of $F_{\mathcal{U}}$ in [7, Example 3.2.3]). In a situation when all endomorphisms of $\mathcal{P}(\mathbb{N})/\mathrm{Fin}$ have additive liftings every complete Boolean algebra embeddable into $\mathcal{P}(\mathbb{N})/\mathrm{Fin}$ would have to be σ-centered (see [7, Proposition 4.11.7]). This conclusion was conjectured to be consistent by A. Dow. It is not known whether it is consistent with the Axiom of Choice (see [7, Question 3.14.2 and Question 4.11.4]). However, every endomorphism of $\mathcal{P}(\mathbb{N})/\mathrm{Fin}$ that has a Baire-measurable lifting has a (completely) additive lifting (essentially by [40], see also Theorem 4.3).

§4. Quotients over Borel ideals. The example given before Problem 1.3 involves a maximal nonprincipal ideal on $\mathbb{N}$. Such ideals are never Borel, as a matter of fact they are not 'definable in a reasonable way.' (We will return to this in §4.1.) In this paper we will restrict our attention to simply definable ideals. Let us first state a conjecture of Todorcevic that has initially inspired the research that resulted in [7].

Conjecture 4.1 (Todorcevic's conjecture, [39, Problem 1]). *Suppose $\mathcal{I}$ is an analytic P-ideal on $\mathbb{N}$ and that Φ is a homomorphism from $\mathcal{P}(\mathbb{N})/\mathrm{Fin}$ into $\mathcal{P}(\mathbb{N})/\mathcal{I}$ with a Baire lifting. Then Φ has a completely additive lifting.*

Definition 4.2 ([7, §1]). An ideal $\mathcal{J}$ has the *Radon–Nikodym property* if every homomorphism $\Phi\colon \mathcal{P}(\mathbb{N})/\mathrm{Fin} \to \mathcal{P}(\mathbb{N})/\mathcal{J}$ with a Baire-measurable lifting has a completely additive lifting.

Thus Todorcevic's conjecture can be restated as 'all analytic P-ideals have the Radon–Nikodym property.' In [7, §1.7] this conjecture was proved to be equivalent to a finite combinatorial statement (see also [8]). This was used to prove the following (the nonpathological ideals were defined in §2.5).

THEOREM 4.3 (Farah, [7, Theorem 1.9.1]). *Every nonpathological analytic P-ideal has the Radon–Nikodym property. Hence Todorcevic's Conjecture is true for all nonpathological analytic P-ideals.* ⊣

This result relied on Ulam-stability of 'approximate homomorphisms.' For more on this, see [8] where it was also shown that similar results can be proved for quotient groups instead of quotient Boolean algebras.

The class of nonpathological ideals contains virtually all examples of analytic P-ideals previously appearing in the literature. In particular, all summable and all density ideals are nonpathological. The proof of Theorem 4.3 used the finitization of Todorcevic's conjecture mentioned before. The observation that this finitization can be avoided has led to the following definition.

DEFINITION 4.4 (Kanovei–Reeken, [20, 21]). An ideal $\mathcal{I}$ is an *F-ideal* if for every $\varepsilon > 0$ and $X \subseteq \mathbb{N} \times 2^{\mathbb{N}}$ such that each vertical section $X_n = \{x : (n, x) \in X\}$ is Haar-measurable and has measure at least ε the set (X^y is the vertical section, $\{n : (n, y) \in X\}$)

$$X^+ = \{y \in 2^{\mathbb{N}} : X^y \notin \mathcal{I}\}$$

is Haar-measurable and has measure at least ε.

A minor modification of the proof of Theorem 4.3 shows that all F-ideals have the Radon–Nikodym property. Also, all nonpathological ideals are F-ideals. This is essentially a consequence of a result of Christensen [4], to the effect that nonpathological submeasures satisfy a variant of Fubini's theorem. But much more is true.

THEOREM 4.5 (Kanovei–Reeken, [20, 21]). *All nonpathological F_σ ideals, as well as all ordinal ideals and all Weiss ideals are F-ideals. In particular, they all have the Radon–Nikodym property.* ⊣

In [7, Theorem 1.9.1] it was shown that some restriction in Todorcevic's conjecture is necessary by constructing certain 'pathological' ideals which violate this conjecture. It should be noted that the status of this conjecture is still unknown if one adds the requirement that Φ is onto, or equivalently, an isomorphism between two analytic quotients. In fact, the following variation on Todorcevic's conjecture is still open (two quotients are *Baire-isomorphic* if there is an isomorphism that has a Baire-measurable—equivalently, continuous—lifting).

CONJECTURE 4.6. *The following are equivalent for any two Borel ideals $\mathcal{I}, \mathcal{J}$:*

(1) *The quotients over $\mathcal{I}$ and $\mathcal{J}$ are Baire-isomorphic.*
(2) $\mathcal{I} \approx_{\mathrm{RK}} \mathcal{J}$.

Moreover, every isomorphism as in (1) has a completely additive lifting.

LEMMA 4.7. *Conjecture 4.6 is true for isomorphisms between quotients over ideals that have the Radon–Nikodym property.*

PROOF. Let $\mathcal{I}$ and $\mathcal{J}$ be ideals with the RNP and let $\Phi\colon \mathcal{P}(\mathbb{N})/\mathcal{I} \to \mathcal{P}(\mathbb{N})/\mathcal{J}$ be an isomorphism. There are partial functions h and g from $\mathbb{N}$ into $\mathbb{N}$ such that $A \mapsto h^{-1}(A)$ is a lifting of Φ and $A \mapsto g^{-1}(A)$ is a lifting of Φ^{-1}. A standard argument shows that the set $X = \{n : g(h(n)) \neq n\}$ belongs to $\mathcal{J}$ (see e.g., the proof of [7, Proposion 1.4.6]). Since h is one-to-one on $\mathbb{N} \setminus X$, it is a Rudin–Keisler isomorphism between $\mathcal{I}$ and $\mathcal{J}$. ⊣

PROPOSITION 4.8. *If $\mathcal{I}$ is an F-ideal and that there is an embedding of $\mathcal{P}(\mathbb{N})/\mathcal{J}$ into $\mathcal{P}(\mathbb{N})/\mathcal{J}$ that has a Baire-measurable lifting, then $\mathcal{J}$ is an F-ideal.*

PROOF. This is very similar to the proof of [7, Proposition 1.10.3]. By the Radon–Nikodym property of F-ideals, the isomorphism has a completely additive lifting, of the form $A \mapsto h^{-1}(A)$ for some $h\colon \mathbb{N} \to \mathbb{N}$. Then h is a Rudin–Keisler reduction between $\mathcal{I}$ and $\mathcal{J}$, i.e., we have $A \in \mathcal{J}$ if and only if $h^{-1}(A) \in \mathcal{I}$. Fix $X \subseteq \mathbb{N} \times 2^{\mathbb{N}}$ such that every vertical section has measure at least ε, for some fixed $\varepsilon > 0$. We need to prove that the set $\{z : X^z \notin \mathcal{J}\}$ has measure at least ε.

Define $Y \subseteq \mathbb{N} \times 2^{\mathbb{N}}$ so that $Y_n = X_{h(n)}$. Since $\mathcal{I}$ is an F-ideal, the set

$$Z = \{z \in Y : Y^z \notin \mathcal{I}\}$$

has measure at least ε. Fix $z \in Z$. Since $(n, z) \in Y$ implies $(h(n), z) \in X$, we conclude that $Y^z \notin \mathcal{I}$ implies $X^z \notin \mathcal{J}$, hence $Z \subseteq \{z \in X : X^z \notin \mathcal{J}\}$, and therefore $\mathcal{J}$ is an F-ideal. ⊣

By Proposition 4.8 and Lemma 4.7 we have a partial answer to Conjecture 4.6.

PROPOSITION 4.9. *Conjecture 4.6 is true for every pair of ideals $\mathcal{I}, \mathcal{J}$ such that at least one of them is an F-ideal.* ⊣

At present it is unclear which analytic ideals have the Radon–Nikodym property, but it is known that not all ideals with the RNP are F-ideals. By [10, Proposition 12.4], NWD($\mathbb{Q}$) has the RNP but it is not an F-ideal.

Several classes of ideals, such as summable ideals or nonpathological ideals, share the property of F-ideals proved in Proposition 4.8 (see [7, §1.10]).

QUESTION 4.10. *Assume $\mathcal{I}$ is a Borel ideal that has the Radon–Nikodym property and that there is an embedding of $\mathcal{P}(\mathbb{N})/\mathcal{J}$ into $\mathcal{P}(\mathbb{N})/\mathcal{I}$ that has a Baire-measurable lifting. Does $\mathcal{J}$ necessarily have the Radon–Nikodym property?*

Similarly to Todorcevic's conjecture, Conjecture 4.6 has an equivalent reformulation in terms of Ulam-stability of *approximate isomorphisms* between finite Boolean algebras (see [7, Question 1.14.4]).

4.1. Quotients over projective ideals and beyond. Consider ideals $\mathcal{I}$ of the form $\mathcal{I} = \{A \subseteq \mathbb{N} : \phi(A)\}$ for a first-order formula ϕ with ordinals and reals as parameters. Under a suitable large cardinal assumption, these are exactly the ideals that belong to $L(\mathbb{R})$, the smallest class containing all reals and all ordinals and closed under the primitively recursive set functions. Following [37], we say that ideals (or, subsets of $\mathcal{P}(\mathbb{N})$) that belong to $L(\mathbb{R})$ are *definable*. By a result of Solovay [35], if there is an inaccessible cardinal κ then after the Levy collapse of κ to ω_1 all definable subsets of $\mathcal{P}(\mathbb{N})$ have the Property of Baire and are Lebesgue-measurable. In particular, there are no nonprincipal ultrafilters on $\mathbb{N}$ hence the example stated before Problem 1.3 does not apply. We say that $L(\mathbb{R})$ of such forcing extension is a *Solovay model.* Moreover, the existence of more substantial large cardinals (like a weakly compact Woodin cardinal) implies that $L(\mathbb{R})$ is elementarily equivalent to some Solovay's model [31, 12]. Hence if sufficiently large cardinals exist the theory of $L(\mathbb{R})$ is unchangeable by forcing and canonical: all definable sets of reals are Lebesgue measurable, have the Property of Baire, are determined, every definable $X \subseteq \mathbb{R}^2$ can be uniformized on a dense G_δ set, and so on (see [31], [2]).

In the following conjecture and elsewhere by 'some Solovay's model' we mean 'a Solovay's model obtained from sufficiently large cardinals.'

Conjecture 4.11. *In some Solovay's model the following are equivalent for any two definable ideals $\mathcal{I}, \mathcal{J}$:*

(1) $\mathcal{P}(\mathbb{N})/\mathcal{I} \approx \mathcal{P}(\mathbb{N})/\mathcal{J}$,

(2) $\mathcal{I} \approx_{\mathrm{RK}} \mathcal{J}$.

Moreover, every isomorphism as in (1) *has a completely additive lifting.*

It is possible that the assumptions of Lemma 4.12 below imply the conclusion of Conjecture 4.11. By Lemma 1.2, the interesting direction is (1) implies (2). Since in Gödels constructible universe and other canonical inner models for moderately large cardinals there is a projective well-ordering of $\mathcal{P}(\mathbb{N})$ [25], and therefore a projective nonprincipal ultrafilter on $\mathbb{N}$, in these models (1) does not imply (2).

The assumptions of the following lemma are true in Solovay's model.

Lemma 4.12. *Assume that every definable relation can be uniformized on a dense G_δ set and that all definable sets of reals have the property of Baire. Then every definable homomorphism between quotients over definable ideals has a continuous lifting.*

Proof. Assume Φ is in $L(\mathbb{R})$, and let

$$X = \{ (A, B) \in \mathcal{P}(\mathbb{N})^2 : \Phi([A]_{\mathcal{I}}) = [B]_{\mathcal{J}} \}.$$

Let F be a function that uniformizes X on a dense G_δ set. We may assume F is Baire-measurable. Hence Φ has a Baire-measurable lifting, and therefore a continuous lifting as well. ⊣

By Lemma 4.12, Conjecture 4.11 is equivalent to:

Conjecture 4.13. *In some Solovay's model every isomorphism between quotients over ideals on $\mathbb{N}$ that has a continuous lifting has a completely additive lifting.*

As an immediate consequence of Lemma 4.12, under its assumptions an isomorphism $\Phi\colon \mathcal{P}(\mathbb{N})/\mathcal{I} \to \mathcal{P}(\mathbb{N})/\mathcal{J}$ has a continuous lifting if and only if its inverse has a continuous lifting.

§5. The influence of the Continuum Hypothesis. The discussion in §4.1 suggests that a 'definable' version of Problem 1.3 may have a satisfactory answer. While our decision to consider only definable ideals is justified by the example given before Problem 1.3, considering only definable isomorphisms seems less natural.

5.1. Saturatedness of quotients. The early study of quotients related to $\mathcal{P}(\mathbb{N})/\mathrm{Fin}$ has concentrated on $\mathbb{R}^{\mathbb{N}}/\mathrm{Fin}$, the structure of real-valued sequences partially ordered by $f \prec g$ if and only if $\lim_{n\to\infty} g(n) - f(n) = \infty$. One of the important features of this structure, discovered by P. du Bois-Reymond and J. Hadamard, is that it has no countable *gaps*. Namely, if $\mathcal{A}, \mathcal{B}$ are two countable subsets of $\mathbb{R}^{\mathbb{N}}/\mathrm{Fin}$ such that for every $f \in \mathcal{A}$ and every $g \in \mathcal{B}$ we have $f \prec g$, then there is h such that $f \prec h \prec g$ for all $f \in \mathcal{A}$ and all $g \in \mathcal{B}$.

Recall that a Boolean algebra $\mathcal{B}$ is *atomless* if for every positive $b \in \mathcal{B}$ there is a positive $c \leq b$ such that $b \setminus c$ is positive. If $\mathcal{I}$ is a proper analytic ideal on $\mathbb{N}$ including Fin, the quotient algebra $\mathcal{P}(\mathbb{N})/\mathcal{I}$ is atomless. Since the theory of atomless Boolean algebras is $\aleph_0$-categorical, all the quotients that we consider in this note are elementarily equivalent (see [3]).

We will use the term *saturated* in the model-theoretic sense (see e.g., [3]). Hence '$\mathcal{B}$ is *countably saturated*' (or '$\mathcal{B}$ is $\aleph_1$-*saturated*') means that every countable finitely satisfiable type with parameters in $\mathcal{B}$ is realized in $\mathcal{B}$. Continuum Hypothesis implies that all countably saturated Boolean algebras of size $2^{\aleph_0}$ (in particular, all quotients $\mathcal{P}(\mathbb{N})/\mathcal{I}$ as above) are saturated, and therefore isomorphic.

THEOREM 5.1 (Just–Krawczyk, [18]). *Every quotient over an F_σ ideal is countably saturated.* ⊣

Thus we have the following result, first proved for summable ideals by Erdös and Monk.

COROLLARY 5.2 (Just–Krawczyk, [18]). *CH implies that all quotients over F_σ ideals are pairwise isomorphic.* ⊣

Note that an isomorphism between two saturated models is constructed via the back-and-forth argument of transfinite length, and an isomorphism constructed in this way is unlikely to have a continuous lifting.

But quotients over many natural ideals are not countably saturated. For example, A_i $(i \in \mathbb{N})$ such that $\lim_n |A_i \cap n|/n = 1/i$ and $A_i \supseteq A_{i+1}$ for all i form a sequence of $\mathcal{Z}_0$-positive sets with no $\mathcal{Z}_0$-positive lower bound. Moreover, a quotient over an analytic P-ideal $\mathcal{I}$ is countably saturated if and only if $\mathcal{I}$ is F_σ (essentially proved in [18], see [7, the end of §1.3]). Still, in [18,

Theorem 3] it was shown that the quotients over many different density ideals, including the ideals $\mathcal{Z}_0$ and $\mathcal{Z}_{\log} = \{A \subseteq \mathbb{N} : \limsup_n (\sum_{i \in A \cap n} 1/i)/\log n = 0\}$, are pairwise isomorphic under CH. It is not difficult to see that $\mathcal{Z}_0 \not\approx_{\mathrm{RK}} \mathcal{Z}_{\log}$ (see [7, Proposition 1.13.13] for a more general result).

Using results of [18], in [7, §3.14] it was shown that there are at least six isomorphism types of analytic quotients in any model of set theory.

Question 5.3 ([7, Question 3.14.3]). *Are there infinitely (or even uncountably) many analytic P-ideals whose quotients are, provably in ZFC, pairwise nonisomorphic?**

This question, as well as the corresponding question for quotients over arbitrary analytic (or definable) ideals is still open. At the moment when [7] was finished, there was a large supply of potential candidates for finding infinitely many pairwise nonisomorphic analytic quotients. For example, it seemed likely that quotients over Weiss ideals and ordinal ideals will turn out to be pairwise nonisomorphic. Also, a large class of analytic P-ideals was constructed by Louveau and Velickovic in [24]. They have constructed a family $\mathcal{J}_A$ ($A \subseteq \mathbb{N}$) of analytic P-ideals such that $\mathcal{P}(\mathbb{N})/\mathcal{J}_A$ is Baire-isomorphic with $\mathcal{P}(\mathbb{N})/\mathcal{J}_B$ (in a rather weak sense) if and only if $A \Delta B$ is finite.

Theorem 5.4 (Farah, [9]). *Assume the Continuum Hypothesis.*

(1) *The quotients over all Louveau–Velickovic ideals are isomorphic.*
(2) *Quotients over all ordinal ideals and all Weiss ideals are countably saturated, and therefore isomorphic to $\mathcal{P}(\mathbb{N})/\mathrm{Fin}$.*
(3) *If $\mathcal{Z}_\mu$ and $\mathcal{Z}_{\mu'}$ are dense and such that*

$$\limsup_{n\to\infty} \mu_n(I_n) = \limsup_{n\to\infty} \mu'_n(I'_n) = \infty,$$

then their quotients are isomorphic. ⊣

Therefore (1) and (2) show that under CH all of the analytic quotients mentioned in this paper or in [7] fall into one of finitely many isomorphism classes.

While Conjecture 7.3 implies that a quotient over an analytic P-ideal cannot be isomorphic to a quotient over an analytic ideal that is not a P-ideal, Theorem 5.4 shows that under CH the situation is quite different. On the other hand, in [11] it was proved that the quotient over $\mathrm{NWD}(\mathbb{Q})$ cannot be isomorphic to a quotient over any analytic P-ideal.

It is worth noting that the Continuum Hypothesis provides the optimal ambient for testing Question 5.3 (see [19] for the definitions).

Theorem 5.5 (Woodin, [43]). *Assume there are class many measurable Woodin cardinals. If $\phi(X)$ is a first-order statement of $L(\mathbb{R})$ and in some forcing*

*Added in proof. This question has been answered by M. R. Oliver in his PhD thesis (UCLA, 2003). He constructed a family of $2^{\ 0}$ analytic P-ideals with pairwise nonisomorphic quotients.

extension there is $X \subseteq \mathbb{R}$ such that $\phi(X)$ holds, then $(\exists X)\phi(X)$ holds in every forcing extension that satisfies CH. $\dashv$

This theorem applies in the case when $\phi(X)$ is saying 'X is a lifting of an isomorphism between $\mathcal{P}(\mathbb{N})/\mathcal{I}$ and $\mathcal{P}(\mathbb{N})/\mathcal{J}$.' Therefore if two Borel ideals have isomorphic quotients in some forcing extension, then they have isomorphic quotients in every forcing extension that satisfies CH, at least assuming large cardinals.

§6. When CH fails. In §5 we have seen that the existence of a well-ordering of the reals all of whose proper initial segments are countable can be used to construct isomorphisms between rather different ideals. But can such isomorphisms be constructed by using a weaker assumption? In general, if the size of the continuum is bigger than $\aleph_1$, an attempt to construct an isomorphism using the back-and-forth method runs into a problem. The following is a generalization of the classical result of Hausdorff [14].

THEOREM 6.1 (Todorcevic, [39]). *Let $\mathcal{I}$ be an analytic ideal. Then $\mathcal{P}(\mathbb{N})/\mathcal{I}$ is not $\aleph_2$-saturated. More precisely, there are two families $\mathcal{A}, \mathcal{B}$ of size $\aleph_1$ such that*

(1) *$A \cap B \in \mathcal{I}$ for all $A \in \mathcal{A}$ and all $B \in \mathcal{B}$, yet*
(2) *there is no C such that $A \setminus C \in \mathcal{I}$ and $C \cap B \in \mathcal{I}$ for all $A \in \mathcal{A}$ and all $B \in \mathcal{B}$.* $\dashv$

We say that families $\mathcal{A}, \mathcal{B}$ satisfing (1) are *orthogonal over $\mathcal{I}$*, or that they form a *pregap in $\mathcal{P}(\mathbb{N})/\mathcal{I}$*. If (2) fails and C is a witness, then C *separates* $\mathcal{A}$ from $\mathcal{B}$ over $\mathcal{I}$. If both (1) and (2) hold, then $(\mathcal{A}, \mathcal{B})$ is a *gap* in $\mathcal{P}(\mathbb{N})/\mathcal{I}$ (for more on gaps in analytic quotients see [39], [37], or [7, Chapter 5]).

Methods of [30] can be used to show that in certain situations when CH fails (in particular, in Cohen's original model for $2^{\aleph_0} = \aleph_2$) all countably saturated quotients over Borel ideals are still isomorphic. The paper [30] was motivated by the study of automorphisms of the Boolean algebra $\mathcal{P}(\mathbb{N})/\mathrm{Fin}$. An automorphism of a quotient algebra $\mathcal{P}(\mathbb{N})/\mathcal{I}$ is *trivial* if and only if it has a completely additive lifting (equivalently, if and only if it is induced by a Rudin–Keisler automorphism of the ideal $\mathcal{I}$). The saturatedness of $\mathcal{P}(\mathbb{N})/\mathrm{Fin}$ implies that it has the maximal number of (mostly) nontrivial automorphisms (a result first proved by W. Rudin, [27]).

The following was the first result in the direction considered in this paper.

THEOREM 6.2 (Shelah, [28]). *There is a forcing extension in which all automorphisms of $\mathcal{P}(\mathbb{N})/\mathrm{Fin}$ are trivial.* $\dashv$

Once Shelah's approach was systematized [40, 16] it became clear that the proof of Theorem 6.2 naturally splits into two parts:

(a) In some forcing extension, all automorphisms of $\mathcal{P}(\mathbb{N})/\mathrm{Fin}$ have continuous liftings.

(b) An automorphism of $\mathcal{P}(\mathbb{N})/\mathrm{Fin}$ has a continuous lifting if and only if it has a completely additive lifting.

Since part (b) belongs to the discussion of §4, let us concentrate on (a). Roughly, its proof proceeds as follows. If Φ is an automorphism with no continuous lifting, then there is a forcing $\mathcal{P}_\Phi$ that adds a set $X \subseteq \mathbb{N}$ such that the following families of ground-model sets (Φ_* is a lifting of Φ):

$$\mathcal{A}_{\Phi,X} = \{\Phi_*(A) : A \in (\mathcal{P}(\mathbb{N}))^V, A \setminus X \in \mathrm{Fin}\},$$
$$\mathcal{B}_{\Phi,X} = \{\Phi_*(B) : B \in (\mathcal{P}(\mathbb{N}))^V, B \cap X \in \mathrm{Fin}\}$$

are forced to form a gap in $\mathcal{P}(\mathbb{N})/\mathrm{Fin}$; hence in the extension Φ cannot be extended to a homomorphism. Such forcings are iterated, taking care (i) that every nontrivial automorphism Φ is destroyed and (ii) that the gap $\mathcal{A}_{\Phi,X}, \mathcal{B}_{\Phi,X}$ remains a gap in the final model. The first task is accomplished by using a standard bookkeeping device; the second one required inventing a new method called *oracle-chain condition* (see [28, §V]).

CONJECTURE 6.3. *Assume there is a weakly compact cardinal. Then there is a forcing extension in which all isomorphisms between quotients over definable ideals have continuous liftings.*

§7. Axiomatic approach. At the end of §5 we have seen that CH implies an extremal answer to Problem 1.3—roughly, any two quotients that could possibly be isomorphic are isomorphic under CH. In this section we attempt to isolate the assumptions under which two (simply definable) ideals have isomorphic quotients if and only if they are isomorphic. This would be an instance of a frequently encountered phenomenon in mathematics, that connecting maps between definable structures are usually definable themselves.

A natural ambient for this program seems to be the Proper Forcing Axiom, PFA. A fragment of the second-order theory of the uncountable turns out to be rather canonical under PFA (see [36, §8], [38]). This phenomenon is explained by Woodin [44] who defines a model in which the Π_2-theory of $\mathcal{P}(\omega_1)$ is maximal. Hence in this model many peculiar objects on ω_1 (like a Suslin tree or a well-ordering of the reals in type ω_1) do not exist. It is not surprising that a large fragment of PFA holds in his model. We are looking for a similar situation for a fragment of the second-order theory of $\mathcal{P}(\mathbb{N})$.

Although the forcing $\mathcal{P}_\Phi$ that destroys a nontrivial automorphism of $\mathcal{P}(\mathbb{N})/\mathrm{Fin}$ used in the proof of Theorem 6.2 is proper, for a while it was unclear whether the PFA implies that all automorphisms of $\mathcal{P}(\mathbb{N})/\mathrm{Fin}$ are trivial. This was finally proved in [29]. One of the novelties introduced in [29] was Lemma 7.1 below. A gap in $\mathcal{P}(\mathbb{N})/\mathcal{I}$ is *indestructible* if it remains a gap in every further $\aleph_1$-preserving extension.

LEMMA 7.1 (Shelah–Steprāns, [29]). *Assume Φ is a nontrivial automorphism of $\mathcal{P}(\mathbb{N})/\mathrm{Fin}$. Then there is a proper forcing that adds $X \subseteq \mathbb{N}$ such that*

$$\mathcal{A}_{\Phi,X} = \{\Phi_*(A) : A \in (\mathcal{P}(\mathbb{N}))^V, A \setminus X \in \mathrm{Fin}\},$$
$$\mathcal{B}_{\Phi,X} = \{\Phi_*(B) : B \in (\mathcal{P}(\mathbb{N}))^V, B \cap X \in \mathrm{Fin}\}$$

form an indestructible gap in $\mathcal{P}(\mathbb{N})/\mathrm{Fin}$. ⊣

The proof of Lemma 7.1 relies on results of [1]. The phenomenon that gaps in $\mathcal{P}(\mathbb{N})/\mathrm{Fin}$ can be made indestructible was discovered by Kunen [22] in the case when both sides of the gap are σ-directed under the inclusion mod Fin (see also [36, p. 74]).

Our motivation for assuming Martin's Maximum (see [12]) instead of the weaker PFA in the following Conjecture is that MM implies that $L(\mathbb{R})$ is elementarily equivalent to some Solovay's model [12], [31], and it therefore provides an ambient for Conjecture 4.11 as well.

CONJECTURE 7.2. *Martin's Maximum implies that every isomorphism between quotients over definable ideals has a continuous lifting.*

We can summarize Conjectures 4.11 and 7.2 as follows.

CONJECTURE 7.3. *Assume Martin's Maximum.*

(a) *The following are equivalent for any two definable ideals* $\mathcal{I}, \mathcal{J}$:
 (1) $\mathcal{P}(\mathbb{N})/\mathcal{I} \approx \mathcal{P}(\mathbb{N})/\mathcal{J}$.
 (2) $\mathcal{I} \approx_{\mathrm{RK}} \mathcal{J}$.

(b) *Moreover, every isomorphism between such quotients has a completely additive lifting.*

In the present paper we have been mainly concerned with isomorphisms between quotients, since both Conjecture 4.6 and Conjecture 7.2 fail for homomorphisms (see [7, Theorem 1.9.5] and [7, §3.2], respectively). The former fails because Todorcevic's conjecture fails outside of class of nonpathological ideals (see §2.5). The reason why the latter fails is much simpler: if $\mathcal{K}$ is a maximal nonprincipal ideal on $\mathbb{N}$, then $\mathcal{P}(\mathbb{N})/\mathcal{K}$ can be embedded into any analytic quotient, and the embedding cannot have a continuous lifting since $\mathcal{K}$ does not have the Property of Baire. However, Conjecture 7.2 does have a plausible generalization for arbitrary homomorphisms.

Let $\mathcal{I}$ be an ideal on $\mathbb{N}$ and let $\mathbb{N} = A \dot\cup B$ be a partition of $\mathbb{N}$ into two disjoint $\mathcal{I}$-positive sets. Let Φ_1 and Φ_2 be homomorphisms of $\mathcal{P}(\mathbb{N})$ into $\mathcal{P}(A)/\mathcal{I}$ and $\mathcal{P}(B)/\mathcal{I}$, respectively. Their *amalgamation* $\Phi = \Phi_1 \oplus \Phi_2$ is defined by the following diagram (id stands for the identity map):

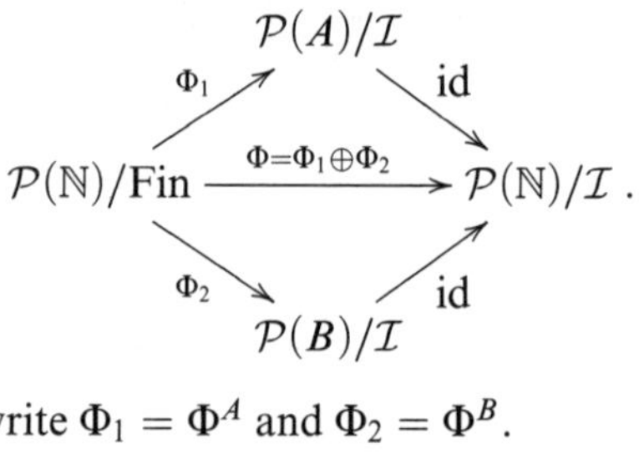

In this situation we write $\Phi_1 = \Phi^A$ and $\Phi_2 = \Phi^B$.

DEFINITION 7.4. The ideal $\mathcal{I}$ has the *continuous lifting property* if for every homomorphism $\Phi\colon \mathcal{P}(\mathbb{N}) \to \mathcal{P}(\mathbb{N})/\mathcal{I}$ is an amalgamation of a homomorphism with a continuous lifting and a homomorphism whose kernel is a nonmeager ideal on $\mathbb{N}$.

LEMMA 7.5. *If $\mathcal{I}$ has the continuous lifting property and $\mathcal{J}$ is an ideal that has the Property of Baire, then every isomorphism between the quotients over $\mathcal{I}$ and $\mathcal{J}$ has a continuous lifting.*

PROOF. Recall that by a result of Jalali-Naini and Talagrand an ideal that contains Fin and has the Property of Baire is nonmeager if and only if it is not proper (see also [7, §3.10]). Therefore if $\Phi\colon \mathcal{P}(\mathbb{N})/\mathcal{J} \to \mathcal{P}(\mathbb{N})/\mathcal{I}$ is an isomorphism and $\ker(\Phi^B)$ is nonmeager, then $B \in \mathcal{I}$. ⊣

Hence the following is a stronger version of Conjecture 7.3.

CONJECTURE 7.6. *Martin's Maximum implies that every definable ideal has the continuous lifting property.*

It should be pointed out that the emphasis of the above conjectures is on the existence of continuous liftings, and not on Martin's Maximum. If there is a model of MM in which there is an isomorphism between quotients over definable ideals with no continuous lifting, this would merely be a suggestion that MM is not the sufficient assumption for Conjecture 7.3. The real refutation of Conjecture 7.2 would be a construction of two definable ideals and an isomorphism between their quotients that has no continuous lifting without using CH or some other additional set-theoretic axiom compatible with large cardinals.

§8. Positive results.

8.1. Quotients over analytic P-ideals. Theorem 8.1 below was proved before the above conjectures were stated; it was actually the starting point for formulating these conjectures. The assumption used in this result is the Open Coloring Axiom, as defined by Todorcevic in [36, §8], taken together with familiar Martin's Axiom, MA. Recall that $[A]^2$ is the family of all two-elements subsets of A and that $B \subseteq A$ is *K-homogeneous* if $[B]^2 \subseteq K$. It is *σ-K-homogeneous* if it can be covered by countably many K-homogeneous sets. A subset K of $[X]^2$ is *open* if $\{(x,y) : \{x,y\} \in K\}$ is an open subset of X^2.

OCA. If X is a separable metric space and $[X]^2 = K_0 \cup K_1$ is a partition such that K_0 is an open subset of $[X]^2 = \{\{x,y\} : x \neq y,\ x,y \in X\}$ then one of the following applies:

(a) X is σ-K_1-homogeneous, or
(b) X has an uncountable K_0-homogeneous subset.

Both OCA and MA are consequences of PFA (for OCA see [36, §8]).

Theorem 8.1 (Farah, [7, Theorem 3.3.5]). OCA *and* MA *imply that all analytic P-ideals have the continuous lifting property. Therefore Conjecture* 7.2 *is true for all analytic P-ideals, and Conjecture* 7.3 *is true for all nonpathological analytic P-ideals.* ⊣

An axiom called OCA was introduced earlier in [1]. This axiom, even if amplified with MA, is not a sufficient assumption for Theorem 8.1, since it holds in Velickovic's model of MA in which there is a nontrivial automorphism of $\mathcal{P}(\mathbb{N})/\mathrm{Fin}$ [41].

Corollary 8.2. *Conjecture* 7.2 *is true for all P-ideals. Conjecture* 7.3 *is true for all nonpathological P-ideals.*

Proof. By [37], in some Solovay's model all P-ideals are analytic, hence the result is an immediate consequence of Theorem 8.1. ⊣

8.2. Quotients over F_σ ideals. OCA_∞ is a strengthening of OCA, introduced in [5], that deals with infinitely many open colorings simultaneously. It was extracted from an earlier unpublished work of Steprāns.

OCA_∞. If X is a separable metric space and $[X]^2 = K_0^n \cup K_1^n$, $n \in \mathbb{N}$, is a sequence of partitions such that each K_0^n is open and that $K_0^{n+1} \subseteq K_0^n$ for all n, then one of the following applies:

(a) $X = \bigcup_{n \in \mathbb{N}} F_n$, where each F_n is K_1^n-homogeneous, or
(b) There is an uncountable $Y \subseteq X$ that is σ-K_0^n-homogeneous for all n.

This axiom was shown to follow from PFA in [5] by following the proof of [36, Theorem 8.0] that PFA implies OCA.

Theorem 8.3 (Farah, 2000). OCA_∞ *and* MA *together imply that all F_σ ideals have the continuous lifting property. Hence Conjecture* 7.2 *is true for all F_σ ideals and Conjecture* 7.3 *is true for all nonpathological F_σ ideals.* ⊣

A 'local' version of this theorem (see §8.4) was proved earlier from OCA and MA by Just [17].

8.3. Using stronger axioms. A more direct and simpler proof of both Theorem 8.1 and Theorem 8.3 from PFA instead of OCA (or OCA_∞) and MA can be given using the following generalization of Shelah–Steprāns Lemma (Lemma 7.1).

Lemma 8.4 (Farah, [10, §6]). *If $\mathcal{I}$ is an analytic P-ideal or an F_σ ideal, $\mathcal{J}$ is an analytic ideal, and $\Phi \colon \mathcal{P}(\mathbb{N})/\mathcal{J} \to \mathcal{P}(\mathbb{N})/\mathcal{I}$ is an isomorphism with no continuous lifting, then there is a proper forcing that adds $X \subseteq \mathbb{N}$ such that*

$$\mathcal{A}_{\Phi,X} = \{\Phi_*(A) : A \in (\mathcal{P}(\mathbb{N}))^V, A \setminus X \in \mathcal{I}\},$$
$$\mathcal{B}_{\Phi,X} = \{\Phi_*(B) : B \in (\mathcal{P}(\mathbb{N}))^V, B \cap X \in \mathcal{I}\}$$

form an indestructible gap in $\mathcal{P}(\mathbb{N})/\mathcal{I}$. ⊣

A positive answer to the following question would be a major step toward confirming Conjecture 7.2.

QUESTION 8.5. *Can Lemma* 8.4 *be proved assuming only that $\mathcal{I}$ is an analytic ideal? Or that $\mathcal{I}$ and $\mathcal{J}$ are definable, under a suitable large cardinal assumption?*

In [10] the author was able to isolate a class of *strongly countably determined by closed approximations* ideals and to verify Conjecture 7.2, and Conjecture 7.6, for the ideals in this class. This class includes all analytic P-ideals, all F_σ ideals, and $\mathrm{NWD}(\mathbb{Q})$. Every ideal that is strongly countably determined by closed approximations is $F_{\sigma\delta}$, and at present it is unknown whether this class of ideals coincides with $F_{\sigma\delta}$ ideals.

The methods used in [10] do not seem to apply to quotients over ideals more complex than $F_{\sigma\delta}$. and at this point we have only partial results on such quotients.

8.4. Quotients over ideals beyond $F_{\sigma\delta}$. An ideal $\mathcal{I}$ is *ccc over* Fin if every family of $\mathcal{I}$-positive sets whose pairwise intersections are finite is countable. Every such ideal is rather large, in particular it is nonmeager (see [7, §3.3]). A homomorphism $\Phi\colon \mathcal{P}(\mathbb{N})/\mathcal{I} \to \mathcal{P}(\mathbb{N})/\mathcal{J}$ *has continuous liftings locally* if the ideal

$$\mathcal{J}_\Phi = \{A \subseteq \mathbb{N} : \Phi \restriction \mathcal{P}(A)/\mathcal{I} \text{ has a continuous lifting}\}$$

is ccc over Fin.

The first step in the proof of Theorem 8.1, Theorem 8.3, as well as all other similar results, is to prove that the homomorphism has continuous liftings locally. In the final step of the proof these liftings are amalgamated into a single continuous lifting.

THEOREM 8.6 (Farah, 2000). *Assume there exists a weakly compact cardinal. Then there is a forcing extension in which every homomorphism between quotients over definable ideals has continuous liftings locally. Moreover, in this model all P-ideals and all F_σ-ideals have the continuous lifting property.* ⊣

The proof of this result, probably relevant to Conjecture 6.3, will appear elsewhere. Corollary 8.7 below shows that a weaker version of Conjecture 7.3 is true for a rather rich class of quotients. Let $\mathcal{C}$ be the class consisting of all nonpathological P-ideals, all nonpathological F_σ-ideals, all ordinal ideals, and all Weiss ideals.

COROLLARY 8.7. *Assume there exists a weakly compact cardinal. Then there is a forcing extension in which for all $\mathcal{I}, \mathcal{J}$ in $\mathcal{C}$ we have $\mathcal{I} \approx_{\mathrm{RK}} \mathcal{J}$ if and only if $\mathcal{P}(\mathbb{N})/\mathcal{I} \approx \mathcal{P}(\mathbb{N})/\mathcal{J}$.*

PROOF. Go to the model of Theorem 8.6. For nonpathological ideals this follows by Proposition 4.9. Recall that ordinal and Weiss ideals are all homogeneous. If $\mathcal{I}$ is homogeneous and an isomorphism between $\mathcal{I}$ and $\mathcal{J}$ has a locally continuous lifting, then $\mathcal{I}$ is isomorphism to $\mathcal{J} \restriction A$ for some $\mathcal{J}$-positive set A. By Theorem 4.5, if $\mathcal{I}$ is an ordinal ideal or a Weiss ideal this implies $\mathcal{I} \approx_{\mathrm{RK}} \mathcal{J} \restriction A$; but by homogeneity this is possible only if $\mathcal{I} \approx_{\mathrm{RK}} \mathcal{J}$. ⊣

REFERENCES

[1] U. Abraham, M. Rubin, and S. Shelah, *On the consistency of some partition theorems for continuous colorings, and the structure of $\aleph_1$-dense real order types*, ***Annals of Pure and Applied Logic***, vol. 29 (1985), pp. 123–206.

[2] A. Andretta, *Notes on descriptive set theory*, in preparation, 2000.

[3] C. C. Chang and H. J. Keisler, ***Model theory***, North–Holland, 1973.

[4] J. P. R. Christensen, *Some results with relation to the control measure problem*, ***Vector space measures and applications II*** (R. M. Aron and S. Dineen, editors), Lecture Notes in Mathematics, vol. 645, Springer, 1978, pp. 27–34.

[5] I. Farah, *Cauchy nets and open colorings*, ***Publications. Institut Mathématique. Nouvelle Série***, vol. 64(78) (1998), pp. 146–152, 50th anniversary of the Mathematical Institute, Serbian Academy of Sciences and Arts (Belgrade, 1996).

[6] ———, *Completely additive liftings*, ***The Bulletin of Symbolic Logic***, vol. 4 (1998), pp. 37–54.

[7] ———, *Analytic quotients: theory of liftings for quotients over analytic ideals on the integers*, ***Memoirs of the American Mathematical Society, vol. 148***, (2000), no. 702, 177 pp.

[8] ———, *Liftings of homomorphisms between quotient structures and Ulam stability*, ***Logic colloquium '98*** (S. Buss, P. Hájek, and P. Pudlák, editors), Lecture notes in logic, vol. 13, A. K. Peters, 2000, pp. 173–196.

[9] ———, *How many Boolean algebras $\mathcal{P}(\mathbb{N})/\mathcal{I}$ are there?*, ***Illinois Journal of Mathematics***, vol. 46 (2003), pp. 999–1033.

[10] ———, *Luzin gaps*, ***Transactions of the American Mathematical Society***, vol. 356 (2004), pp. 2197–2239.

[11] I. Farah and S. Solecki, *Two $F_{\sigma\delta}$ ideals*, ***Proceedings of the American Mathematical Society***, vol. 131 (2003), pp. 1971–1975.

[12] M. Foreman, M. Magidor, and S. Shelah, *Martin's maximum, saturated ideals and nonregular ultrafilters, I*, ***Annals of Mathematics***, vol. 127 (1988), pp. 1–47.

[13] D. H. Fremlin, *Notes on* Farah *P99*, preprint, University of Essex, June 1999.

[14] F. Hausdorff, *Die Graduierung nach dem Endverlauf*, ***Abhandlungen der Königlich Sächsischen Gesellschaft der Wissenschaften; Mathematisch–Physische Klasse***, vol. 31 (1909), pp. 296–334.

[15] W. Just, *Repercussions on a problem of Erdös and Ulam about density ideals*, ***Canadian Journal of Mathematics***, vol. 42 (1990), pp. 902–914.

[16] ———, *A modification of Shelah's oracle chain condition with applications*, ***Transactions of the American Mathematical Society***, vol. 329 (1992), pp. 325–341.

[17] ———, *A weak version of AT from* OCA, ***Mathematical Science Research Institute Publications***, vol. 26 (1992), pp. 281–291.

[18] W. Just and A. Krawczyk, *On certain Boolean algebras $\mathcal{P}(\omega)/I$*, ***Transactions of the American Mathematical Society***, vol. 285 (1984), pp. 411–429.

[19] A. Kanamori, ***The higher infinite: large cardinals in set theory from their beginnings***, Perspectives in Mathematical Logic, Springer–Verlag, Berlin–Heidelberg–New York, 1995.

[20] V. Kanovei and M. Reeken, *On Ulam's problem concerning the stability of approximate homomorphisms*, ***Trudy Matematicheskogo Instituta Imeni V. A. Steklova. Rossiĭskaya Akademiya Nauk***, vol. 231 (2000), no. Din. Sist., Avtom. i Beskon. Gruppy, pp. 249–283.

[21] ———, *New Radon–Nikodym ideals*, ***Mathematika***, vol. 47 (2002), pp. 219–227.

[22] K. Kunen, *$\langle\kappa, \lambda\rangle$-gaps under* MA, preprint, 1976.

[23] C. Laflamme, *Combinatorial aspects of F_σ filters with an application to $\mathcal{N}$-sets*, ***Proceedings of the American Mathematical Society***, vol. 125 (1997), pp. 3019–3025.

[24] A. Louveau and B. Velickovic, *A note on Borel equivalence relations*, ***Proceedings of the American Mathematical Society***, vol. 120 (1994), pp. 255–259.

[25] B. LÖWE and J. STEEL, *An introduction to core model theory*, ***Sets and proofs, Logic Colloquium 1997, volume 1***, London Mathematical Society Lecture Note Series, no. 258, Cambridge University Press, 1999.

[26] K. MAZUR, *F_σ-ideals and $\omega_1\omega_1$-gaps in the Boolean algebra $\mathcal{P}(\omega)/I$*, ***Fundamenta Mathematicae***, vol. 138 (1991), pp. 103–111.

[27] W. RUDIN, *Homogeneity problems in the theory of Čech compactifications*, ***Duke Mathematics Journal***, vol. 23 (1956), pp. 409–419.

[28] S. SHELAH, ***Proper forcing***, Lecture Notes in Mathematics 940, Springer, 1982.

[29] S. SHELAH and J. STEPRĀNS, *PFA implies all automorphisms are trivial*, ***Proceedings of the American Mathematical Society***, vol. 104 (1988), pp. 1220–1225.

[30] ———, *Non-trivial homeomorphisms of $\beta\mathbb{N}\setminus\mathbb{N}$ without the continuum hypothesis*, ***Fundamenta Mathematicae***, vol. 132 (1989), pp. 135–141.

[31] S. SHELAH and W. H. WOODIN, *Large cardinals imply that every reasonably definable set of reals is Lebesgue measurable*, ***Israel Journal of Mathematics***, vol. 70 (1990), pp. 381–394.

[32] S. SOLECKI, *Analytic ideals*, ***The Bulletin of Symbolic Logic***, vol. 2 (1996), pp. 339–348.

[33] ———, *Analytic ideals and their applications*, ***Annals of Pure and Applied Logic***, vol. 99 (1999), pp. 51–72.

[34] ———, *Filters and sequences*, ***Fundamenta Mathematicae***, vol. 163 (2000), pp. 215–228.

[35] R. SOLOVAY, *A model of set theory in which every set of reals is Lebesgue measurable*, ***Annals of Mathematics***, vol. 92 (1970), pp. 1–56.

[36] S. TODORČEVIĆ, ***Partition problems in topology***, Contemporary Mathematics, vol. 84, American Mathematical Society, Providence, Rhode Island, 1989.

[37] ———, *Definable ideals and gaps in their quotients*, ***Set theory: Techniques and applications*** (C. A. DiPrisco et al., editors), Kluwer Academic Press, 1997, pp. 213–226.

[38] ———, *Basis problems in combinatorial set theory*, ***Proceedings of the international congress of mathematicians, vol. II (Berlin, 1998)***, Documenta Mathematica, 1998, pp. 43–52.

[39] ———, *Gaps in analytic quotients*, ***Fundamenta Mathematicae***, vol. 156 (1998), pp. 85–97.

[40] B. VELICKOVIC, *Definable automorphisms of $\mathcal{P}(\omega)/\mathrm{Fin}$*, ***Proceedings of the American Mathematical Society***, vol. 96 (1986), pp. 130–135.

[41] ———, OCA *and automorphisms of $\mathcal{P}(\omega)/\mathrm{Fin}$*, ***Topology and its Applications***, vol. 49 (1992), pp. 1–12.

[42] W. WEISS, *Partitioning topological spaces*, ***Mathematics of Ramsey Theory*** (J. Nešetřil and V. Rödl, editors), Algorithms and Combinatorics, vol. 5, Springer–Verlag, Berlin, 1990, pp. 154–171.

[43] W. H. WOODIN, *Σ^2_1-absoluteness*, handwritten note of May 1985.

[44] ———, ***The axiom of determinacy, forcing axioms and the nonstationary ideal***, de Gruyter Series in Logic and Its Applications, vol. 1, de Gruyter, 1999.

DEPARTMENT OF MATHEMATICS AND STATISTICS
4700 KEELE STREET
TORONTO, ON M2J 1P3, CANADA
and
MATEMATICKI INSTITUT
KNEZA MIHAILA 35
BELGRADE, SERBIA AND MONTENEGRO
E-mail: ifarah@yorku.ca
URL: http://www.math.yorku.ca/ ifar

METAPREDICATIVE AND EXPLICIT MAHLO: A PROOF-THEORETIC PERSPECTIVE

GERHARD JÄGER

Abstract. After briefly discussing the concepts of *predicativity*, *metapredicativity* and *impredicativity*, we turn to the notion of *Mahloness* as it is treated in various contexts. Afterwards the appropriate Mahlo axioms for the framework of explicit mathematics are presented. The article concludes with relating explicit Mahlo to certain nonmonotone inductive definitions.

§1. Introduction. More than 100 years ago Cantor developed the theory of infinite sets (Cantor's paradise). Shortly afterwards, Russell found his famous paradox, and, as a consequence, many mathematicians became very concerned about the foundations of mathematics, and the expression *foundational crisis* was coined.

To overcome this crisis, Hilbert proposed the program of *Beweistheorie* as a method of rescuing *Cantor's paradise*. A few years later, however, Gödel showed that Hilbert's program—at least in its original strong form—cannot work. Again, after only a short while a first new idea was brought in by Gentzen, and a break-through along these lines was obtained by his proof-theoretic analysis of first order arithmetic. Then, during the last decades, Gentzen's work has been extended to stronger and stronger subsystems of second order arithmetic and set theory, most prominently by the schools of Schütte and Takeuti, leading to what today is denoted as *infinitary* and *finitary proof theory*, respectively.

A position completely different from Hilbert's was taken by Brouwer who advocated the restriction of mathematics to those principles which could be justified on constructive grounds. Starting off from his pioneering work various "dialects" of constructive mathematics have been put forward (e.g., in the Netherlands directly following Brouwer and Heyting, the Russian form(s) of constructivism, Bishop's approach, Martin-Löf type theory, Feferman's explicit mathematics).

In a certain sense the research directions originating from Hilbert's and Brouwer's original ideas come together again under the heading of *reductive proof theory* which tries to justify classical theories and classical principles by

Logic Colloquium 2000
Edited by R. Cori, A. Razborov, S. Todorčević, and C. Wood
Lecture Notes in Logic, 19

reducing them to a (more) constructive framework. For further reading and detailed information about this topic we refer, for example, to Beeson [6], Feferman [17] and Troelstra and van Dalen [61].

§2. The predicative, impredicative and metapredicative. The general picture described so far is, however, oversimplified in that it leaves out many important intermediate approaches. A particularly interesting line of thought was initiated by Poincaré's conviction that many foundational problems are caused by making use of so called *impredicative* definitions.[1] On the other hand, he did not consider the use of classical logic as being critical.

Poincaré was followed by Weyl, and they focused on the *arithmetical foundations of mathematics* (using their own terminology): their starting point being the usual structure $(\mathbb{N}, \dots)$ of the natural numbers with the schema of complete induction; moreover, all predicatively definable subsets of $\mathbb{N}$ are permitted.

This informal Poincaré-Weyl program was later brought into precise mathematical and logical terms by Feferman; see Feferman [13]. A further guiding line is his attempt to answer the following question: *What is implicit in the structure of the natural numbers together with the principle of induction?*

During the sixties Feferman and Schütte independently characterized *predicative mathematics* and showed that the associated ordinal is the famous Feferman-Schütte ordinal Γ_0. They achieved this by employing autonomous progressions of theories or ramified systems of second order arithmetic; for details see, for example, Feferman [8] and Schütte [51]. The theories capturing exactly predicative mathematics have their least standard model at

$$(\mathbb{N}, L_{\Gamma_0} \cap Pow(\mathbb{N}), \dots)$$

and are equivalent to the system of second order arithmetic $\mathsf{AUT}(\Pi^0_1) + (\mathsf{BR})$ which comprises autonomously iterated Π^0_1 comprehension and the bar rule. Starting off from the Feferman-Schütte notion of predicativity, we can now distinguish the following three collections of theories:

A. Predicatively reducible systems. They comprise all those theories which are (finitely) reducible to a predicative system, i.e., whose proof-theoretic ordinal is less than or equal to Γ_0.

Trivially, all predicative theories are predicatively reducible. On the other hand, the least standard model of a predicatively reducible but not predicative theory can well be of the form

$$(\mathbb{N}, L_\alpha \cap Pow(\mathbb{N}), \dots) \quad \text{or} \quad (L_\alpha, \in, \dots)$$

[1] The definition of a set S is called impredicative if it refers to a totality of sets to which S itself belongs to.

for some $\alpha > \Gamma_0$. A typical such theory is the system Σ^1_1-AC of second order arithmetic with the Σ^1_1 axiom of choice; its proof-theoretic ordinal is the ordinal $\varphi\varepsilon_0 0 < \Gamma_0$, although its least standard model is only reached at the first non-recursive ordinal ω_1^{CK}.

In most cases the proof-theoretic analysis of a predicatively reducible theory can be obtained by forms of partial cut elimination and consecutive asymmetric interpretations or use of Skolem operators.

Examples of further important predicatively reducible systems are the theory Σ^1_1-AC + (BR), Friedman's theory ATR_0 of arithmetic transfinite recursion (cf. e.g., [20]), Avigad fixed point theory FP_0 (cf. e.g., [5]), Feferman's theory $\widehat{\mathsf{ID}}_{<\omega}$ of finitely many iterated fixed point axioms (cf. e.g., [12]) and Jäger's theory KPi^0 for a recursively inaccessible universe without foundation (cf. e.g., [27]).

B. Impredicative systems. Traditionally, all theories which are not predicatively reducible have been subsumed under the heading *impredicative*. Such an approach, however, has some undesired consequences.

Friedman's theory ATR_0, mentioned above, has proof-theoretic ordinal Γ_0, thus is predicatively reducible. But for obtaining this result it is essential that in ATR_0 complete induction on the natural numbers is restricted to sets. If complete induction on the natural numbers is permitted for arbitrary formulas, then results of Friedman (see Simpson [53]) and Jäger [24] show that the corresponding theory, called ATR, is of proof-theoretic strength Γ_{ε_0}. As a consequence, ATR is not predicatively reducible.

Recall, in addition, that the schema of complete induction on the natural numbers is at the core of predicativity a la Poincaré, Weyl and Feferman. Thus the predicatively reducible theory ATR_0 would be shifted into the impredicative by adding a purely predicative principle.

Moreover, the proof-theoretic analysis of ATR requires more of less the same concepts and machinery as the proof-theoretic analysis of ATR_0. Hence also from the point of view of methods involved, a very sharp dividing line between ATR_0 and ATR seems out of place.

To overcome such atrocities, we suggest to use the proof-theoretic techniques involved as criterion for structuring the range of systems which are not predicatively reducible. This then leads to the following "definition" of impredicativity: *The collection of impredicative systems comprises all those theories whose proof-theoretic (ordinal) analysis requires the use of impredicative methods.*

Of course, this is far from being a formal definition since we refer to the notion of *impredicative method*, and it is nowhere exactly pinned down what that means. However, given a specific ordinal analysis of a theory, we are convinced that all proof-theorists would agree on whether this analysis is done via impredicative techniques or not.

Our experience shows that typical impredicative methods always refer to some sort of collapsing techniques and collapsing functions, either directly applied to infinitary proofs or to the ordinals assigned to proofs or to both.

The first system for whose proof-theoretic analysis such impredicative methods have been used is the famous first order theory ID_1 for non-iterated positive arithmetic inductive definitions. On the other hand, recent work reveals (see below) that also an alternative approach is possible.

C. Metapredicative systems. The division into predicatively reducible and impredicative systems provided so far leaves some space between these two collections which is filled by the so-called metapredicative systems introduced now: *The collection of metapredicative systems comprises all those theories which are not predicatively reducible and whose proof-theoretic analysis can be carried through without making use of any impredicative methods.*

This description of metapredicativity is as informal as that of impredicativity. Therefore it would be very interesting to answer the following two questions:

(1) Is there a formal counterpart of this informal notion?
(2) If so, what is the limit of metapredicativity?

A satisfactory answer to question (1) has not yet been given, thus also (2) is still open. A first and necessary step in the direction of learning more about metapredicativity certainly consists of

- analyzing typical metapredicative systems and
- identifying their structural properties.

Years ago theories like ATR—lying in strength between Γ_0 and ID_1—have been considered as exceptional cases; today we know that many interesting systems can be found in this area.

§3. More about metapredicative systems. The first metapredicative system which called for attention was the above mentioned theory ATR with proof-theoretic ordinal Γ_{ε_0} and was considered to be a rather isolated phenomenon at the time of its analysis. It was only later that more and more related theories turned up, and the systematic approach to metapredicativity began with the proof-theoretic analysis of the transfinitely iterated fixed point theories $\widehat{\mathsf{ID}}_\alpha$ and $\widehat{\mathsf{ID}}_{<\beta}$ for (recursive) ordinals $\alpha \geq \omega$ and $\beta > \omega$ in Jäger, Kahle, Setzer and Strahm [32].

These first order systems extend Feferman's theories $\widehat{\mathsf{ID}}_n$ into the transfinite and are a very good tool for calibrating the initial part of metapredicativity. ATR, for example, is proof-theoretically equivalent to $\widehat{\mathsf{ID}}_\omega$. In addition, Jäger and Strahm [35] established the role of the schema of Σ^1_1 dependent choice in the contexts of ATR_0 and ATR and proved that $\mathsf{ATR}_0 + (\Sigma^1_1\text{-DC})$

has the same proof-theoretic strength as $\widehat{\mathsf{ID}}_{<\omega^\omega}$ whereas $\mathsf{ATR} + (\Sigma^1_1\text{-}\mathsf{DC})$ is proof-theoretically equivalent to $\widehat{\mathsf{ID}}_{<\varepsilon_0}$.

By using this result we could also answer a question of Simpson [54] about the strength of some of his second order systems for transfinite Σ^1_1 and Π^1_1 induction:

$$\Sigma^1_1\text{-}\mathsf{TI}_0 + \Pi^1_1\text{-}\mathsf{TI}_0 \equiv \widehat{\mathsf{ID}}_{<\omega^\omega} \quad \text{and} \quad \Sigma^1_1\text{-}\mathsf{TI} + \Pi^1_1\text{-}\mathsf{TI} \equiv \widehat{\mathsf{ID}}_{<\varepsilon_0}.$$

Afterwards Strahm studied autonomously iterated fixed point theories; see Strahm [56]. In his article he also establishes the relationship between autonomous fixed point iteration and transfinite fixed point recursion in the spirit of Friedman's arithmetic transfinite recursion.

In the context of admissible set theory without foundation, we obtain an interesting metapredicative system, if complete induction on the natural numbers is added to the previously mentioned theory KPi^0,

$$\mathsf{KPi}^0 + (\text{full induction on } \mathbb{N}) \equiv \widehat{\mathsf{ID}}_{<\varepsilon_0}.$$

To obtain this result we first have to show in KPi^0 that for each set a there exists a least admissible containing a (cf. Jäger [30]). Moreover, because of full complete induction on $\mathbb{N}$, we have transfinite induction for all formulas up to each $\alpha < \varepsilon_0$ and can thus build hierarchies of admissibles of the same height. Then it is easy to embed $\widehat{\mathsf{ID}}_{<\varepsilon_0}$. The upper bound is established by an extension of the methods in Jäger [27]. It is interesting to see that these arguments are sensitive to the question of whether in KPi^0 the admissible sets are linearly ordered or not.

Later in this article the formalism of explicit mathematics will be (very) briefly introduced. However, for the reader already familiar with this approach we include some remarks about metapredicative systems of explicit mathematics.

In the following we write EETJ for the theory comprising of the basic first order axioms plus elementary comprehension and join. Furthermore, we have a so-called *limit axiom*,

$$(\mathsf{Lim}) \qquad (\forall x)\big[\Re(x) \to (x \mathrel{\dot{\in}} \ell x \wedge \mathcal{U}(x))\big]$$

stating that each name x of a type is contained in a universe named ℓx. Then $\mathsf{EETJ} + (\mathsf{Lim})$ plus complete induction on $\mathbb{N}$ for types is a predicatively reducible theory of strength Γ_0; for details see Marzetta [41] and Kahle [39]. According to Strahm [55], the addition of compete induction on $\mathbb{N}$ for arbitrary formulas gives a metapredicative theory, again proof-theoretically equivalent to $\widehat{\mathsf{ID}}_{<\varepsilon_0}$. A non-uniform version of the limit axiom (Lim) is considered in Marzetta [41].

Although all theories discussed so far are strictly weaker than *metapredicative Mahlo*,[2] we must point out that this is by no means the limit of the area

[2]This notion will be described below.

of metapredicativity. Recent work of Jäger and Strahm has shown that all instances of Π_n reflection $(n < \omega)$ can safely be added to KPi^0 without surpassing metapredicativity. This yields, among other things, a metapredicative justification for ID_1.

Axioms for explicit mathematics corresponding to the set-theoretic Π_n reflections have also been studied by Jäger and Strahm. And there are certain stability properties for higher order operations in explicit mathematics which all give rise to metapredicative theories and reach far beyond metapredicative Mahlo.

The least standard models of all these metapredicative systems are based on sets L_α for comparatively large ordinals α, and these theories formulate (represent) many aspects of these sets which are sufficient for mathematical practice. On the other hand, since the available induction principles are very restricted, their proof-theoretic strengths stays rather low. So we can summarize some *charatieristics* of of metapredicative systems as follows:

- They cover a good deal of ordinary mathematics, for example large parts of analysis, discrete mathematics, category theory.
- They allow a philosophically careful (justified) proof-theoretic treatment from below in the sense that no collapsing techniques have to be used.
- In connection with metapredicative theories we can have *large* (*complicated*) *sets* but have to deal with *low consistency strength* only.

Further evidence for these remarks will now be given by looking closer at the role of Mahloness in various formal frameworks.

§4. A short survey of Mahloness. Mahlo axioms play an important role in present day proof theory. They have been studied during the last years from different perspectives, and some of these will be sketched below. In a sense, Mahloness draws the borderline of the part of proof theory that is so well understood, that the interaction between different standpoints becomes clear.

The Mahlo axioms go back to Mahlo's pioneering work from around 1911; see Mahlo [40]. Today an ordinal α is called a *Mahlo ordinal* if and only if

$$(\forall f : \alpha \to \alpha)\ (\exists \beta \in \mathrm{Reg})\ \big[\beta < \alpha \wedge f : \beta \to \beta\big],$$

Reg denoting the class of all regular cardinal numbers. The least Mahlo ordinal M_0 outgrows all inaccessible, hyperinaccessible, ... ordinals. M_0 cannot be reached from below by any sort of iteration of inaccessibility.

The usual approach of obtaining the *recursive analogues* of classical cardinal numbers also was applied to Mahlo ordinals by simply replacing

- *regular cardinals* by *admissible sets* and
- *arbitrary* functions by *recursive* functions.

Following the tradition in recursion theory, we will not directly use the corresponding reformulation of the definition of Mahlo ordinal as above, but instead work in the context of admissible set theory.

Let KPu^0 be the system of *Kripke-Platek set theory* above the natural numbers as urelements with induction on the natural numbers restricted to Δ_0 formulas and without $\in$ induction.[3] In this context, the recursive version of Mahlo can be characterized via Π_2 reflection on admissibles.

Accordingly, KPm^0 is defined to be the set theory which extends KPu^0 by the following schema

$$A(\vec{u}) \rightarrow (\exists x)\big[\mathsf{Ad}(x) \wedge \vec{u} \in x \wedge A^x(\vec{u})\big]$$

for all Π_2 formulas $A(\vec{u})$ with the parameters shown. An ordinal α is called *recursively Mahlo* if L_α, or more precisely the structure $(\mathbb{N}, L_\alpha(\mathbb{N}), \in, \dots)$, is a model of KPm^0, and we write μ_0 for the least recursively Mahlo ordinal.

Now we are going to mention two alternative characterizations of μ_0. The first one is in terms of Gandy's *superjump* $\mathbb{S}$, introduced in Gandy [21]. The superjump is an important type 3 functional whose associated closure ordinal has been studied in Aczel and Hinman [4] and Harrington [22].

THEOREM 1 (Aczel, Harrington, Hinman). *μ_0 is the least ordinal which is not recursive in Gandy's superjump $\mathbb{S}$, i.e.,*

$$\mu_0 = \omega_1^{\mathbb{S}}.$$

For the next description of μ_0 we turn to nonmonotone inductive definitions. Let Φ be some arbitrary operator which maps the power set $Pow(\mathbb{N})$ to itself. Then Φ can be considered as (inducing) an inductive definition whose stages are introduced by recursion on the ordinals as follows:

$$I_\Phi^\alpha := I_\Phi^{<\alpha} \cup \Phi\left(I_\Phi^{<\alpha}\right) \quad \text{and} \quad I_\Phi^{<\alpha} := \bigcup \{I_\Phi^\xi : \xi < \alpha\}.$$

Obviously we have $I_\Phi^\alpha \subset I_\Phi^\beta$ for $\alpha \leq \beta$. A simple cardinality argument thus implies the existence of a least ordinal α such that I_Φ^α and $I_\Phi^{<\alpha}$ are identical. This ordinal is often called the *closure ordinal* of Φ and denoted by $\|\Phi\|$,

$$\|\Phi\| := \text{least } \alpha \text{ such that } I_\Phi^{<\alpha} = I_\Phi^\alpha.$$

Correspondingly, if $\mathcal{C}$ is a collection of operators, then the closure ordinal $\|\mathcal{C}\|$ of $\mathcal{C}$ is defined to be the ordinal $\sup\{\|\Phi\| : \Phi \in \mathcal{C}\}$.

An interesting way of combining two operators Φ and Ψ was introduced in Richter [47]. A new *combined operator* $[\Phi, \Psi]$ is generated from Φ and Ψ by setting for all subsets X of $\mathbb{N}$:

$$[\Phi, \Psi](X) := \begin{cases} \Phi(X) & \text{if } \Phi(X) \not\subset X, \\ \Psi(X) & \text{if } \Phi(X) \subset X. \end{cases}$$

[3]The omission of $\in$ induction will be crucial for obtaining metapredicative systems; it has no effect, of course, if transitive standard models are considered.

Some further notation: let $[\text{POS-}\Pi^0_\infty, \Pi^0_1]$ be the collection of all combined operators whose first component is definable by an X-positive arithmetical formula and whose second component by a (not necessarily X-positive) Π^0_1 formula. Then Richter [47] contains the following theorem.

THEOREM 2 (Richter). $\mu_0 = \|[POS\text{-}\Pi^0_\infty, \Pi^0_1]\|$.

Mahloness in explicit mathematics has a natural classical and recursion-theoretic interpretation. We omit its discussion now and refer to Section 6, in which we will review it in detail and give the exact formulations.

Mahlo axioms are presently also of much interest in connection with constructive set theories and constructive type theories. Older approaches towards constructive versions of Zermelo-Fraenkel set theory are due to Friedman, Myhill and Scott, among others, and deal with (sub)systems of ZF, but with intuitionistic logic instead of classical logic (cf. [42]).

More recently, Aczel, in a series of papers [1, 2, 3], propagates alternative systems of constructive set theory CZF which incorporate (constructive) variants of the usual set theoretic principles, although their consistency strength stays comparatively small. His work has been extended in Rathjen [45] so that the constructive versions of large cardinal axioms, including Mahlo, find their place. Crosilla [7] deals with an extension of CZF without foundation for inaccessibility.

CZF and its extensions use intuitionistic logic as well, and we can obtain their constructive justification by interpretations into *Martin-Löf type theory* MLTT, which provides a philosophically motivated framework for constructive reasoning. The axioms reflecting the idea of Mahlo sets in MLTT are originally due to Setzer [52].

Let us end this section with recapitulating some of the most important proof-theoretic results about Mahlo in its various settings.

A. Full recursive Mahlo. The ordinal analysis of the canonical formalization of full recursive Mahlo, i.e., of the set theory for Π_2 reflection on admissibles

$$\mathsf{KPm} := \mathsf{KPm}^0 + (\text{full induction}),$$

has been given by Rathjen [43], making use of methods of traditional impredicative proof theory. The corresponding system $\mathsf{FID}([\text{POS-}\Pi^0_\infty, \Pi^0_1])$ for first order inductive definitions treating combined operators from $[\text{POS-}\Pi^0_\infty, \Pi^0_1]$ was studied in Jäger [31] and Jäger and Studer [38].

B. Full and metapredicative explicit Mahlo. Full and metapredicative explicit Mahlo will be introduced in Section 6. Then we also state the respective results concerning their proof-theoretic strength.

In a nutshell: full explicit Mahlo is obtained from Feferman's T_0 by adding the Mahlo axioms (M1) and (M2) formulated below; metapredicative explicit Mahlo is obtained from full explicit Mahlo by deleting inductive generation (for details see Section 6).

C. Constructive Mahlo. As mentioned above, the design and analysis of the first extension of Martin-Löf type theory with one Mahlo universe was given by Setzer. Related formalizations in MLTT and CZF and their treatment are due to Rathjen. In the context of Martin-Löf type theory a sort of Mahlo rule is formulated in Rathjen [46].

D. Metapredicative Mahlo in set theory. KPm^0 is a natural formalism for metapredicative Mahlo in a set-theoretic context. Even if the schema $(\mathcal{L}^*\text{-}\mathsf{I_N})$ of complete induction on the natural numbers for all formulas of the language $\mathcal{L}^*$ of KPm^0 is added, we do not leave the area of metapredicativity.

THEOREM 3 (Jäger and Strahm). *We have the following proof-theoretic ordinals for metapredicative Mahlo in set theory*:

$$|\mathsf{KPm}^0| = \varphi\omega 00 \quad \textit{and} \quad |\mathsf{KPm}^0 + (\mathcal{L}^*\text{-}\mathsf{I_N})| = \varphi\varepsilon_0 00.$$

This theorem is proved in Jäger and Strahm [36] and Strahm [57]. The lower bounds are established by well-ordering proofs, the upper bounds by interpretations of the respective theories into suitable ordinal theories with fixed point operators; the treatment of those is via partial cut elimination and asymmetric interpretations.

§5. The basics of explicit mathematics. Explicit mathematics was introduced by Feferman around 1975. The three basic papers which illuminate explicit mathematics from various angles are Feferman [9, 10, 11]. The original aim of explicit mathematics was to provide a natural formal framework for Bishop-style constructive mathematics. Soon it turned out, however, that the range of applications of explicit mathematics is much wider and includes, for example, also the following subjects:

Reductive proof theory. Systems of explicit mathematics play an important role in studying the relationship between subsystems of analysis, subsystems of set theory and theories for inductive definitions and for the reduction of classical theories to constructively (better) justified formalisms.

Abstract recursion theory. Several of its basic first order features (λ abstraction, fixed point theorem) are recursion-theoretic in nature and are not tied to any specific structure; it is also a good tool in developing a proof theory of higher order functionals (cf. e.g., Feferman and Jäger [18, 19] and Jäger and Strahm [37]).

Type systems. Flexible (polymorphic) type systems find a natural place in explicit mathematics; the most practically needed type constructs can be modeled in systems of low proof-theoretic strength (cf. e.g., Feferman [14], Jäger [29]).

Programming. Feferman [15, 16] deals with properties of functional programs; Studer [59, 60] employs explicit mathematics for foundational questions in object-oriented programming.

In the following we do not work with Feferman's original formulation of explicit mathematics but use instead the framework of theories of types and names introduced in Jäger [28]. Their general "ontology" can then be described as follows:

- *individuals* are explicitly given and can be interpreted as objects, operations, (constructive) functions, programs and the like;
- self-application is possible; we define new operations (terms) by means of principles such as λ abstraction and the fixed point theorem;
- induction is then often used in order to show that these new operations have the desired properties;
- *types* are abstractly defined collections of operations; they have names and are addressed via these names.

The focus of explicit mathematics is on the explicit presentation of operations rather than their constructive justification; it is possible to be explicit without being constructive (and vice versa).

Explicit mathematics starts off from a language $\mathbb{L}$ of two sorts, those being individuals $(a, b, c, x, y, z, \dots)$ and types $(U, V, W, X, Y, Z, \dots)$. There are several constants $\mathsf{k}, \mathsf{s}, \mathsf{p}, \mathsf{p}_0, \mathsf{p}_1, \mathsf{p}_\mathsf{N}, \mathsf{s}_\mathsf{N}, \dots$ whose meaning will be explained later plus one binary function symbol Ap for application. *Terms* are generated from the individual variables and constants by this form of application,

$$\text{Terms } (r, s, t, \dots)\text{: variables} \mid \text{constants} \mid \mathsf{Ap}(s, t).$$

In the following we often abbreviate $\mathsf{Ap}(s, t)$ as $(s \cdot t)$ or simply as (st) or st. We also adopt the convention of association to the left so that $s_1 s_2 \dots s_n$ stands for $(\dots(s_1 s_2)\dots s_n)$. In addition, we often write $s(t_1, \dots, t_n)$ for $st_1 \dots t_n$. Further we put $t' := \mathsf{s}_\mathsf{N} t$ and $1 := 0'$.

In addition, we have two unary relation symbols $\downarrow$ and N where $r\downarrow$ and $\mathsf{N}(r)$ express that r is defined (has a value) and r is a natural number, respectively. The only further relation symbols of our language of explicit mathematics are the binary $=$ for equality between individuals and between types, $\in$ for elementhood of individuals in types and $\Re$ for the naming relation; if $\Re(r, U)$ then we say that the individual r represents (is a name of) the type U. Therefore we have the atomic formulas

$$r\downarrow, \quad \mathsf{N}(r), \quad r = s, \quad U = V, \quad \Re(r, U),$$

and from those our formulas are generated as usual. A formula is called *elementary* if it contains neither the relation symbol $\Re$ nor bound type variables.

Finally, our logic is the classical *Beeson-Feferman logic of partial terms* with equality in both sorts as described, for example, in Beeson [6] and Troelstra and van Dalen [61]. Since it is not guaranteed that terms have values, a *partial equality* $\simeq$ à la Kleene is introduced by

$$(s \simeq t) := (s\downarrow \vee\, t\downarrow) \to (s = t).$$

To simplify the notation, we frequently also use the following abbreviations concerning the predicate N:

$$\begin{aligned} t \in \mathsf{N} &:= \mathsf{N}(t), \\ t \notin \mathsf{N} &:= \neg\mathsf{N}(t), \\ (\exists x \in \mathsf{N})A &:= (\exists x)(x \in \mathsf{N} \wedge A), \\ (\forall x \in \mathsf{N})A &:= (\forall x)(x \in \mathsf{N} \rightarrow A), \\ t \in (\mathsf{N} \rightarrow \mathsf{N}) &:= (\forall x \in \mathsf{N})(tx \in \mathsf{N}), \\ t \in (\mathsf{N}^{k+1} \rightarrow \mathsf{N}) &:= (\forall x \in \mathsf{N})(tx \in (\mathsf{N}^k \rightarrow \mathsf{N})). \end{aligned}$$

5.1. Basic theory BON **of operations and numbers.** BON was introduced in Feferman and Jäger [18]. The nonlogical axioms of BON formalize that the individuals form a partial combinatory algebra, that we have pairing and projection and the usual closure conditions on the natural numbers as well as definition by numerical cases. We divide the axioms into the following five groups:

I. Partial combinatory algebra.
 (1) $\mathsf{k}ab = a$,
 (2) $\mathsf{s}ab\downarrow \wedge \mathsf{s}abc \simeq (ac)(bc)$.

II. Pairing and projection.
 (3) $\mathsf{p}_0(\mathsf{p}ab) = a \wedge \mathsf{p}_1(\mathsf{p}ab) = b$.

III. Natural numbers.
 (4) $0 \in \mathsf{N} \wedge (\forall x \in \mathsf{N})(x' \in \mathsf{N})$,
 (5) $(\forall x \in \mathsf{N})(x' \neq 0 \wedge \mathsf{p_N}(x') = x)$,
 (6) $(\forall x \in \mathsf{N})(x \neq 0 \rightarrow \mathsf{p_N}x \in \mathsf{N} \wedge (\mathsf{p_N}x)' = x)$.

IV. Definition by numerical cases.
 (7) $a \in \mathsf{N} \wedge b \in \mathsf{N} \wedge a = b \rightarrow \mathsf{d_N}uvab = u$,
 (8) $a \in \mathsf{N} \wedge b \in \mathsf{N} \wedge a \neq b \rightarrow \mathsf{d_N}uvab = v$.

V. Primitive recursion on N.
 (9) $f \in (\mathsf{N}^2 \rightarrow \mathsf{N}) \wedge a \in \mathsf{N} \rightarrow \mathsf{r_N}fa \in (\mathsf{N} \rightarrow \mathsf{N})$,
 (10) $f \in (\mathsf{N}^2 \rightarrow \mathsf{N}) \wedge a \in \mathsf{N} \wedge b \in \mathsf{N} \wedge h = \mathsf{r_N}fa \rightarrow$
 $h0 = a \wedge h(b') = fb(hb)$.

There are two crucial principles which follow already from the the axioms of a partial combinatory algebra, i.e., from axioms (1) and (2) of BON: *λ abstraction* and the *fixed point* (*recursion*) *theorem*. These are of course standard results which have been discussed in the relevant literature a long time ago; cf. e.g., Beeson [6], Feferman [9] or Troelstra and van Dalen [61].

The existence of $\mathsf{r_N}$ as claimed in (9) and (10) surely allows us to introduce representing terms for all primitive recursive functions. The defining equations and totality of (the representing terms for) these functions are derivable in BON.

In view of the availability of the fixed point theorem in BON one might even suspect that $\mathsf{r_N}$ and the axioms (9) and (10) are superfluous. Unfortunately, this is only the case if sufficiently strong induction principles are available. Actually, all induction principles formulated below would suffice. Nevertheless we decided to include (9) and (10) since these two axioms belong to the now "official" formulation of BON.

5.2. Basic axioms about types. Our next step is to formulate some basic axioms about types and their names, which will be included in all our further systems of explicit mathematics. We first claim that each type has a name, that types with the same name are identical and that the equality of types is extensional.

Naming and extensionality axioms (N&E).

(11) $(\forall X)(\exists a)\Re(a, X)$,
(12) $\Re(a, X) \wedge \Re(a, Y) \rightarrow X = Y$,
(13) $(\forall a)(a \in X \leftrightarrow a \in Y) \leftrightarrow X = Y$.

Our systems of explicit mathematics combine intensionality and extensionality: on the level of types we are extensional, and types may be considered as objects in a Platonistic universe, given by abstract definitions. On the other hand, on the level of names we are intensional, and names have to be concretely given (introduced) terms. This idea also manifests itself by the following treatment of elementary comprehension.

Elementary comprehension (ECA). Nowadays we prefer to work with a finite axiomatization (f-ECA) of elementary comprehension. That means that we add further constants to our language corresponding to several basic operations on types so that the following theorem can be proved.

THEOREM 4 (ECA). *For every elementary formula $A(x, \vec{y}, \vec{Z})$ with all its free variables indicated we can define a term t_A so that*

$$\Re(\vec{v}, \vec{V}) \rightarrow (\exists X)\big[X = \{x : A(x, \vec{u}, \vec{V})\} \wedge \Re\left(t_A\left(\vec{u}, \vec{v}\right), X\right)\big].$$

Here we assume that $\vec{u}$ is a finite string $u_1, \dots, u_n$ of individual variables, $\vec{V}$ is a finite string $V_1, \dots, V_n$ of type variables and $\Re(\vec{v}, \vec{V})$ is short for the conjunction of the formulas $\Re(v_i, V_i)$ (for $i = 1, \dots, n$). This form of elementary comprehension is uniform in the individual and type parameters of the formula involved.

Join (J). The join (J) is the way in which explicit mathematics treats disjoint unions. Suppose that we have a type A and an operation f which maps each element x of A to the name fx of a type, say, B_x. Then we write $\Sigma\{fx : x \in A\}$ for the disjoint union of the types B_x, indexed by A.

We want a uniform formulation of join and therefore choose a new constant j which names the intended disjoint union depending on a name of the index type and the operation from this index type to names; hence the axiom (J) can be written as

$$\Re(a, A) \wedge (\forall x \in A)(\exists X)\Re(fx, X) \rightarrow$$
$$(\exists Y)\big[Y = \Sigma\{fx : x \in A\} \wedge \Re(\mathsf{j}(a, f), Y)\big].$$

This finishes the description of the basic type-theoretic axioms of explicit mathematics. From now on we will write EETJ (elementary explicit typing with join) for the extension of BON by the naming and extensionality axioms (N&E), elementary comprehension (ECA) and join (J),

$$\mathsf{EETJ} := \mathsf{BON} + (\mathsf{N\&E}) + (\mathsf{f\text{-}ECA}) + (\mathsf{J}).$$

5.3. Induction in explicit mathematics. There are many induction principles which are studied in the context of explicit mathematics. In the following we confine ourselves here to type induction and $\mathbb{L}$ induction with respect to the natural numbers. The first is the axiom

$$(\mathsf{T\text{-}I_N}) \quad (\forall X)\big[0 \in X \wedge (\forall x \in \mathsf{N})\,(x \in X \rightarrow x' \in X) \rightarrow (\forall x \in \mathsf{N})(x \in X)\big].$$

Obviously, type induction is a subcase of the schema of $\mathbb{L}$ induction stating for all formulas A of $\mathbb{L}$

$$(\mathbb{L}\text{-}\mathsf{I_N}) \qquad A(0) \wedge (\forall x \in \mathsf{N})(A(x) \rightarrow A(x')) \rightarrow (\forall x \in \mathsf{N})A(x).$$

Weaker forms of complete induction on the natural numbers, referring to the first order part of explicit mathematics, are studied at length in Jäger and Feferman [18].

5.4. Marriage of convenience. There are two main roads leading to models of explicit mathematics: one, in which the individuals are interpreted as (codes of) partial functions in the sense of classical set theory, and a second, which restricts itself to (codes of) partial *recursive* functions for dealing with the individuals. Details about such model constructions can be found, for example, in Feferman [11].

If we write $\mathrm{Gen}(V_{\aleph_1})$ and $\mathrm{Gen}(L_{\omega_1^{CK}})$ for the set-theoretic and recursion-theoretic model generated from the structure $(V_{\aleph_1}, \in)$ and $(L_{\omega_1^{CK}}, \in)$, respectively, the following observation can be easily established:

$$\mathrm{Gen}(V_{\aleph_1}) \models \mathsf{EETJ} + (\mathbb{L}\text{-}\mathsf{I_N}) \quad \text{and} \quad \mathrm{Gen}(L_{\omega_1^{CK}}) \models \mathsf{EETJ} + (\mathbb{L}\text{-}\mathsf{I_N}).$$

The (full) set-theoretic and the recursion-theoretic interpretations of explicit mathematics are connected by what Feferman [10] calls a *marriage of convenience*. Consider one of the usual extensions S of EETJ studied so far and a formula A provable in S. Then we may interpret A in the full set-theoretic model(s) of S—thus yielding the classical meaning $A^{(\mathsf{set})}$ of A—and in the recursion-theoretic models of S for obtaining the recursive reflection $A^{(\mathsf{rec})}$ of the classical assertion $A^{(\mathsf{set})}$.

5.5. Universes in explicit mathematics. Universes were introduced into explicit mathematics in Feferman [12], Marzetta [41] and Jäger, Kahle and Studer [33] as a powerful method for increasing its expressive and proof-theoretic strength. Informally speaking, universes play a similar role in explicit mathematics as admissible sets in weak set theory and the sets V_κ (for regular cardinals κ) in full classical set theory; explicit universes are also related to universes in Martin-Löf type theory. More formally, universes in explicit mathematics are types which consist of names only and reflect the theory EETJ.

In the following we write $U \models \mathsf{EETJ}$ for the conjunction of the finitely many formulas of $\mathbb{L}$ which express that the type U validates all type-theoretic axioms of EETJ; see Jäger, Kahle and Studer [33] for the exact formulation. Furthermore, the following shorthand notations are convenient:

$$\begin{aligned} \Re(a) &:= (\exists X)\Re(a, X), \\ a \mathrel{\dot\in} b &:= (\exists X)(\Re(b, X) \wedge a \in X), \\ \mathsf{U}(W) &:= W \models \mathsf{EETJ} \wedge (\forall x \in W)\Re(x), \\ \mathcal{U}(a) &:= (\exists X)\big[\mathsf{U}(X) \wedge \Re(a, X)\big]. \end{aligned}$$

Thus $\Re(a)$ means that the individual a names some type; the formula $a \mathrel{\dot\in} b$ expresses that the individual a is an element of the type named by b; $\mathsf{U}(W)$ and $\mathcal{U}(a)$ say that the type W is a universe and the individual a a name of a universe, respectively.

The first important axiom in connection with universes, also studied in Jäger, Kahle and Studer [33], is the earlier mentioned *limit axiom*

$$(\mathsf{Lim}) \qquad (\forall x)\big[\Re(x) \rightarrow (x \mathrel{\dot\in} \ell x \wedge \mathcal{U}(x))\big]$$

which states that the individual ℓ uniformly picks for each name x of a type the name ℓx of a universe containing x.

For several proof-theoretic aspects of (Lim) see the above mentioned Jäger, Kahle and Studer [33]; the proof-theoretic strength of (Lim) in a metapredicative context is analyzed in Strahm [55].

§6. The Mahlo axioms in explicit mathematics. The limit axiom (Lim) together with EETJ describes the explicit analogue of (recursive) inaccessibility. Now we go an important step further and adapt the formulation of Mahloness to our explicit context. To simplify the notation we set

$$\begin{aligned} f \in (\Re \rightarrow \Re) &:= (\forall x)\,(\Re(x) \rightarrow \Re(fx)), \\ f \in (a \rightarrow a) &:= (\forall x)\,(x \mathrel{\dot\in} a \rightarrow fx \mathrel{\dot\in} a) \end{aligned}$$

to express that the individual f is an operation form names to names and the type named by a to itself, respectively. Then the *Mahlo axioms* are as follows:

$$\text{(M1)} \qquad \Re(a) \wedge f \in (\Re \to \Re) \to \mathcal{U}(\mathsf{m}(a, f)) \wedge a \,\dot{\in}\, \mathsf{m}(a, f),$$

$$\text{(M2)} \qquad \Re(a) \wedge f \in (\Re \to \Re) \to f \in (\mathsf{m}(a, f) \to \mathsf{m}(a, f)).$$

m is a fresh individual constant for obtaining a formulation of these two axioms which is uniform in the name a and the operation f from names to names. From now on the theory $\mathsf{EETJ} + (\mathsf{M1}) + (\mathsf{M2})$ is usually written as $\mathsf{EETJ(M)}$.

Let M_0 be the first Mahlo cardinal and μ_0 the first recursively Mahlo ordinal. Then natural models of explicit Mahlo are generated from the full set-theoretic and the corresponding recursion-theoretic model of Mahloness.

$$\text{Gen}\,(V_{M_0}) \models \mathsf{EETJ(M)} + (\mathbb{L}\text{-}\mathsf{I_N}) \quad \text{and} \quad \text{Gen}\,(L_{\mu_0}) \models \mathsf{EETJ(M)} + (\mathbb{L}\text{-}\mathsf{I_N}).$$

The proof-theoretic analysis of $\mathsf{EETJ(M)}$ with type and formula induction on the natural numbers is carried through in Jäger and Strahm [36] and in Strahm [57]. Related results can also be obtained for corresponding systems of explicit mathematics with intuitionistic logic.

THEOREM 5 (Jäger and Strahm). *We have the following two proof-theoretic equivalences*:

1. $\mathsf{EETJ(M)} + (\mathsf{T}\text{-}\mathsf{I_N}) \equiv \mathsf{KPm}^0$,
2. $\mathsf{EETJ(M)} + (\mathbb{L}\text{-}\mathsf{I_N}) \equiv \mathsf{KPm}^0 + (\mathcal{L}^*\text{-}\mathsf{I_N})$.

All induction principles in these "metapredicative Mahlo" theories are restricted to the natural numbers. Later in this article stronger theories will be considered as well.

§7. Metapredicative Mahlo in second order arithmetic. This section is a short insertion turning to the problem of metapredicative Mahlo in the context of subsystems of second order arithmetic. The basic reference is Rüede's recent PhD thesis [48].

A subsystem of second order arithmetic which is proof-theoretically equivalent to KPm, i.e., KPm^0 with full $\in$ induction, is introduced and analyzed in Rathjen [44]. For obtaining systems of the same strength as KPm^0, Rüede had to proceed differently.

The role of universes is played in his approach by countable coded ω-models of $(\Sigma^1_1\text{-DC})$, and for such universes M, the "elements" of M are the sets which can be written as projections of M, i.e., for all subsets X of the natural numbers we simply define

$$X \,\varepsilon\, M := (\exists y)(X = (M)_y).$$

Countable coded ω-models of $(\Sigma^1_1\text{-AC})$ would not suffice since for all limits λ of admissible ordinals, the set $L_\lambda \cap Pow(\mathbb{N})$ is a countable ω-model of $(\Sigma^1_1\text{-AC})$, but not necessarily admissible.

Two schemas are central. The first is Π^1_2 *reflection on countable coded ω-models of* $(\Sigma^1_1\text{-DC})$ and consists of

$$(\Pi^1_2\text{-REF})^{(\Sigma^1_1\text{-DC})} \qquad A(X) \to (\exists M)\big[X \in M \wedge M \models_\omega (\Sigma^1_1\text{-DC}) \wedge A^M(X)\big]$$

for all Π^1_2 formulas $A(X)$ with the only free set variable being X; of course finite strings of set parameters could be permitted as well. The second important schema is Σ^1_1 *transfinite dependent choice*, consisting of

$$(\Sigma^1_1\text{-TDC}) \quad \mathsf{WO}(\prec) \wedge (\forall x)(\forall X)(\exists Y)A(x, X, Y) \to$$
$$(\forall X)(\exists Z)\ \big[(Z)_0 = X \wedge (\forall x)(0 \prec x \to A(x, (Z)_{\prec \restriction x}, (Z)_x))\big]$$

for all Σ^1_1 formulas $A(x, X, Y)$. As it turns out, $(\Pi^1_2\text{-REF})^{(\Sigma^1_1\text{-DC})}$ is equivalent to $(\Sigma^1_1\text{-TDC})$ and has the desired proof-theoretic strength.

THEOREM 6 (Rüede). *The two schemas* $(\Pi^1_2\text{-REF})^{(\Sigma^1_1\text{-DC})}$ *and* $(\Sigma^1_1\text{-TDC})$ *are equivalent over* ACA_0 *and have the same proof-theoretic strength as metapredicative Mahlo*; *i.e., we have*

1. $\mathsf{ACA}_0 + (\Pi^1_2\text{-REF})^{(\Sigma^1_1\text{-DC})} = \mathsf{ACA}_0 + (\Sigma^1_1\text{-TDC})$,
2. $\mathsf{ACA}_0 + (\Sigma^1_1\text{-TDC}) \equiv \mathsf{KPm}^0$.

For a proof of these two results see the above mentioned PhD thesis Rüede [48] or Rüede [49, 50].

§8. Mahlo beyond Feferman's T_0. Feferman's famous theory T_0 was the starting point of explicit mathematics; it extends the theory $\mathsf{EETJ} + (\mathbb{L}\text{-}\mathsf{I_N})$ by the powerful principle of *inductive generation* (IG): for every type A named a and every binary relation R on A with name r there exists the type of the R-accessible elements of A and is named $\mathsf{i}(a, r)$. So we set

$$\mathsf{T}_0 := \mathsf{EETJ} + (\mathsf{IG}) + (\mathbb{L}\text{-}\mathsf{I_N}).$$

Originally, T_0 was formulated within intuitionistic logic, but for some time classical logic has been used. The intuitionistic version of T_0 is called T^i_0 nowadays and provides an elegant framework for Bishop-style constructive mathematics. The constructive justification of T^i_0 is via a realizability interpretation.

The proof-theoretic strength of T_0 and T^i_0 is substantial and has been established in the following four articles: Jäger [26] presents a well-ordering proof for T^i_0; Feferman [11] shows that T_0 can be embedded into $(\Delta^1_2\text{-CA}) + (\mathsf{BI})$; Jäger [23, 25] contain proofs that $(\Delta^1_2\text{-CA}) + (\mathsf{BI})$ is contained in KPi; Jäger and Pohlers [34], finally, deals with the upper proof-theoretic limit of KPi.

THEOREM 7 (Feferman, Jäger and Pohlers). *The theory* T_0 *and its intuitionistic version* T_0^i *possess the same proof-theoretic strength, which is determined by the following equivalences*:

$$\mathsf{T}_0^i \equiv \mathsf{T}_0 \equiv (\Delta^1_2\text{-}\mathsf{CA}) + (\mathsf{BI}) \equiv \mathsf{KPi}.$$

Conceptually, T_0 and T_0^i go beyond what is reachable via iterated monotone inductive definability. Both systems provide a first step towards nonmonotone inductive definitions. This becomes very perspicuous in the context of the new model constructions for explicit mathematics given the help of certain classes of nonmonotone inductive definitions in Jäger [31], Jäger and Studer [38] and Studer [58].

Nonmonotone inductive definability is even more important if the Mahlo axioms $(\mathsf{M1})$ and $(\mathsf{M2})$ are added to T_0; call the resulting theory $\mathsf{T}_0(\mathsf{M})$ for simplicity. In the next section we will try to convey an idea how nonmonotone inductive definitions can be used for modeling $\mathsf{EETJ}(\mathsf{M})$ and $\mathsf{T}_0(\mathsf{M})$.

The rest of this section is dedicated to some recent results about the proof-theoretic strength of $\mathsf{T}_0(\mathsf{M})$, $\mathsf{T}_0^i(\mathsf{M})$ and other related theories. For the upper bound of $\mathsf{T}_0(\mathsf{M})$ we only have to refer to Jäger and Studer [38].

THEOREM 8 (Jäger and Studer). *The theory* $\mathsf{T}_0(\mathsf{M})$ *can be interpreted in the theory* KPm, *i.e.*, $\mathsf{T}_0(\mathsf{M}) \subset \mathsf{KPm}$.

Recent work of Tupailo is concerned with providing realizability interpretations of subsystems of second order arithmetic and extensions of Aczel's system CZF into explicit mathematics. The article Tupailo [62] embeds a subsystem of second order arithmetic of the same strength as the familiar $\Delta^1_2\text{-}\mathsf{CA} + (\mathsf{BI})$ into T_0^i whereas Tupailo [63] deals with the extension of CZF by a form of the Mahlo axiom (Mahlo) suited for the constructive setting.

In view of Rathjen [43] and some of his unpublished observations about constructive set theory it follows that the well-ordering proof for the proof-theoretic ordinal of KPm can be carried through in $\mathsf{CZF}+(\mathsf{Mahlo})$. Combining all these results we thus have the following theorem.

THEOREM 9. *The theories* $\mathsf{T}_0(\mathsf{M})$ *and* $\mathsf{T}_0^i(\mathsf{M})$ *have the same proof-theoretic strength as the theory* KPm,

$$\mathsf{CZF} + (\mathsf{Mahlo}) \equiv \mathsf{T}_0^i(\mathsf{M}) \equiv \mathsf{T}_0(\mathsf{M}) \equiv \mathsf{KPm}.$$

Setzer's approach to Mahloness within the framework of Martin-Löf type theory has also to be mentioned in this context; in [52] he studies a type theory of comparable strength.

§9. Modeling $\mathsf{EETJ}(\mathsf{M}) + (\mathbb{L}\text{-}\mathsf{I_N})$ **and** $\mathsf{T}_0(\mathsf{M})$**.** We conclude this survey with a brief description of how explicit Mahlo can be modeled in systems of nonmonotone inductive definitions of the same strength. We confine ourselves to the basic ideas; all further details concerning the theory $\mathsf{T}_0(\mathsf{M})$ for full recursive Mahlo can be found in Jäger and Studer [38]; the metapredicative variant is in Jäger and Strahm [36].

The treatment of the applicative part of $\mathsf{EETJ(M)} + (\mathbb{L}\text{-}\mathsf{I_N})$ and $\mathsf{T_0(M)}$ is as usual: the individuals are supposed to range over the set of natural numbers $\mathbb{N}$, we assume that $\{e\}$ for $e \in \mathbb{N}$ is a usual indexing of the partial recursive functions and let individual application be translated by setting

$$(e \bullet n) :\simeq \{e\}(n).$$

Then standard applications of the well-known S-m-n theorem provide natural numbers so that the axioms of BON are satisfied. The interpretation of types and names is more interesting. For this purpose we choose suitable formulas

$$\mathfrak{A}(X, a, b, c) \in \mathsf{POS}\text{-}\Pi^0_\infty \text{ and } \mathfrak{B}(X, a, b, c) \in \Pi^0_1,$$

with the corresponding operators $\Phi_{\mathfrak{A}}$ and $\Phi_{\mathfrak{B}}$ and work with the combined operator $\Theta := [\Phi_{\mathfrak{A}}, \Phi_{\mathfrak{B}}]$ generated from $\Phi_{\mathfrak{A}}$ and $\Phi_{\mathfrak{B}}$. The sets I^α_Θ are the stages of the inductive definition induced by Θ, and I_Θ is defined as

$$I_\Theta := \bigcup \{I^\alpha_\Theta : \alpha < \|\Theta\|\},$$

with α ranging—in the formalized versions—over ordinals or linear orderings, depending on whether we want to treat, respectively, the stronger or the weaker (metapredicative) system.

The set I_Θ consists of triples of natural numbers which code names and their elements in the following sense:

$$(e, 0, 0) \in I_\Theta \sim e \text{ is (name of) a type},$$
$$(e, n, 1) \in I_\Theta \sim e \text{ is a type and } n \mathrel{\dot\in} a,$$
$$(e, n, 2) \in I_\Theta \sim e \text{ is a type and } n \mathrel{\dot\notin} a.$$

The operator form $\mathfrak{A}(X, a, b, c)$ is reminiscent of the definition of Kleene's $\mathcal{O}$ and is used to deal with

- elementary comprehension and join,
- *elements* of universes (and of accessible parts).

The operator form $\mathfrak{B}(X, a, b, c)$, which contains negative occurrences in an essential way, provides for

- *names* of universes (and accessible parts),
- *extensions* of the complements of universes (and accessible parts).

Formalization of this approach in the suitable systems of nonmonotone inductive definitions immediately provides the sharp upper proof-theoretic bounds (cf. [38, 36]).

According to Theorem 9, our system $\mathsf{T_0(M)}$ has the same proof-theoretic strength as the system of constructive set theory $\mathsf{CZF} + (\mathsf{Mahlo})$ and thus is justified in the sense of reductive proof theory. However, the question remains whether there is a direct constructive justification of the system $\mathsf{T}_0^i(\mathsf{M})$ and, if so, what such a justification would mean.

It may be an interesting general approach to analyze whether nonmonotone inductive definitions of the form discussed above or even more powerful ones could be instrumentalized as tool for the constructive justification of strong systems. It could be the role of nonmonotone operators to provide for the possibility of carrying through *sufficiently long* iterations (of such operators) and the corresponding ordinals. More research in this direction is left for future publications.

REFERENCES

[1] Peter Aczel, *The type-theoretic interpretation of constructive set theory*, ***Logic Colloquium '77*** (A. MacIntyre, L. Pacholski, and J. Paris, editors), North-Holland, 1978, pp. 55–66.

[2] ———, *The type-theoretic interpretation of constructive set theory: choice principles*, ***The L. E. J. Brouwer Centenary Symposium*** (A. S. Troelstra and D. van Dalen, editors), North-Holland, 1982, pp. 1–40.

[3] ———, *The type-theoretic interpretation of constructive set theory: inductive definitions*, ***Logic, Methodology and Philosophy of Science VII*** (R. B. Marcus, G. J. W. Dorn, and P. Weingartner, editors), North-Holland, 1986, pp. 17–49.

[4] Peter Aczel and Peter G. Hinman, *Recursion in the superjump*, ***Generalized Recursion Theory, Proceedings of the 1972 Symposium, Oslo*** (J. E. Fenstad and P. G. Hinman, editors), vol. 94, North-Holland, 1974, pp. 3–41.

[5] Jeremy Avigad, *On the relationship between ATR_0 and $\widehat{ID}_{<\omega}$*, ***The Journal of Symbolic Logic***, vol. 61 (1996), no. 3, pp. 768–779.

[6] Michael J. Beeson, ***Foundations of Constructive Mathematics: Metamathematical Studies***, Springer, 1985.

[7] Laura Crosilla, *Realizability Models for Constructive Set Theories with Restricted Induction*, ***Ph.D. thesis***, School of Mathematics, University of Leeds, 2000.

[8] Solomon Feferman, *Systems of predicative analysis*, ***The Journal of Symbolic Logic***, vol. 29 (1964), no. 1, pp. 1–30.

[9] ———, *A language and axioms for explicit mathematics*, ***Algebra and Logic*** (J. N. Crossley, editor), Lecture Notes in Mathematics, vol. 450, Springer, 1975, pp. 87–139.

[10] ———, *Recursion theory and set theory: a marriage of convenience*, ***Generalized Recursion Theory II, Oslo 1977*** (J. E. Fenstad, R. O. Gandy, and G. E. Sacks, editors), North-Holland, 1978, pp. 55–98.

[11] ———, *Constructive theories of functions and classes*, ***Logic Colloquium '78*** (M. Boffa, D. van Dalen, and K. McAloon, editors), North-Holland, 1979, pp. 159–224.

[12] ———, *Iterated inductive fixed-point theories: application to Hancock's conjecture*, ***The Patras Symposion*** (G. Metakides, editor), North-Holland, 1982, pp. 171–196.

[13] ———, *Weyl vindicated: "Das Kontinuum" 70 years later*, ***Temi e Prospettive della Logica e della Filosofia della Scienza Contemporanee*** (C. Celluci and G. Sambin, editors), CLUEB, 1988, pp. 59–93.

[14] ———, *Polymorphic typed lambda-calculi in a type-free axiomatic framework*, ***Logic and Computation*** (W. Sieg, editor), Contemporary Mathematics, vol. 106, American Mathematical Society, 1990, pp. 101–136.

[15] ———, *Logics for termination and correctness of functional programs*, ***Logic from Computer Science*** (Y. N. Moschovakis, editor), Springer, 1991, pp. 95–127.

[16] ———, *Logics for termination and correctness of functional programs II: Logics of strength PRA*, ***Proof Theory*** (P. Aczel, H. Simmons, and S. S. Wainer, editors), Cambridge University Press, 1992, pp. 195–225.

[17] ———, *Does reductive proof theory have a viable rationale?*, ***Erkenntnis***, vol. 53 (2000), pp. 63–96.

[18] SOLOMON FEFERMAN and GERHARD JÄGER, *Systems of explicit mathematics with non-constructive μ-operator. Part I*, ***Annals of Pure and Applied Logic***, vol. 65 (1993), no. 3, pp. 243–263.

[19] ———, *Systems of explicit mathematics with non-constructive μ-operator. Part II*, ***Annals of Pure and Applied Logic***, vol. 79 (1996), no. 1, pp. 37–52.

[20] HARVEY FRIEDMAN, *Some systems of second order arithmetic and their use*, ***Proceedings of the International Congress of Mathematicians, Vancouver 1974***, vol. 1, Canadian Mathematical Congress, 1975, pp. 235–242.

[21] ROBIN O. GANDY, *General recursive functionals of finite type and hierarchies of functions*, ***Annales de la Faculté des Sciences de l'Université de Clermont Ferrant***, vol. 35 (1967), pp. 5–24.

[22] LEO HARRINGTON, *The superjump and the first recursively Mahlo ordinal*, ***Generalized Recursion Theory, Proceedings of the 1972 Symposium, Oslo*** (J. E. Fenstad and P. G. Hinman, editors), vol. 94, North-Holland, 1974, pp. 3–41.

[23] GERHARD JÄGER, *Die konstruktible Hierarchie als Hilfsmittel zur beweistheoretischen Untersuchung von Teilsystemen der Mengenlehre und Analysis*, ***Ph.D. thesis***, Universität München, 1979.

[24] ———, *Theories for iterated jumps*, ***Technical notes***, Mathematical Institute, Oxford University, 1980.

[25] ———, *Iterating admissibility in proof theory*, ***Logic Colloquium '81. Proceedings of the Herbrand Symposion***, North-Holland, 1982, pp. 137–146.

[26] ———, *A well-ordering proof for Feferman's theory* T_0, ***Archiv für mathematische Logik und Grundlagenforschung***, vol. 23 (1983), pp. 65–77.

[27] ———, *The strength of admissibility without foundation*, ***The Journal of Symbolic Logic***, vol. 49 (1984), no. 3, pp. 867–879.

[28] ———, *Induction in the elementary theory of types and names*, ***Computer Science Logic '87*** (E. Börger, H. Kleine Büning, and M. M. Richter, editors), Lecture Notes in Computer Science, vol. 329, Springer, 1988, pp. 118–128.

[29] ———, *Type theory and explicit mathematics*, ***Logic Colloquium '87*** (H.-D. Ebbinghaus, J. Fernandez-Prida, M. Garrido, M. Lascar, and M. Rodriguez Artalejo, editors), North-Holland, 1989, pp. 117–135.

[30] ———, *The next admissible in set theories without foundation*, ***Technical notes***, Institut für Informatik und angewandte Mathematik, Universität Bern, 1995.

[31] ———, *First order theories for nonmonotone inductive definitions: recursively inaccessible and Mahlo*, ***The Journal of Symbolic Logic***, vol. 66 (2001), pp. 1073–1089.

[32] GERHARD JÄGER, REINHARD KAHLE, ANTON SETZER, and THOMAS STRAHM, *The proof-theoretic analysis of transfinitely iterated fixed point theories*, ***The Journal of Symbolic Logic***, vol. 64 (1999), no. 1, pp. 53–67.

[33] GERHARD JÄGER, REINHARD KAHLE, and THOMAS STUDER, *Universes in explicit mathematics*, ***Annals of Pure and Applied Logic***, vol. 109 (2001), pp. 141–162.

[34] GERHARD JÄGER and WOLFRAM POHLERS, *Eine beweistheoretische Untersuchung von* $(\Delta^1_2\text{-CA}) + (\text{BI})$ *und verwandter Systeme*, ***Sitzungsberichte der Bayerischen Akademie der Wissenschaften***, Mathematisch-naturwissenschaftliche Klasse, 1982, pp. 1–28.

[35] GERHARD JÄGER and THOMAS STRAHM, *Fixed point theories and dependent choice*, ***Archive for Mathematical Logic***, vol. 39 (2000), no. 7, pp. 493–508.

[36] ———, *Upper bounds for metapredicative Mahlo in explicit mathematics and admissible set theory*, ***The Journal of Symbolic Logic***, vol. 66 (2001), pp. 935–958.

[37] ———, *The proof-theoretic strength of the Suslin operator in applicative theories*, ***Reflections on the foundations of mathematics*** (W. Sieg, R. Sommer, and C. Talcott, editors), Lecture Notes in Logic, vol. 15, A K Peters, 2002, pp. 270–292.

[38] Gerhard Jäger and Thomas Studer, *Extending the system* T_0 *of explicit mathematics: the limit and Mahlo axioms*, ***Annals of Pure and Applied Logic***, vol. 114 (2002), pp. 79–101.

[39] Reinhard Kahle, *Uniform limit in explicit mathematics with universes*, ***Technical report, IAM-97-002***, Institut für Informatik und angewandte Mathematik, Universität Bern, 1997.

[40] Paul Mahlo, *Über lineare transfinite Mengen*, ***Berichte über die Verhandlungen der königlich sächsischen Gesellschaft der Wissenschaften zu Leipzig***, Mathematisch-Physische Klasse, vol. 63, 1911, pp. 187–225.

[41] Markus Marzetta, *Predicative Theories of Types and Names*, ***Ph.D. thesis***, Institut für Informatik und angewandte Mathematik, Universität Bern, 1993.

[42] John Myhill, *Constructive set theory*, ***The Journal of Symbolic Logic***, vol. 40 (1975), no. 3, pp. 347–382.

[43] Michael Rathjen, *Proof-theoretic analysis of* KPM, ***Archive for Mathematical Logic***, vol. 30 (1991), pp. 377–403.

[44] ———, *The recursively Mahlo property in second order arithmetic*, ***Mathematical Logic Quaterly***, vol. 42 (1996), pp. 59–66.

[45] ———, *The higher infinite in proof theory*, ***Logic Colloquium '95*** (J. Makowsky and E. Ravve, editors), Lecture Notes in Logic, vol. 11, Springer, 1998, pp. 275–304.

[46] ———, *The superjump in Martin-Löf type theory*, ***Logic Colloquium '98*** (S. Buss, P. Hájek, and P. Pudlák, editors), Lecture Notes in Logic, vol. 13, Association for Symbolic Logic, 2000, pp. 363–386.

[47] W. Richter, *Recursively Mahlo ordinals and inductive definitions*, ***Logic Colloquium '69*** (R. O. Gandy and C. E. M. Yates, editors), North-Holland, 1971, pp. 273–288.

[48] Christian Rüede, *Metapredicative Subsystems of Analysis*, ***Ph.D. thesis***, Institut für Informatik und angewandte Mathematik, Univeristät Bern, 2000.

[49] ———, *Transfinite dependent choice and* ω*-model reflection*, ***The Journal of Symbolic Logic***, vol. 67 (2002), no. 3, pp. 1153–1168.

[50] ———, *The proof-theoretic analysis of* Σ^1_1 *transfinite dependent choice*, ***Annals of Pure and Applied Logic***, vol. 122 (2003), pp. 195–234.

[51] Kurt Schütte, *Eine Grenze für die Beweisbarkeit der transfiniten Induktion in der verzweigten Typenlogik*, ***Archiv für Mathematische Logik und Grundlagen der Mathematik***, vol. 7 (1964), pp. 45–60.

[52] Anton Setzer, *Extending Martin-Löf type theory by one Mahlo universe*, ***Archive for Mathematical Logic***, vol. 39 (2000), no. 3, pp. 155–181.

[53] Stephen G. Simpson, *Set-theoretic aspects of* ATR_0, ***Logic Colloquium '80*** (D. van Dalen, D. Lascar, and J. Smiley, editors), North-Holland, Amsterdam, 1982, pp. 255–271.

[54] ———, Σ^1_1 *and* Π^1_1 *transfinite induction*, ***Logic Colloquium '80*** (D. van Dalen, D. Lascar, and J. Smiley, editors), North-Holland, Amsterdam, 1982, pp. 239–253.

[55] Thomas Strahm, *First steps into metapredicativity in explicit mathematics*, ***Sets and Proofs*** (S. Barry Cooper and John Truss, editors), Cambridge University Press, 1999, pp. 383–402.

[56] ———, *Autonomous fixed point progressions and fixed point transfinite recursion*, ***Logic Colloquium '98*** (S. Buss, P. Hájek, and P. Pudlák, editors), Lecture Notes in Logic, vol. 13, Association for Symbolic Logic, 2000, pp. 449–464.

[57] ———, *Wellordering proofs for metapredicative Mahlo*, ***The Journal of Symbolic Logic***, vol. 67 (2002), pp. 260–278.

[58] Thomas Studer, *Explicit mathematics: W-type, models*, ***Master's thesis***, Institut für Informatik und angewandte Mathematik, 1997.

[59] ———, *Constructive foundations for featherweight java*, ***Proof Theory in Computer Science*** (R. Kahle, P. Schroeder-Heister, and R. Stärk, editors), Lecture Notes in Computer Science, vol.

2183, Springer, 2001, pp. 202–238.

[60] ———, *A semantics for $\lambda^{\{\}}_{str}$: a calculus with overloading and late-binding*, ***Journal of Logic and Computation***, vol. 11 (2001), pp. 527–544.

[61] Anne S. Troelstra and D. van Dalen, ***Constructivism in Mathematics, vol. I and II***, North-Holland, 1988.

[62] Sergei Tupailo, *Realization of analysis into explicit mathematics*, ***The Journal of Symbolic Logic***, vol. 66 (2001), pp. 1848–1864.

[63] ———, *Realization of constructive mathematics into explicit mathematics: a lower bound for an impredicative Mahlo universe*, ***Annals of Pure and Applied Logic***, vol. 120 (2003), pp. 165–196.

INSTITUT FÜR INFORMATIK UND ANGEWANDTE MATHEMATIK
UNIVERSITÄT BERN
NEUBRÜCKSTRASSE 10
CH-3012 BERN, SWITZERLAND
E-mail: jaeger@iam.unibe.ch

A TWO-DIMENSIONAL TREE IDEAL

SVEN JOSSEN AND OTMAR SPINAS

Introduction. Each of the classical tree forcings, such as Sacks forcing $\mathbb{S}$, Laver Forcing $\mathbb{L}$ or Miller forcing $\mathbb{M}$, is associated with a σ-ideal on 2^ω (in the case of $\mathbb{S}$) or on ω^ω (in the case of $\mathbb{L}$ or $\mathbb{M}$). These can be considered forcing ideals in the sense that a real (in 2^ω or ω^ω) is generic for the respective forcing iff it avoids all the sets from the associated ideal defined in the ground model. Such ideals were first studied by Marczewski [9], later by Veličković [16], and Judah, Miller, Shelah [5]. They all studied the Sacks ideal. The ideals associated with Laver or Miller forcing were investigated by Brendle, Goldstern, Johnson, Repický, Shelah and Spinas (see [3], [4], [12] and [2] (chronological order)). In all these papers, among other things the additivity and covering coefficients were studied. Let $J(Q)$ denote the ideal associated with the forcing Q. The typical problem was whether for $Q \in \{\mathbb{S}, \mathbb{L}, \mathbb{M}\}$, in the trivial chain of inequalities

$$\omega_1 \leq \mathbf{add}(J(Q)) \leq \mathbf{cov}(J(Q)) \leq 2^\omega$$

any of the inequalities could consistently be strict. Here $\mathbf{add}(J)$, $\mathbf{cov}(J)$ denotes the additivity, covering coefficient of the ideal J, respectively. One of the main results of [5] is that, letting $V \models$ZFC+GCH and $\mathbb{S}_{\omega_2}$ a countable support iteration of $\mathbb{S}$ of length ω_2, then $V^{\mathbb{S}_{\omega_2}}$ is a model for $\mathbf{add}(J(\mathbb{S})) < \mathbf{cov}(J(\mathbb{S}))$. The hard part of the argument is to show that $V^{\mathbb{S}_{\omega_2}}$ is a model for $\mathbf{add}(J(\mathbb{S})) = \omega_1$. Building on these arguments, analogues of this result were proved in [4] for $J(\mathbb{M})$ and $J(\mathbb{L})$: For $Q \in \{\mathbb{L}, \mathbb{M}\}$, letting Q_{ω_2} denote a countable support iteration of length ω_2 of Q and $V \models$ZFC+GCH, $V^{Q_{\omega_2}}$ is a model for $\mathbf{add}(J(Q)) < \mathbf{cov}(J(Q))$. Actually, for $Q = \mathbb{M}$ this result needs a lemma from [12] which was not yet available in [4]. Again, the hard part was to show that $\mathbf{add}(J(Q)) = \omega_1$ holds in the extension. For this, it was proved that the cardinal invariant $\mathfrak{h}$ which is defined as the distributivity number of $\mathcal{P}(\omega)/\text{fin}$ is an upper bound of $\mathbf{add}(J(Q))$ in $V^{Q_{\omega_2}}$. Then this was combined with, first, the difficult result from [6] that for any $Q \in \{\mathbb{L}, \mathbb{M}\}$, in the model $V^{Q_{\omega_2}}$ there exists a nonmeasurable set of reals of size $\aleph_1$ and, secondly, with the easy result that such a set must have size at least $\mathfrak{h}$.

In the present paper we study the ideals associated with finite powers Q^n, for any $Q \in \{\mathbb{L}, \mathbb{M}\}$. We observe that $J(\mathbb{L}^n)$ is not a σ-ideal for any $n \geq 2$ and

Logic Colloquium 2000
Edited by R. Cori, A. Razborov, S. Todorčević, and C. Wood
Lecture Notes in Logic, 19

that $J(\mathbb{M}^m)$ is not a σ-ideal for any $m \geq 3$. A deeper result says that $J(\mathbb{M}^2)$ *is* a σ-ideal. In the line of work outlined above we proceed to show that if $(\mathbb{M}^2)_{\omega_2}$ is a countable support iteration of $\mathbb{M}^2$ and $V \models$ZFC+CH, then $V^{(\mathbb{M}^2)_{\omega_2}}$ is a model for $\mathbf{add}(J(\mathbb{M}^2)) < \mathbf{cov}(J(\mathbb{M}^2))$. Part of the proof is completely analogous to [4], in particular that $\mathbf{add}(J(\mathbb{M}^2)) \leq \mathfrak{h}$ and $\mathbf{cov}(J(\mathbb{M}^2)) = \omega_2$ hold in $V^{(\mathbb{M}^2)_{\omega_2}}$. But there is a new point. Forcing with $\mathbb{M}^2$ makes the old reals a Lebesgue null set because it adds a nonsplit real, i.e., a set $a \in [\omega]^\omega$ for which there is no $b \in [\omega]^\omega \cap V$ with $b \cap a$ and $a \setminus b$ both infinite (b does not split a). This implies that in $V^{(\mathbb{M}^2)_{\omega_2}}$ the splitting number $\mathfrak{s}$ is $\aleph_2$ and hence every set of reals of size $\aleph_1$ is a null set.

Nevertheless we are able to show that $\mathfrak{h} = \omega_1$ holds in $V^{(\mathbb{M}^2)_{\omega_2}}$ and therefore $\mathbf{add}(J(\mathbb{M}^2)) = \omega_1$ as desired. This is the core of the paper. The proof is a continuation of the difficult analysis of the forcing $\mathbb{M}^2$ which was carried out in [14], [13]. As a side-product we obtain a relatively simple model for $\mathfrak{h} < \min\{\mathfrak{b}, \mathfrak{s}\}$. So far only one model for this had been known (see [11]), which is a very involved construction of Shelah using creature forcing.

§1. **Preliminaries.** Let us first fix our notation. Given a tree $p \subseteq \omega^{<\omega}$, the set of its infinite branches is denoted with $[p]$. We let $|\sigma|$ denote the length of $\sigma \in \omega^{<\omega}$. For $\sigma \in \omega^{<\omega}$ and $n \in \omega$, we let $\sigma^\frown n$ be the sequence of length $|\sigma| + 1$ with initial segment σ and last coordinate n. By $\mathrm{succ}_p(\sigma)$ we denote the set of all $n < \omega$ with $\sigma^\frown n \in p$. Then σ is a *splitnode* of p iff $\mathrm{succ}_p(\sigma)$ has at least two elements. A tree $p \subseteq \omega^{<\omega}$ is called *superperfect* (or *Miller*), if $p \neq \emptyset$ and for every $\sigma \in p$ there exists $\tau \in p$ such that $\sigma \subseteq \tau$ and $\mathrm{succ}_p(\tau)$ is infinite. Such τ are called *infinite splitnodes*. The set of all infinite splitnodes of p is denoted by $\mathrm{split}(p)$. For $\sigma \in \mathrm{split}(p)$, by $\mathrm{Succ}_p(\sigma)$ we denote the set of infinite successor splitnodes of σ, i.e., those $\tau \in \mathrm{split}(p)$ such that $\sigma \subsetneq \tau$ and for no ϱ with $\sigma \subsetneq \varrho \subsetneq \tau$ do we have $\varrho \in \mathrm{split}(p)$. By $\mathrm{Lev}_n(p)$ we denote the set of those $\sigma \in \mathrm{split}(p)$ which have precisely n proper initial segments which belong to $\mathrm{split}(p)$. For convenience we shall always assume that a superperfect tree has only infinite splitnodes. Then $\mathrm{st}(p)$ denotes the shortest splitnode of p, and p^- denotes the set of nodes of p which extend $\mathrm{st}(p)$. If $\sigma \in p$ then $p(\sigma)$ consists of all $\nu \in p$ which are comparable with σ. A superperfect tree is called a *Laver tree* if for every $\sigma \in p^-$ the set $\mathrm{succ}_p(\sigma)$ is infinite. We let $[\sigma]$ denote the set of all reals with initial segment σ. If $p \subseteq \omega^{<\omega}$ is a finite tree, we denote the set of all its terminal nodes by $\mathrm{leaves}(p)$.

Let $\prec$ be the following well-ordering of $\omega^{<\omega}$ in type ω:

$$\begin{aligned}
\sigma \prec \tau \iff & \max\{|\sigma|, \max\mathrm{ran}(\sigma)\} < \max\{|\tau|, \max\mathrm{ran}(\tau)\} \\
& \vee \left[\max\{|\sigma|, \max\mathrm{ran}(\sigma)\} = \max\{|\tau|, \max\mathrm{ran}(\tau)\} \wedge |\sigma| < |\tau|\right] \\
& \vee \Big[\max\{|\sigma|, \max\mathrm{ran}(\sigma)\} = \max\{|\tau|, \max\mathrm{ran}(\tau)\} \wedge |\sigma| = |\tau| \\
& \qquad \wedge \sigma \text{ precedes } \tau \text{ lexicographically}\Big].
\end{aligned}$$

Letting $\langle \varrho_n : n < \omega \rangle$ be the $\prec$-increasing enumeration of $\omega^{<\omega}$ we write $\sharp \varrho_n = n$. Note that $\sharp \varrho \geq \max\{|\varrho|, \max \operatorname{ran}(\varrho)\}$. We shall often construct a superperfect tree p by constructing inductively $\langle \sigma_n : n < \omega \rangle$, the set of splitnodes of p. This will be done so that for every n, the set of splitnodes and leaves of the tree generated by $\langle \sigma_i : i < n \rangle$ is isomorphic to the tree of the n first (with respect to $\prec$) members of $\omega^{<\omega}$. Then we shall say that $\langle \sigma_i : i < n \rangle$ is an *initial segment of the splitnodes of some superperfect tree* or that $\langle \sigma_n : n < \omega \rangle$ is the *increasing enumeration of the splitnodes of some superperfect tree*. Note this does not imply that $\langle \sigma_i : i < n \rangle$ is $\prec$-increasing; however it can of course be arranged that it is.

The set of all superperfect trees will be denoted by $\mathbb{M}$, the set of all Laver trees by $\mathbb{L}$. Then $(\mathbb{L}, \subseteq)$ is Laver forcing (see [8]), and $(\mathbb{M}, \subseteq)$ is Miller forcing (see [10]). Let Q be $\mathbb{L}$ or $\mathbb{M}$. For $p, q \in Q$ we shall write $p \leq q$ instead of $p \subseteq q$, and by $p \leq^n q$ we mean that $p \leq q$ and p and q have the same $n + 1$ first splitnodes in the sense mentioned above.

Then Q^m carries the coordinatewise ordering. For sequences $\langle p_0, \dots, p_{m-1} \rangle$, $\langle q_0, \dots, q_{m-1} \rangle \in Q^m$, by writing $\langle p_0, \dots, p_{m-1} \rangle \leq^n \langle q_0, \dots, q_{m-1} \rangle$ we mean $p_i \leq^n q_i$ for all $i < m$. A set $S \subseteq \omega^\omega$ is called *superperfect* iff it equals $[p]$ for some superperfect tree p. By $[p] \times^+ [q]$ we denote the upper half of the superperfect rectangle $[p] \times [q]$, i.e., the set of all $\langle x, y \rangle \in [p] \times [q]$ with $x(|\mathrm{st} p|) < y(|\mathrm{st} q|)$. Similarly, $[p] \times^- [q]$ denotes the lower half of $[p] \times [q]$.

In [14, Corollary 1.4] it has been shown that for every $\langle p, q \rangle \in \mathbb{M}^2$ there exists $\langle p', q' \rangle \leq^0 \langle p, q \rangle$ such that for every $\langle x, y \rangle \in [p'] \times^+ [q']$ there exists an increasing sequence $\langle k_i : i < \omega \rangle$ such that $k_0 = |\mathrm{st} q|$, $k_1 > |\mathrm{st} p|$ and for all $n < \omega$ we have that the following hold:

1. $k_{2n} = \min\{i < \omega : y(i) > \sharp x \restriction k_{2n+1}\}$,
2. $k_{2n+1} = \min\{i < \omega : x(i) > \sharp y \restriction k_{2n+2}\}$.
3. $k_{2n+1} < y(k_{2n}) < k_{2n+2} < x(k_{2n+1}) < k_{2n+3}$,
4. $x \restriction k_{2n+1} \in \operatorname{split}(p')$, $y \restriction k_{2n} \in \operatorname{split}(q')$.

Let $\sigma = \mathrm{st} p$ and $\tau = \mathrm{st} q$. In this case we say that $\langle x, y \rangle$ *oscillates infinitely often above* $\langle \sigma, \tau \rangle$. Hence every $\langle x, y \rangle \in [p'] \times^+ [q']$ has a unique associated sequence $\langle k_i : i < \omega \rangle$ which is determined solely by $\langle x, y \rangle$ and $\langle \sigma, \tau \rangle$. The sequence

$$\langle \sigma, \tau, x \restriction k_1, y \restriction k_2, x \restriction k_3, y \restriction k_4, \dots \rangle$$

is called the *type*$_{\sigma,\tau}$*-sequence of* $\langle x, y \rangle$. Let $\mathrm{tp}_{\sigma,\tau}$-0-pair$(x, y) = \langle \sigma, \tau \rangle$, and for all $n > 0$, let $\mathrm{tp}_{\sigma,\tau}$-$(2n + 1)$-pair$(x, y) = \langle x \restriction k_{2n+1}, y \restriction k_{2n} \rangle$ and $\mathrm{tp}_{\sigma,\tau}$-$(2n + 2)$-pair$(x, y) = \langle x \restriction k_{2n+1}, y \restriction k_{2n+2} \rangle$.

Using this we can define a partial function

$$\mathrm{tp}^{p',q'}_{\sigma,\tau} : (\omega^{<\omega})^2 \to \omega$$

by letting $\mathrm{tp}^{p',q'}_{\sigma,\tau}(\mu, \nu) = n$ iff there exists $\langle x, y\rangle \in [p'] \times^+ [q']$ such that $\langle \mu, \nu\rangle = \mathrm{tp}_{\sigma,\tau}\text{-}n\text{-pair}(x, y)$.

Suppose that $\mathrm{tp}^{p',q'}_{\sigma,\tau}(\mu, \nu) = 2n$ for some $\langle \mu, \nu\rangle \in p' \times q'$. Then there exists a unique sequence $\langle \mu_0, \nu_0, \ldots, \mu_n = \mu, \nu_n = \nu\rangle$, which is the initial sequence of length $2n+2$ of the $\mathrm{type}_{\sigma,\tau}$-sequence of $\langle x, y\rangle$ for some $\langle x, y\rangle \in [p'] \times^+ [q']$ that is infinitely oscillating over $\langle \sigma, \tau\rangle$. We will call this sequence the *$\mathrm{type}_{\sigma,\tau}$-sequence of* $\langle \mu, \nu\rangle$ and let $\mathrm{tp}_{\sigma,\tau}\text{-}i\text{-pair}(\mu, \nu) = \mathrm{tp}_{\sigma,\tau}\text{-}i\text{-pair}(x, y)$ for all $i \leq 2n$. A similar remark applies when $\mathrm{tp}_{\sigma,\tau}(\mu, \nu)$ is odd.

If $\langle \sigma, \tau\rangle$ or $\langle p', q'\rangle$ are clear from context, we omit them in the above notation.

For any $\langle p, q\rangle \in \mathbb{M}^2$, $\langle \sigma, \tau\rangle \in p \times q$, let

$$\mathrm{TP}_{\sigma,\tau}(p, q) = \mathrm{dom}\left(\mathrm{tp}^{p,q}_{\sigma,\tau}\right) \cap \mathrm{split}(p(\sigma)) \times^+ \mathrm{split}(q(\tau)),$$

the set of all the type pairs of $\langle p, q\rangle$, and for $n < \omega$ let $\mathrm{TP}^n_{\sigma,\tau}(p, q)$ denote the set of all pairs in $TP_{\sigma,\tau}(p, q)$ with type n. Then for some $\langle \mu, \nu\rangle \in \mathrm{TP}_{\sigma,\tau}(p, q)$ with $\mathrm{tp}^{p,q}_{\sigma,\tau}(\mu, \nu) = n$, say, n even, we denote the set of possible successive oscillation points as

$$\mathrm{Sop}^{p,q}_{\sigma,\tau}(\mu, \nu) = \{\mu' \in \mathrm{split}(p(\mu)^-) \colon \mathrm{tp}^{p,q}_{\sigma,\tau}(\mu', \nu) = n + 1 \\ \wedge\ \mathrm{tp}_{\sigma,\tau}\text{-}n\text{-pair}\,(\mu', \nu) = \langle \mu, \nu\rangle\},$$

and for odd n we define $\mathrm{Sop}^{p,q}_{\sigma,\tau}(\mu, \nu)$ symmetrically. If it is clear from the context, we will omit some or all of the indices.

Note that if $\langle p', q'\rangle$, $\langle \sigma, \tau\rangle$ are as above and $\langle u, v\rangle \leq \langle p', q'\rangle$, then in general we have that $\mathrm{TP}_{\sigma,\tau}(p', q') \cap (u \times v) \neq \mathrm{TP}_{\sigma,\tau}(u, v)$. However, there exists $\langle u', v'\rangle \leq^0 \langle u, v\rangle$ such that for almost all $n < \omega$ and all $\langle x, y\rangle \in [u'] \times^+ [v']$, we have $\mathrm{tp}_{\sigma,\tau}\text{-}n\text{-pair}(x, y) \in \mathrm{TP}_{\sigma,\tau}(u', v')$.

For the rest of the paper we tacitly assume that we always work with elements $\langle p, q\rangle$ which have the property of the above $\langle p', q'\rangle$. Since the set of all such $\langle p, q\rangle$ is dense in $\mathbb{M}^2$, forcing with this partial order is isomorphic to forcing with $\mathbb{M}^2$.

Let $\langle p, q\rangle \in \mathbb{M}^2$, ϕ a statement in the forcing language of $\mathbb{M}^2$ and $* \in \{+, -\}$. Define

$$\langle p, q\rangle \Vdash^*_{\mathbb{M}^2} \phi$$

iff $V[\langle g_0, g_1\rangle] \models \phi$, for all pairs of $\mathbb{M}^2$-generic reals $\langle g_0, g_1\rangle \in [p] \times^* [q]$. By [14, Main Lemma 4.2], $\mathbb{M}^2$ has the *weak decision property*, i.e., if $p \in \mathbb{M}^2$ and ϕ_0 and ϕ_1 are statements in the forcing language of $\mathbb{M}^2$ with $p \Vdash \phi_0 \vee \phi_1$, then there exist $q \leq^0 p$ and $i_+, i_- \in 2$ (which are not necessarily equal) with $q \Vdash^+ \phi_{i_+}$ and $q \Vdash^- \phi_{i_-}$.

We will often use the following result from [10]:

LEMMA 1 (Miller). *For every coloring of the splitnodes of a superperfect tree, there exists a superperfect subtree such that all splitnodes have the same color.*

From [13], we have the following

Fusion Property of $\mathbb{M}^2$**.** Suppose that P is a property of elements of $\mathbb{M}^2$ such that for every $\langle u, v\rangle \in \mathbb{M}^2$ there exists $\langle u', v'\rangle \leq^0 \langle u, v\rangle$ which has P. Then for every $\langle u, v\rangle \in \mathbb{M}^2$ there exists $\langle u', v'\rangle \leq^0 \langle u, v\rangle$ such that for every $\langle \mu, \nu\rangle \in \mathrm{TP}(u', v')$,

- if $\mathrm{tp}_{\mathrm{st}u,\mathrm{st}v}(\mu, \nu)$ is even and if $\langle u'', v''\rangle \leq^0 \langle u'(\mu), v'(\nu)\rangle$ is defined implicitly by letting $u'' \times v''$ be the downward closure of
$$\left\{\langle \varrho, \xi\rangle \in u'(\mu)^- \times v'(\nu) \colon \varrho(|\mu|) > \sharp \nu\right\},$$
then $\langle u'', v''\rangle$ has P, and
- if $\mathrm{tp}_{\mathrm{st}u,\mathrm{st}v}(\mu, \nu)$ is odd and if $\langle u'', v''\rangle \leq^0 \langle u'(\mu), v'(\nu)\rangle$ is defined implicitly by letting $u'' \times v''$ be the downward closure of
$$\left\{\langle \varrho, \xi\rangle \in u'(\mu) \times v'(\nu)^- \colon \xi(|\nu|) > \sharp \mu\right\},$$
then $\langle u'', v''\rangle$ has P.

Let $\langle P_\alpha, \dot{Q}_\beta \colon \alpha \leq \omega_2, \beta < \omega_2\rangle$ be a countable support iteration of $\mathbb{M}^2$ of length ω_2, hence for all $\beta < \omega_2$, $\Vdash_{P_\beta}$ " $\dot{Q}_\beta$ is $\mathbb{M}^2$ ". For $\bar{p} \in P_{\omega_2}$ we write
$$\bar{p} = \langle \langle \bar{p}(\alpha)^0, \bar{p}(\alpha)^1\rangle : \alpha \in \omega_2\rangle$$
with
$$\bar{p} \restriction \alpha \Vdash_{P_\alpha} \langle \bar{p}(\alpha)^0, p(\alpha)^1\rangle \in \mathbb{M}^2.$$
If G is a P_{ω_2}-generic filter over V, for $\alpha \in [1, \omega_2]$, let $G_\alpha = \{p \restriction \alpha \colon p \in G\}$ (which by [1, Theorem 1.2.] is a P_α-generic filter), and let
$$\left\langle \left\langle g_0^\beta, g_1^\beta\right\rangle : \beta < \alpha\right\rangle$$
be the pairs of generic reals determined by G_α.

Definition 2. *Let $F \in [\omega_2]^{<\omega}$, $\bar{p}, \bar{q} \in P_{\omega_2}$, $*_\alpha \in \{+, -\}$ for $\alpha \in F$, $n \in \omega$ and ϕ a statement in the forcing language of P_{ω_2}.*

1. *$\bar{p} \leq_F^n \bar{q}$ iff $\bar{p} \leq \bar{q}$ and $\forall \alpha \in F\, \bar{p} \restriction \alpha \Vdash_{P_\alpha} \bar{p}(\alpha) \leq^n \bar{q}(\alpha)$,*
2. *$\bar{p} \Vdash^{\langle *_\alpha \colon \alpha \in F\rangle} \phi$ iff for all P_{ω_2}-generic G with $\bar{p} \in G$ and such that*
$$g_0^\alpha\left(\left|\mathrm{st}\bar{p}(\alpha)^0[G_\alpha]\right|\right) < g_1^\alpha\left(\left|\mathrm{st}\bar{p}(\alpha)^1[G_\alpha]\right|\right) \iff *_\alpha = +$$
holds for every $\alpha \in F$, we have $V[G] \models \phi$.

Lemma 3. *Suppose $\bar{p} \in P_{\omega_2}$, $F \in [\omega_2]^{<\omega}$, ϕ_i statements in the forcing language of P_{ω_2} for $i < 2$ such that*
$$\bar{p} \Vdash_{P_{\omega_2}} \phi_0 \vee \phi_1.$$
Also let $*_\alpha \in \{+, -\}$ for all $\alpha \in F$. There exists $\bar{q} \leq_F^0 \bar{p}$ and $i < 2$ such that
$$\bar{q} \Vdash^{\langle *_\alpha \colon \alpha \in F\rangle} \phi_i.$$

PROOF. By induction on $\max(F)$. Let $\alpha = \max(F)$. Let $G_{\alpha+1}$ be $P_{\alpha+1}$-generic over V such that $\bar{p} \restriction \alpha + 1 \in G_{\alpha+1}$. In $V[G_{\alpha+1}]$ find q^α and $l < 2$ such that

$$q^\alpha \leq \bar{p} \restriction [\alpha + 1, \omega_2)\,[G_{\alpha+1}] \text{ and } q^\alpha \Vdash_{P_{\omega_2}/G_{\alpha+1}} \phi_l.$$

As $G_{\alpha+1}$ was arbitrary, in $V[G_\alpha]$ there exist $P_{\alpha+1}/G_\alpha$-names $\dot{l}$, $\dot{q}^\alpha$ such that

$$\bar{p}(\alpha)\,[G_\alpha] \Vdash_{P_{\alpha+1}/G_\alpha} \text{“}\dot{q}^\alpha \Vdash_{P_{\omega_2}/G_{\alpha+1}} \phi_{\dot{l}}\text{”}.$$

(We use that $P_{\alpha+1}/G_\alpha \cong Q_\alpha[G_\alpha] = (\mathbb{M}^2)^{V[G_\alpha]}$). Since $\mathbb{M}^2$ has the weak decision property, in $V[G_\alpha]$ there is $k < 2$ and $q(\alpha) \leq^0 p(\alpha)$ with $q(\alpha) \in (\mathbb{M}^2)^{V[G_\alpha]}$ such that

$$q(\alpha) \Vdash^{\langle *_\alpha \rangle} \dot{l} = k.$$

In V there are P_α-names $\dot{q}(\alpha)$ and $\dot{k}$ such that

$$\bar{p} \restriction \alpha \Vdash_{P_\alpha} \text{“}\dot{k} \in 2 \wedge \dot{q}(\alpha) \Vdash^{\langle *_\alpha \rangle} \dot{l} = \dot{k}\text{”}.$$

If $F \cap \alpha \neq 0$ then by induction hypothesis (and trivially otherwise) there exist $i < 2$ and $q_\alpha \leq^0_{F \cap \alpha} \bar{p} \restriction \alpha$ such that

$$q_\alpha \Vdash^{\langle *_\beta :\, \beta \in F \cap \alpha \rangle} \dot{k} = i.$$

Finally, let

$$\bar{q} = q_\alpha{}^\frown \langle \dot{q}(\alpha) \rangle^\frown \dot{q}^\alpha,$$

then $\bar{q}$ is as desired. ⊣

For $a, b \in [\omega]^{<\omega}$ we will write $a \lhd b$ iff a is an initial segment of b, i.e., $a \subseteq b$ and $\min(b \setminus a) > \max(a)$.

DEFINITION 4. *Let $J \subseteq \omega$ be infinite and $\langle s_j : j \in J \rangle$ a family of finite sets in ω. Then*

$$\lim_{j \in J} s_j = s$$

for some $s \subseteq \omega$, iff the following holds:

1. $\forall l < \omega\, \exists j_0 < \omega \forall j \in J \setminus j_0\, (s \cap l = s_j \cap l)$,
2. *if s is finite, additionally* $s = \displaystyle\bigcap_{j \in J} s_j$.

Thus s is the limit point of the characteristic functions of the s_j in 2^ω, where 2^ω is equipped with the product topology of the discrete topology on 2.

REMARK 5. (a) By König's Lemma, for every sequence $\langle s_j : j \in J \rangle$ as above there exists some infinite $J' \subseteq J$ such that $\lim_{j \in J'} s_j$ exists.

(b) If s is finite, then we have

$$\forall l > \max(s)\, \exists j_0 \forall j \in J \setminus j_0 \quad l \notin s_j.$$

§2. The n-dimensional Laver and Miller ideal. For $Q \in \{\mathbb{L}, \mathbb{M}\}$, let $J(Q^n)$ be the set of all $X \subseteq (\omega^\omega)^n$ with the property that for every $\langle p_i : i < n \rangle \in Q^n$ there exists $\langle q_i : i < n \rangle \in Q^n$ extending $\langle p_i : i < n \rangle$ such that $X \cap \prod_{i<n}[q_i] = \emptyset$. So $J(Q^n)$ is an ideal in $(\omega^\omega)^n$, which we call the n-dimensional Q-ideal. It is easy to see that $J(\mathbb{L})$ and $J(\mathbb{M})$ are σ-ideals. In this section we shall show that also $J(\mathbb{M}^2)$ is a σ-ideal, but $J(\mathbb{L}^2)$ and $J(\mathbb{M}^3)$ are not.

The additivity (**add**) of any ideal is defined as the minimal cardinality of a family of sets belonging to the ideal whose union does not. The covering number (**cov**) is defined as the least cardinality of a family of sets from the ideal whose union is the whole set on which the ideal is defined.

THEOREM 6. *Suppose $n \geq 2$ and $m \geq 3$, then $J(\mathbb{L}^n)$ and $J(\mathbb{M}^m)$ are not σ-ideals.*

PROOF. We show first that $\mathbf{add}(J(\mathbb{L}^2)) = \omega_0$, secondly that $\mathbf{add}(J(\mathbb{M}^3)) = \omega_0$. The same proofs hold for $m > 2$ and $n > 3$.

1. For any $\sigma, \tau \in \omega^{<\omega}$ and $\langle x, y \rangle \in [\sigma] \times [\tau]$, we define $\mathrm{Osc}_{\sigma,\tau}(x, y) \in 3^\omega$ such that

$$\mathrm{Osc}_{\sigma,\tau}(x,y)(n) = \begin{cases} 0 & \text{if } x(n+|\sigma|) < y(n+|\tau|), \\ 1 & \text{if } y(n+|\tau|) < x(n+|\sigma|), \\ 2 & \text{if } x(n+|\sigma|) = y(n+|\tau|). \end{cases}$$

Fix $\alpha \in 2^\omega$ and let

$$X^\alpha_{\sigma,\tau} = \left\{ \langle x, y \rangle \in (\omega^\omega)^2 \colon \mathrm{Osc}_{\sigma,\tau}(x,y) = \alpha \right\}.$$

For $\sigma, \tau \in \omega^{<\omega}$ we show $X^\alpha_{\sigma,\tau} \in J(\mathbb{L}^2)$. For $\langle p, q \rangle \in \mathbb{L}^2$, if $\langle \sigma, \tau \rangle \in p \times q$, then choose $m \in \mathrm{succ}_p(\sigma)$ and $n \in \mathrm{succ}_q(\tau)$ such that if $\alpha(0) = 0$ then $m > n$ else $n > m$, hence $[p(\sigma^\frown m)] \times [q(\tau^\frown n)] \cap X^\alpha_{\sigma,\tau} = \emptyset$.

On the other hand $X = \bigcup_{\sigma,\tau} X^\alpha_{\sigma,\tau}$ is not in $J(\mathbb{L}^2)$. Indeed, for $\langle p, q \rangle \in \mathbb{L}^2$ one can easily construct some $\langle x, y \rangle \in [p] \times [q]$ with $\mathrm{Osc}_{\mathrm{st}p,\mathrm{st}q}(x, y) = \alpha$.

2. We define the partial function $f : (\omega^\omega)^3 \to 2^\omega$ essentially as in the proof of Theorem 1 of [17]: For $x, y, z \in \mathbb{M}$, wherever it is defined let

$$\begin{aligned} l_0 &= \min\left\{i \colon z(i) > \max\{x(i), y(i)\}\right\}, \\ m_0 &= \min\left\{i \colon x(i) > z(l_0)\right\}, \\ n_0 &= \min\left\{i \colon y(i) > z(l_0)\right\}, \\ l_{k+1} &= \min\left\{i \colon z(i) > \max\{x(m_k), y(n_k)\}\right\}, \\ m_{k+1} &= \min\left\{i \colon x(i) > z(l_{k+1})\right\}, \\ n_{k+1} &= \min\left\{i \colon y(i) > z(l_{k+1})\right\}, \end{aligned}$$

and

$$f(x, y, z)(k) = \begin{cases} 0, & x(m_k) \le y(n_k), \\ 1, & \text{otherwise.} \end{cases}$$

Now for $\varrho, \sigma, \tau \in \omega^{<\omega}$ and $\langle x, y, z\rangle \in [\varrho] \times [\sigma] \times [\tau]$ we define

$$f_{\varrho,\sigma,\tau}(x, y, z) = f(x_\varrho, y_\sigma, z_\tau)$$

where $x_\varrho(n) = x(n + |\varrho|)$, $y_\sigma(n) = y(n + |\sigma|)$ and $z_\tau(n) = z(n + |\tau|)$.

We fix $\alpha \in 2^\omega$ and let

$$X^\alpha_{\varrho,\sigma,\tau} = \{\langle x, y, z\rangle \in [\varrho] \times [\sigma] \times [\tau] \colon f_{\varrho,\sigma,\tau}(x, y, z) = \alpha\}.$$

Then on one hand, for $\varrho, \sigma, \tau \in \omega^{<\omega}$, we show $X^\alpha_{\varrho,\sigma,\tau} \in J(\mathbb{M}^3)$. Let $\langle r, s, t\rangle \in \mathbb{M}^3$. Suppose $\langle \varrho, \sigma, \tau\rangle \in \text{split}(r) \times \text{split}(s) \times \text{split}(t)$. Choose $\varrho_0 \in \text{Succ}_r(\varrho)$ and $\sigma_0 \in \text{Succ}_s(\sigma)$ arbitrarily and $\tau_0 \in \text{Succ}_t(\tau)$ such that

$$\tau_0(|\tau|) > \max\{\sharp\varrho_0, \sharp\sigma_0\}.$$

Choose $\varrho_1 \in \text{Succ}_r(\varrho_0)$ and $\sigma_1 \in \text{Succ}_s(\sigma_0)$ such that if $\alpha(0) = 0$ then

$$\varrho_1(|\varrho_0|) > \sigma_1(|\sigma_0|) > \sharp\tau_0.$$

So for any $\langle x, y, z\rangle \in [\varrho_1] \times [\sigma_1] \times [\tau_0]$ we have $f_{\varrho,\sigma,\tau}(x, y, z)(0) = 1 \neq \alpha(0)$ and therefore $[r(\varrho_1)] \times [s(\sigma_1)] \times [t(\tau_0)] \cap X^\alpha_{\varrho,\sigma,\tau} = \emptyset$.

If $\langle \varrho, \sigma, \tau\rangle \in r \times s \times t \setminus \text{split}(r) \times \text{split}(s) \times \text{split}(t)$, analogously we find some longer $\varrho_1, \sigma_1, \tau_0$ and $k < \omega$ such that for any $\langle x, y, z\rangle \in [\varrho_1] \times [\sigma_1] \times [\tau_0]$ we have $f_{\varrho,\sigma,\tau}(k) \neq \alpha(k)$.

On the other hand, with the coding argument in the proof of Theorem 1 of [17], we see that $\bigcup_{\varrho,\sigma,\tau} X^\alpha_{\varrho,\sigma,\tau}$ is not in $J(\mathbb{M}^3)$. ⊣

Remark 7. For any $Q \in \{\mathbb{L}, \mathbb{M}\}$ and $n < \omega$, we have $\mathbf{cov}(J(Q^n)) \geq \omega_1$. Indeed, for a sequence $\langle X_n \colon n < \omega\rangle$ in $J(Q^n)$, let $\langle p_i^0 \colon i < n\rangle = (\omega^{<\omega})^n$, and for $k < \omega$ let $\langle p_i^{k+1} \colon i < n\rangle \in Q^n$ be an extension of $\langle p_i^k \colon i < n\rangle$ with $\text{st}(p_i^{k+1}) \supsetneq \text{st}(p_i^k)$, all i, and $\Pi_{i<n}[p_i^{k+1}] \cap X_k = \emptyset$. Let x_i be the unique elements of $\bigcap_k [p_i^k]$ for $i < n$, hence $\langle x_i \colon i < n\rangle \notin \bigcup_k X_k$.

Theorem 8. *$J(\mathbb{M}^2)$ is a σ-ideal.*

Lemma 9. *For any $\langle p, q\rangle \in \mathbb{M}^2$ and $X \in J(\mathbb{M}^2)$,*

(a) *there exists a $\mathbb{M}^2$-extension $\langle p', q'\rangle \leq^0 \langle p, q\rangle$ such that $[p'] \times^+ [q'] \cap X = \emptyset$,*

(b) *there exists a $\mathbb{M}^2$-extension $\langle p', q'\rangle \leq^0 \langle p, q\rangle$ such that $[p'] \times^- [q'] \cap X = \emptyset$.*

Corollary 10. *For any $\langle p, q\rangle \in \mathbb{M}^2$ and $X \in J(\mathbb{M}^2)$ there is a $\mathbb{M}^2$-extension $\langle p', q'\rangle \leq^0 \langle p, q\rangle$ such that $[p'] \times [q']$ is disjoint from X.*

Proof of Theorem 8 from Corollary 10. Suppose $\{X_n : n < \omega\} \subseteq J(\mathbb{M}^2)$. For any $\langle p, q\rangle \in \mathbb{M}^2$, let $\mu_0 = \operatorname{st} p$ and $\nu_0 = \operatorname{st} q$, let μ_1 and ν_1 be the $\prec$-minimal members of $\operatorname{Succ}_p(\operatorname{st} p)$ and $\operatorname{Succ}_q(\operatorname{st} q)$, respectively. By applying Corollary 10 four times, we get $\langle p^0, q^0\rangle \leq^0 \langle p, q\rangle$ with $\mu_1 \in \operatorname{split}(p^0)$, $\nu_1 \in \operatorname{split}(q^0)$ such that the following are disjoint from X_0:

$$[p^0(\mu_1)] \times [q^0(\nu_1)], \qquad [p^0 \setminus p^0(\mu_1)^-] \times [q^0(\nu_1)],$$
$$[p^0(\mu_1)] \times [q^0 \setminus q^0(\nu_1)^-], \qquad [p^0 \setminus p^0(\mu_1)^-] \times [q^0 \setminus q^0(\nu_1)^-],$$

and therefore $[p^0] \times [q^0] \cap X_0 = \emptyset$. In this way one can easily construct a fusion sequence $\langle \langle p^n, q^n\rangle : n < \omega\rangle$ in $\mathbb{M}^2$ together with $\langle \mu_n : n < \omega\rangle$ and $\langle \nu_n : n < \omega\rangle$ with the following properties:

1. $\langle \mu_n : n < \omega\rangle$ and $\langle \nu_n : n < \omega\rangle$ are increasing enumerations of the splitnodes of two superperfect trees,
2. $\langle p^{n+1}, q^{n+1}\rangle \leq^0 \langle p^n, q^n\rangle \leq^0 \langle p, q\rangle$,
3. $\langle \mu_i, \nu_i\rangle \in p^n \times q^n$, for all $i \leq n$,
4. $[p^n] \times [q^n] \cap X_n = \emptyset$.

Note that at each step we have to extend our trees only finitely many times.

Finally let p' and q' be the trees determined by their sets of splitnodes $\{\mu_n : n < \omega\}$ and $\{\nu_n : n < \omega\}$, respectively. Hence $\langle p', q'\rangle \leq^0 \langle p, q\rangle$ and $[p'] \times [q'] \cap \bigcup_{n<\omega} X_n = \emptyset$ and therefore $\bigcup_{n<\omega} X_n \in J(\mathbb{M}^2)$. ⊣

Proof of Lemma 9. Let $\sigma = \operatorname{st} p$ and $\tau = \operatorname{st} q$. Recall that $\langle p, q\rangle$ is such that all $\langle x, y\rangle \in [p] \times^+ [q]$ are infinitely oscillating above $\langle \sigma, \tau\rangle$.

Suppose that there does not exist any $\langle p', q'\rangle \leq^0 \langle p, q\rangle$ such that $[p] \times^+ [q] \cap X = \emptyset$. We shall construct some $\langle p_1, q_1\rangle \leq^0 \langle p, q\rangle$ such that for every $\mu \in \operatorname{split}(p_1)$ there does not exist $\langle p', q'\rangle \leq^0 \langle p_1(\mu), q_1\rangle$ with $[p'] \times^- [q'] \cap X = \emptyset$. Note that this is equivalent to saying that for every $\langle \mu, \nu\rangle \in \mathrm{TP}^1_{\sigma,\tau}(p_1, q_1)$ for no $\langle p', q'\rangle \leq^0 \langle p_1(\mu), q_1(\nu)\rangle$ we have $[p'] \times^- [q'] \cap X = \emptyset$.

We recursively construct sequences $\langle \sigma_n : n < \omega\rangle$, $\langle \xi_n : n < \omega\rangle$ in $\omega^{<\omega}$ and $\langle u_n : n < \omega\rangle$, $\langle v_n : n < \omega\rangle$ in $\mathbb{M}$, such that the following holds for all n:

1. $\langle \sigma_n : n < \omega\rangle$ is the increasing enumeration of the splitnodes of some superperfect tree,
2. $\sharp\sigma_n > \xi_{n-1}(|\tau|)$ and if $k < n$ is maximal such that $\sigma_k \subsetneq \sigma_n$, then $\sigma_n \in \operatorname{split}(u_k)$ and $u_n \leq^0 u_k(\sigma_n)$ (hence $\operatorname{st}(u_n) = \sigma_n$),
3. $v_{n+1} \leq^0 v_n \leq^0 q$,
4. ξ_n is the leftmost element of $\operatorname{Succ}_{v_n}(\tau)$ and $\xi_n(|\tau|) < \xi_{n+1}(|\tau|)$,
5. if $k < n$ is maximal such that $\sigma_k \subseteq \sigma_n$ and there exists $\langle p', q'\rangle \leq^0 \langle u_k(\sigma_n), v_{n-1}\rangle$ such that $[p'] \times^- [q'] \cap X = \emptyset$, then $\langle u_n, v_n\rangle$ equals such $\langle p', q'\rangle$.

Let $\sigma_0 = \sigma$, $u_0 = p$, $v_0 = q$ and let ξ_0 be the $\prec$-minimal element of $\operatorname{Succ}_\tau(q)$. At stage $n \geq 1$ suppose σ_k needs an extension. Choose $\sigma_n \in \operatorname{split}(u_k^-)$ such that $\sharp\sigma_n > \xi_{n-1}(|\tau|)$. Let v' be the downward closure of all $\xi \in \operatorname{split}(v_{n-1})$

with $\xi(|\tau|) > \xi_{n-1}(|\tau|)$ and find $\langle u_n, v_n\rangle \leq^0 \langle u_k(\sigma_n), v'\rangle$ with possibly $[u_n] \times^- [v_n] \cap X = \emptyset$. Let ξ_n be the leftmost member of $\mathrm{Succ}_{v_n}(\tau)$.

Finally let $u \in \mathbb{M}$ be determined by $\mathrm{split}(u) = \{\sigma_n \colon n < \omega\}$ and let $v = \bigcup_{n<\omega} v_n(\xi_n) \in \mathbb{M}$.

Now for all $n < \omega$ we color σ_n by "yes" if at step n we could find $\langle u_n, v_n\rangle$ such that $[u_n] \times^- [v_n] \cap X = \emptyset$, otherwise color σ_n by "no". By Lemma 1 there exists $u' \leq^0 u$ such that every splitnode of $(u')^-$ has the same color. Choose $\langle p_1, q_1\rangle \leq^0 \langle u', v\rangle$ with all branches in the upper half infinitely oscillating.

In order to get a contradiction suppose that every splitnode of p_1 has color "yes". Fix $\langle x, y\rangle \in [p_1] \times^+ [q_1]$. Then we get some $n < \omega$ such that $\mathrm{tp}_{\sigma,\tau}\text{-1-pair}(x, y) = \langle \sigma_n, \tau\rangle$. Thus $y(|\tau|) > \sharp\sigma_n$, which, by 2, is $> \xi_{n-1}(|\tau|)$. Hence $\langle x, y\rangle \in [u_n] \times^- [v_n]$, and therefore we have $\langle x, y\rangle \notin X$, since σ_n is colored "yes". As $\langle x, y\rangle$ was arbitrary, we have shown that $[p_1] \times^+ [q_1] \cap X = \emptyset$. So since $\langle p_1, q_1\rangle \leq^0 \langle p, q\rangle$ we get a contradiction to our assumption.

So we have proved part (a) of the following:

CLAIM 10.1. *Suppose $X \in J(\mathbb{M}^2)$ and $\langle p, q\rangle \in \mathbb{M}^2$ and let $\sigma = \mathrm{st}p$, $\tau = \mathrm{st}q$.*

(a) *Suppose that for no $\langle p', q'\rangle \leq^0 \langle p, q\rangle$ do we have $[p'] \times^+ [q'] \cap X = \emptyset$. Then there exists $\langle p_1, q_1\rangle \leq^0 \langle p, q\rangle$ such that for every $\langle \mu, \tau\rangle \in \mathrm{TP}^1_{\sigma,\tau}(p_1, q_1)$ there is no $\langle p', q'\rangle \leq^0 \langle p_1(\mu), q_1\rangle$ with $[p'] \times^- [q'] \cap X = \emptyset$.*

(b) *Suppose that for no $\langle p', q'\rangle \leq^0 \langle p, q\rangle$ do we have $[p'] \times^- [q'] \cap X = \emptyset$. Then there exists $\langle p_1, q_1\rangle \leq^0 \langle p, q\rangle$ such that for every $\langle \mu, \tau\rangle \in \mathrm{TP}^1_{\sigma,\tau}(p_1, q_1)$ there is no $\langle p', q'\rangle \leq^0 \langle p_1(\mu), q_1\rangle$ with $[p'] \times^+ [q'] \cap X = \emptyset$.*

The proof for part (b) is the same as for part (a) with "+" and "−" exchanged.

Now by recursion we build a fusion sequence $\langle\langle p_n, q_n\rangle \colon n < \omega\rangle$ such that for every odd $n < \omega$ we get

$(-)_n$ for every $\langle \mu, \nu\rangle \in \mathrm{TP}^n_{\sigma,\tau}(p_n, q_n)$, for no $\langle p', q'\rangle \leq^0 \langle p_n(\mu), q_n(\nu)\rangle$ we have $[p'] \times^- [q'] \cap X = \emptyset$,

and for every even n we get

$(+)_n$ for every $\langle \mu, \nu\rangle \in \mathrm{TP}^n_{\sigma,\tau}(p_n, q_n)$, for no $\langle p', q'\rangle \leq^0 \langle p_n(\mu), q_n(\nu)\rangle$ we have $[p'] \times^+ [q'] \cap X = \emptyset$.

Let $\langle p', q'\rangle$ be the fusion, i.e., $p' = \bigcap_{n<\omega} p_n$ and $q' = \bigcap_{n<\omega} q_n$, so $\langle p', q'\rangle \leq^0 \langle p, q\rangle$. Now choose $i \in \mathrm{succ}_{p'}(\sigma)$, $j \in \mathrm{succ}_{q'}(\tau)$ with $i < j$. Find $\langle u, v\rangle \leq \langle p'(\sigma^\frown i), q'(\tau^\frown j)\rangle$ with $[u] \times [v] \cap X = \emptyset$. Then $\langle \mathrm{st}u, \mathrm{st}v\rangle$ has some $\mathrm{type}_{\sigma,\tau}$, say n. As $\langle u, v\rangle \leq^0 \langle p_n(\mathrm{st}u), q_n(\mathrm{st}v)\rangle$ we get a contradiction to $(*)_n$, where $* \in \{+, -\}$ depending on the parity of n.

The construction of the fusion sequence is a recursion of length ω^ω, in the way it was explicitly done in [14]. The reason why we get length ω^ω for the recursion is that we have to go through all type-pairs and for each n, the type-n-pairs are canonically wellordered in type ω^n.

We show how to get $\langle p_2, q_2\rangle$ with $(+)_2$. Suppose $\langle \mu, \tau\rangle \in \mathrm{TP}^1_{\sigma,\tau}(p_1, q_1)$. So by $(-)_1$, for no $\langle p', q'\rangle \leq^0 \langle p_1(\mu), q_1\rangle$ do we have $[p'] \times^- [q'] \cap X = \emptyset$ and equivalently $[q'] \times^+ [p'] \cap X^{-1} = \emptyset$. By Claim 10.1 we obtain $\langle q^\mu, p^\mu\rangle \leq^0 \langle q_1, p_1(\mu)\rangle$ such that for all $\langle \nu, \mu\rangle \in \mathrm{TP}^1_{\tau,\mu}(q^\mu, p^\mu)$ we do not have $\langle q', p'\rangle \leq^0 \langle q^\mu(\nu), p^\mu\rangle$ with $[q'] \times^- [p'] \cap X^{-1} = \emptyset$. So in particular we have that for all pairs $\langle \mu', \nu'\rangle \in \mathrm{TP}^2_{\sigma,\tau}(p^\mu, q^\mu)$ with $\mu' = \mu$, for no $\langle p', q'\rangle \leq^0 \langle p^\mu(\mu), q^\mu(\nu')\rangle$, $[p'] \times^+ [q'] \cap X = \emptyset$. So with a fusion over all $\mu \in \mathrm{split}(p_1)$ we get $\langle p_2, q_2\rangle$ such that $(+)_2$ holds. ⊣

§3. **Bounds for add**$(J(\mathbb{M}^2))$. We show that under the assumption $\mathfrak{d} = \mathfrak{c}$, **add**$(J(\mathbb{M}^2))$ is less or equal to $\mathfrak{h}$, the distributivity number of $\mathcal{P}(\omega)/\mathrm{fin}$.

LEMMA 11.

(1) **add**$(J(\mathbb{M}^2)) \leq$ **add**$(J(\mathbb{M}))$,
(2) **cov**$(J(\mathbb{M}^2)) \leq$ **cov**$(J(\mathbb{M}))$.

PROOF. (1) Suppose $\langle X_\alpha : \alpha < \kappa\rangle \subseteq J(\mathbb{M})$ with $\kappa <$ **add**$(J(\mathbb{M}^2))$. We have to show that $\bigcup_{\alpha<\kappa} X_\alpha \in J(\mathbb{M})$. Let $X'_\alpha = X_\alpha \times \omega^\omega$. Clearly $X'_\alpha \in J(\mathbb{M}^2)$. Hence $\bigcup_{\alpha<\kappa} X'_\alpha \in J(\mathbb{M}^2)$. Let $p \in \mathbb{M}$. Find $q_0, q_1 \leq p$ such that $[q_0] \times [q_1] \cap \bigcup_{\alpha<\kappa} X'_\alpha = \emptyset$. As $\bigcup_{\alpha<\kappa} X'_\alpha = (\bigcup_{\alpha<\kappa} X_\alpha) \times \omega^\omega$, we have $[q_0] \cap \bigcup_{\alpha<\kappa} X_\alpha = \emptyset$.
The proof of (2) is similar. ⊣

REMARK 12. (a) The consistency of **add**$(J(\mathbb{M}^2)) <$ **add**$(J(\mathbb{M}))$ is open. A natural conjecture is that a model for this is obtained by iteratively forcing with amoeba forcing associated with $\mathbb{M}$ (see [12]).

(b) A model for **cov**$(J(\mathbb{M}^2)) <$ **cov**$(J(\mathbb{M}))$ is $V^{\mathbb{M}_{\omega_2}}$, where $\mathbb{M}_{\omega_2}$ is a countable support iteration of $\mathbb{M}$ of length ω_2. This follows from the ZFC version of Lemma 17 below which says that **add**$(J(\mathbb{M}^2)) \leq \mathfrak{b}$, together with the result from [10] that that $V^{\mathbb{M}_{\omega_2}} \models \mathfrak{b} = \omega_1$.

COROLLARY 13. $\mathfrak{d} = \mathfrak{c}$. **add**$(J(\mathbb{M}^2)) \leq \mathfrak{h}$.

PROOF. Lemma 11 shows **add**$(J(\mathbb{M}^2)) \leq$ **add**$(J(\mathbb{M}))$. By results from [4] and [12] and the fact that $\mathfrak{d} = \mathfrak{c}$, the latter is less or equal to the collapsing number $\kappa(\mathbb{M})$, which is defined as the least cardinal to which forcing with $\mathbb{M}^2$ collapses the continuum. That $\kappa(\mathbb{M})$ is at most $\mathfrak{h}$ was shown in [4]. ⊣

§4. **The consistency of add**$(J(\mathbb{M}^2)) <$ **cov**$(J(\mathbb{M}^2))$.

THEOREM 14. *Suppose that P_{ω_2} is a countable support iteration of length ω_2 of $\mathbb{M}^2$ and $V \models$ ZFC+GCH. Then $V^{P_{\omega_2}} \models$* **cov**$(J(\mathbb{M}^2)) = \omega_2$.

PROOF. In $V^{P_{\omega_2}}$ let $\langle X_\alpha : \alpha < \omega_1\rangle \in V^{P_{\omega_2}}$ be a sequence of sets in $J(\mathbb{M}^2)$. For every $\alpha < \omega_2$, there exists a function $f_\alpha : \mathbb{M}^2 \longrightarrow \mathbb{M}^2$ such that $f_\alpha(p) \leq p$ and $[f_\alpha(p)^0] \times [f_\alpha(p)^1] \cap X_\alpha = \emptyset$, for all $p \in \mathbb{M}^2$. In V we have a P_{ω_2}-name $\dot{f}_\alpha$ for f_α, so $\dot{f}_\alpha$ is a set of pairs $\langle \dot{p}, \dot{q}\rangle$ such that $\Vdash_{P_{\omega_2}} \dot{p}, \dot{q} \in \mathbb{M}^2$. Because P_{ω_2}

is proper we can assume without loss of generality that $\dot{p}$ and $\dot{q}$ are hereditarily countable, for all $\langle \dot{p}, \dot{q} \rangle \in \dot{f}_\alpha$. Let

$$C_\alpha = \left\{\beta < \omega_2 \colon \text{if } \langle \dot{p}, \dot{q} \rangle \in \dot{f}_\alpha \text{ and } \dot{p} \in V^{P_\beta}, \text{ then } \dot{q} \in V^{P_\beta}\right\}.$$

Then C_α is ω_1-club, i.e., unbounded in ω_2 and closed under increasing ω_1-sequences. Hence, letting $C = \bigcap_{\alpha<\omega_1} C_\alpha$, C is ω_1-club as well. Fix $\beta \in C$ and let $\dot{f}^\beta_\alpha = \{\langle \dot{p}, \dot{q} \rangle \in \dot{f}_\alpha \colon \dot{p} \in V^{P_\beta}\}$. Hence $f_\alpha \restriction \mathbb{M}^{2(V^{P_\beta})} = f^\beta_\alpha \in V^{P_\beta}$, for all $\alpha < \omega_1$. Let $\langle \dot{g}^\beta_0, \dot{g}^\beta_1 \rangle$ be the P_β-name for the pair of generic reals determined by $\dot{Q}_\beta$. We claim

$$\Vdash_{P_{\omega_2}} \left\langle \dot{g}^\beta_0, \dot{g}^\beta_1 \right\rangle \notin \bigcup_{\alpha<\omega_1} X_\alpha.$$

Suppose we have $p \in P_{\omega_2}$ and $\alpha < \omega_1$ such that $p \Vdash_{P_{\omega_2}} \langle \dot{g}^\beta_0, \dot{g}^\beta_1 \rangle \in X_\alpha$. Let G be a P_{ω_2}-generic filter containing p. We work in $V[G_\beta]$, where $p \restriction [\beta, \omega_2) \Vdash_{P_{\omega_2}/G_\beta} \langle \dot{g}^\beta_0, \dot{g}^\beta_1 \rangle \in X_\alpha$. Let $q = f^\beta_\alpha(p(\beta))$ and let $p' = q^\frown p \restriction (\beta, \omega_2)$, so $p' \leq p \restriction [\beta, \omega_2)$ and since $p' \Vdash \langle \dot{g}^\beta_0, \dot{g}^\beta_1 \rangle \in [q^0] \times [q^1]$, we have $p' \Vdash \langle \dot{g}^\beta_0, \dot{g}^\beta_1 \rangle \notin X_\alpha$. Hence we get a contradiction. ⊣

THEOREM 15. *Suppose that P_{ω_2} and V are as in Theorem 14. Then $V^{P_{\omega_2}} \models$ $\mathbf{add}(J(\mathbb{M}^2)) = \omega_1$.*

The proof consists of a series of lemmas after the following result.

THEOREM 16. *Forcing with $\mathbb{M}^2$ adds a nonsplit real, i.e., in $V^{\mathbb{M}^2}$ there exists a set $a \in [\omega]^\omega$ such that for all $b \in [\omega]^\omega \cap V$ either $a \subseteq^* b$ or $|a \cap b| < \omega$. Hence $V^{P_{\omega_2}} \models \mathfrak{s} = \omega_2$.*

PROOF. Let G be an $\mathbb{M}^2$-generic filter over V and $\langle g_0, g_1 \rangle$ the generic pair of reals determined by G. By genericity there exists $\langle p, q \rangle \in G$ and $\sigma \in p, \tau \in q$ such that every $\langle x, y \rangle \in \langle p, q \rangle$ oscillates infinitely often above $\langle \sigma, \tau \rangle$, hence the same holds for $\langle g_0, g_1 \rangle$. Let

$$a = \langle g_0 \restriction k_{2n+1} \colon n < \omega \rangle,$$

the sequence of the left oscillation points of $\langle g_0, g_1 \rangle$ above $\langle \sigma, \tau \rangle$. Fix $b \in [\omega^{<\omega}]^\omega$. By Lemma 1 the set of $\langle p', q' \rangle$ such that either $\text{split}(p') \subseteq b$ or $\text{split}(p') \cap b = \emptyset$ is dense below $\langle p, q \rangle$. Hence such $\langle p', q' \rangle$ belongs to G. Since $\{g_0 \restriction k_{2n+1} \colon n < \omega\} \subseteq^* \text{split}(p')$, we get that a is as desired.

Now assume $X \subseteq [\omega]^\omega$ is a splitting family in $V^{P_{\omega_2}}$ with $|X| = \omega_1$. For some $\alpha < \omega_2$ we have that $X \in V^{P_\alpha}$. So $\dot{Q}_\alpha$ adds some real that is not split by any $b \in X$. ⊣

LEMMA 17. *Forcing with $\mathbb{M}^2$ adds a dominating real, i.e., some function f such that for any $g \in \omega^\omega \cap V$ we have $f \geq^* g$. Hence $V^{P_{\omega_2}} \models \mathfrak{b} = \mathfrak{c}$.*

PROOF. Let $\langle g_0, g_1 \rangle$ be an $\mathbb{M}^2$-generic pair of reals. As in the proof of Theorem 16, find $\langle p, q \rangle$ and $\sigma \in p$, $\tau \in q$ such that $\langle g_0, g_1 \rangle \in [p(\sigma)] \times [q(\tau)]$

and every $\langle x, y\rangle \in [p(\sigma)] \times [q(\tau)]$ oscillates infinitely often above σ, τ. Let $\langle k_n : n < \omega\rangle$ be the type-sequence of $\langle g_0, g_1\rangle$. Define $f = \langle k_{2n+1} : n < \omega\rangle$. Note that for every $g \in \omega^\omega \cap V$, the set of all $\langle p', q'\rangle \leq \langle p, q\rangle$ such that

$$\forall n \forall \varrho \in \mathrm{Lev}_n(p') |\varrho| > g(n)$$

is dense below $\langle p, q\rangle$. It is easy to see that by genericity this implies $g \leq^* f$. ⊣

REMARK 18. Theorem 16 and Lemma 17 easily translate to the ZFC-inequalities $\mathbf{cov}(J(\mathbb{M}^2)) \leq \mathfrak{b}$ and $\mathbf{cov}(J(\mathbb{M}^2)) \leq \mathfrak{s}$.

Combining the previous result with the results in Section 3, we get

$$V^{P_{\omega_2}} \models \mathbf{add}(J(\mathbb{M}^2)) \leq \mathfrak{h}.$$

In order to show that the additivity number stays small in the iterated forcing extension by $\mathbb{M}^2$, we want to show that $\mathfrak{h} = \omega_1$ holds in that model. In the following lemma we establish the main construction for the proof.

LEMMA 19. *Suppose $\dot{a}$ is a P_{ω_2}-name and $\widetilde{p} \in P_{\omega_2}$ with $\widetilde{p} \Vdash \dot{a} \in [\omega]^\omega$. Then for every finite $\widetilde{F} \subset \omega_2 \setminus \{0\}$, there exist $r \in P_{\omega_2}$ with*

$$r \leq^0_{\{0\} \cup \widetilde{F}} \widetilde{p}$$

and refining finite partitions

$$\dot{\Gamma}_n = \{\dot{r}^n_i : i < n+1\}, \quad n < \omega$$

of the Boolean completion of P_{ω_2} below $r \restriction [1, \omega_2)$, and, letting $\langle u, v\rangle = r(0)$, $\sigma = \mathrm{st}u$, $\tau = \mathrm{st}v$, there exists a set

$$\left\{w^{\mu,\nu}_i, y^{\mu,\nu}_{i'} \in \mathcal{P}(\omega) : \exists n < \omega \left(\langle \mu, \nu\rangle \in \mathrm{TP}^n_{\sigma,\tau}(u, v) \wedge i < n+2 \wedge i' < n+1\right)\right\},$$

and a countable set $\mathcal{D}$ of disjoint families in $[\omega]^\omega$ such that for any $\langle \mu, \nu\rangle \in \mathrm{TP}^n(u, v)$, $n < \omega$, and any $i < n+2$ and $i' < n+1$, letting $J = \mathrm{succ}_u(\mu) \setminus \sharp\nu$ and $J' = \mathrm{succ}_v(\nu) \setminus \sharp\mu$, the following properties hold:

1. *Suppose n is even. For any sequence $\langle G^j : j \in J\rangle$ of P_{ω_2}-generic filters over V such that $\langle u(\mu^\frown j), v\rangle ^\frown \dot{r}^{n+1}_i \in G^j$ and $\mathrm{tp}_{\sigma,\tau}$-$n$-pair$(g^j_0, g^j_1) = \langle \mu, \nu\rangle$ where $G^j_1 = \langle g^j_0, g^j_1\rangle$ and $\langle \mu^j, \nu\rangle$ is the type-$(n+1)$-pair of $\langle g^j_0, g^j_1\rangle$, we have*

$$w^{\mu,\nu}_i = \lim_{j \in J} \dot{a}\left[G^j\right] \cap \left|\mu^j\right|.$$

Suppose n is odd, for any sequence $\langle G^j : j \in J'\rangle$ of P_{ω_2}-generic filters over V such that $\langle u, v(\nu^\frown j)\rangle ^\frown \dot{r}^{n+1}_i \in G^j$ and $\mathrm{tp}_{\sigma,\tau}$-$n$-pair$(g^j_0, g^j_1) = \langle \mu, \nu\rangle$ where $G^j_1 = \langle g^j_0, g^j_1\rangle$ and $\langle \mu, \nu^j\rangle = \mathrm{tp}_{\sigma,\tau}$-$(n+1)$-pair$(g^j_0, g^j_1)$,

$$w^{\mu,\nu}_i = \lim_{j \in J'} \dot{a}\left[G^j\right] \cap \left|\nu^j\right|.$$

2. *Suppose $n > 0$ and even. For any sequence $\langle G^j : j \in J' \rangle$ of P_{ω_2}-generic filters over V such that $\langle u, v(v^\frown j) \rangle ^\frown \dot{r}^n_{i'} \in G^j$ and $\mathrm{tp}_{\sigma,\tau}$-$n$-$\mathrm{pair}(g^j_0, g^j_1) = \langle \mu, v^j \rangle$ with $G^j_1 = \langle g^j_0, g^j_1 \rangle$ and some $v^j \supsetneq v$ (hence $v^\frown j \subseteq v^j$), we have*

$$y^{\mu,v}_{i'} = \lim_{j \in J'} \dot{a}\left[G^j\right] \cap j.$$

If n is odd, for any sequence $\langle G^j : j \in J \rangle$ of P_{ω_2}-generic filters over V such that $\langle u(\mu^\frown j), v \rangle ^\frown \dot{r}^n_{i'} \in G^j$ and $\mathrm{tp}_{\sigma,\tau}$-$n$-$\mathrm{pair}(g^j_0, g^j_1) = \langle \mu^j, v \rangle$ where $G^j_1 = \langle g^j_0, g^j_1 \rangle$ and some $\mu^j \supsetneq \mu$ (hence $\mu^\frown j \subseteq \mu^j$), we have

$$y^{\mu,v}_{i'} = \lim_{j \in J} \dot{a}\left[G^j\right] \cap j.$$

3. *Suppose $w^{\mu,v}_i$ is finite. If n is even, then all $y^{\mu',v}_i$ with $\mu' \in \mathrm{Sop}^{u,v}_{\sigma,\tau}(\mu, v)$ are finite or they are all infinite. In the finite case we have either one of the following cases:*
   (a) (empty case) *For all μ' as above, $y^{\mu',v}_i = w^{\mu,v}_i$. Let $\{\dot{r}^{n+2}_{i_0}, \dot{r}^{n+2}_{i_1}\}$ consist of those elements of $\dot{\Gamma}_{n+2}$ which are below $\dot{r}^{n+1}_i$, so possibly $i_0 = i_1$. Then for $j < 2$, we have either*
      (i) *for all $\mu' \in \mathrm{Sop}(\mu, v)$, $w^{\mu',v}_{i_j} \setminus w^{\mu,v}_i \neq \emptyset$, or*
      (ii) *for all such μ', $w^{\mu',v}_{i_j} \setminus w^{\mu,v}_i = \emptyset$.*
   Moreover, for all $k \in \mathrm{succ}_u(\mu)$, letting

$$d^i_k(\mu, v) = \bigcup \left\{ w^{\mu',v}_{i_j} j < 2 \wedge \backslash w^{\mu,v}_i : \mu' \in u(\mu^\frown k) \cap \mathrm{Sop}(\mu, v) \right\},$$

   we have $\{d^i_k(\mu, v) : k \in \mathrm{succ}_u(\mu)\} \in \mathcal{D}$.
   (b) (constant case) *For every $\mu', \mu'' \in \mathrm{Sop}(\mu, v)$ with $\mu' \subseteq \mu''$, we have $y^{\mu',v}_i \setminus w^{\mu,v}_i = y^{\mu'',v}_i \setminus w^{\mu,v}_i \neq \emptyset$. Let*

$$d^i(\mu, v) = \bigcup \left\{ y^{\mu',v}_i \setminus w^{\mu,v}_i : \mu' \in \mathrm{Sop}(\mu, v) \right\}.$$

   (c) (tree case) *For every $\mu', \mu'' \in \mathrm{Sop}(\mu, v)$, $y^{\mu',v}_i \setminus w^{\mu,v}_i \neq \emptyset$ and if $\mu' \subsetneq \mu''$, $y^{\mu'',v}_i \setminus y^{\mu',v}_i \neq \emptyset$. Then, for all $k \in \mathrm{succ}_u(\mu)$, letting*

$$d^i_k(\mu, v) = \bigcup \left\{ y^{\mu',v}_i \setminus w^{\mu,v}_i : \mu' \in u(\mu^\frown k) \cap \mathrm{Sop}(\mu, v) \right\},$$

   we have $\{d^i_k(\mu, v) : k \in \mathrm{succ}_u(\mu)\} \in \mathcal{D}$.

   For odd n, we have the symmetric properties and $d^i(\mu, v)$ and disjoint families $\{d^i_k(\mu, v) : k \in \mathrm{succ}_v(v)\} \in \mathcal{D}$.

Moreover, for any $\mu \in \operatorname{split}(u)$ *and* $\nu \in \operatorname{split}(v)$, *letting*

$$d_\mu = \bigcup \Big\{ d^i\,(\mu, \hat{\nu}) : \hat{\nu} \in \operatorname{split}(v) \wedge \operatorname{tp}(\mu, \hat{\nu}) = n \text{ is even} \\ \wedge \langle \mu, \hat{\nu} \rangle \text{ of constant case for } i < n+2 \Big\},$$

$$d'_\nu = \bigcup \Big\{ d^i\,(\hat{\mu}, \nu) : \hat{\mu} \in \operatorname{split}(u) \wedge \operatorname{tp}(\hat{\mu}, \nu) = n \text{ is odd} \\ \wedge \langle \hat{\mu}, \nu \rangle \text{ of constant case for } i < n+2 \Big\},$$

then we have

$$\{d_\mu, d'_\nu \colon \mu \in \operatorname{split}(u) \wedge \nu \in \operatorname{split}(v)\} \in \mathcal{D}.$$

Proof. Fix $G_1 = \langle g_0, g_1 \rangle$, a Q_0-generic filter over V containing $\widetilde{p}(0)$. Let $\langle p, q \rangle = \widetilde{p}(0)$, $\sigma = \operatorname{st} p$ and $\tau = \operatorname{st} q$; for the rest of the proof, the types will always refer to σ, τ. Let $\langle k_n \colon n \in \omega \rangle$ enumerate the lengths of the oscillation points of $\langle g_0, g_1 \rangle$, i.e., for all $n \in \omega$, if $\langle \mu, \nu \rangle$ is the type-n-pair of $\langle g_0, g_1 \rangle$, then let $k_n = \max\{|\mu|, |\nu|\}$.

In $V[G_1]$ we consider $\langle \dot{a}[G_1] \cap k_i \colon i < \omega \rangle$, which is a P_{ω_2}/G_1-name for a member in $\Pi_{i<\omega} 2^{k_i}$. (We identify $\mathcal{P}(k_i)$ with 2^{k_i}). By [1, Theorem 5.2.], P_{ω_2}/G_1 is isomorphic to $(P_{\omega_2})^{V[G_1]}$. It is straightforward to apply Lemma 3 to build a fusion sequence $\langle p_j \colon j < \omega \rangle$ with $p_j \leq^0_{\widetilde{F}} \widetilde{p} \restriction [1, \omega_2)$ such that

1. $p_0 = \widetilde{p} \restriction [1, \omega_2)$,
2. for every n there exists a partition

   $$\Gamma_n = \{r^n_i \colon i < n+1\}$$

   of the Boolean completion of P_{ω_2}/G_1 below p_{n+1} and

   $$B_n = \{b^n_i \colon i < n+1\} \subseteq 2^{k_n}$$

   such that

   $$\forall i < n+1 \quad r^n_i \Vdash_{P_{\omega_2}/G_1} \dot{a} \restriction k_n = b^n_i,$$

   and if $n > 0$, then Γ_n refines Γ_{n-1}.

Now let $\bar{q}$ be the infimum of $\langle p_n \colon n < \omega \rangle$. As G_1 was arbitrary, in V there exist Q_0-names $\langle \dot{k}_n \colon n < \omega \rangle$, $\dot{\Gamma}_n = \{\dot{r}^n_i \colon i < n+1\}$, $\dot{B}_n = \{\dot{b}^n_i \colon i < n+1\}$, $\langle \dot{p}_n \colon n < \omega \rangle$ and $\dot{q}$ such that $\widetilde{p}(0)$ Q_0-forces that $\langle \dot{k}_n \colon n < \omega \rangle$ are the lengths of the oscillation points of $\langle \dot{g}^0_0, \dot{g}^0_1 \rangle$, $\langle \dot{p}_j \colon j < \omega \rangle$ is a fusion sequence in $P_{\omega_2}/\dot{G}_1$ with $\dot{p}_j \leq^0_{\widetilde{F}} \widetilde{p} \restriction [1, \omega_2)$ and infimum $\dot{q}$ and $\dot{\Gamma}_n$ is a partition of the Boolean completion of $P_{\omega_2}/\dot{G}_1$ below $\dot{p}_{n+1}$, such that $\dot{\Gamma}_{n+1}$ refines $\dot{\Gamma}_n$, and $\dot{B}_n = \{\dot{b}^n_i \colon i < n+1\} \subseteq 2^{\dot{k}_n}$ such that

$$\forall i < n+1 \quad \dot{r}^n_i \Vdash_{P_{\omega_2}/\dot{G}_1} \dot{a} \restriction \dot{k}_n = \dot{b}^n_i,$$

for all $n < \omega$, so $\widetilde{p}(0)$ forces that for all n, $\dot{\Gamma}_n$ is a partition below $\dot{q}$ and actually $\dot{\Gamma}_{n+1}$ refines $\dot{\Gamma}_n$.

Since $\mathbb{M}^2$ has the weak decision property, for every $\langle \mu, \nu \rangle \in \mathrm{TP}^n(p,q)$, $n < \omega$, there exists an extension $\langle u', v' \rangle \leq^0 \langle p(\mu), q(\nu) \rangle$ such that every generic $\langle g_0, g_1 \rangle$ thru $\langle u', v' \rangle$ with $\langle \mu, \nu \rangle$ as type pair gives $\dot{B}_n$ the same value, say

$$\langle b_i^n(\mu, \nu) \colon i < n+1 \rangle.$$

Also if $n > 0$ and even, then there exists $\langle u'', v'' \rangle \leq \langle u', v' \rangle$ such that every generic $\langle g_0, g_1 \rangle$ thru $\langle u'', v'' \rangle$ which has $\langle \mu, \nu' \rangle$ as type-n-pair for some $\nu' \supseteq \nu^\frown j$, gives $\dot{B}_n \restriction j$ the same value, say

$$\langle s_i^{n,j}(\mu, \nu) \colon i < n+1 \rangle.$$

To show the latter, again we use the weak decision property to decide $\dot{B}_n \restriction j$ above every $\langle \mu, \nu' \rangle$ with $\nu' \supseteq \nu^\frown j$ together with Lemma 1 applied on the trees $v'(\nu^\frown j)$. Symmetrically we get $\langle s_i^{n,j}(\mu, \nu), i < n+1 \rangle$ in case n is odd. So by the Fusion Property of $\mathbb{M}^2$, we can assume without loss of generality that for every $\langle \mu, \nu \rangle \in \mathrm{TP}^n(p,q)$, $\langle p(\mu), q(\nu) \rangle$ decides $\dot{B}_n$ and $\dot{B}_n \restriction j$ with $j \in \mathrm{succ}_q(\nu)$ for even n and $j \in \mathrm{succ}_p(\mu)$ for odd n, as explained above.

By Remark 5(a) and the Fusion Property, without loss of generality we may assume that for every $\langle \mu, \nu \rangle \in \mathrm{TP}^n(p,q)$, $i < n+2$ and $i' < n+1$, we have $w_i^{\mu,\nu}$ and $y_{i'}^{\mu,\nu}$, such that if n is even, then

1. for every $\langle \mu^j \colon j \in \mathrm{succ}_p(\mu) \rangle$ with $\mathrm{tp}(\mu^j, \nu) = n+1$ and $\mu^j \supseteq \mu^\frown j$, $w_i^{\mu,\nu} = \lim_{j \in \mathrm{succ}_p(\mu)} b_i^{n+1}(\mu^j, \nu)$,
2. $y_0^{\sigma,\tau} = \emptyset$ and if $n > 0$ then $y_{i'}^{\mu,\nu} = \lim_{j \in \mathrm{succ}_q(\nu)} s_{i'}^{n,j}(\mu, \nu)$,

and symmetrically for odd n. So 1 and 2 of Lemma 19 hold. Note that if $\langle \mu, \nu \rangle \in \mathrm{TP}^n(p,q)$, $i < n+2$ and $i' < \mathrm{tp}(\mu, \nu) + 3$ such that $\dot{r}_i^n$ is compatible with $\dot{r}_{i'}^{n+1}$, and $\mu', \mu'', \mu''' \in \mathrm{Sop}(\mu, \nu)$ with $\mu' \subseteq \mu'' \subsetneq \mu'''$, then

$$w_i^{\mu,\nu} \lhd w_{i'}^{\mu',\nu} \quad \text{and} \quad w_i^{\mu,\nu} \lhd y_i^{\mu',\nu} \lhd y_i^{\mu'',\nu} \lhd w_{i'}^{\mu''',\nu}.$$

Now we fix $\langle \mu, \nu \rangle \in \mathrm{TP}^n(p,q)$ with, say, even type n, also fix $i < n+2$. If $w_i^{\mu,\nu}$ is finite, by Lemma 1 we can find $u^0 \leq^0 p(\mu)$, such that either for all $\mu' \in \mathrm{Sop}^{u^0,q}(\mu, \nu)$ we have that $y_i^{\mu',\nu}$ is infinite or for all such μ', $y_i^{\mu',\nu}$ is finite. In the second case, again by Lemma 1 we can prepare u^0 (i.e., extend u^0 to some $u' \leq^0 u^0$) such that we are either in the empty, constant or tree case of Lemma 19.

In the empty case, we look at $\dot{r}_i^{n+1} \in \dot{\Gamma}_{n+1}$. Since $\dot{\Gamma}_{n+2}$ is a refinement of $\dot{\Gamma}_{n+1}$, at most one member of $\dot{\Gamma}_{n+1}$ gets split into two members of $\dot{\Gamma}_{n+2}$, and all other members of $\dot{\Gamma}_{n+1}$ belong to $\dot{\Gamma}_{n+2}$. So there are $i_0, i_1 < n+3$ such that $\{\dot{r}_{i_0}^{n+2}, \dot{r}_{i_1}^{n+2}\}$ is a partition of $\dot{r}_i^{n+1}$. In particular, for $j < 2$, below $\dot{r}_{i_j}^{n+2}$ we have that $\dot{b}_i^{n+1}$ is an initial segment of $\dot{b}_{i_j}^{n+2}$ and therefore $w_i^{\mu,\nu} \lhd w_{i_j}^{\mu',\nu}$ for all $\mu' \in \mathrm{Sop}(\mu, \nu)$.

CLAIM 19.1. *There exists $u^1 \leq^0 u^0$ such that the following hold:*

1. *for $j < 2$, we have either $w_{i_j}^{\mu',\nu} \setminus w_i^{\mu,\nu} \neq \emptyset$ for all $\mu' \in \mathrm{Sop}^{u^1,q}(\mu,\nu)$, or $w_{i_j}^{\mu',\nu} \setminus w_i^{\mu,\nu} = \emptyset$ for all such μ',*
2. *for all $k \in \mathrm{succ}_{u^1}(\mu)$, letting*
$$\underline{d}_k^i(\mu,\nu) = \bigcup\left\{w_{i_j}^{\mu',\nu} \setminus w_i^{\mu,\nu} \colon j < 2 \wedge \mu' \in u^1(\mu^\frown k) \cap \mathrm{Sop}(\mu,\nu)\right\},$$
then for all $k, l \in \mathrm{succ}_{u^1}(\mu)$, if $k \neq l$ then $\underline{d}_k^i(\mu,\nu) \cap \underline{d}_l^i(\mu,\nu) = \emptyset$.

Note that either all $\underline{d}_k^i(\mu,\nu)$ are infinite (if the new pieces are not empty for at least one $j < 2$) or all $\underline{d}_k^i(\mu,\nu)$ are finite (if all the new pieces are empty for both $j = 0$ and $j = 1$).

PROOF OF CLAIM 19.1. By Lemma 1, we can find $u' \leq^0 u^0$ such that 1 holds. For the disjointness, by induction we construct the splitnodes $\langle \sigma_l : l < \omega \rangle$ of u^1 such that

1. $\sigma_l \in \mathrm{Sop}^{u',q}(\mu,\nu)$,
2. if $0 < k < l$, then for $j < 2$ with $w_{i_j}^{\sigma_l,\nu} \setminus w_i^{\mu,\nu} \neq \emptyset$ and for any $j' < 2$ we have
$$\min\left(w_{i_j}^{\sigma_l,\nu} \setminus w_i^{\mu,\nu}\right) > \max\left(w_{i_{j'}}^{\sigma_k,\nu} \setminus w_i^{\mu,\nu}\right).$$

Fix $\sigma_0 \in \mathrm{Sop}^{u',q}(\mu,\nu)$ arbitrarily. At stage l, assume $\sigma_{l'}$ needs an extension. For $j < 2$, suppose $w_{i_j}^{\mu',\nu} \setminus w_i^{\mu,\nu} \neq \emptyset$ for all $\mu' \in \mathrm{Sop}(\mu,\nu)$. Suppose $\sigma_{l'} = \mu$. As $w_i^{\mu,\nu}$ is finite, we can apply Remark 5(b), hence by construction we can choose $k \in \mathrm{succ}_{u'}(\mu)$ large enough, such that for every $\mu' \in u'(\mu^\frown k) \cap \mathrm{Sop}(\mu,\nu)$, $\min(w_{i_j}^{\mu',\nu} \setminus w_i^{\mu,\nu})$ is as large as we like. For $\sigma_{l'} \supsetneq \mu$, note that for all $\mu' \in \mathrm{split}(u'(\sigma_{l'}))$, $w_{i_j}^{\mu',\nu} \setminus w_i^{\mu,\nu} = w_{i_j}^{\mu',\nu} \setminus y_i^{\mu'',\nu}$, where μ'' is the predecessor splitnode of μ'. So with the same argument as before we can make the new piece of some $w_{i_j}^{\mu',\nu}$ start as late as we like. Therefore we can choose $\sigma_l \in \mathrm{Succ}_{u'}(\sigma_{l'})$ as desired. ⊣

In the tree case we have the following:

CLAIM 19.2. *There exists $u^2 \leq^0 u^1$ such that for all $k \in \mathrm{succ}_{u^2}(\mu)$, letting*
$$\underline{d}_k^i(\mu,\nu) = \bigcup\left\{y_i^{\mu',\nu} \setminus w_i^{\mu,\nu} \colon \mu' \in u^2(\mu^\frown k) \cap \mathrm{Sop}(\mu,\nu)\right\},$$
we have $\underline{d}_k^i(\mu,\nu) \cap \underline{d}_l^i(\mu,\nu) = \emptyset$ for $k \neq l$ and all $\underline{d}_k^i(\mu,\nu)$ are infinite.

PROOF OF CLAIM 19.2. Note that for such u^2 and $\underline{d}_k^i(\mu,\nu)$, if $\mu' \in \mathrm{Sop}(\mu,\nu) \cap u(\mu^\frown k) \cap \mathrm{Succ}_{u^2}(\mu)$, then
$$\underline{d}_k^i(\mu,\nu) = \bigcup\left\{y_i^{\mu',\nu} \setminus w_i^{\mu',\nu}, y_i^{\varrho',\nu} \setminus y_i^{\varrho,\nu} \colon \varrho \in \mathrm{split}(u^2(\mu')) \wedge \varrho' \in \mathrm{Succ}_{u^2}(\varrho)\right\}$$
where μ' is the single element of $\mathrm{Succ}_{u^2}(\mu) \cap \mathrm{Sop}^{u^2,v}(\mu,\nu)$ with $\mu'(|\mu|) = k$,

and with the arguments from the proof of Claim 19.1 we can make all these new pieces of the y's pairwise disjoint finite sets. ⊣

For odd-type pairs, we proceed symmetrically. So, by applying the Fusion Property, we obtain some $\langle \widehat{u}, \widehat{v} \rangle \leq^0 \langle p, q \rangle$ such that part 3 of the lemma holds.

Finally we need the following:

Claim 19.3. *There exists* $\langle u, v \rangle \leq^0 \langle \widehat{u}, \widehat{v} \rangle \leq \langle p, q \rangle$ *such that* $\{d_\mu, d'_\nu \colon \mu \in \operatorname{split}(u) \wedge \nu \in \operatorname{split}(v)\}$ *is a disjoint family.*

Proof of Claim 19.3. We construct $\langle u, v \rangle$ by a simultaneous induction on splitnodes: Suppose we got the initial segments $\langle \sigma_l \colon l < k \rangle$, $\langle \tau_l \colon l < k \rangle$ of the splitnodes of u, v, respectively, and σ_j needs an extension (and therefore also τ_j). Fix τ_l such that $\operatorname{tp}(\sigma_j, \tau_l) = n$ is even and also fix $i < n + 2$. Since $w_i^{\sigma_j, \tau_l}$ is finite, we can choose $\sigma_k \in \operatorname{Sop}(\sigma_j, \tau_l) \cap \operatorname{Succ}_{\hat{u}}(\sigma_j)$ so much right that $\min(y_i^{\sigma_k, \tau_l} \setminus w_i^{\sigma_j, \tau_l})$ is as large as we like, say m_*. So we make sure $m_* > \max(y_{i'}^{\sigma_{k'}, \tau_{l'}} \setminus w_{i'}^{\sigma_{k''}, \tau_{l'}})$ for all $k'' < k' \leq k$ such that $\sigma_{k''} \subsetneq \sigma_{k'}$ and $\operatorname{tp}(\sigma_{k''}, \tau_{l'}) = n'$ is even and all $i' < n' + 2$, and also $m_* > \max(y_{i'}^{\sigma_{k'}, \tau_{l'}} \setminus w_{i'}^{\sigma_{k'}, \tau_{l''}})$ for all $l'' < l' < k$ and $\tau_{l''} \subsetneq \tau_{l'}$ and $\operatorname{tp}(\sigma_{k'}, \tau_{l''}) = n'$ is odd and $i' < n' + 2$. Get τ_k analogously. ⊣

Note that for all $\langle \mu, \nu \rangle \in \operatorname{TP}^n(u, v)$ and $i < n + 2$ as in the empty case or tree case, all $k \in \operatorname{succ}_u(\mu)$ for even type or $k \in \operatorname{succ}_v(\nu)$ for odd type, $\underline{d}_k^i(\mu, \nu) \supseteq d_k^i(\mu, \nu)$. Hence we get

$$r = \langle u, v \rangle^\frown \dot{q} \leq^0_{\{0\} \cup \widetilde{F}} \widetilde{p}$$

as desired, which finishes the proof of Lemma 19. ⊣

Lemma 20. *Suppose that* $\langle a_\alpha \colon \alpha < \kappa \rangle$ *is an almost disjoint family on* ω *with* κ *infinite and* $\kappa < \mathfrak{b}$. *Let* $p \in \mathbb{M}$ *and let* $\langle k(\mu) \colon \mu \in \operatorname{split}(p) \rangle$ *be a one-to-one family such that* $k(\mu) \in \omega$. *There exists* $q \leq^0 p$, $q \in \mathbb{M}$, *such that one of the following holds*:

1. *for every* $\alpha < \kappa$, *the set*

$$\{k(\mu) \colon \mu \in \operatorname{split}(q)\} \cap a_\alpha$$

is finite;

2. *for every* $i \in \operatorname{succ}_q(\operatorname{st} q)$ *there exists* $\alpha_i < \kappa$ *such that*

$$\{k(\mu) \colon \mu \in \operatorname{split}(q(\operatorname{st}(q)^\frown i))\} \subseteq a_{\alpha_i}.$$

Proof. The proof is given in the following two claims:

Claim 20.1. *For every* $\mu \in \operatorname{split}(p)$ *there exist a countable set* $C \subseteq \kappa$ *and a sequence* $\langle \nu_n \colon n \in \omega \rangle$ *in* $\operatorname{split}(p)$ *such that for all* $n < m < \omega$,

$$\mu \subsetneq \nu_n, \quad \nu_n(|\mu|) \neq \nu_m(|\mu|),$$

and

$$\{k(\nu_n) \colon n \in \omega\} \cap a_\alpha$$

is finite for every $\alpha \in \kappa \setminus C$.

PROOF OF CLAIM 20.1. Choose $\langle x_n : n \in \omega \rangle \subseteq [p]$ such that for every $n < m$

$$\mu \subseteq x_n \text{ and } x_n(|\mu|) \neq x_m(|\mu|).$$

Let $n < \omega$. If possible choose $\alpha_n < \kappa$ such that

$$\{k(\sigma) : \sigma \in \operatorname{split}(p) \wedge \sigma \subseteq x_n\} \cap a_{\alpha_n}$$

is finite. In this case let $\langle \sigma_i^n : i < \omega \rangle$ be such that for every $i < j < \omega$

$$\mu \subsetneq \sigma_i^n \subsetneq \sigma_j^n \subsetneq x$$

and

$$k(\sigma_i^n) \in a_{\alpha_n}.$$

Otherwise choose $\langle \sigma_i^n : i < \omega \rangle$ arbitrarily such that for every $i < j < \omega$

$$\mu \subsetneq \sigma_i^n \subsetneq \sigma_j^n \subseteq x.$$

Let $C = \{\alpha_n : n < \omega\}$.

For $\alpha \in \kappa \setminus C$ define $f_\alpha \in \omega^\omega$ such that

$$k(\sigma_i^n) \notin a_\alpha$$

for every $i \geq f_\alpha(n)$. By $\kappa < \mathfrak{b}$ find $f \in \omega^\omega$ with $f \geq^* f_\alpha$, for all $\alpha \in \kappa \setminus C$. Now let $v_n = \sigma_{f(n)}^n$. Then $\langle v_n : n \in \omega \rangle$ is as desired. ⊣

Using Claim 20.1 it is straightforward to construct $r \leq^0 p$, $r \in \mathbb{M}$, and $\langle C_\mu : \mu \in \operatorname{split}(r) \rangle$ such that $C_\mu \subseteq \kappa$ is countable and

$$\{k(v) : v \in \operatorname{Succ}_r(\mu)\} \cap a_\alpha$$

is finite for every $\alpha \in \kappa \setminus C_\mu$, for every $\mu \in \operatorname{split}(r)$. Let $C = \bigcup\{C_\mu : \mu \in \operatorname{split}(r)\}$.

CLAIM 20.2. *There exists $w \leq^0 r$, $w \in \mathbb{M}$, such that*

$$\{k(\mu) : \mu \in \operatorname{split}(w)\} \cap a_\alpha$$

is finite for every $\alpha \in \kappa \setminus C$.

PROOF OF CLAIM 20.2. First we construct $u \in \mathbb{M}$ by inductively defining $\operatorname{Lev}_n(u)$, $n \in \omega$. We stipulate that $\operatorname{Lev}_0(u) = \operatorname{Lev}_0(r)$ and $\operatorname{Lev}_1(u) = \operatorname{Lev}_1(r)$. Let $\langle \mu_n : n \in \omega \rangle$ canonically enumerate $\operatorname{Lev}_1(u)$. By construction of r, for every $\alpha \in \kappa \setminus C$ there exists $f_\alpha \in \omega^\omega$ such that for every $v \in \operatorname{Succ}_r(\mu_n)$, if $v(|\mu_n|) \geq f_\alpha(n)$, then $k(v) \notin a_\alpha$. Let $f \in \omega^\omega$ dominate every f_α, $\alpha \in \kappa \setminus C$. Now $\operatorname{Lev}_2(u)$ contains precisely those $v \in \operatorname{Succ}_r(\mu_n)$ with $v(|\mu_n|) \geq f(n)$, for every n. Note that $\{k(v) : v \in \bigcup_{i<3} \operatorname{Lev}_i(u)\} \cap a_\alpha$ is finite for every $\alpha \in \kappa \setminus C$. Suppose we got $\operatorname{Lev}_n(u)$. Fix $\mu_m \in \operatorname{Lev}_1(u)$. By repeating above μ_m the construction of $\operatorname{Lev}_n(u)$ we just did (so not changing any of $\operatorname{Lev}_i(u)$, $i \leq n$) we obtain $\operatorname{Lev}_n(u')$ for some $u' \leq^0 u(\mu_m)$, such that if v denotes the tree generated by $\operatorname{Lev}_n(u')$, then

$$\{k(v) : v \in \operatorname{split}(v)\} \cap a_\alpha$$

is finite for every $\alpha \in \kappa \setminus C$. For such α define $f_\alpha(m) \in \omega$ large enough, such that if v' is the finite subtree of v containing $\mathrm{st}(v)$ and with every $\nu \supseteq \mathrm{st}(v)$ also precisely the $f_\alpha(m)$ leftmost successor splitnodes of ν in v (if there are such ones), then

$$\{k(\nu) \colon \nu \in \mathrm{split}(v)\} \cap a_\alpha \subseteq \{k(\nu) \colon \nu \in \mathrm{split}\,(v') \cup \mathrm{leaves}\,(v')\}.$$

Let f dominate every f_α, $\alpha \in \kappa \setminus C$.

Now let $m \in \omega$ and let v' be defined as above, except that in its definition $f_\alpha(m)$ is replaced by $f(m)$. Let $\mu \in \mathrm{Lev}_n(u)$ such that $\mu_m \subseteq \mu$. In case $\mu \in v'$ we stipulate that $\mathrm{Succ}_u(\mu)$ equals the set $\mathrm{Succ}_r(\mu)$ with its $f(m)$ leftmost members being removed. Otherwise $\mathrm{Succ}_u(\mu) = \mathrm{Succ}_r(\mu)$. This finishes the construction of $\mathrm{Lev}_{n+1}(u)$, and hence of u.

By construction we have that $M(n, \alpha) = \{k(\nu) \colon \nu \in \bigcup_{i<n} \mathrm{Lev}_i(u)\} \cap a_\alpha$ is finite for every $\alpha \in \kappa \setminus C$ and $n \in \omega$. Let $g_\alpha(n)$ be large enough that if v' is the finite subtree of the tree generated by $\mathrm{Lev}_{n+1}(u)$ containing $\mathrm{st}(u)$, the $g_\alpha(n)$ leftmost elements of $\mathrm{Succ}_u(\mathrm{st}(u))$, for any $\nu \in \bigcup_{i<n} \mathrm{Lev}_i(u)$ the $g_\alpha(n)$ leftmost elements of $\mathrm{Succ}_u(\nu)$ etc., then

$$M(n, \alpha) \subseteq \{k(\nu) \colon \nu \in \mathrm{split}\,(v') \cup \mathrm{leaves}\,(v')\}.$$

Let $g \in \omega^\omega$ dominate every g_α, $\alpha \in \kappa \setminus C$. Obtain $w \leq^0 u$ as desired as follows: Inductively define $\mathrm{Lev}_n(w)$: $\mathrm{Lev}_0(w) = \{\mathrm{st}(u)\}$, $\mathrm{Lev}_1(w)$ equals the set $\mathrm{Lev}_1(u)$ with its $g(0)$ leftmost elements being removed. Suppose $\mathrm{Lev}_n(w)$ has been defined. Let v' be defined as above except that in its definition $g(n)$ is used instead of $g_\alpha(n)$.

Let $\mu \in \mathrm{Lev}_n(w)$. In case $\mu \in v'$, let $\mathrm{Succ}_w(\mu)$ equal $\mathrm{Succ}_u(\mu)$ with its $g(n)$ leftmost elements removed. Otherwise $\mathrm{Succ}_w(\mu) = \mathrm{Succ}_u(\mu)$.

It is straightforward to check that w is as desired. ⊣

Now it is an easy matter to construct $q \leq^0 w$ as in the Theorem. If there are only finitely many $i \in \mathrm{succ}_w(\mathrm{st}(w))$ for which there is $v \leq w(\mathrm{st}(w)^\frown i)$ and $\alpha \in C$ such that $\{k(\mu) \colon \mu \in \mathrm{split}(v)\} \subseteq a_\alpha$ we get alternative 1 by an easy fusion, otherwise we have alternative 2. ⊣

Corollary 21. *Let $\mathcal{A}_n$, $n < \omega$, be almost disjoint families on ω of size less than $\mathfrak{b}$, and let p and $\langle k(\mu) \colon \mu \in \mathrm{split}(p)\rangle$ be as in Lemma 20. There exist $q \leq^0 p$, $q \in \mathbb{M}$ and some countable $C \subseteq \bigcup_{n<\omega} \mathcal{A}_n$ such that*

$$\{k(\mu) \colon \mu \in \mathrm{split}(q)\} \cap a$$

is finite for every $a \in \bigcup_{n<\omega} \mathcal{A}_n \setminus C$.

Proof. Essentially the same construction as in the proof of Lemma 20 works. The only new point is that in Claim 20.1 and also in the construction of r after Claim 20.1, when we have chosen $\langle x_n : n < \omega\rangle$ we choose $\langle \sigma_i^n : i < \omega\rangle$ such that for every $m < \omega$,

$$\{k\,(\sigma_i^n) : n < \omega\}$$

is either disjoint with every member of $\mathcal{A}_m$ or almost contained in some member of $\mathcal{A}_m$. Then C will consist of all members of $\bigcup_{m<\omega} \mathcal{A}_m$ as in the latter case. ⊣

REMARK 22. Let $\mathcal{A}_n, q, C$ as in Corollary 21, and suppose that there does not exist any $r \leq^0 q$, $r \in \mathbb{M}$ such that for every $a \in C$ the set $\{k(\mu) \colon \mu \in \text{split}(r)\}$ is either almost contained in a or almost disjoint with a. Note that this is a Π^1_1-property of q and C, hence it is absolute. Now let $b \in [\omega]^\omega$ such that either $b \subseteq^* a$ or $|b \cap a| < \omega$ for every $a \in C$. It is easy to see that we can find $r \leq^* q$, $r \in \mathbb{M}$, such that $k(\mu) \notin b$ for every $\mu \in \text{split}(r^-)$. Here b may belong to some extension of the universe and still we can find r (in the extension) as above.

THEOREM 23 (**GCH**). *There exists a P_{ω_2}-name $\dot{\mathcal{M}} = \langle \dot{\mathcal{M}}_\gamma \colon \gamma < \omega_1 \rangle$ for a family of $\aleph_1$ mad families on ω which does not have a refinement in $V^{P_{\omega_2}}$. Hence $V^{P_{\omega_2}} \models \mathfrak{h} = \omega_1$.*

PROOF. Recursively we construct $\langle \dot{\mathcal{A}}^\alpha \colon \alpha < \omega_2 \rangle$ such that for every $\alpha < \omega_2$

$$\dot{\mathcal{A}}^\alpha = \langle \dot{\mathcal{A}}^\alpha_\gamma \colon \gamma < \omega_1 \rangle$$

is a family of P_α-names with the following properties in V^{P_α}:

1. $\dot{\mathcal{A}}^\alpha_\gamma$ is a mad family on ω for all $\gamma < \omega_1$,
2. $\dot{\mathcal{A}}^\alpha$ is refining,
3. $\forall \beta < \alpha \forall \gamma < \omega_1 (\dot{\mathcal{A}}^\beta_\gamma \subseteq \dot{\mathcal{A}}^\alpha_\gamma)$.

Suppose that $[\omega]^\omega \cap V^{P_\alpha} \setminus \bigcup_{\beta<\alpha} V^{P_\beta} \neq \emptyset$. Note that this happens precisely if α has countable cofinality, and that then by Lemma 17 , in V^{P_α} there exists a dominating real over $\bigcup_{\beta<\alpha} V^{P_\beta}$. Hence for every $x \in [\omega]^\omega \cap V^{P_\alpha}$ there exists $y \in [x]^\omega$ such that either for every $\gamma < \omega_1$ there is $b \in \bigcup \dot{\mathcal{A}}^\beta_\gamma$ with $y \subseteq^* b$, or else there exists $\gamma < \omega_1$ such that for every $\gamma \leq \delta < \omega_1$ and every $b \in \bigcup_{\beta<\alpha} \dot{\mathcal{A}}^\beta_\delta$, $y \cap b$ is finite. Hence we can find $\langle \dot{x}^\alpha_\nu \colon \nu < \omega_1 \rangle \subseteq [\omega]^\omega \cap V^{P_\alpha}$ such that $b \cap \dot{x}^\alpha_\nu$ is finite for every $b \in \bigcup_{\beta<\alpha} \dot{\mathcal{A}}^\beta_\nu$ and for every $y \in [\omega]^\omega \cap V^{P_\alpha}$ either there is $\nu < \omega_1$ such that $\dot{x}^\alpha_\nu \subseteq^* y$ or there is $z \in [y]^\omega$ and for every $\nu < \omega$ there is $b \in \bigcup_{\beta<\alpha} \dot{\mathcal{A}}^\beta_\nu$ with $z \subseteq^* b$. Moreover we may assume that either $\dot{x}^\alpha_\nu \cap \dot{x}^\alpha_\mu$ is finite or $\dot{x}^\alpha_\mu \subseteq^* \dot{x}^\alpha_\nu$ and $\dot{x}^\alpha_\nu \setminus \dot{x}^\alpha_\mu$ is infinite for every $\nu < \mu < \omega_1$.

Suppose we got $\langle \dot{\mathcal{A}}^\beta \colon \beta < \alpha \rangle$ for some $\alpha < \omega_2$. Let $\dot{\mathcal{A}}^{<\alpha}_\gamma = \bigcup_{\beta<\alpha} \dot{\mathcal{A}}^\beta_\gamma$ and $\dot{\mathcal{A}}^{<\alpha} = \langle \dot{\mathcal{A}}^{<\alpha}_\gamma \colon \gamma < \omega_1 \rangle$.

If $\text{cf}(\alpha) = \omega_1$ we let $\dot{\mathcal{A}}^\alpha_\gamma = \dot{\mathcal{A}}^{<\alpha}_\gamma$. If $\text{cf}(\alpha)$ is countable, the construction of $\dot{\mathcal{A}}^\alpha$ will be such that the following properties hold:

4. $\forall \gamma < \omega_1 \forall \nu < \gamma \forall b \in \dot{\mathcal{A}}^\alpha_\gamma \setminus \dot{\mathcal{A}}^{<\alpha}_\gamma (|\dot{x}^\alpha_\nu \setminus b| = \omega)$;
5. $\forall \gamma < \omega_1 \forall \nu \leq \gamma \forall b \in \dot{\mathcal{A}}^\alpha_\gamma \setminus \dot{\mathcal{A}}^{<\alpha}_\gamma (b \subseteq^* \dot{x}^\alpha_\nu \vee |b \cap \dot{x}^\alpha_\nu| < \omega)$.

Finally suppose that in V^{P_α}

$$\left\langle \left(\dot{p}_\nu^\alpha, \left\langle \dot{k}_\nu^\alpha(\mu) \colon \mu \in \operatorname{split}\left(\dot{p}_\nu^\alpha\right) \right\rangle \right) \colon \nu < \omega_1 \right\rangle$$

enumerates all pairs of a superperfect tree and a one-to-one family of natural numbers. Then we shall ensure that

6. for every $\nu < \omega_1$ there exists $q \leq^0 \dot{p}_\nu^\alpha$ such that one of the following holds:
   - there exists $\gamma < \omega_1$ such that for every $i \in \operatorname{succ}_q(\operatorname{st}(q))$
   $$\left\{ \dot{k}_\nu^\alpha(\mu) \colon \mu \in \operatorname{split}\left(q\left(st(q)^\frown i\right)\right) \right\} \in \dot{A}_\gamma^\alpha \setminus \dot{A}_\gamma^{<\alpha},$$
   - there exists $\gamma < \omega_1$ and some countable $C \subseteq \bigcup_{\delta<\gamma} \dot{A}_\delta^\alpha$ such that for no $r \leq^0 q$ is it the case that for every $a \in C$ the set
   $$\{k(\mu) \colon \mu \in \operatorname{split}(r)\}$$
   is either almost contained in a or almost disjoint from a,
   - for every $\delta < \omega_1$ and $i \in \operatorname{succ}_q(\operatorname{st}(q))$ there exists $a \in \bigcup_{\delta<\gamma} \dot{A}_\delta^\alpha$ such that
   $$\left\{ \dot{k}_\nu^\alpha(\mu) \colon \mu \in \operatorname{split}\left(q\left(\operatorname{st}(q)^\frown i\right)\right) \right\} \subseteq^* a.$$

For the construction of $\dot{A}^\alpha$ in V^{P_α} we apply the following:

Claim 23.1. *Let $p \in \mathbb{M}$ and let $\langle k(\mu) \colon \mu \in \operatorname{split}(p) \rangle$ be a one-to-one family such that $k(\mu) \in \omega$. There exists $q \leq^0 p$, $q \in \mathbb{M}$, such that one of the following holds:*

i. *There exists $\gamma < \omega_1$ such that $\{k(\mu) \colon \mu \in \operatorname{split}(q)\} \cap a$ is finite for every $a \in \dot{A}_\gamma^{<\alpha}$;*
ii. *For every $\delta < \omega_1$ and $i \in \operatorname{succ}(\operatorname{st}(q))$ there exists $a \in \dot{A}_\delta^{<\alpha}$ such that*
$$\left\{ k(\mu) \colon \mu \in \operatorname{split}\left(q(\operatorname{st}(q)^\frown i)\right) \right\} \subseteq^* a.$$

Remark 24. We shall show below that the second alternative never occurs.

Proof. It is based on the proof of Lemma 20, hence we only give a sketch. If possible we choose $\gamma_0 < \omega_1$, $\langle x_n \colon n \in \omega \rangle \subseteq [p]$ such that $\langle x_n(|\operatorname{st}(p)|) \colon n \in \omega \rangle$ is one-to-one, and a one-to-one family $\langle a_{\langle n \rangle} \colon n \in \omega \rangle \subseteq \dot{A}_{\gamma_0}^{<\alpha}$ such that

$$\{k(\mu) \colon \mu \in \operatorname{split}(p) \wedge \mu \subseteq x_n\} \cap a_{\langle n \rangle}$$

is infinite.

As in V^{P_α} we have dominating reals over $\bigcup_{\beta<\alpha} V^{P_\alpha}$, we can pick $\mu_{\langle n \rangle} \in \operatorname{split}(p)$ such that $\operatorname{st}(p) \subsetneq \mu_{\langle n \rangle} \subseteq x_n$, $k(\mu_{\langle n \rangle}) \in a_{\langle n \rangle}$ and $\{k(\mu_{\langle n \rangle}) \colon n \in \omega\} \cap a$ is finite for every $a \in \mathcal{A}_{\gamma_0}^{<\alpha}$.

Above each $\mu_{\langle n \rangle}$ we proceed in the same way, if possible, constructing $\gamma_{\langle n \rangle} < \omega_1$ and $\mu_{\langle n,m \rangle}$, $m \in \omega$, such that

$$\left\{ k\left(\mu_{\langle n,m \rangle}\right) \colon m \in \omega \right\} \cap a$$

is finite for every $a \in \mathcal{A}^{<\alpha}_{\gamma\langle n\rangle}$. We continue similarly. Suppose that for all $n \geq n_0$ for some n_0, above $\mu_{\langle n\rangle}$ we can proceed forever, thus constructing μ_σ for every $\sigma \in \omega^{<\omega}$ with $\sigma(0) \geq n_0$. Let $\gamma = \sup_{\sigma\in\omega^{<\omega}} \gamma_\sigma$ and let $v \leq^0 p$ be the tree generated by all μ_σ, $\sigma(0) \geq n_0$. Hence for every $\mu \in \operatorname{split}(v)$ and $a \in \mathcal{A}^{<\alpha}_\gamma$,

$$\{k(\nu) \colon \nu \in \operatorname{Succ}_v(\mu)\} \cap a$$

is finite. By the proof Claim 20.2 we may obtain $q \leq^0 v$ such that

$$\{k(\mu) \colon \mu \in \operatorname{split}(q)\} \cap a$$

is finite for every $a \in \mathcal{A}^{<\alpha}_\gamma$, thus we have obtained alternative i.

Suppose now that this construction breaks down already at $\operatorname{st}(p)$, i.e., for every $\delta < \omega_1$ there is n_δ and finite $F_\delta \subseteq \mathcal{A}^{<\alpha}_\delta$ such that

$$\{k(\mu) \colon \mu \in \operatorname{split}(p) \wedge \mu \subseteq x\} \subseteq^* \bigcup_{\delta<\omega_1} F_\delta$$

for every $x \in [p]$ with $x(|\operatorname{st}(p)|) \geq n_\delta$. Without loss of generality n_δ is constantly n and F_δ is constantly of size m for every $\delta < \omega_1$. Fix $\delta < \omega_1$. A straightforward application of the Theorem of Kechris and St. Raymond [7, 15] yields a single $a \in F_\delta$ and $v \leq^0 p$ such that

$$\{k(\mu) \colon \mu \in \operatorname{split}(p) \wedge \mu \subseteq x\} \subseteq^* a$$

for every $x \in [v]$. It is now easy to get $q \leq^0 v$ such that

$$\{k(\mu) \colon \mu \in \operatorname{split}(q^-)\} \subseteq a.$$

As $\langle \mathcal{A}^{<\alpha}_\beta \colon \beta < \omega_1\rangle$ is refining, we have a more special form of alternative ii.

Finally suppose that the construction at the beginning of the present proof does not break down already at $\operatorname{st}(p)$ but only at μ_{σ_n} for $\sigma_n \in \omega^{<\omega}$ with $\operatorname{st}(p) \subsetneq \sigma_n$ and $\sigma_n(|\operatorname{st}(p)|) < \sigma_{n+1}(|\operatorname{st}(p)|)$. We use the previous argument above each μ_{σ_n} and obtain alternative ii. again. ⊣

Inductively we construct $\langle \gamma_\nu \colon \nu < \omega_1\rangle$ and $\langle \mathcal{C}^\nu \colon \nu < \omega_1\rangle$ where $\mathcal{C}^\nu = \langle \mathcal{C}^\nu_\delta \colon \delta \leq \gamma_\nu\rangle$ such that

(a) $\nu < \nu' < \omega_1$ implies $\gamma_\nu < \gamma_{\nu'} < \omega_1$;

(b) $\mathcal{C}^\nu_\delta$ is a countable almost disjoint family such that for all $a \in \mathcal{C}^\nu_\delta$ and all $b \in \mathcal{A}^{<\alpha}_\delta$, $|a \cap b| < \omega$; every $a \in \mathcal{C}^\nu_\delta$ satisfies requirements 4 and 5; $\mathcal{C}^\nu$ is refining and $\nu < \nu' < \omega_1$ implies $\mathcal{C}^\nu_\delta \subseteq \mathcal{C}^{\nu'}_\delta$ for all $\delta \leq \gamma_\nu$.

Suppose we got $\langle \gamma_\nu \colon \nu < \varrho\rangle$ and $\langle \mathcal{C}^\nu \colon \nu < \varrho\rangle$ for some $\varrho < \omega_1$. Let $\bar\gamma = \sup_{\nu<\varrho} \gamma_\nu$. We apply Claim 23.1 to $\dot p^\alpha_\varrho$, $\langle \dot k^\alpha_\varrho(\mu) \colon \mu \in \operatorname{split}(\dot p^\alpha_\varrho)\rangle$ and obtain $q' \leq^0 \dot p^\alpha_\varrho$. Suppose we have the first alternative. In this case we get $\gamma_\varrho > \bar\gamma$ such that $\{\dot k^\alpha_\varrho(\mu) \colon \mu \in \operatorname{split}(q')\} \cap a$ is finite for every $a \in \dot{\mathcal{A}}^{<\alpha}_{\gamma_\varrho}$. By Corollary 21 we obtain $q \leq^0 q'$ and some countable $\mathcal{C} \subseteq \bigcup_{\nu\leq\gamma_\varrho} \dot{\mathcal{A}}^{<\alpha}_\nu$ such that

$$\{\dot k^\alpha_\varrho(\mu) \colon \mu \in \operatorname{split}(q)\} \cap a$$

is finite for every $a \in \bigcup_{v \leq \gamma_\varrho} \dot{\mathcal{A}}_\gamma^{<\alpha} \setminus C$.

Let $C' = C \cup \bigcup\{C_\delta^v : \delta \leq \gamma_v, v < \varrho\}$, so C' is countable. Suppose first there does not exist $r \leq^0 q$ such that, letting

$$b(r) = \{\dot{k}_\varrho^\alpha(\mu) : \mu \in \text{split}(r)\},$$

for every $a \in C'$ we have either $b(r) \subseteq^* a$ or $|b(r) \cap a| < \omega$. In this case we define C^v for $\bar{\gamma} \leq v \leq \gamma_\varrho$ arbitrarily, except for requirements (a) and (b).

Otherwise we choose such r and $b(r)$. Note that if both $b(r) \cap \dot{x}_\gamma^\alpha$ and $b(r) \setminus \dot{x}_\gamma^\alpha$ are infinite for some $\gamma < \gamma_\varrho$, then from the construction of $\langle \dot{x}_v^\alpha : v < \omega_1 \rangle$ and 5 it follows that $\dot{x}_\gamma^\alpha \cap a$ is finite for every $a \in \dot{\mathcal{A}}_\gamma^{<\alpha} \cup \bigcup_{v<\varrho} C_\gamma^v$. Let D be the set of all these γ.

Suppose first that there exists $r' \leq^0 r$ such that for every $\gamma \in D$ either $b(r') \cap \dot{x}_\gamma^\alpha$ is finite or $b(r') \subseteq^* \dot{x}_\gamma^\alpha$. Then we put $b(r(st(r)^\frown i)) \in C_{\gamma_\varrho}^\varrho$ for every $i \in \text{succ}_q(st(q))$. If there does not exist such r' we put $b(r) \cap \dot{x}_\gamma^\alpha \in C_\gamma^\varrho$ for every $\gamma \in D$ and besides that, for $\gamma \leq \gamma_\varrho$ we define C_γ^ϱ such that (a) and (b) hold. Note that then there is no $r' \leq^0 r$ such that for every $a \in \{b \cap \dot{x}_\gamma^\alpha : \gamma \in D\}$ we have that either $b(r') \subseteq^* a$ or $b(r') \cap a$ is finite.

If after applying Claim 23.1 we had the second alternative, we define C^ϱ arbitrarily, satisfying (a) and (b).

In the end we let

$$\dot{\mathcal{A}}_\gamma^\alpha \setminus \dot{\mathcal{A}}_\gamma^{<\alpha} = \bigcup_{v<\omega_1} C_\gamma^v.$$

This finishes the construction of $\langle \dot{\mathcal{A}}^\alpha : \alpha < \omega_2 \rangle$. Let $\dot{\mathcal{M}}_\gamma = \bigcup_{\alpha<\omega_2} \dot{\mathcal{A}}_\gamma^\alpha$, for $\gamma < \omega_1$.

Suppose that for some P_{ω_2}-name $\dot{a}$ and $p \in P_{\omega_2}$ we have

$$p \Vdash_{P_{\omega_2}} \dot{a} \in [\omega]^\omega.$$

We shall find $q \leq p$ and $\gamma < \omega_1$ such that

$$q \Vdash_{P_{\omega_2}} \forall b \in \dot{\mathcal{M}}_\gamma \dot{a} \not\subseteq^* b.$$

We may assume that there exists $\alpha \in \omega_2$ with countable cofinality such that

$$p \Vdash \dot{a} \in V\left[\dot{G}_\alpha\right] \setminus \bigcup_{\beta<\alpha} V\left[\dot{G}_\beta\right].$$

Hence we may assume that $\dot{a}$ is a P_α-name, but not a P_β-name for any $\beta < \alpha$. By construction of $\dot{\mathcal{M}}$, we know

$$\Vdash_{P_{\omega_2}} \forall \gamma < \omega_1 \forall \dot{b} \in \dot{\mathcal{M}}_\gamma \setminus \dot{\mathcal{A}}_\gamma^\alpha \dot{a} \not\subseteq^* \dot{b}.$$

Hence if for some $q \leq p$, $\gamma < \omega_1$ and $\dot{b} \in \dot{\mathcal{M}}_\gamma$ we have

$$q \Vdash_{P_{\omega_2}} \dot{a} \subseteq^* \dot{b},$$

then in fact

$$q \restriction \alpha \Vdash_{P_\alpha} \dot{a} \subseteq^* \dot{b} \wedge \dot{b} \in \dot{A}^\alpha_\gamma.$$

We may therefore assume that $p \in P_\alpha$.

Using Lemma 19, we construct a fusion sequence $\langle\langle p_k, F_k\rangle : k < \omega\rangle$ in P_α such that the following holds for all $\beta \leq \alpha$:

1. We have $p_0 \leq p$ and there exists $\nu_0 < \omega_1$ such that either $p_0 \Vdash_{P_\alpha} \forall \nu < \omega_1\ \exists b \in \dot{A}^{<\alpha}_\nu \dot{a} \subseteq^* b$, or $p_0 \Vdash_{P_\alpha} \dot{x}^\alpha_{\nu_0} \subseteq^* \dot{a}$.
2. If $\beta \in \operatorname{supp}(p_k)$, then there exists $m > k$ such that the following hold:
   (a) For every $n < \omega$, we have hereditarily countable P_β-names $\dot{\Gamma}_n(\beta)$ for a finite partition of $p \restriction [\beta + 1, \alpha)$ in V^{P_β}, P_β-names $\dot{w}^{\mu,\nu}_i(\beta)$, $\dot{y}^{\mu,\nu}_{i'}(\beta)$ for every P_β-name $\langle \mu, \nu\rangle$ for an element in $\mathrm{TP}^n(p_m(\beta))$ and $i < n+2$, $i' < n+1$, which are all obtained together with p_m by applying Lemma 19 in V^{P_β} to $\dot{a}[\dot{G}_\beta]$ and $p_{m-1} \restriction [\beta, \alpha)$ and $F_k \setminus \beta$.
   (b) There exists $\gamma_\beta < \omega_1$ such that
      (i) $p_m \restriction \beta \Vdash_{P_\beta}$ "If $a \in \{\dot{w}^{\mu,\nu}_i(\beta), \dot{y}^{\mu,\nu}_{i'}(\beta) : \langle \mu, \nu\rangle \in \mathrm{TP}^n(p(\beta)) \wedge i < n+2, i' < n+1\} \cap [\omega]^\omega$, then either $\forall \nu < \omega_1\ \exists b \in \dot{A}^{<\alpha}_\nu\ \dot{a} \subseteq^* b$ or else $x^\beta_\nu \subseteq^* \dot{a}$ for some $\nu < \gamma_\beta$."
      (ii) Let $p_m(\beta) = (\dot{u}, \dot{v})$ and (μ, ν) a type-pair of $(\dot{u}, \dot{v})$, say of even type (for odd type-pairs a symmetric requirement must hold). If we have the empty case at (μ, ν), let $\dot{k}^j(\varrho) = \min(\dot{w}^{\varrho,\nu}_{i_j}(\beta) \setminus \dot{w}^{\mu,\nu}_i(\beta))$ for every $\varrho \in \mathrm{Sop}^{\dot{u},\dot{v}}(\mu, \nu)$ and $j < 2$. Then for every $j < 2$, letting $\nu < \omega_1$ such that

$$\begin{aligned}(\dot{u}(\mu), \langle \dot{k}^j(\varrho) : \varrho \in \operatorname{split}(\dot{u}(\mu))\rangle) \\ = (\dot{p}^\beta_\nu, \langle \dot{k}^\beta_\nu(\varrho) : \varrho \in \operatorname{split}(\dot{p}^\beta_\nu)\rangle),\end{aligned}$$

the row γ at which this pair has been taken care of in the construction of the matrix $\dot{A}^\beta$ (denoted γ_ν in that construction) is below γ_β. Moreover, if $p_{m+1}(\beta) = (\dot{u}', \dot{v}')$, then either

$$\{\dot{k}^j(\varrho) : \varrho \in \operatorname{split}(\dot{u}'(\mu^\frown i))\} \in \dot{A}^\beta_\gamma \setminus \bigcup_{\delta<\beta} \dot{A}^\delta_\gamma$$

for some $\gamma < \gamma_\beta$, for every $i \in \operatorname{succ}_{u'(\mu)}(\mu)$, or there exists some countable $C \subseteq \bigcup_{\gamma<\gamma_\beta} \dot{A}^\beta_\gamma$ such that for no $r \leq^0 \dot{u}'(\mu)$ is it the case that for every $a \in C$ the set $\{\dot{k}^j(\varrho) : \varrho \in \operatorname{split}(r)\}$ is either almost contained in a or almost disjoint from a, or for every $\delta < \omega_1$ and $i \in \operatorname{succ}_{\dot{u}'}(\operatorname{st}(\dot{u}'))$ there exists $a \in \dot{A}^\beta_\delta$ such that

$$\{\dot{k}^j(\varrho) : \varrho \in \operatorname{split}(\dot{u}'(\operatorname{st}(\dot{u}')^\frown i))\} \subseteq^* a.$$

Now suppose that we have the constant case at (μ, ν). Let $(\dot{u}, \dot{v})$ and $(\dot{u}', \dot{v}')$ be as above. Let $\langle \mu_l : l < \omega\rangle$ enumerate

$\mathrm{Succ}_{\dot{u}(\mu)}(\mu) \cap \mathrm{Sop}^{\dot{u},\dot{v}}(\mu, \nu)$. For every $r \leq^0 \dot{u}(\mu)$ let

$$d(r, \mu, \nu, i) = \left\{\min(y_i^{\mu_l,\nu} \setminus w_i^{\mu,\nu}) \colon \mu_l \in r\right\}.$$

Then either $\forall \nu < \omega_1 \, \exists b \in \dot{\mathcal{A}}_\nu^\alpha \, d(\dot{u}', \mu, \nu, i) \subseteq^* b$, or else we have $d(\dot{u}', \mu, \nu, i) \in \dot{\mathcal{A}}_\nu^{<\beta}$ for some $\nu < \gamma_\beta$. Moreover we make sure that no μ_l is a splitnode of $\dot{u}'$.

Finally suppose that we have the tree case at (μ, ν). let $\dot{k}(\varrho) = \max(\dot{y}_i^{\varrho,\nu}(\beta))$ for every $\varrho \in \mathrm{Sop}^{\dot{u},\dot{v}}(\mu, \nu)$. Letting $\nu < \omega_1$ such that

$$\left(\dot{u}(\mu), \left\langle \dot{k}(\varrho) \colon \varrho \in \mathrm{split}(\dot{u}(\mu))\right\rangle\right) = \left(\dot{p}_\nu^\beta, \left\langle \dot{k}_\nu^\beta(\varrho) \colon \varrho \in \mathrm{split}\left(\dot{p}_\nu^\beta\right)\right\rangle\right),$$

then the row γ at which this pair has been taken care of in the construction of the matrix $\dot{\mathcal{A}}^\beta$ (denoted γ_ν in that construction) is before γ_β. Moreover, if $p_{m+1}(\beta) = (\dot{u}', \dot{v}')$, then either

$$\left\{\dot{k}(\varrho) \colon \varrho \in \mathrm{split}\left(\dot{u}'(\mu^\frown i)\right)\right\} \in \dot{\mathcal{A}}_\gamma^\beta \setminus \dot{\mathcal{A}}_\gamma^{<\beta}$$

for some $\gamma < \gamma_\beta$, for every $i \in \mathrm{succ}_{u'(\mu)}(\mu)$, or there exists some countable $C \subseteq \bigcup_{\gamma<\gamma_\beta} \dot{\mathcal{A}}_\gamma^\beta$ such that for no $r \leq^0 \dot{u}'(\mu)$ is it the case that for every $a \in C$ the set $\{\dot{k}(\varrho) \colon \varrho \in \mathrm{split}(r)\}$ is either almost contained in a or almost disjoint from a, or for every $\delta < \omega_1$ and $i \in \mathrm{succ}_{\dot{u}'}(\mathrm{st}(\dot{u}'))$ there exists $a \in \mathcal{A}_\delta^\beta$ such that

$$\left\{\dot{k}(\varrho) \colon \varrho \in \mathrm{split}\left(\dot{u}'\left(\mathrm{st}\left(\dot{u}'\right)^\frown i\right)\right)\right\} \subseteq^* a.$$

3. For every $\beta \in \mathrm{supp}(p_k)$, all the (countably many) coordinates needed to evaluate $p_k(\beta)$ and the families

$$\left\langle \dot{w}_i^{\mu,\nu}(\beta), \dot{y}_{i'}^{\mu,\nu}(\beta) \colon \langle \mu, \nu\rangle \in \mathrm{TP}^n(p(\beta)) \wedge i < n+2, i' < n+1\right\rangle$$

and $\left\langle \dot{\Gamma}_n(\beta) \colon n < \omega\right\rangle$, and also their supremum belong to $\mathrm{supp}(p_m)$, for some $m > k$. Hence if $\bar{\beta}$ is this supremum and $\bar{\beta} < \beta$, then all these objects are $P_{\bar{\beta}}$-names. Moreover we require that there exists $\gamma'_\beta < \omega_1$ such that the precise analogues of 2a and 2b hold about these objects at stage $\bar{\beta}$ before row γ'_β of the matrix $\dot{\mathcal{A}}^{\bar{\beta}}$. In particular, $p_{m+1}(\beta)$ will be obtained by shrinking $p_m(\beta)$ accordingly.

Now suppose q is the infimum of $\{p_k \colon k < \omega\}$ and let

$$\gamma = \sup\left\{\nu_0, \gamma_\beta, \gamma'_\beta \colon \beta \in \mathrm{supp}(q)\right\}.$$

In order to get a contradiction, suppose that for some P_{ω_2}-extension $r \leq q$, $\beta < \omega_2$, $\dot{b} \in \dot{\mathcal{A}}_\gamma^\beta \setminus \dot{\mathcal{A}}_\gamma^{<\beta}$ and $n_0 < \omega$, we have

$$(\dagger) \qquad r \Vdash_{P_{\omega_2}} \dot{a} \setminus n_0 \subseteq \dot{b}.$$

Case 1. Suppose $\beta \in \operatorname{supp}(q)$. We work in V^{P_β} below $r \restriction \beta$. Let $r(\beta) = (\dot{u}, \dot{v})$. We say that $\dot{w}_i^{\mu,\nu}(\beta)$ or $\dot{y}_{i'}^{\mu,\nu}(\beta)$ is *relevant* (*for* $r \restriction [\beta, \omega_2)$), iff $\langle \mu, \nu \rangle \in \mathrm{TP}^n(\dot{u}, \dot{v})$ for some n with $i < n+2$, $i' < n+1$ and $\langle \dot{u}(\mu), \dot{v}(\nu) \rangle^\frown \dot{r}_i^{n+1} \| r'$ or $\langle \dot{u}(\mu), \dot{v}(\nu) \rangle^\frown \dot{r}_{i'}^{n} \| r'$, respectively. We investigate two subcases:

Subcase 1*a*. There exists an infinite $\dot{w}_i^{\mu,\nu}(\beta)$ or $\dot{y}_i^{\mu,\nu}(\beta)$ that is relevant for $r \restriction [\beta, \omega_2)$. Suppose for example that $\dot{w}_i^{\mu,\nu}(\beta)$ is relevant and infinite for some $\langle \mu, \nu \rangle \in \mathrm{TP}^n(r'(0))$ with even type n and some $i < n+2$. By 2.(b)(i) (fusion) and the construction of $\dot{\mathcal{M}}$ we conclude that $\dot{w}_i^{\mu,\nu}(\beta) \setminus b$ is infinite. Pick $k \in (\dot{w}_i^{\mu,\nu}(\beta) \setminus b) \setminus n_0$. By the definition of $\dot{w}_i^{\mu,\nu}(\beta)$ we can get $\dot{u}' \leq^0 \dot{u}$ by pruning $\dot{u}$ at μ only, such that

$$(\dot{u}', \dot{v})^\frown r \restriction [\beta+1, \omega_2) \Vdash k \in \dot{a}.$$

This is a contradiction.

Subcase 1*b*. All relevant $\dot{w}_i^{\mu,\nu}(\beta)$ and $\dot{y}_i^{\mu,\nu}(\beta)$ are finite. Therefore we are always in the finite case of 3 of Lemma 19. By applying the fusion property of $\mathbb{M}^2$, 2(b)ii and the construction of $\dot{\mathcal{M}}$, we can easily find $(\dot{u}', \dot{v}') \leq^0 (\dot{u}, \dot{v})$ such that for every $(\mu, \nu) \in \mathrm{TP}(\dot{u}', \dot{v}')$ the following hold (we only state it for even type-pairs):

Suppose we have the empty case at (μ, ν). Let

$$\dot{k}^j(\varrho) = \min\left(\dot{w}_{i_j}^{\varrho,\nu}(\beta) \setminus \dot{w}_i^{\mu,\nu}(\beta)\right)$$

for every $\varrho \in \mathrm{Sop}^{\dot{u}',\dot{v}'}(\mu, \nu)$ and $j < 2$. Then

$$b \cap \left\{\dot{k}^j(\varrho) \colon \varrho \in \mathrm{Sop}^{\dot{u}',\dot{v}'}(\mu, \nu)\right\}$$

is empty for every $j < 2$.

If we have the constant case at (μ, ν), then either $b \cap d(\dot{u}'(\mu), \mu, \nu, i)$ is empty or else $b \subseteq^* d(\dot{u}'(\mu), \mu, \nu, i)$.. Note that by Lemma 19, by 2(b)ii and the construction of $\dot{\mathcal{M}}$ the second alternative holds for at most one μ.

Finally suppose we have the tree case at (μ, ν). Let

$$\dot{k}(\varrho) = \max\left(\dot{y}_i^{\varrho,\nu}(\beta)\right)$$

for every $\varrho \in \mathrm{Sop}^{\dot{u}',\dot{v}'}(\mu, \nu)$. Then

$$b \cap \left\{\dot{k}(\varrho) \colon \varrho \in \mathrm{Sop}^{\dot{u}',\dot{v}'}(\mu, \nu)\right\}$$

is empty.

Now fix a P_{ω_2}-generic filter G containing

$$r \restriction \beta^\frown (\dot{u}', \dot{v}')^\frown r \restriction [\beta+1, \omega_2),$$

and let $G(\beta)$ determine the generic reals (g_0, g_1). We write $\dot{w}_i^{\mu,\nu}(\beta)[G] = w_i^{\mu,\nu}$, $\dot{u}'[G] = u'$ etc. For any $(\mu, \nu) \in \mathrm{TP}^n(u', v')$, let $i < n+2$ be the unique number such that we have $r \restriction \beta^\frown (\dot{u}', \dot{v}')^\frown \dot{r}_i^{n+1} \in G$ and let $w^{\mu,\nu} = w_i^{\mu,\nu}$, similarly for $y^{\mu,\nu}$.

Let

$$I = \{\langle \mu, \nu \rangle : \exists n \exists \langle \mu', \nu' \rangle \, [\langle \mu, \nu \rangle = \text{tp-}n\text{-pair}\,(g_0, g_1) \wedge \langle \mu', \nu' \rangle = \text{tp-}(n+1)\text{-pair}\,(g_0, g_1) \wedge w^{\mu',\nu'} \setminus w^{\mu,\nu} \neq \emptyset]\},$$

the set of type-pairs where the new part of w gives a contribution to $\dot{a}[G]$. Note that I is infinite.

Subcase $1b\alpha$. We have the empty case for infinitely many $(\mu, \nu) \in I$. We may assume without loss of generality that infinitely many of them have even type. Suppose (μ, ν) is one of them, say $\text{tp}(\mu, \nu) = 2n$, and suppose that (μ', ν) is the type-$2n+1$-pair of (g_0, g_1). By construction we have that $k(\mu') := \min(w^{\mu',\nu} \setminus w^{\mu,\nu})$ belongs to $\dot{a}[G] \setminus b$. As clearly $k(\mu') \neq k(\mu'')$ for $\mu' \subsetneq \mu''$ we conclude that $\dot{a}[G] \setminus b$ is infinite, a contradiction.

Subcase $1b\beta$. We have the constant case for infinitely many $(\mu, \nu) \in I$. Again we may assume that infinitely many of them have even type. Let (μ, ν) be one of them and $\text{tp}(\mu, \nu) = 2n$. Let (μ', ν) be the type-$2n+1$-pair of (g_0, g_1). By construction there exists $\mu \subsetneq \varrho \subsetneq \mu'$ such that $\text{tp}(\varrho, \nu) = 2n+1$, and hence $k(\varrho) := \max(y^{\varrho,\nu})$ belongs to $\dot{a}[G] \setminus b$. As for different ϱ the $k(\varrho)$ are different, we conclude that $\dot{a}[G] \setminus b$ is infinite, a contradiction.

Subcase $1b\gamma$. We have the tree case for infinitely many $(\mu, \nu) \in I$. We get a contradiction as in Subcase $1b\alpha$, using the $k(\varrho)$ defined above.

Case 2. Suppose $\beta > \alpha$. As $\dot{a} \in V^{P_\alpha}$ and $b \in \dot{\mathcal{A}}_\gamma^\beta \setminus \dot{\mathcal{A}}_\gamma^{<\beta}$, this contradicts the maximality of $\mathcal{A}_\gamma^\alpha$ in V^{P_α}.

Case 3. Suppose $\beta = \alpha$. By 1 of the fusion and as $\nu_0 < \gamma$, we have that $q \Vdash |\dot{a} \setminus b| = \omega$, a contradiction.

Case 4. Suppose $\beta \notin \text{supp}(q)$ and $\beta < \alpha$. Choose $\beta' \in \text{supp}(q)$ minimal such that $\beta' > \beta$.

By 3 of the fusion, letting $\bar{\beta}$ the supremum of all coordinates needed to evaluate $p_k(\beta')$ and the families

$$\langle \dot{w}_i^{\mu,\nu}(\beta'), \dot{y}_{i'}^{\mu,\nu}(\beta') : \langle \mu, \nu \rangle \in \text{TP}^n(p(\beta')) \wedge i < n+2, i' < \text{tp}(\mu, \nu) + 1 \rangle,$$

$\langle \dot{\Gamma}_n(\beta') : n < \omega \rangle$, we have $\bar{\beta} < \beta$. We use the term *relevant* as above.

Subcase $4a$. For at least one $\langle \mu, \nu \rangle \in r(\beta')$ and $i < \omega$ we have an infinite $\dot{w}_i^{\mu,\nu}(\beta')$ or $\dot{y}_i^{\mu,\nu}(\beta')$ that is relevant for $r(\beta')$. As $\dot{w}_i^{\mu,\nu}(\beta'), \dot{y}_i^{\mu,\nu}(\beta') \in V^{P_{\bar{\beta}}}$ and $\dot{b}$ is added at the later stage β, by construction of $\dot{\mathcal{M}}$ we know that $\dot{w}_i^{\mu,\nu}(\beta') \setminus \dot{b}$ or $\dot{y}_i^{\mu,\nu}(\beta') \setminus \dot{b}$ is infinite. So since $\dot{a}$ contains more and more of $\dot{w}_i^{\mu,\nu}(\beta')$ or $\dot{y}_i^{\mu,\nu}(\beta')$ as we go right with the succeeding splitnode of $\langle \mu, \nu \rangle$, we can force some point k above n_0 into $\dot{a}$ which is outside $\dot{b}$.

Subcase $4b$. All relevant $\dot{w}_i^{\mu,\nu}(\beta')$ and $\dot{y}_i^{\mu,\nu}(\beta')$ are finite. Essentially as in Case 1, Subcase 1b, working in $V^{P_{\beta'}}$, we construct $\dot{r}' \leq^0 r(\beta')$ such that

$$r \restriction \beta^\frown \dot{r}'^\frown r \restriction [\beta + 1, \omega_2) \Vdash_{P_{\omega_2}} |\dot{a} \setminus b| = \omega.$$

For this we use the same arguments together with Remark 22.

This completes the proof of Theorem 23 and hence of Theorem 15. ⊣

REFERENCES

[1] J. E. BAUMGARTNER, *Iterated forcing*, ***Surveys in set theory*** (A. R. D. Mathias, editor), London Mathematical Society Lecture Notes, vol. 8, Cambridge University Press, 1983, pp. 1–59.

[2] J. BRENDLE, *Combinatorial aspects of the meager and null ideals and of other ideals on the reals*, 1994/95, Thesis for Habilitation, Eberhard-Karls-Universität Tuebingen.

[3] M. GOLDSTERN, M. JOHNSON, and O. SPINAS, *Towers on trees*, ***Proceedings of the American Mathematical Society***, vol. 122 (1994), pp. 7557–564.

[4] M. GOLDSTERN, M. REPICKÝ, S. SHELAH, and O. SPINAS, *On tree ideals*, ***Proceedings of the American Mathematical Society***, vol. 123 (1995), pp. 1573–1581.

[5] H. JUDAH, A. MILLER, and S. SHELAH, *Sacks forcing, Miller forcing and Martin's axiom*, ***Archive for Mathematical Logic***, vol. 31 (1992), pp. 145–161.

[6] H. JUDAH and S. SHELAH, *The Kunen-Miller chart*, ***The Journal of Symbolic Logic***, vol. 55 (1990), pp. 909–927.

[7] A. S. KECHRIS, *On a notion of smallness for subsets of the Baire space*, ***Transactions of the American Mathematical Society***, vol. 229 (1977), pp. 191–207.

[8] L. LAVER, *On the consistency of Borel's conjecture*, ***Acta Mathematica***, vol. 137 (1976), pp. 151–169.

[9] E. MARCZEWSKI, *Sur une classe de fonctions de W. Sierpiński et la classe correspondente d'ensembles*, ***Fundamentae Mathematicae***, vol. 24 (1935), pp. 17–34.

[10] A. MILLER, *Rational perfect set forcing*, ***Contemporary Mathematics***, vol. 31 (1984), pp. 143–159.

[11] S. SHELAH, *On cardinal invariants of the continuum*, ***Contemporary Mathematics***, vol. 31 (1984), pp. 183–207.

[12] O. SPINAS, *Generic trees*, ***The Journal of Symbolic Logic***, vol. 60 (1995), pp. 705–726.

[13] ———, *Canonical behaviour of Borel functions on superperfect rectangles*, ***Journal of Mathematical Logic***, vol. 1 (2001), no. 2, pp. 173–220.

[14] ———, *Ramsey and freeness properties of Polish planes*, ***Proceedings of the London Mathematical Society***, vol. 82 (2001), no. 3, pp. 31–63.

[15] J. ST. RAYMOND, *Approximation des sous-ensembles analytiques par l'inérieur*, ***Comptes Rendus de l'Académie des Sciences. Série I. Mathématique***, vol. 281 (1975), pp. 85–87.

[16] B. VELIČKOVIĆ, *CCC posets of perfect trees*, ***Compositio Mathematicae***, vol. 79 (1991), pp. 279–294.

[17] B. VELIČKOVIĆ and H. WOODIN, *Complexity of reals in inner models of set theory*, ***Annals of Pure and Applied Logic***, vol. 92 (1998), pp. 283–295.

MATHEMATIK DEPARTEMENT
ETH ZÜRICH
8092 ZÜRICH, SWITZERLAND
E-mail: jossen@math.ethz.ch

MATHEMATISCHES SEMINAR
CHRISTIAN-ALBRECHTS-UNIVERSITÄT ZU KIEL
24098 KIEL, GERMANY
E-mail: spinas@math.uni-kiel.de

PSYCHOLOGY LOOKS HOPEFULLY TO LOGIC

DANIEL N. OSHERSON AND ERIC MARTIN

Abstract. The development of science is among the most distinctive accomplishments of the human species. To help clarify how science is achieved, psychologists need a perspective on inductive logic that does not invoke the subjective probability of rival theories. One alternative starts from the all-or-none concept of acceptance as opposed to graded belief. The inductive logic of acceptance is governed by principles of hypothesis selection and revision rather than probability. This theory of inductive logic has already seen development, but many questions remain.

§1. The psychologist's problem. Since its inception in the 19th century, psychological science has made steady progress investigating perceptual and motoric abilities—how the visual system encodes color, for example, or how we shift our gaze to peripheral events.[1] Much less is understood about abilities that make us distinctly human. Some information is available about the mechanisms of natural language.[2] But there is hardly any insight into how people create scientific theories about the world. This is an embarassing gap for Psychology since scientific achievement is the most distinctive and remarkable feature of our species.

What's blocking progress is that the most natural account of this ability seems to face an insuperable difficulty. According to the account in question, most everyone has an innate disposition to reason in rough conformity with normatively correct principles of deductive and inductive logic—just as most everyone is endowed with perceptual mechanisms that give us a roughly accurate picture of the environment. How else could our ancestors have met the challenges of survival? It is the twin pillars of natural reasoning—deductive and inductive—that allow people to draw out the consequences of rival scientific theories and assign sensible credibilities to each in the light of data.

These vague remarks are just an attempt to prepare for sharper theories. But we stumble even at this initial step, because one of the twin pillars seems to be absent. The problem is not so much with deductive logic. It can be challenging to communicate the informal concept of logical necessity. But

---

Thanks to Richard Grandy and Moshe Vardi for comments on an earlier version of this paper.

[1] For an entry to the literature on visual perception and action, see [17], [31, Chapters 10–11].

[2] An introduction to the study of natural language is available in [8] and [5, Ch. 8].

Logic Colloquium 2000
Edited by R. Cori, A. Razborov, S. Todorčević, and C. Wood
Lecture Notes in Logic, 19

once this is achieved, most people distinguish validity from invalidity on an intuitive basis across a broad class of arguments.[3]

The situation appears to be different for the other pillar. We usually conceive of inductive logic as governing numerical measures of credibility, and as imposing constraints on their assignment to events or propositions. Credibilities governed by the usual postulates are called "probabilities." The probabilities in question would be subjective, since a frequency conception seems not to apply to the accreditation of scientific theories.

Now there is no doubt that people enjoy rich intuitions about chances. Most everyone expects the internet to expand next year, and Tony Blair to be in the news. It doesn't seem to be an accident that we are often right about such cases.[4] But notice that the events just mentioned—involving the internet and Tony Blair—are logically unrelated. In contrast, when people are called upon to distribute credibilities over logically connected events we find cause for alarm about naive inductive logic.

Indeed, numerous experiments have documented the tendency of both college students and professionals to supply incoherent estimates when asked to judge the probability of logically related events.[5] The "incoherence" in question consists in citing numerical chances that violate elementary axioms of probability theory. Even in the simple case of a conjunction of sentences versus one of its conjuncts, people willingly assign greater chance to the former compared to the latter when certain conditions are met.[6] This so-called "conjunction fallacy" has given rise to a voluminous and occasionally acrimonious literature, but there is little doubt that it survives arduous attempts to clarify the problem to respondents.[7]

So this is the psychologist's problem. Probability seems not to ground the inductive logic necessary for scientific discovery. Yet signs of human scientific prowess abound.

§2. Inductive logic based on acceptance. One possibility is that inductive competence is reserved for a small elite. Perhaps science can be explained to the masses, but can only be created by the few who are sensitive to probabilistic coherence. Episodes of theory-evaluation in the history of science have in fact been reconstructed in Bayesian terms.[8] But it seems hard to picture pre-modern

[3]Alternative theories of the psychology of deductive inference are offered in [13, 30]. Commentary on the theoretical options is provided in [25]. The functional neuroanatomy of deductive reasoning is discussed in [1, 26, 27].

[4]For the accuracy of subjectively assessed probability, see [14] and [28, Ch. 19].

[5]See [24, 12] and references cited there.

[6]The conditions in question are discussed in [32].

[7]For opposing positions, see [15, 7]. The psychological difficulty of maintaining probabilistic coherence should perhaps be no cause for surprise. It is known to be an intractable problem even for computers. See [6].

[8]See [3], and [22] for an opposing view.

scientists like Copernicus or Newton calculating the odds that various theories are right, even implicitly.[9]

Anyway, on the assumption that scientific competence is more democratically distributed, what's needed is an inductive logic that doesn't rely on probability. An alternative foundation starts with the concept of *hypothesis acceptance*. Instead of scattering degrees-of-belief over all the hypotheses in play, only one is accepted at a time. This is the hypothesis designated for elaboration and test, and for public defense as the most interesting and viable idea at a particular moment of investigation.[10] Acceptance is provisional, of course. New observations may lead the scientist to an alternative view.

Within the acceptance perspective, scientific inquiry proceeds on the basis of *theory selection and revision*. Suppose that plausible schemes for these processes could in fact be shown to embody successful inductive strategies. Then we would be able to see, at least in principle, how a creature not born to maintain probabilistic coherence might nonetheless find out so much about the world.[11]

Recent studies have begun to formalize inductive logic in these terms. We'll now review a few results and indicate gaps in our understanding.[12] Rational strategies for discovery are considered first. Then we return to the human case.

§3. A formal model of scientific inquiry. Our analysis of inquiry consists of three parts. First, a class of inductive problems is specified, along with criteria for success and failure in a given problem. Next, schemes for theory-choice and revision are introduced. Last, the success of the schemes for solving the inductive problems is assessed.

Turning to the first part of our approach, inductive problems can be conceived as a guessing game between two players. Call them "Nature" and "the scientist." The following game-pieces are used:

1. a countable, first-order language $\mathcal{L}$, with identity,
2. a collection of countable structures that interpret $\mathcal{L}$, and
3. a partition of the collection into cells.

To illustrate, let the nonlogical vocabulary be limited to a single unary predicate, say, H. Let the countable structures be all those in which the extension of H is finite. And let the partition be based on the cardinality of

[9]For example, the chance of a theory being correct is not mentioned in Newton's "Rules of Reasoning" [23]. A similar absence is evident in Copernicus' preface to *De Revolutionibus* (as presented in [18, Ch. 5]).

[10]For analysis of the concept of acceptance, see [2]. Karl Popper's philosophy of science [29] contains similar ideas.

[11]For comparison of probabilistic versus acceptance based inquiry, see [4, 16].

[12] For proofs and extensions of the results cited below, see [21, 20, 19]. Other approaches to acceptance-based discovery include [16, 11].

H's extension. In one cell, the extension has cardinality 0; in another, it has cardinality 1, and so forth (there is no cell for infinite cardinality).

Nature prepares herself for the game as follows. First she chooses one structure $\mathcal{S}$ from some cell of the partition. This structure is "reality." Pursuing our illustration, she may choose the structure consisting of the positive integers with H interpreted as the set $\{1, 2, 3\}$. She then maps the variables onto the domain of $\mathcal{S}$, thereby using the variables as names for all the objects in the countable universe of $\mathcal{S}$. Nature is allowed to choose any surjective mapping she pleases.

Next, Nature chooses an ω-sequence of all the basic formulas true in her choices of structure and variable-mapping. For example, the ω-sequence might start this way:

$$\neg Hv_0 \quad Hv_1 \quad Hv_2 \quad v_1 \neq v_2 \ldots.$$

These are the data made available to the scientist. It's as if Nature says:

> "H is false of the first thing I'm showing you, H is true of the second and third things. The second and third things are distinct . . . ."

and so forth. The scientist examines ever longer, finite initial segments of Nature's ω-sequence of data. After each segment, he designates a cell of the partition as his guess about the origin of the structure Nature chose at the outset. The scientist wins the game just in case all but finitely many of his conjectures are correct. So, in our illustrative game, the scientist's conjectures must ultimately stabilize to the cell in which the cardinality of H's extension is three.

Formally, a scientist is a mapping from the set of finite initial segments of any ω-sequence of data to the collection of classes of structures (i.e., potential cells of a partition). A scientist is considered to *converge* to cell C of a partition just in case cofinitely many of his conjectures are nonempty subsets of C (hence, cofinitely often, the scientist's conjectures are consistent and imply C).

If you are willing to think of classes of structures as *propositions*, then the scientist is confronted with a family of mutually incompatible propositions, and his guesses are supposed to converge to the correct one. In our illustration, the propositions correspond to the claims that there are n things satisfying H.

The data available to the scientist in the finite initial sequences of Nature's ω-sequence might never imply the correct cell of the partition. In our illustration at no stage of the game is the cardinality of H's extension made certain to the scientist. But success in the game may nevertheless be possible.

Indeed, in our illustration, the scientist has a reliable strategy. If at each round of the game he guesses the cell corresponding to the smallest cardinality for H that is consistent with his data, then he is guaranteed to converge to the correct cell. A scientist following this strategy is thus *reliable* in the sense of succeeding regardless of Nature's choices.

In contrast, suppose that we add a cell made up of structures in which the extension of H is infinite. Then it can be shown that no scientist succeeds reliably in the game based on this expanded partition. In other words, no function from data to cells of the expanded partition is guaranteed to converge to the right cell.

Let us give a name to reliable success.

DEFINITION 3.1. Let partition $\mathcal{P}$ and scientist Ψ be given. Ψ *solves* $\mathcal{P}$ just in case for every cell $C \in \mathcal{P}$, every structure $\mathcal{S} \in C$, every variable-assignment h onto $|\mathcal{S}|$, and every ω-sequence O of the basic formulas true in $(\mathcal{S}, h)$, Ψ converges on O to C.

Thus, we say that the scientist *solves* a partition if no matter what structure Nature chooses, no matter how she maps variables onto the structure's domain, and no matter how she lists the basic facts, the scientist's guesses converge to the cell from which the structure was drawn. Intuitively, the scientist discovers which proposition is true of his world. Thus, our first partition is *solvable* whereas the expanded partition is not solvable.

To describe a different kind of partition, call a total order *one-sided* if it has a least point or a greatest point but not both. Suppose that the nonlogical vocabulary of $\mathcal{L}$ contains just the binary relation symbol R, and consider a partition of all countable structures that interpret R as a one-sided order. In one cell lie the one-sided orders with a least point. In the other cell are the one-sided orders with a greatest point. There are no other cells. Call this partition "the one-sided split." It is solved by alternating between subconjectures of the form *variable* v_i *is the least point* and *variable* v_i *is the greatest point*. A given subconjecture implies one of the cells of the partition, and it is retained until contradicted by the data. The partition becomes unsolvable if a third cell is added containing the *two-sided* orders, namely, total orders with both a least and greatest point.

The one-sided split and its expansion have the particularity of being specifiable within the first-order language $\mathcal{L}$. To state the matter precisely, we rely on the following definition.

DEFINITION 3.2. A partition $\mathcal{P}$ is *first-order* iff there is a set $T \subseteq \mathcal{L}$ of sentences, and further sentences $\theta_0 \ldots \theta_n$ such that:

1. every model of T satisfies exactly one of the θ_i's, and
2. the cells of $\mathcal{P}$ have the form $MOD(T \cup \{\theta_i\})$,

where $MOD(X)$ are the countable models of X.

In such partition, the set T is said to *underlie* the partition. First-order partitions resemble scientific settings in which an underlying theory T delimits the collection of possible realities, and research is oriented towards extending the theory by selecting a new axiom θ_i from among a finite set of them.[13] We have already seen that first-order definability is neither necessary nor sufficient for solvability. Thus, our first example—involving different finite cardinalities for the extension of the predicate H—is solvable but not first-order. And expanding the partition of one-sided orders by adding a cell for the two-sided orders yields a partition that is first-order but not solvable. As our examples merely hint, it is not trivial to trace the boundary between the solvable and unsolvable partitions. Combinatorial conditions for solvability have been discovered in recent years.[14]

The inductive games described above can be modified in various ways. For example, the data made available to scientists can be enriched or impoverished, the restriction to countable structures can be dropped, and success can be made a more graded affair. These variants and others have been discussed in the literature devoted to acceptance-based inductive logic. Rather than review these developments, we will stick with the simple paradigm presented here, and turn to the second component of our analysis—schemes of theory-selection and revision.

§4. **Theory-selection and revision.** Here we look for inspiration to well-known work on belief revision in the tradition of Alchourrón, Gärdenfors and Makinson, and especially to the more recent work of Sven Hansson.[15] In the context of scientific discovery, however, revision is conceived somewhat differently compared to earlier work.

Within our analysis, revision-based scientists have the form $\lambda\sigma \,.\, X \dot{+} \sigma$, where X is a starting theory, consisting of an arbitrary set of formulas, and $\dot{+}$ is a *belief-revision* function (to be explained shortly). Thus, scientists based on belief revision have two working parts. The first is a starting theory, which the scientist chooses prior to encountering data. The second is a scheme for revising the starting theory under the impact of data. Such a scheme is called a "revision function," often denoted by the dotted plus. It maps a given theory X and a given sequence σ of data into another theory, $X \dot{+} \sigma$. The latter theory will consist of an appropriately chosen subset of X (the fragment that survives the confrontation with fact) along with the formulas appearing in σ (the facts themselves).

To fill in the picture, imagine the scientist working his way along the data-enumeration provided by Nature. At each stage, the scientist views an increasing sequence σ of basic formulas. In response, the scientist consults his

[13]By the compactness of first-order logic, there is no theory T whose models are partitioned by an infinite collection of logically distinct sentences θ_i.

[14]See the references in note 12.

[15]See, for example, [9]. A very useful survey of theories is [10].

starting theory X and his revision operator $\dot{+}$. He computes $X \dot{+} \sigma$, then asks: "Does the class of (countable) structures that satisfy the new theory fit inside a unique cell of the game's partition?" If so, then this cell is the scientist's conjecture. In the absence of such a cell, no conjecture is made. Note that the same starting theory X is used at each stage of the game. Only the data change, through accumulation. Call scientists who operate in this way *revision based.*

It remains to describe the kind of revision operators that equip revision based scientists. A preliminary definition is needed.

DEFINITION 4.1. Let $X \subseteq \mathcal{L}$ and data-sequence σ be given. A formula $\varphi \in X$ is called *σ-innocent* (with respect to X) iff there is no $A \subseteq X$ such that:

1. A contradicts σ.
2. Every proper subset of A is consistent with σ.
3. $\varphi \in A$.

Thus, a formula of X is called *σ-innocent* if it is a member of no $\subseteq$-minimal part of X contradicted by σ. It seems clear that every σ-innocent formula in X should survive X's confrontation with σ, since such formulas are not involved in contradictions between X and σ. This is because no such formula is involved in a contradiction between the theory X and the data σ. Preserving the σ-innocent members of X is the principal condition that we impose on theory revision. Officially:

DEFINITION 4.2. A *revision* of a starting theory X in the face of data σ is any set of the form $Y \cup range(\sigma)$, where Y is a subset of X that is consistent with σ and includes all of X's σ-innocent members.

Thus, it is required that the theory resulting from the collision between starting theory X and data σ include the data σ, be internally consistent, embrace X's σ-innocent members, and contain no formula outside of X and σ. Any mapping of theories X and data σ into a successor theory having the foregoing properties counts as a *revision function*. In turn (as noted earlier), any function of the form $\lambda\sigma \,.\, X \dot{+} \sigma$, where $X \subseteq \mathcal{L}$ and $\dot{+}$ is a revision function, is a revision based scientist.

§5. Discovery through belief revision. Now we can consider the inductive powers of revision based scientists. What partitions do they solve? The following theorem illustrates findings in this area.

PROPOSITION 5.1. *Let $T \subseteq \mathcal{L}$ be an arbitrary set of sentences, and let $\mathcal{P}$ be any partition of the collection of T's countable models. If $\mathcal{P}$ is solvable (by any kind of scientist) then there is a starting theory $X \subseteq \mathcal{L}$ such that:*

1. *X is a consistent extension of T, and*
2. *every revision based scientist $\lambda\sigma \,.\, X \dot{+} \sigma$ using X as starting theory solves the partition (the revision function $\dot{+}$ can be chosen arbitrarily).*

On these kinds of partitions, if successful inquiry is possible at all, then it can be carried out using any revision function whatsoever. Since the class of revision functions is broad, the theorem seems to put inductive proficiency within common reach—without the use of any probability concepts, just as we wanted. But such a cheerful interpretation encounters difficulties. Only a few will be discussed here.

§6. Efficient inquiry. Suppose we are given a partition. Recall that after Nature chooses a structure from one of its cells, she goes on to choose a variable-mapping for that structure, and a listing of all the true basic formulas that result. The structure is thus associated with uncountably many potential streams of data. The scientist is required to succeed on all of them, eventually converging to the correct guess about the cell from which Nature drew her choice.

On any given data stream, the successful scientist reaches an earliest stage after which all of his conjectures are accurate. Call this the *convergence point* for the scientist and the data stream. Now consider another successful scientist whose convergence points never come later than those for the first scientist, and sometimes come earlier. This second scientist may be said to *dominate* the first, in the sense of being a strictly more efficient inductive agent for the partition in question. We consider a scientist to be *efficient* for a partition if it solves the partition and is dominated by no other scientist that solves it.

In this weak sense of efficiency, revision functions are capable of efficient induction on every solvable problem. But belief revision is no guarantee of efficiency. There are revision functions that cannot be used to efficiently solve certain solvable partitions, no matter what starting theory is employed. This fact can be formulated as follows.

PROPOSITION 6.1. *Suppose that the nonlogical vocabulary of $\mathcal{L}$ consists of countably many constants. Then there is a decidable set T of sentences and a sentence θ with the following properties.*

1. *The partition*

$$\mathcal{P} = \{\ MOD(T \cup \{\theta\}), \quad MOD(T \cup \{\neg\theta\})\ \}$$

is solvable efficiently.

2. *There are revision functions $\dotplus$ such that for every $B \subseteq \mathcal{L}$, the revision based scientist $\lambda\sigma\,.\,B \dotplus \sigma$ fails to solve $\mathcal{P}$ efficiently. (That is, some other scientist dominates $\lambda\sigma\,.\,B \dotplus \sigma$ on $\mathcal{P}$.)*

Thus, the revision based scientists using the revision functions evoked in the proposition are dominated by other scientists who solve the same partitions strictly faster.

We seem to be faced with a defect in our conception of belief revision, which is broad enough to include pathological functions. For creatures with finite lives, it seems unreasonable to employ a dominated method of hypothesis choice. So, we seek an additional property of revision functions that is intuitively rational and guarantees efficiency.

An attractive additional constraint on belief revision is to require the successor theory to be a $\subseteq$-maximal choice among potential candidates. More precisely, a revision function $\dotplus$ is called *maxichoice* iff for every starting theory X and data-sequence σ, there is no revision function $\dot{\oplus}$ with $X \dotplus \sigma \subset X \dot{\oplus} \sigma$. Maxichoice revision functions implement the conservative policy of inflicting minimal change on existing theories, since a largest possible subset of the original theory is preserved.

For a yet narrower kind of revision function, consider the set Y in the successor $Y \cup range(\sigma)$ to X and σ. We can require it to be the first admissible candidate in a preestablished total ordering of all sets of formulas. The ordering is thought of as coding *a priori* preferences over theories. Revision functions that meet this additional constraint are called *definite*. Officially:

DEFINITION 6.2. Recall that a revision of a starting theory X in the face of data σ is any set of the form $Y \cup range(\sigma)$ where Y is a subset of X that is consistent with σ and includes all of X's σ-innocent members.

If a revision function $\dotplus$ is such that Y is always first with these properties in some pre-established total ordering of all sets of formulas, then $\dotplus$ is *definite*.

It can be proved that definite revision functions exist, and that some of them are maxichoice. Definite, maxichoice revision satisfies many criteria of rationality. It manifests transitive and connected preferences among theories, and revises beliefs so as to minimize change and maximize acceptability.

Unhappily, these properties do not guarantee efficiency—for, the culprit revision function in our theorem about inefficiency can be chosen to be definite and maxichoice. That is, even some maxichoice and definite revision functions are dominated by other revision functions no matter what starting theory is chosen. Some other property of revision functions is thus called for, to naturally delimit the efficient ones. We do not know what this property might be.[16]

§7. Other lacunae. Additional gaps in our understanding appear when we attempt to define revision functions with more human demeanor. For example, in the paradigm discussed so far, the scientist carries along the same starting theory throughout his career, modifying it over and over with accumulating data. This is not how real scientists operate. Rather, the theory tested against the data at stage $n + 1$ is often the one emerging from the

---

[16]It is known, however, that belief revision on a solvable partition with underlying (sentential) theory T can always be carried out efficiently if the revision operator is biased towards retaining T at the expense of open formulas (such as those of $X - T$ in the notation of Proposition 5.1). It remains unclear whether such a constraint on belief revision can be independently motivated.

encounter with the data of stage n. Such use of a revision function may be called *iterative*. The difference iteration makes to revision based scientists may be pictured as follows. Let $d_1, d_2, d_3, d_4 \ldots$ be Nature's data-sequence. A non-iterative revision based scientist operates like this:

$$X \dotplus (d_1),\ X \dotplus (d_1, d_2),\ X \dotplus (d_1, d_2, d_3)\ \ldots.$$

In contrast, an iterative, revision based scientist proceeds this way:

$$X \dotplus (d_1),\ (X \dotplus d_1) \dotplus d_2,\ ((X \dotplus d_1) \dotplus d_2) \dotplus d_3\ \ldots.$$

Iterated revision has some drawbacks compared to repeated revision of the starting theory. For example, some revision functions used iteratively cannot be made to work on certain partitions (very simple ones), whereas they all work non-iteratively. Even imposing the definiteness and maxichoice properties does not lift this limitation. But we are lacking an illuminating characterization of which revision functions show such iterative weakness.

For another gap, notice that all our revision functions implement uncomputable processes since they must determine consistency in order to spot "innocent" formulas. If human cognition is Turing simulable (as widely supposed in Psychology), the relevance of Proposition 5.1 to scientific achievement is cast further in doubt. Natural ways of weakening the computability requirement on belief revision have yet to be investigated. In contrast, some information is available about the powers of computable scientists who do not rely on belief revision.

The computable scientists may be defined in the following way. Let f be a computable function from the (decidable) set of finite data-sequences to indexes for r.e. sets W_i of formulas. Then $\lambda\sigma\,.\,MOD(W_{f(\sigma)})$ is a computable scientist. [As usual, $MOD(X)$ denotes the countable models of X.] Given input data σ, such a scientist is considered to conjecture the class of structures that satisfy the indexed sentences.

Call a partition *r.e.* iff it partitions the class of (countable) models of an r.e. underlying theory $T \subseteq \mathcal{L}$. It can be shown that every solvable r.e. first-order partition is solved by some computable scientist. (That is, no solvable r.e. first-order partition escapes the competence of machine simulable scientists.) The theorem breaks down if the collection is not axiomatizable. Indeed, there is a more general fact.

PROPOSITION 7.1. *Suppose the nonlogical vocabulary of $\mathcal{L}$ is limited to a binary predicate, two constants, and a unary function symbol. Then for every countable collection Σ of scientists there is a partition $\mathcal{P}$ with the following properties.*

1. *Every member of $\mathcal{P}$ is an elementary class (in other words, definable by a single sentence of $\mathcal{L}$).*
2. *$\mathcal{P}$ is solvable.*
3. *No member of Σ solves $\mathcal{P}$.*

That is, every countable collection of scientists fails on some partition made up of elementary classes. Since the computable scientists form a countable set, they also fail on some such partition. The same is true of the countable collection of scientists simulable by Turing Machines with oracle O (for a fixed choice of O). Note that these failures do not derive from inexpressibility of the cells of the partition (since each is an elementary class). Beyond these crude distinctions, not much is known. The most pressing issue is to elucidate the inductive range of computable scientists operating under time and space constraints, or implementing neural nets.

§8. Concluding remarks. Many additional issues can be raised, but perhaps this is enough detail. Let us recall the overall aim of our enterprise. It is to specify a broad class of revision functions that are inductively proficient. They should support efficient solution of solvable partitions, and at least some of them should resemble the kinds of mechanisms that people use in choosing a theory to accept. If this aim can be achieved, human scientific capacity will begin to seem less mysterious. It will issue from membership of human revision functions in a larger class with proven inductive powers.

Bringing this programme to fruition requires insight into human psychology, concerning how people move from sparse information about the environment to a provisional theory. The programme also requires mathematical analysis of inductive logic, to determine which hypotheses about human induction bestow sufficiently generous scientific capacity on people. A new alliance between logic and psychology may thus yield interesting results for both sides.

It might turn out, of course, that what is discovered about how people choose hypotheses is incompatible with any successful model of induction. In this case, the research programme will have failed to cast light on the remarkable scientific capacity of our species. Human scientific achievement will need explanation on some other basis.

REFERENCES

[1] S. C. Baker, R. J. Dolan, and C. D. Frith, *The functional anatomy of logic: A PET study of inferential reasoning*, ***NeuroImage***, vol. 3 (1996), p. S218.

[2] L. J. Cohen, ***An essay on belief and acceptance***, Oxford University Press, Oxford, UK, 1992.

[3] J. Dorling, *Bayesian personalism, the methodology of scientific research programs, and Duhem's problem*, ***Studies in the History and Philosophy of Science***, vol. 10 (1979), pp. 177–187.

[4] J. Earman, ***Bayes or bust?***, MIT Press, Cambridge MA, 1992.

[5] M. S. Gazzaniga, R. B. Ivry, and G. R. Mangun, ***Cognitive neuroscience***, W. W. Norton, New York NY, 1998.

[6] G. Georgakopoulos, D. Kavvadias, and C. Papadimitriou, *Probabilistic satisfiability*, ***Journal of Complexity***, vol. 4 (1988), pp. 1–11.

[7] G. GIGERENZER, *Reply to Tversky and Kahneman*, ***Psychological Review***, vol. 103 (1996), no. 3, pp. 592–593.

[8] L. R. Gleitman and M. Liberman (editors), ***Language (An invitation to cognitive science)***, second ed., vol. 1, MIT Press, Cambridge MA, 1995.

[9] S. O. HANSSON, *Kernel contraction*, ***The Journal of Symbolic Logic***, vol. 59 (1994), no. 3, pp. 845–859.

[10] ———, ***A textbook of belief dynamics: Theory change and database updating***, Kluwer Academic Publishers, Norwell MA, 1999.

[11] V. F. HENDRICKS, ***The convergence of scientific knowledge***, Kluwer Academic Publishers, Dordrecht, the Netherlands, 2000.

[12] L. C. IDSON, D. H. KRANTZ, D. OSHERSON, and N. BONINI, *The relation between probabilitiy and evidence judgment: An extension of support theory*, ***Journal of Risk and Uncertainty***, vol. 22 (2001), pp. 227–250.

[13] P. N. JOHNSON-LAIRD and R. M. J. BYRNE, ***Deduction***, Erlbaum, Hillsdale, NJ, 1991.

[14] S. K. JONES, K. T. JONES, and D. FRISCH, *Biases of probability assessment: A comparison of frequency and single-case judgments*, ***Organizational Behavior and Human Decision Processes***, vol. 61 (1995), pp. 109–122.

[15] D. KAHNEMAN and A. TVERSKY, *On the reality of cognitive illusions*, ***Psychological Review***, vol. 103 (1996), no. 3, pp. 582–591.

[16] K. T. KELLY, ***The logic of reliable inquiry***, Oxford University Press, New York, NY, 1996.

[17] S. M. Kosslyn and D. N. Osherson (editors), ***Visual cognition (An invitation to cognitive science)***, second ed., vol. 2, MIT Press, Cambridge MA, 1995.

[18] T. S. KUHN, ***The copernican revolution***, Harvard University Press, Cambridge MA, 1957.

[19] E. MARTIN and D. OSHERSON, *Scientific discovery on positive data via belief revision*, ***Journal of Philosophical Logic***, vol. 29 (2000), pp. 483–506.

[20] ———, *Induction by enumeration*, ***Information and Computation***, vol. 171 (2001), pp. 50–68.

[21] ERIC MARTIN and DANIEL OSHERSON, ***Elements of scientific inquiry***, M.I.T. Press.

[22] D. G. MAYO, ***Error and the growth of experimental knowledge***, University of Chicago Press, Chicago IL, 1996.

[23] ISAAC NEWTON, ***Newton's philosophy of nature: Selections from his writings***, Hafner, New York, 1953.

[24] D. OSHERSON, *Probability judgment*, ***Thinking (An invitation to cognitive science)***, vol. 3, MIT Press, Cambridge MA, second ed., 1995.

[25] D. OSHERSON, L. MACCHI, and W. HODGES, *Models, rules, and deductive reasoning*, manuscript, 1998.

[26] D. OSHERSON, D. PERANI, S. CAPPA, T. SCHNUR, F. GRASSI, and F. FAZIO, *Distinct brain loci in deductive versus probabilistic reasoning*, ***Neuropsychologia***, vol. 36 (1998), no. 4, pp. 369–376.

[27] L. PARSONS and D. OSHERSON, *New evidence for distinct right and left brain systems for deductive vs probabilistic reasoning*, manuscript, 2000.

[28] S. PLOUS, ***The psychology of judgment and decision making***, McGraw-Hill, 1993.

[29] K. POPPER, ***The logic of scientific discovery***, Hutchinson, London, 1959.

[30] L. RIPS, ***The psychology of proof***, MIT Press, Cambridge, MA, 1994.

[31] M. R. ROSENZWEIG, A. L. LEIMAN, and S. M. BREEDLOVE, ***Biological psychology***, second ed., Sinauer Associates, Sunderland, MA, 1999.

[32] E. SHAFIR, E. E. SMITH, and D. OSHERSON, *Typicality and reasoning fallacies*, ***Memory and Cognition***, vol. 18 (1990), no. 3, pp. 229–239.

DEPARTMENT OF PSYCHOLOGY
PRINCETON UNIVERSITY
PRINCETON, NJ 08544, USA
E-mail: osherson@princeton.edu

SCHOOL OF COMP. SCI. AND ENGINEERING
UNIVERSITY OF NEW SOUTH WALES
SYDNEY, AUSTRALIA
E-mail: emartin@cse.unsw.edu.au

RUSSELL'S LOGICS

PHILIPPE DE ROUILHAN

Abstract. In 1903, in Appendix B of *The Principles of Mathematics*, Russell described a paradox he had discovered late in 1902. This paradox is more difficult than the better known paradox of 1901, which depends upon the notion of *set*. The later paradox concerns what we should call today the "hyperintensional" notion of *proposition*, understood as the meaning of a possible sentence, and it reveals that this notion is subject to difficulties analogous to those that the earlier paradox uncovered in the concept of set. Russell seems to have forgotten or ignored the 1903 paradox, for he never mentioned it again, at least in works published in his lifetime. John Myhill rediscovered it, in a slightly different form, in 1958; consequently it attracted the attention of logicians, notably Church, in the seventies.

The hyperintensional paradox, and others of the same sort, have been unjustly neglected. When they and their implications are taken into account, the development of Russell's thought, and the history of logic in general, from *The Principles* to *Principia Mathematica*, appear in a new light. If we look with care at the succession of logics that Russell conceived, it becomes evident that *logical* paradoxes of this little known kind played an important role. Indeed, Russell himself underestimated their significance and gave undue weight to the *epistemological* paradoxes. This has led many to overemphasize the importance of the *semantical* paradoxes.

There are at least three ways of telling the story of Russellian logic, which may be qualified as *popular*, *scholarly*, and *rational*. I'll consider them in turn. It is in terms of the last one that I have tried, over the course of last years, to contribute to the renewal of Russellian studies in logic. The results of my efforts have essentially been consigned to a book [17], whose ideas are echoed here and to which I refer the interested reader for fuller details and references.

§1. The popular story. Here is the popular story, recounted as simply as possible.

Like Frege, Russell thought that he could solve the problem of the foundations of mathematics (from elementary arithmetic to classical mathematical analysis) by showing that they were reducible to logic (this was the fundamental thesis of "logicism"). Logic was then understood in a broad sense, covering notably the Cantorian theory of sets. But, as Russell himself was to discover in 1901, the notion of *set* (or of *class*, as he would say) was contradictory. Russell proved this in his famous paradox of non-self-membership, thereby

Logic Colloquium 2000
Edited by R. Cori, A. Razborov, S. Todorčević, and C. Wood
Lecture Notes in Logic, 19

joining a tradition dating at least from the ancient antinomy of the Liar and including such recent instances as the Burali–Forti paradox.

Following Poincaré, Russell believed that he had detected a subtle form of circularity as being the source of all the paradoxes. This led him to formulate for the first time the "vicious circle principle", designed to counteract precisely this kind of circularity. It was this principle, he thought, that would enable a thorough reconstruction of logic and an eventual end to the famous "crisis of foundations." Thus was born the Russellian theory of types, the most completely developed form of which was to find expression in the first volume of the first edition of *Principia Mathematica* in 1910 [29] and which would become known retrospectively as "the ramified theory of types."

As regards the theory of types—the story continues—if Russell had wanted only to solve the *logical* paradoxes (such as his own and Burali–Forti's), a much simpler theory of types would have sufficed. The hierarchy of types could have been reduced to one type for individuals, another for classes of individuals, another for two-place (three-place, etc.) relations (in extension)[1] among individuals, another for classes of classes of individuals, and so on. But Russell wanted to solve *all* the main paradoxes in one fell swoop, not only the logical paradoxes, but in addition and notably, the *semantical* paradoxes (such as the Liar's paradox). That is why he needed the vicious circle principle, and that is what led him to such a complex theory of types.

The theory actually was extremely complex, but that was the least of its failings. It was just too weak to allow the reconstruction of classical mathematics without resorting to the expedient of an *ad hoc* axiom void of any intuitive self-evidence, the notorious "axiom of reducibility", not to mention the axiom of infinity and the axiom of choice, whose logical nature was more than dubious, and whose presence in any case marked the failure of the logicist enterprise.

Finally, Ramsey came—the story is drawing to a close. He actually simplified matters and put them in order. Russell had put all the paradoxes into the same bag. Ramsey would distinguish among them. The job of solving the *logical* paradoxes, but not the *extra-logical* (semantical or other) paradoxes, was to fall to the logician as such. Ramsey abandoned the vicious circle principle, which was useless for his ends, and even wrong. He proposed a simplified, or, as some have said, a "Ramsified" version of the ramified theory of types: whence the "simple theory of types", in which one could reconstruct mathematics much more simply (with neither ramification nor the axiom of reducibility). This did not however save the logicist enterprise since the axiom of infinity and the axiom of choice were still needed. In the process of this reconstruction, Russellian ramified type theory had been at best a useless detour.

[1]I assume this qualification in what follows.

Subsequently, in spite of the relative "simplicity" of the new theory of types, mathematicians ultimately preferred Zermelo–Frænkel set theory, or some variant of it, as the framework within which to reconstruct the whole of mathematics and shelter it from the paradoxes, as the *Éléments de Mathématique* of Nicolas Bourbaki would show in exemplary fashion.

§2. The scholarly story. The popular story is coherent, but it's wrong. It is not so much that it is false, though much of it is false, as that it is oversimplified and misleading. The true story is more complex and more profound. The scholarly story that I sketch below is faithful to Russell's understanding and also to the ways in which historians of logic, scholars, have understood the facts in his wake. In the interests of avoiding repetition, I'll indicate only certain aspects of the popular story that need correction. In two initial sections I set forth two essential motifs of Russellian thought about which the popular story keeps strangely silent. In a third and final one I'll touch upon the possibility of completing the new picture.

2.1. First motif: the idea of "incomplete symbol." In his 1959 account of his philosophical development, and again in 1967 in his autobiography, Russell would state in a sibylline manner that his 1905 theory of descriptions [20] represented a first step towards a solution of the main paradoxes, meaning in fact towards the ramified theory of types sketched in 1908 and developed in the first edition of *Principia* in 1910.[2] The hidden middle term would be the idea of "incomplete symbol", omnipresent throughout all his work since 1905 and whose importance is impossible to overestimate.

From 1905 on, the primary, principal, and fundamental strategy for solving the paradoxes that accompany certain notions did not, in Russell's eyes, consist in detecting any kind of vicious circle in their use or in establishing type distinctions among the corresponding entities. It consisted in defining these notions contextually, in terms of less problematical notions, and thereby reducing the corresponding entities to simple "*façons de parler*", or, as he also said, to "incomplete symbols", "logical fictions", "logical constructions", "linguistic constructions", "false abstractions." The best historical precedent for this method was the reconstruction of the (infinitesimal) Calculus by Cauchy, Bolzano and Weierstrass, in which Leibniz' infinitesimals no longer appeared except as simple *façons de parler*. In this setting, statements which apparently concerned these dubious entities had been systematically paraphrased into statements that clearly no longer referred to them.

The first use of this strategy in Russell's work can be detected, as Russell himself said later, as early as 1905, in his famous analysis of definite descriptions [20]. The main paradoxes were not at issue there. The subject was

[2]Until I expressly say otherwise, I am for the time being setting to one side the revised, corrected version of the theory outlined in 1925 in the second edition of *Principia*.

simply the logical analysis of denoting expressions in general and of definite descriptions in particular. In order to solve certain puzzles that troubled the analyses accepted at the time, Russell proposed a new analysis. The fundamental idea of this can be expressed in a few words: denoting expressions in general, and definite descriptions in particular, are only incomplete symbols; the sentences containing these expressions must be paraphrased into sentences that no longer contain them.

Russell was, however, still bothered by the paradoxes and in particular by the paradox about the notion of class[3] that he had discovered in 1901. In 1905, he envisioned several solutions [21]. He called one of these the "no-classes theory": expressions standing for classes would only be incomplete symbols. In other words, classes would not be genuine entities; they would be logical fictions.

But what about the vicious circle principle? According to the popular story this principle together with the distinguishing of types form the major components of Russell's solution of the paradoxes. The reduction strategy that I am setting forth seems to play no role, or at least no explicit role, in this solution. In fact, in 1906, the following year, when Russell espoused Poincaré's diagnosis of the paradoxes, according to which they all resulted from a vicious circle, he declared that it was precisely to avoid vicious circles that he had invented the no-classes theory [23]. As if the principle did not contribute anything new! He was exaggerating, either out of bad faith or lack of lucidity, as the rest of the story will make clear. Indeed, the reduction strategy and the requirement of "predicativity"(the terminology later assigned it) expressed in the vicious circle principle would work together. Ideally, they would reduce the problematical entities to simple *façons de parler* on the basis of an irreducible ontology satisfying the requirement in question. What is true in Russell's declaration is that once this requirement is satisfied on the basic level, meaning in the so to speak real world, it is also automatically satisfied in the realm of logical fictions, without further precautionary measure. In 1908–1910, the ramified theory of types itself would be a no-classes theory.[4] Classes would appear there as logical fictions, and they would *ipso facto* obey the requirement of predicativity.

Indeed, classes are but a paradigmatic example of entities reduced to logical fictions. Relations of any degree meet the same fate. To size up the operation, it must be remembered that the logic to which the logicist Russell wanted then to reduce[5] mathematics must be understood in an extremely

[3]More precisely: "class *as a whole*", "class *as one*."

[4]Meaning a theory without classes "*as one*", but that would, as a matter of fact, also be a theory without classes "*as many*."

[5]Need it be said? The two reductions in question must not be confused. Reducing a discipline considered until then to be specifically mathematical, *extra-logical*, like elementary arithmetic, for example, to *logic* (and therefore its objects, whole numbers in this case, to purely logical

broad sense: besides individuals, it countenanced extensional objects such as classes and relations, with which mathematics classically deals, but also propositions and (propositional)[6] functions. All those objects—individuals, propositions, functions, classes and relations—be it as entities or as logical fictions, figure in the ramified theory of types. Propositions and functions are not extensional, they are intensional, or better, *hyperintensional* objects. They satisfy more severe criteria of identity than simple intensions. Propositions, for example, may be different even though the sentences expressing them are logically equivalent.

The notions of class, relation, proposition, and function are all problematical. As astonishing as it may seem, in the ramified theory of types it is the classes and relations that are contextually defined in terms of propositions and functions, and not the other way around. It is the classes and relations that are reduced to the state of logical fictions, while the propositions and functions retain their ontological status of entities, and not the other way around.[7] It is as if Russell had found the latter less puzzling than the former. Was that really the case, and, if so, why? To my knowledge, Russell never confronted this question head on, and thus it is hard to determine his views. Quine ([13], § 35) contended that, a victim of confusing use and mention, Russell had attributed to propositions and functions the clarity that he had found in the expression of them. That thesis is hardly convincing. Of course, Russell all too often expressed himself in an off-hand way, but his thought was always rigorous, and Quine's reproach is undeserved. In any event, the question remains: Why should hyperintensional objects be less problematic than extensional ones? It seems to me that this question loses much of its point when one realizes that, in Russell's mind, the ramified theory of types represented only one temporary stage in carrying out an analytical program at the end of which propositions and functions as entities were themselves called upon to disappear. Viewed in this light, that Russell first analyzed classes and relations in terms of propositions and functions, rather than the latter in terms of the former, does not mean that he found one category intrinsically more mysterious than the other.

The fact remains that the only entities in the ramified theory of types are individuals, propositions and functions. Classes and relations are but logical

---

entities) is one thing; reducing logical or extra-logical *entities* to *logical fictions* is another. The first reduction defines logicism; the second, the fundamental Russellian strategy for solving the paradoxes. The two are not mutually exclusive. Frege had reduced the whole numbers to purely logical entities (namely to certain classes—"extensions of concepts"—defined in a purely logical way) and left matters there; Russell reduced these logical entities in turn, along with all classes and relations, to logical fictions.

[6] I assume this qualification in what follows.

[7] In this paper, I neglect the possibility signaled and exploited by Russell of contextually defining propositions in terms of functions.

fictions. The theory thus resolves the paradoxes about the notions of class and relation. The paradoxes about the notions of proposition and of function remain, however. It was only after this reduction that Russell was obliged to resort expressly to the vicious circle principle and the requirement of predicativity. Technically, Russell embraced this principle in the following form: "Whatever involves [or contains] an apparent[8] variable must not be among the possible value of this variable" [23]. Russell is not thinking here of linguistic entities, as one might believe, but of extra-linguistic entities, for example general propositions, or functions.[9] And, while he applies the vicious circle principle to them, this is not, as one might think, because he has taken them to be mental constructions. Constructivism of this sort aside, the vicious circle principle turns out to be neither more nor less natural for the hyperintensional entities of the ramified theory of types than is the principle of foundation for the extensional entities of set theory.

2.2. Second motif: the ideal of universality. Russell's preferred strategy for solving the paradoxes did not invoke the vicious circle principle at all, even though Russell believed in this principle. Neither did it have anything to do with setting up type differences among the entities corresponding to the notions in question. As 1908 approached, the year of the first draft of the 1910 theory, he even ruled out recourse to type distinctions to solve the paradoxes. He had, let us recall, considered resorting to such distinctions in 1903 in the appendix of *The Principles of Mathematics* devoted to the "doctrine of types", but he had immediately given up on the idea—not just because he had discovered that the doctrine was inconsistent, but also because, in a general way, he thought that type distinctions impinged upon the universality of logic.

Russell was a *universalist* about logic. In his view, it was impossible to speak about anything at all, and about logic in particular, from *outside* logic. Frege was also a universalist, but he admitted the possibility of a sort of "absolute metalanguage", as it were. He spoke of a "language of exposition" ("*Darlegungssprache*") (see "Logische Allgemeinheit", in [7]); void of any properly theoretical legitimacy, but which might be heuristically useful like

[8]I.e., bound.

[9]I cannot resist correcting Russell here (anticipating the rational story). Unable to recognize the variables characteristic of functions (those in the argument positions) as apparent, he could not directly apply the vicious circle principle to the latter. He thus found himself obliged to formulate an additional, special principle for them. Inversely, moreover, when the ontological status of extensions—classes and relations—of those functions had not yet been ratified, he was ready to say that these objects, except perhaps those which are finite and can therefore be given by simple enumeration, "involve [or contain] an apparent variable", and to apply the vicious circle principle to them, while only a version of this principle *other* than the one he had embraced in the beginning would have been applicable to them. For, even if an extension is "determined" (or "defined") by a function, and the latter is acknowledged as "involving" (or "containing") an apparent variable, one wonders in what rather substantial sense that extension "involves" (or "contains") itself an apparent variable.

Wittgenstein's ladder, at the end of the *Tractatus* [30], to be pushed away after one climbed it, when one finally has an "accurate picture of the world." Russell, for once (but once is not a habit!), was more rigorous than Frege and refused this kind of comfort. In my terminology, Frege was universalist *in the weak sense*; Russell was universalist *in the strong sense*.[10]

However, while a typed logic was compatible with Frege's weak universalism (and the logic of Frege was actually typed), it was not, it seemed, with Russell's strong universalism. The exposition of such a logic seemed condemned to paradoxical declarations like: "There are different types of entities, and one cannot say anything about these that would not be limited to entities of a definite type", transgressing by their very formulation the prohibitions that they presumed to impose on any possible theoretical language.[11] Frege and Russell were both sensitive to this paradox, but they drew different lessons from it. Frege concluded that the expository langage could not be properly theoretical. Russell concluded that type distinctions were unacceptable. Frege's logic used several sorts of variables, each one relative, specific to a type. Russell's was to use but one. The variables had to be absolute, universal in range. Frege's logic was typed. Russell's was to be type-free.

Thus, the strong universalist requirement on the Russellian enterprise excluded the fundamental distinctions of the ramified theory of types, the very theory that Russell is famous for having discovered or invented. That discovery was certainly a great accomplishment, but it presupposed the abandonment of the universalist position, and, paradoxically, it must have had for Russell the bitter taste of failure. How had this come about?

In my view, Russell did not so much abandon the universalist program in logic as put it aside provisionally and, he hoped, temporarily. As ingenious and monumental as it was, the ramified theory of types was in his eyes only a temporary response to the problem of the paradoxes, a strategic withdrawal, while waiting for better times to come when a genuine solution to the paradoxes complying with the universalist requirement would finally be possible.

Before constructing the ramified theory of types, and before he had even heard of the vicious circle principle, Russell first tried to develop a universal (in the strong sense of type-free) logic, the simple[12] "substitutional theory" ([21], [22]). After adopting the vicious circle principle, he concentrated on a predicative logic, but still universal (in the same sense), logic: the ramified[13]

---

[10]In his *Recherches sur l'universalisme logique*, F. Rivenc calls Russell's and Carnap's universalism "positive", and Frege's and early Wittgenstein's "negative".

[11]Like many others, I believed it, but I do not believe it any more. I explain this in "On What There Are", 2002.

[12]The phrase "substitutional theory" is Russell's, but I borrow the qualification "simple" (vs "ramified") from Hylton [9](1980).

[13]See note 12. In fact, Hylton fails to distinguish, as I do, the so called ramified substitutional theory from what I'll call "intermediate theory", see section 2.3 below.

"substitutional theory" [23]. He gave up on the first logic because he believed it incapable of solving the *epistemological* paradoxes.[14] He gave up on the second one because it was manifestly too weak to express all the notions of elementary logic, *a fortiori* those of mathematics.[15]

Such a logic is indeed predicative; whatever contains an apparent variable cannot be a value of this variable. It is however also universal in the strong sense that all variables are universal. Therefore, whatever contains an apparent variable does not exist. Hence there can be no general propositions, no functions, no classes or relations, except if need be those that are finite and can be given by simple enumeration. In such a logic, all these objects must be reconstructed as simple *façons de parler*. Unable to see how to realize such a reconstruction plan, in 1908 Russell gave up a portion of his ambitions. He had to be satisfied with a logic that was predicative and universal *in the weak sense*. This was the ramified theory of types.[16] Thus was born the ramified theory of types: Far from crowning the success of the Russellian enterprise, it was a by-product of its failure.

In this theory, a hierarchy of individuals, propositions, and functions satisfying the vicious circle principle is presumed at the outset. Classes and relations are then reconstructed on this basis as *façons de parler*. Curiously, the original hierarchy was not the one to which the straightforward application of the vicious circle principle would have led. It is rather based on two principles. The first, analogous to the principle that had led Frege to his own hierarchy of functions, would have yielded a simple hierarchy and a simple theory of types;[17] the second principle was the vicious circle principle, which led to the ramified hierarchy and the ramified theory of types. As for classes and relations, just by virtue of being defined step-by-step in terms of propositions and functions, they were arranged in a simple hierarchy. All things considered, therefore, there was a ramified hierarchy of entities (individuals, propositions and functions) and a simple hierarchy of logical fictions (classes and relations).

2.3. To complete the picture. With these themes in mind it is possible to retrace the history within which the 1910 ramified theory of types developed, as Russell himself saw it and as scholars have seen it after him. This can be done with a formal care that neither Russell nor, as far as I know, any historian ever took. One gets a sequence of formal theories, as in the chart below, going

---

[14]I will reconsider the reason for this abandonment at length in section 3, below. Notice that Russell's concern is not with the *semantical* paradoxes.

[15]Gregory Landini, in *Russell's Hidden Substitutional Theory*, taking account of a number of Russell's unpublished manuscripts on the substitutional theory, comes to different conclusions than those of my 1996 book on this matter. Let me not go into the details here. See [1] where the two approaches are briefly compared.

[16]Here I pass over what I call the "intermediate theory." See section 2.3.

[17]I will come back to this theory, in section 3.1 (see. n. 19).

from the 1903 doctrine of types (in *The Principles*) to the 1925 form of the ramified theory of types (in the second edition of *Principia*).

(1) The 1903 doctrine of types
(2) The (type-free, hyperintensional) simple subtitutional theory, **SST** (1905, 1906)
(3) The (type-free, hyperintensional) ramified substitutional theory, **RST** (1906)
(4) The (hyperintensional) intermediate theory, **IT** (1908)
(5) The hyperintensional, ramified theory of types, **RTT** (1908, 1910)
(6) The partially extensional, ramified theory of types (1925)

After Russell abandoned the 1903 doctrine of types because of its inconsistency (let alone its non-universality), he devised the simple substitutional theory (**SST**, 1905, 1906), which was universal, but impredicative, and thus had to be abandoned in turn. Whence arose the ramified substitutional theory (**RST**, 1906). But that theory appeared much too weak to fill the bill. It was followed by a theory that I call the "intermediate theory" (**IT**, 1908). Russell alluded in passing to this theory just prior to outlining the ramified theory of types for the first time in his 1908 paper. Like **SRT**, it was predicative and substitutional, but it enforced a hierarchy of types. Russell renounced it because of its "technical inconvenience." The ramified theory of types (**RTT**, 1908, 1910) was developed for the first time in the first edition of *Principia*. Russell renounced it because of the problematical axiom of reducibility, without which the theory seemed too weak to enable one to reconstruct classical mathematics. Finally, it is important to distinguish the ramified theory of types of the first edition of *Principia*, my preferred reference up to this point, from the later (1925) version outlined in the Introduction to the second edition. There the axiom of reducibility was replaced by Wittgenstein's thesis of extensionality, which comes down to identifying materially equivalent propositions or coextensive functions *of the same order*,[18] thus eliminating the hyperintensionality which had been an essential feature of all the previous theories. The resulting theory is extensional only in part, not completely extensional as would be a theory of individuals, classes, and relations. (I'll say more about all this below.)

§3. The rational story.

3.1. If one is to believe the popular story, the ramified theory of types was a *vain*, if heroic, effort to solve the paradoxes and preserve classical mathematics. The effort was vain because it achieved its end (to the extent

[18]Or, what is equivalent here, *of the same type*. The type of a proposition is determined by its order. The type of a function is determined by the order of the function and the types of its arguments. Functions of the same type are hence always of the same order, but not, contrary to what is often said, conversely: Two functions of the same order may be of different types, unless they have the same arguments, which is of course the case if they are coextensive.

it did this) only by resorting to the disputable axiom of reducibility. Above all, the popular story continues, the ramified theory was a *waste of time*. On the one hand, as Ramsey showed, it aimed to solve all the main paradoxes, the extra-logical as well as the logical, not merely the logical paradoxes. But non-logical problems call for non-logical solutions (semantical solutions to the semantical paradoxes, for example). Let logic keep its house in order and not meddle in the affairs of its non-logical neighbors. Type theory, ramified or not, can but be a logical theory and it is thus completely out of its element in tackling problems of semantics or epistemology. The vicious circle principle, ramification of the hierarchy of types, the axiom of reducibility, these are all just useless complications. When the extra-logical paradoxes are ignored, on the other hand, it is clear that a simplified version of the theory of types—not ramified, not complicated—gives the right logical solution to the logical paradoxes.

On this account, essentially Ramsey's, the ramified theory was a dead end. The simple theory does the logical job and it is, of course, simpler. And the account is blessed by the best logicians. In all of this, it is clearly the ordinary, *extensional* simple theory to which the popular story refers. It is as if that theory were the result of an operation of simplification applied to the ramified theory. But the theory of the first edition of *Principia* is not only ramified, it is also hyperintensional, as the scholarly story makes clear, and simplifying that theory yields a *hyperintensional* simple theory of types,[19] not the ordinary, extensional theory. To get to the latter theory requires in addition the introduction of extensionality which brings with it a radical shift in ontology; from a world of individuals, propositions, and functions, to one of individuals, classes, and relations.

The above chart can be expanded, as below, beyond Russell's logics, mentioning those different simple theories to which the Ramseyan simplification leads, according to how it is applied to the different ramified theories of the first and second editions of *Principia*.[20]

(1) The 1903 doctrine of types
(2) The (type-free, hyperintensional) simple subtitutional theory, **SST** (1905)
(3) The (type-free, hyperintensional) ramified substitutional theory, **RST** (1906)
(4) The (hyperintensional) intermediate theory, **IT** (1908)
(5) The hyperintensional, ramified theory of types, **RTT** (1908, 1910)
(6) The partially extensional, ramified theory of types (1925)
(7) [The hyperintensional, simple theory of types, $\mathbf{STT}_h$]
(8) [The extensional, simple theory of types, $\mathbf{STT}_e$]

[19]This theory is just the one mentioned at the end of section 2.2, above. See note 17.

[20]I have bracketed the last two theories to recall that they are not properly Russellian.

The hyperintensional ontology of the theory **RTT** of the first edition of *Principia* is commonly and mistakenly not distinguished from the partially extensional ontology of the theory of the second edition, or worse it is taken to be extensional, thus assuring that Ramsey's simplification will be misunderstood.[21] Applied to the earlier, hyperintensional theory **RTT**, the Ramseyan simplification yields the theory $\mathbf{STT}_h$. Applied to the theory of the second edition it yields a theory in which the residual ontological, non-typological, distinctions of order are without point, and from there by an obvious move to the theory $\mathbf{STT}_e$, the *terminus ad quem* of the popular story. Applied to an extensional ramified theory of types, the Ramseyian simplification would lead directly, without passing by the theory described above, to $\mathbf{STT}_e$. But Russell never envisioned an extensional ramified theory of types.[22]

3.2. The move from the ramified theory of types of the first edition of *Principia* (**RTT**) to the extensional, simple theory of types ($\mathbf{STT}_e$) is not merely a matter of "simplification." It requires also "extensionalization." If one imagines this extensionalization accomplished, meaning that one envisages henceforth an ontology of individuals, classes and relations, then no logical paradox seems to stand in the way of simplification, yielding $\mathbf{STT}_e$. There is every reason to believe that $\mathbf{STT}_e$ is consistent. On the other hand, it would be possible to simplify the theory **RTT** without affecting its hyperintensionality; without extensionalizing it. The result would be the theory $\mathbf{STT}_h$: a hyperintensional, simplified theory of types, carrying with it a hyperintensional ontology of individuals, propositions, and functions. And the obvious question is now whether $\mathbf{STT}_h$ is, or would be, consistent.

I suspect that not so very long ago most logicians would have answered yes to this question, and I am sure that is how Russell would have responded. Such a theory, even assuming its consistency, would not have satisfied him, however, since it can be shown, and this would have been obvious to Russell, that no such theory could resolve the extra-logical paradoxes like the Liar. And Russell insisted upon a uniform solution to all of the paradoxes. That is why he devised the ramified theory **RTT**. But, if we accept Ramsey's division of labor and leave it to the special sciences to solve the extra-logical paradoxes that come with the use of certain of extra-logical concepts, then the theory $\mathbf{STT}_h$

[21]The misinterpretation dates back at least to Quine [12]. For a recent example, see Feferman [5], reprinted in [6]. Ramsey himself failed to be always clear about the distinction in question.

[22]This is not to say that Ramsey's simplification of the theory of the second edition of *Principia* is the historical source of the extensional simple theory of types ($\mathbf{STT}_e$). As Solomon Feferman pointed out in his discussion of the spoken form of this paper, the simple extensional theory appeared in Tarski's 1923 dissertation, three years before Ramsey's 1926 article. Church [2] finds an even earlier development of the idea in Schröder's 1890 *Algebra der Logik*. I've shown elsewhere [16] to what extent the idea can also be found in Frege's work as well as in Russell's 1903 doctrine of types.

and the question of its consistency assume a new and increased importance. And the question of consistency is relevant not only to that theory, but also to the very novel theory **SST** that Russell outlined in 1905 and abandoned afterward for the same bad reason that would have led him to reject $\mathbf{SST}_{\mathrm{h}}$: it looked not to be able to deal with the extra-logical paradoxes. In fact, the theory **SST** is more interesting than the theory $\mathbf{STT}_{\mathrm{h}}$, since it is type-free and for this reason does not contravene the Russellian principle of logical universalism. The investigation of **SST** and of its hypothetical descendant $\mathbf{STT}_{\mathrm{h}}$ is thus of some interest. Are these theories consistent?

No, these theories are not consistent. They are inconsistent. They fall prey to certain purely *logical* paradoxes, practically unnoticed by logicians all throughout the last century, and in the first place by Russell himself, at least at the time that he was at work on the simple substitutional theory **SST** as presented in his [21] and [22]. But here the story has a major surprise in store for us. For who was the first logician to have dealt with a paradox of this not well known kind? It was Russell himself, once again, at the end of 1902, and he presented it in 1903, in an appendix to *Principles*, in order to show that his doctrine of types was inconsistent. But how could he have forgotten a paradox that is so unforgettable once one has understood it? It must be concluded that he had not really understood it. Russell had discovered the paradox but rendered it obscurely, in terms of the problematical notions of truth and class. Its import was thus disguised, like that of gold mixed in ore. Here is his formulation (my emphasis):

> If m be a *class* of *propositions*, the *proposition* "every m is *true*" may or may not be itself an m. But there is a one-one relation of this *proposition* to m: if n be different from m, "every n is *true*" is not the same *proposition* as "every m is *true*." Consider now the whole *class* of *propositions* of the form "every m is *true*", and having the property of not being members of their respective m's. Let this class be w, and let p be the *proposition* "every w is *true*." If p is a w, it must possess the defining property of w; but this property demands that p should not be a w. On the other hand, if p be not a w, then p does not possess the defining property of w, and therefore is a w. Thus the contradiction appears unavoidable.

Russell never, to my knowledge, identified the paradox in its pure state as the paradox of *propositions*. Or, more precisely, let us say that, since simply intensional propositions would not let such a paradox take hold, Russell did not identify it as a paradox of their *hyperintensionality*. Not surprising that it slipped his mind.

The paradox in question seems not to have escaped Chwistek's vigilant eye, if one is to believe Gödel [8]. It was rediscovered by Myhill in 1958, and once again captured the attention of some logicians, Church among them, beginning in the 1970s. It is better known today as the Russell–Myhill paradox. I have already given the historical reason why this paradox has remained in the shadows, while the other, Russell's paradox about classes, shone in the bright light of day: While the theory of types was being simplified, it was also being extensionalized. One had gone from the hyperintensional, ramified theory of types to an extensional, simple theory of types, and there was, and still is, every reason to believe that the latter theory is consistent.

I first appreciated the significance of the propositional paradox when reading a 1980 article by P. Hylton about the simple substitutional theory, in which he described the theory in the same informal mode as Russell had in 1906 and gave, curiously without mentioning the Russellian source, the idea of an inconsistency proof. Afterward, I endeavored to formalize the theory and the proof of its inconsistency. As for the inconsistency of the hyperintensional, simple theory of types, I soon discovered that it had already been formally proved by Church in 1984, with explicit reference made to Russell's ideas as expressed at the end of *The Principles*.

3.3. There is a paradox of propositions, and not surprinsingly there are other similar paradoxes for functions with an individual argument and for entities for any type of the hyperintensional, simple theory of types. All these paradoxes are analogous, and, in the context in which they appear, they deserve being called *logical* paradoxes. Each of them proves the inconsistency of the hyperintensional, simple theory of types. And it is they, and not some extra-logical paradoxes (semantical, epistemological, or whatever) that are the true reason for the ramification, no matter what Russell, and others after him, may have thought.

It should moreover be said that the popular story goes wrong in at least *two* ways. The scholarly story corrects the first error. Contrary to appearance, Russell was never seriously interested in the *semantical* paradoxes, in the semantical versions of the Liar, for example, which involve the semantical notion of truth, relative to *sentences*. He was interested in the *epistemological* paradoxes, in the epistemological versions of the Liar, for example, which involve the logical notion of truth, relative to *propositions*, and the epistemological notion of assertion, connecting the knowing subjects and the propositions (whence the adjective "epistemological", which came to them from Ramsey). The epistemological paradoxes in question are logically very different from semantical paradoxes: the ramified theory of types solves the former, while it is impotent against the latter. The rational story corrects the second error. It was not the epistemological paradoxes that could justify recourse to the ramified theory of types without any possible way out. It was the logical paradox of propositions and others of the same kind, in short the logical paradoxes of hyperintensionality.

The great Russellian logical enterprise was subject to powerful constraints quite exogenous to Russell's thought and mind. Russell may not have been aware of them, but their effects were in no way diminished by his ignorance, and these same forces apply, as is better realized today, to any possible effort to construct a hyperintensional logic.[23]

REFERENCES

[1] SERGE BOZON, *Review of Philippe de Rouilhan's "Russell et le cercle des paradoxes"*, ***Dialogue***, vol. 40 (2001), pp. 820–824.

[2] ALONZO CHURCH, *Schröder's anticipation of the simple theory of types*, ***Erkenntnis***, vol. X (1976), pp. 407–411, "This paper was presented at the Fifth International Congress for the Unity of Science in Cambridge, Massachusetts, in 1939. Preprints of the paper were distributed to the members of the Congress and the paper was to have been published in ***The Journal of Unified Science*** (***Erkenntnis***), Volume IX, pp. 149–152. But this volume never appeared and the paper has not otherwise had publication", p. 411.

[3] ———, *Outline of a revised formulation of the logic of sense and denotation, part I*, ***Noûs***, vol. 7 (1973), pp. 24–33, *part II*, ***Noûs***, vol. 8 (1974), pp. 135–136.

[4] ———, *Russell's theory of identity of propositions*, ***Philosophia Naturalis***, vol. 21 (1984), pp. 513–522.

[5] SOLOMON FEFERMAN, *Infinity in mathematics: Is Cantor necessary?*, ***L'infinito nella scienza—Infinity in Science*** (Giuliano Toraldo di Francia, editor), Istituto della Enciclopedia Italiana, Roma, 1987, Reprinted in [7].

[6] ———, ***In the light of logic***, Oxford University Press, Oxford, 1998.

[7] GOTTLOB FREGE, ***Nachgelassene Schriften*** (H. Hermes, F. Kambartel, and F. Kaulbach, editors), Felix Meiner, Hamburg, 1969.

[8] KURT GÖDEL, *Russell's mathematical logic*, ***The philosophy of Bertrand Russell*** (P.A. Schilpp, editor), Northwestern University, Chicago, 1944.

[9] PETER HYLTON, *Russell's substitutional theory*, ***Synthese***, vol. 45-1 (1980), pp. 1–31.

[10] GREGORY LANDINI, ***Russell's hidden substitutional theory***, Oxford University Press, Oxford, 1998.

[11] JOHN MYHILL, *Problems arising in the formalization of intensional logic*, ***Logique et Analyse***, vol. 1 (1958), pp. 74–83.

[12] WILLARD VAN ORMAN QUINE, *On the axiom of reducibility*, ***Mind***, vol. 45 (1936), pp. 498–500.

[13] ———, ***Set theory and its logic***, Harvard University Press, Cambridge, Massachusetts, 1963, second edition: 1969.

[14] FRANK PLUMPTON RAMSEY, *The foundations of mathematics*, ***Proceedings of the London Mathematical Society***, vol. 25 (1926), pp. 338–384, paper read before the London Mathematical Society in 1925.

[15] FRANÇOIS RIVENC, ***Recherches sur l'universalisme logique—Russell et Carnap***, Payot, Paris, 1993.

[16] PHILIPPE DE ROUILHAN, ***Frege—les paradoxes de la représentation***, Éditions de Minuit, Paris, France, 1988.

[17] ———, ***Russell et le cercle des paradoxes***, Presses Universitaires de France, Paris, France, 1996.

[23] I am grateful to Claire Ortiz Hill, who translated a previous draft of this paper into English, and to John M. Vickers, whose helpful comments influenced the definitive version.

[18] ———, *On what there are*, ***Proceedings of the Aristotelian Society***, vol. 102 (2001–2002), pp. 183–200.

[19] Bertrand Russell, ***The principles of mathematics***, Cambridge University Press, Cambridge, 1903, 2d ed: G. Allen and Unwin, 1937.

[20] ———, *On denoting*, ***Mind***, vol. 14 (1905), pp. 479–493.

[21] ———, *On some difficulties in the theory of transfinite numbers and order types*, ***Proceedings of the London Mathematical Society***, vol. 4 (1906), pp. 183–200, paper read before the London Mathematical Society in 1905.

[22] ———, *On the substitutional theory of classes and relations*, ***Essays in analysis*** (D. Lackey, editor), Allen and Unwin, London, 1973, paper read before the London Mathematical Society in 1906.

[23] ———, *Les paradoxes de la logique*, ***Revue de Métaphysique et de Morale***, vol. 14 (1906), pp. 627–650, (original english version, "On 'Insolubilia' and their Solution by Symbolic logic", published in [11]).

[24] ———, *Mathematical logic as based on the theory of types*, ***American Journal of Mathematics***, vol. 30 (1908), pp. 222–262.

[25] ———, ***My philosophical development***, Allen and Unwin, London, 1959.

[26] ———, ***The autobiography of Bertrand Russell***, Allen and Unwin, London, Vol. I (1872–1914): 1967; Vol. II (1914–1944): 1968; Vol. III (1944–1967): 1969.

[27] ———, ***Essays in analysis***, Allen and Unwin, London, 1973, (D. Lackey, editor).

[28] Paul Arthur Schilpp (editor), ***The philosophy of Bertrand Russell***, Northwestern University, Chicago, 1994.

[29] Alfred North Whitehead and Bertrand Russell, ***Principia mathematica***, Cambridge University Press, Cambridge, Vol. I, 1910; Vol. II, 1912; Vol. III, 1913; 2d edition: Vol. I, 1925; Vol. II & III, 1927.

[30] Ludwig Wittgenstein, *Logisch-Philosophische Abhandlung*, ***Annalen der Naturphilosophie***, vol. 14 (1921).

INSTITUT D'HISTOIRE ET DE PHILOSOPHIE DES SCIENCES ET DES TECHNIQUE
(CNRS AND UNIVERSITÉ PARIS I — PANTHÉON-SORBONNE)
13, RUE DU FOUR
75006 PARIS, FRANCE
E-mail: rouilhan@ext.jussieu.fr

PARTITIONING PAIRS OF UNCOUNTABLE SETS

MASAHIRO SHIOYA

Abstract. Let κ be a regular cardinal, λ a cardinal $>\kappa$ and $\kappa^{\mathrm{cf}\,\lambda} = \kappa$. We color pairs of sets from $[\lambda]^\kappa$ with λ^+ colors so that an unbounded subset of $[\lambda]^\kappa$ contains some pair of a given color.

§1. Introduction. In [9] Todorčević introduced the method of minimal walks and established among other things $\omega_1 \not\to [\omega_1]^2_{\omega_1}$: There is a coloring $c\colon [\omega_1]^2 \to \omega_1$ such that $c``[X]^2 = \omega_1$ for $X \subset \omega_1$ unbounded. With a simplified argument avoiding the use of walks (see [9]), Velleman [11] obtained a proper class of cardinals λ such that $[\lambda]^\omega \not\to [\mathrm{unbdd}]^2_\lambda$, i.e., there is $c\colon [[\lambda]^\omega]^2_\subset \to \lambda$ such that $c``[X]^2_\subset = \lambda$ for $X \subset [\lambda]^\omega$ unbounded. Todorčević [10] soon extended the result for an arbitrary $\lambda > \omega$, and even indicated $[\lambda]^\kappa \not\to [\mathrm{unbdd}]^2_\lambda$ for κ regular $<\lambda$ with the full exercise of his method.

Naturally one wonders if the number of colors can be larger than λ for $\lambda > \kappa$ of cofinality $\leq \kappa$, in which case an unbounded subset of $[\lambda]^\kappa$ has size $>\lambda$. As shown by Todorčević [10], this is indeed the case for $\kappa = \omega$ in the constructible universe. This paper provides a positive answer for λ of cofinality much smaller than κ:

THEOREM. *Let κ be regular, $\lambda > \kappa$ and $\kappa^{\mathrm{cf}\,\lambda} = \kappa$. Then $[\lambda]^\kappa \not\to [\mathrm{unbdd}]^2_{\lambda^+}$, i.e., there is $c\colon [[\lambda]^\kappa]^2_\subset \to \lambda^+$ such that $c``[X]^2_\subset = \lambda^+$ for $X \subset [\lambda]^\kappa$ unbounded.*

§2. Preliminaries. Throughout the paper, κ denotes a regular cardinal and λ a cardinal $>\kappa$ respectively. We generally follow the terminology of Kanamori [3]. As an exception, we let $[\lambda]^\kappa = \{x \subset \lambda\colon |x| = \kappa\}$, to which the notions of unboundedness and stationarity are transferred from $\mathcal{P}_{\kappa^+}\lambda$. We understand $\max\emptyset = 0$. We also let $S^\kappa_\lambda = \{\alpha < \lambda\colon \mathrm{cf}\,\alpha = \kappa\}$ and $\lim A = \{\alpha < \sup A\colon \sup(A \cap \alpha) = \alpha > 0\}$ for a set A of ordinals.

This work was partially supported by Grant-in-Aid for Scientific Research (No. 10640099), Ministry of Education, Culture and Science. The author would like to thank Professor Stevo Todorčević for answering questions during and after International Conference on Topology and its Applications held in Yokohama, 1999.

Logic Colloquium 2000
Edited by R. Cori, A. Razborov, S. Todorčević, and C. Wood
Lecture Notes in Logic, 19

Building on Shelah's idea [4], we (hopefully) simplify the original proof of the Todorčević theorem $[\lambda]^\kappa \not\to [\text{unbdd}]^2_\lambda$ in the case λ is regular. By Solovay's theorem, we have $\pi\colon \lambda \to \lambda$ such that $S^\kappa_\lambda \cap \pi^{-1}\{\xi\}$ is stationary in λ for $\xi < \lambda$. It is then simple to see that the composition of f below and π witnesses the theorem:

Proposition. *Let $\kappa < \lambda$ be both regular. Then there is $f\colon [[\lambda]^\kappa]^2_{\subset} \to \lambda$ such that for $X \subset [\lambda]^\kappa$ unbounded there is a club $D \subset \lambda$ with $S^\kappa_\lambda \cap D \subset f``[X]^2_{\subset}$.*

Proof. Let $y \neq \emptyset$ be a set of ordinals of size $\leq \kappa$. Set

$$C_y = \{\min y, \sup y\}$$

if $\sup y \in y$. Otherwise take $C_y \subset y$ so that $\sup C_y = \sup y$, $\operatorname{ot} C_y \leq \kappa$ and $\min(y - \sup a) \in C_y$ for $a \subset C_y$ with $\sup a < \sup y$. Fix $\alpha \in y$. Define inductively

$$y^\alpha_i \in y \cup \{\sup y\}$$

by

$$y^\alpha_0 = \sup y \text{ and } y^\alpha_{i+1} = \min(C_{y \cap y^\alpha_i} - \alpha) < y^\alpha_i$$

as long as $\alpha < y^\alpha_i$. In the end we have $n(\alpha, y) < \omega$ with $y^\alpha_{n(\alpha,y)} = \alpha$. Define

$$\rho(\alpha, y)\colon n(\alpha, y) \to y \cap (\alpha + 1)$$

by

$$\rho(\alpha, y)(i) = \min(y - \sup(C_{y \cap y^\alpha_i} \cap \alpha)).$$

(For the proof in Section 4, note that $\rho(\alpha, y)(i) \in C_{y \cap y^\alpha_i}$ by the choice of $C_{y \cap y^\alpha_i}$.)

We next establish some basic facts. Assume further $\alpha < \delta \in y$, $j \leq n(\delta, y)$ and $\max \rho(\delta, y)``j < \alpha$. Starting from $y^\alpha_0 = y^\delta_0 = \sup y$, we have inductively

$$y^\alpha_{i+1} = \min(C_{y \cap y^\alpha_i} - \alpha) = \min(C_{y \cap y^\delta_i} - \alpha) = \min(C_{y \cap y^\delta_i} - \delta) = y^\delta_{i+1}$$

for $i < j$, since $C_{y \cap y^\delta_i} \cap \delta - \alpha = \emptyset$ by

$$\sup(C_{y \cap y^\delta_i} \cap \delta) \leq \min(y - \sup(C_{y \cap y^\delta_i} \cap \delta)) = \rho(\delta, y)(i) < \alpha.$$

Hence $j < n(\alpha, y)$ by

$$y^\alpha_j = y^\delta_j \geq \delta > \alpha.$$

Also

$$\begin{aligned}\rho(\alpha, y)(i) &= \min(y - \sup(C_{y \cap y^\alpha_i} \cap \alpha)) \\ &= \min(y - \sup(C_{y \cap y^\delta_i} \cap \delta)) = \rho(\delta, y)(i)\end{aligned}$$

for $i < j$, since

$$C_{y \cap y^\alpha_i} \cap \alpha = C_{y \cap y^\delta_i} \cap \alpha = C_{y \cap y^\delta_i} \cap \delta$$

by $y^\alpha_i = y^\delta_i$ and $C_{y \cap y^\delta_i} \cap \delta - \alpha = \emptyset$.

Now fix $(x, y) \in [[\lambda]^\kappa]^2_\subset$ with $\sup x \in y$. For $j \le n(\sup x, y)$, let $\varepsilon(x, y, j)$ be the minimal $\varepsilon < \lambda$ such that $\max \rho(\sup x, y)``j < \varepsilon \in x$, if there is one. Set

$$f(x, y) = \max\Big\{y_j^{\sup x}\colon j < n(\sup x, y) \wedge \varepsilon(x, y, j) \text{ exists} \wedge \\ j \le n(\varepsilon(x, y, j), x) \wedge x_j^{\varepsilon(x,y,j)} \le \rho(\sup x, y)(j)\Big\}.$$

We claim that f works.

Let $X \subset [\lambda]^\kappa$ be unbounded. Fix $t \in \lambda^{<\omega}$. Set

$$X_{t\gamma} = \Big\{x \in X : \gamma \in x \wedge \rho(\gamma, x) \in \prod_{i<|t|} t(i)\Big\}$$

for $\gamma < \lambda$, and

$$A_t = \Big\{\gamma < \lambda\colon \sup\{\sup x : x \in X_{t\gamma}\} = \lambda\Big\}.$$

If $\sup A_t = \lambda$, take $\max \operatorname{ran} t < \gamma_t \in A_t$, and $g_t(\alpha) \in X_{t\gamma_t}$ for $\alpha < \lambda$ so that $\alpha < \sup g_t(\alpha)$. We show that

$$D = \Big\{\delta < \lambda\colon \forall t \in \delta^{<\omega}\Big(\sup A_t < \delta \vee \Big(\sup A_t = \lambda \wedge \gamma_t < \delta \wedge \\ \forall \alpha < \delta\Big(\sup g_t(\alpha) < \delta\Big)\Big)\Big)\Big\}$$

is the desired club set.

Fix

$$\delta \in S^\kappa_\lambda \cap D.$$

Build $z \in [\delta]^\kappa$ with $\sup z = \delta$ so that

$$g_t(\alpha) \cup \{\sup g_t(\alpha)\} \subset z$$

for $t \in z^{<\omega}$ with $\sup A_t = \lambda$, and $\alpha \in z$. Recall that

$$\sup\{\sup x : x \in X_{t\delta}\} < \lambda$$

for $t \in \lambda^{<\omega}$ with $\delta \notin A_t$. Hence we have $z \cup \{\delta\} \subset y \in X$ such that $\delta, \sup\{\sup x : x \in X_{t\delta}\} < \sup y$ for $t \in z^{<\omega}$ with $\delta \notin A_t$, since $X \subset [\lambda]^\kappa$ is unbounded.

Set $l = n(\delta, y) < \omega$. Then $l > 0$ by $\delta < \sup y$. Also $\sup(C_{y \cap y_i^\delta} \cap \delta) < \delta$ for $i < l$, since $\operatorname{ot}(C_{y\cap y_i^\delta} \cap \delta) < \kappa = \operatorname{cf}\delta$. Hence

$$\rho(\delta, y)(i) = \min(y - \sup(C_{y\cap y_i^\delta} \cap \delta)) < \delta$$

for $i < l$, since $\sup(y \cap \delta) = \delta$. Thus we can define

$$t\colon l \to z$$

by

$$t(i) = \min(z - (\rho(\delta, y)(i) + 1)),$$

since $\sup z = \delta$. Then $y \in X_{t\delta}$ by $\delta \in y$ and the choice of t. Note that $\sup A_t = \lambda$: Otherwise we would have $\sup A_t < \delta$ by $\delta \in D$ and $t \in \delta^{<\omega}$, hence $\delta \notin A_t$. Thus $\sup\{\sup x : x \in X_{t\delta}\} < \sup y$ by $t \in z^{<\omega}$ and the choice of y, contradicting $y \in X_{t\delta}$.

Set $\gamma = \gamma_t$. Then $\gamma < \delta$ by $\delta \in D$ and $\sup A_t = \lambda$. We have $\alpha \in z$ with $C_{y\cap\delta} \cap \alpha - \gamma \neq \emptyset$, since $\sup C_{y\cap\delta} = \sup z = \delta$. Set $x = g_t(\alpha) \in X_{t\gamma} \subset X$. Then $x \cup \{\sup x\} \subset z \subset y \cap \delta$ by $t \in z^{<\omega}$, $\sup A_t = \lambda$ and $\alpha \in z$. We claim that $f(x, y) = \delta$, as desired.

Note first

$$|\rho(\delta, y)| = n(\delta, y) = l$$

and

$$\max \operatorname{ran} \rho(\delta, y) < \max \operatorname{ran} t < \gamma \in x$$

by the choice of t and $\gamma_t = \gamma$, and $x \in X_{t\gamma}$. Hence

$$\max \rho(\delta, y)``l < \sup x < \delta \in y.$$

By the basic facts

$$l < n(\sup x, y), \quad y_l^{\sup x} = y_l^{\delta} = \delta$$

and

$$\rho(\sup x, y) \mid l = \rho(\delta, y) \mid l = \rho(\delta, y).$$

Also

$$\max \rho(\sup x, y)``l = \max \operatorname{ran} \rho(\delta, y) < \gamma \in x$$

by the claims above. Thus $\varepsilon(x, y, j)$ (exists and) $\leq \gamma$ for $j \leq l$.

Note next

$$n(\gamma, x) = |\rho(\gamma, x)| = |t| = l \text{ and } \rho(\gamma, x)(i) < t(i)$$

for $i < l$, since $x \in X_{t\gamma}$. Hence

$$\rho(\gamma, x)(i) \leq \rho(\delta, y)(i)$$

for $i < l$, since

$$\rho(\gamma, x)(i) \in x \cap t(i) \subset z \cap t(i) \subset \rho(\delta, y)(i) + 1$$

by the choice of $t(i)$. Thus

$$\max \rho(\gamma, x)``j \leq \max \rho(\delta, y)``j = \max \rho(\sup x, y)``j < \varepsilon(x, y, j) \leq \gamma \in x$$

for $j \leq l$ by the claims above and the choice of $\varepsilon(x, y, j)$. Again by the basic facts

$$j \leq n(\varepsilon(x, y, j), x) \text{ and } x_j^{\varepsilon(x,y,j)} = x_j^{\gamma} \text{ for } j \leq l.$$

Finally

$$\begin{aligned} x_l^{\varepsilon(x,y,l)} = x_l^\gamma = \gamma &\le \sup(C_{y\cap\delta} \cap \sup x) \\ &= \sup(C_{y\cap y_l^{\sup x}} \cap \sup x) \\ &\le \min(y - \sup(C_{y\cap y_l^{\sup x}} \cap \sup x)) \\ &= \rho(\sup x, y)(l), \end{aligned}$$

since

$$C_{y\cap\delta} \cap \sup x - \gamma \neq \emptyset$$

by

$$C_{y\cap\delta} \cap \alpha - \gamma \neq \emptyset \quad \text{and} \quad \alpha < \sup g_t(\alpha) = \sup x.$$

On the other hand

$$x_i^{\varepsilon(x,y,i)} = x_i^\gamma > x_l^\gamma = \gamma > \rho(\sup x, y)(i)$$

for $i < l$ by the claims above. ⊣

To show $[\lambda]^\omega \not\to [\mathrm{unbdd}]^2_\lambda$ for λ singular, Todorčević [10] introduced a new splitting of $[\lambda]^\omega$ into λ stationary sets (see also [8]). In the next section, we invoke the Foreman–Magidor argument [1], which enables us to split $[\lambda]^\kappa$ into λ^+ stationary sets when $\operatorname{cf}\lambda < \kappa$ (see also [8]). One of the crucial facts is the existence of a square-like sequence due to Shelah [5], whose proof is sketched here for convenience. We need a sequence $\langle e_\gamma : \gamma \in S^\kappa_{\kappa^{++}}\rangle$ which is diamond-like [7] (see also [8]): e_γ is club in γ, $\operatorname{ot} e_\gamma = \kappa$ and $\{\gamma \in S^\kappa_{\kappa^{++}} : e_\gamma \subset D\}$ is stationary for a club $D \subset \kappa^{++}$.

Lemma 1. *Let κ and $\nu > \kappa^{++}$ be both regular. Then there is an increasing and continuous $\langle B_\xi : \xi < \nu\rangle \in \prod_{\xi<\nu}[[\xi]^{<\kappa}]^{<\nu}$ such that*

$$S = \left\{\xi \in S^\kappa_\nu : \exists d \subset \xi(\sup d = \xi \wedge \operatorname{ot} d = \kappa \wedge \forall\zeta < \xi(d \cap \zeta \in B_\xi))\right\}$$

is stationary.

Proof. Fix $\langle e_\gamma : \gamma \in S^\kappa_{\kappa^{++}}\rangle$ as above and an increasing, continuous and cofinal $g_\beta : \kappa^{++} \to \beta$ for $\beta \in S^{\kappa^{++}}_\nu$. We claim that

$$B_\xi = \left\{g_\beta``(e_\gamma \cap \alpha) : \alpha < \gamma \in S^\kappa_{\kappa^{++}} \wedge \beta \in S^{\kappa^{++}}_\nu \cap \xi\right\}$$

for $\xi < \nu$ is as desired.

To see that S is stationary, fix a club $D \subset \nu$. Build an increasing and continuous $f : \kappa^{++} \to D$ so that $f``(e_\gamma \cap \alpha) \in B_{f(\alpha+1)}$ for $\alpha < \gamma \in S^\kappa_{\kappa^{++}}$ with $f``(e_\gamma \cap \alpha) \in \bigcup_{\xi<\nu} B_\xi$. Set $\beta = \sup\operatorname{ran} f \in S^{\kappa^{++}}_\nu$. Take a club $E \subset \kappa^{++}$ with $f \restriction E = g_\beta \restriction E$, and then $\gamma \in S^\kappa_{\kappa^{++}}$ with $e_\gamma \subset E$. Now $f``e_\gamma$ witnesses $f(\gamma) \in S$, as desired: Fix $\alpha < \gamma$. Then $f``(e_\gamma \cap \alpha) = g_\beta``(e_\gamma \cap \alpha) \in B_{\beta+1}$, hence $f``(e_\gamma \cap \alpha) \in B_{f(\alpha+1)} \subset B_{f(\gamma)}$. ⊣

Let a be a set of cardinals. By "$\forall^* \mu \in a\ \varphi(\mu)$" or "$\varphi(\mu)$ for a.a. $\mu \in a$", we mean $\{\mu \in a : \varphi(\mu)\}$ is cobounded. We let $k <^* k'$ iff $\forall^* \mu \in a (k(\mu) < k'(\mu))$ for $k, k' \in \prod a$, and $=^*$, $\leq^*$ have the derived definitions. We conclude this section with Shelah's scale theorem [6], which is another ingredient of the Foreman–Magidor argument:

LEMMA 2. *Let λ be singular. Then there are a set a of regular cardinals $> \operatorname{cf} \lambda$ with $\sup a = \lambda$ and $\operatorname{ot} a = \operatorname{cf} \lambda$, and a $<^*$-increasing and $<^*$-cofinal $h \colon \lambda^+ \to \prod a$.*

A $\leq^*$-upper bound $k^* \in \prod a$ of $K \subset \prod a$ is called exact if for $k <^* k^*$ there is $k' \in K$ with $k <^* k'$. Note that an exact upper bound is a $\leq^*$-least upper bound, hence unique modulo $=^*$. In Lemma 2, we can require the scale h to be continuous: For $\delta < \lambda^+$ limit, $h(\delta)$ is an exact upper bound of $h``\delta$, if there is one.

§3. Proof of Theorem.

In this section we freely exploit the material in the previous proof of Proposition for the following

PROOF OF THEOREM. By Lemmas 1 and 2, we have an increasing and continuous $\langle B_\xi : \xi < \lambda^+ \rangle \in \prod_{\xi < \lambda^+} [[\xi]^{<\kappa}]^{<\lambda^+}$ such that

$$S = \left\{ \xi \in S^\kappa_{\lambda^+} : \exists d \subset \xi\, (\sup d = \xi \wedge \operatorname{ot} d = \kappa \wedge \forall \zeta < \xi\, (d \cap \zeta \in B_\xi)) \right\}$$

is stationary, a set a of regular cardinals $>\kappa$ with $\sup a = \lambda$ and $\operatorname{ot} a = \operatorname{cf} \lambda < \kappa$, and a continuous scale $h \colon \lambda^+ \to \prod a$.

Fix $(x, y) \in [[\lambda]^\kappa]^2_\subset$ with $\{\sup(x \cap \mu) : \mu \in a\} \subset y$. For

$$j \leq^* \langle n(\sup(x \cap \mu), y \cap \mu) : \mu \in a \rangle,$$

let $\varepsilon(x, y, j)$ be the minimal $\varepsilon < \lambda^+$ such that

$$\max \rho(\sup(x \cap \mu), y \cap \mu)``j(\mu) < h(\varepsilon)(\mu) \in x$$

for a.a. $\mu \in a$, if there is one. Set

$$\begin{aligned} f(x, y) = \sup \Big\{ \xi < \lambda^+ : \exists j <^* \big\langle n(\sup(x \cap \mu), y \cap \mu) : \mu \in a \big\rangle \\ \Big(h(\xi) =^* \big\langle (y \cap \mu)^{\sup(x \cap \mu)}_{j(\mu)} : \mu \in a \big\rangle \wedge \\ \varepsilon(x, y, j) \text{ exists} \wedge \\ j \leq^* \big\langle n(h(\varepsilon(x, y, j))(\mu), x \cap \mu) : \mu \in a \big\rangle \wedge \\ \big\langle (x \cap \mu)^{h(\varepsilon(x,y,j))(\mu)}_{j(\mu)} : \mu \in a \big\rangle \\ \leq^* \big\langle \rho(\sup(x \cap \mu), y \cap \mu)(j(\mu)) : \mu \in a \big\rangle \Big) \Big\}, \end{aligned}$$

which is the maximum of the set if it exists. We claim that for $X \subset [\lambda]^\kappa$ unbounded there is a club $D \subset \lambda^+$ with $S \cap D \subset f``[X]^2_{\subset}$, which suffices as before.

Let $X \subset [\lambda]^\kappa$ be unbounded. Fix $t \in \prod_{\mu \in a} \mu^{<\omega}$. Set

$$X_{t\gamma} = \Big\{x \in X : \forall^* \mu \in a\Big(h(\gamma)(\mu) \in x \wedge \rho(h(\gamma)(\mu), x \cap \mu) \in \prod_{i<|t(\mu)|} t(\mu)(i)\Big)\Big\}$$

for $\gamma < \lambda^+$, and

$$A_t = \Big\{\gamma < \lambda^+ : \sup\Big\{\xi < \lambda^+ : \exists x \in X_{t\gamma}(\operatorname{ran} h(\xi) \subset x)\Big\} = \lambda^+\Big\}.$$

If $\sup A_t = \lambda^+$, take $\gamma_t \in A_t$ with $\langle \max \operatorname{ran} t(\mu) : \mu \in a\rangle <^* h(\gamma_t)$, and $g_t(\alpha) \in X_{t\gamma_t}$ for $\alpha < \lambda^+$ so that $\operatorname{ran} h(\xi) \subset g_t(\alpha)$ for some $\alpha < \xi < \lambda^+$. Otherwise take $g_t(\alpha) \in [\lambda]^\kappa$ for $\alpha < \lambda^+$ so that $\operatorname{ran} h(\xi) \subset g_t(\alpha)$ for some $\sup A_t < \xi < \lambda^+$.

By induction on the order type of $b \in [\lambda^+]^{<\kappa}$, let $\operatorname{cl} b \subset \lambda$ be the smallest

$$w \supset \bigcup_{\xi \in b} \operatorname{cl}(b \cap \xi) \cup \big\{\sup(\operatorname{cl}(b \cap \xi) \cap \mu) + 1 : \mu \in a\big\} \cup \operatorname{ran} h(\xi)$$

such that

$$\bigcup_{\xi \in b} g_t(\xi) \cup \big\{\sup(g_t(\xi) \cap \mu) : \mu \in a\big\} \subset w$$

for $t \in \prod_{\mu \in a}(w \cap \mu)^{<\omega}$. Then $|\operatorname{cl} b| \leq \kappa^{\operatorname{cf} \lambda} = \kappa$ by induction on $b \in [\lambda^+]^{<\kappa}$. We show that

$$D = \Big\{\delta < \lambda^+ : \forall \zeta < \delta \forall b \in B_\zeta\ \exists \xi < \delta\Big(\langle \sup(\operatorname{cl} b \cap \mu) : \mu \in a\rangle <^* h(\xi)\Big)\Big\}$$

is the desired club set.

Fix $\delta \in S \cap D$. By $\delta \in S$ we have an unbounded $d \subset \delta$ with $\operatorname{ot} d = \kappa$ and $d \cap \xi \in B_\delta$ for $\xi \in d$. Set $z = \bigcup_{\xi \in d} \operatorname{cl}(d \cap \xi) \in [\lambda]^\kappa$. Note that

$$\big\{\sup\left(\operatorname{cl}(d \cap \xi) \cap \mu\right) : \xi \in d\big\} \subset \sup(z \cap \mu)$$

is unbounded, hence $\operatorname{cf} \sup(z \cap \mu) = \kappa$ for $\mu \in a$. Also $\langle \sup(z \cap \mu) : \mu \in a\rangle$ is an upper bound of $h``\delta$, since $\operatorname{ran} h(\xi) \subset \operatorname{cl}(d \cap (\xi + 1)) \subset z$ for $\xi \in d$, and $\sup d = \delta$. We claim that the upper bound is exact, hence $\langle \sup(z \cap \mu) : \mu \in a\rangle =^* h(\delta)$ by the choice of h.

Fix $k \in \prod_{\mu \in a} \sup(z \cap \mu)$. We have $\beta \in d$ such that

$$k(\mu) \leq \sup(\operatorname{cl}(d \cap \beta) \cap \mu)$$

for $\mu \in a$ by the note above and $\operatorname{ot} a = \operatorname{cf} \lambda < \kappa = \operatorname{ot} d$. Then

$$\langle \sup(\operatorname{cl}(d \cap \beta) \cap \mu) : \mu \in a\rangle <^* h(\xi)$$

for some $\xi < \delta$, as desired, since $\delta \in D$ and $d \cap \beta \in B_\delta = \bigcup_{\zeta < \delta} B_\zeta$.

Recall that

$$\sup\{\xi < \lambda^+ : \exists x \in X_{t\delta}(\operatorname{ran} h(\xi) \subset x)\} < \lambda^+$$

for $t \in \prod_{\mu \in a} \mu^{<\omega}$ with $\delta \notin A_t$. Hence we have $\delta < \delta^* < \lambda^+$ such that $\sup\{\xi < \lambda^+ : \exists x \in X_{t\delta}(\operatorname{ran} h(\xi) \subset x)\} < \delta^*$ for $t \in \prod_{\mu \in a}(z \cap \mu)^{<\omega}$ with $\delta \notin A_t$, since $|\{t \in \prod_{\mu \in a}(z \cap \mu)^{<\omega} : \delta \notin A_t\}| \leq \kappa^{\operatorname{cf} \lambda} = \kappa$. Thus we have

$$z \cup \{\sup(z \cap \mu) : \mu \in a\} \cup \operatorname{ran} h(\delta^*) \subset y \in X$$

such that $\sup(z \cap \mu) < \sup(y \cap \mu)$ for $\mu \in a$, since $X \subset [\lambda]^\kappa$ is unbounded.

Fix $\mu \in a$. Set $l(\mu) = n(\sup(z \cap \mu), y \cap \mu) < \omega$. Then $l(\mu) > 0$ by $\sup(z \cap \mu) < \sup(y \cap \mu)$. Also $\sup(C_{y \cap (y \cap \mu)_i^{\sup(z \cap \mu)}} \cap \sup(z \cap \mu)) < \sup(z \cap \mu)$ for $i < l(\mu)$, since $\operatorname{ot}(C_{y \cap (y \cap \mu)_i^{\sup(z \cap \mu)}} \cap \sup(z \cap \mu)) < \kappa = \operatorname{cf} \sup(z \cap \mu)$. Hence

$$\begin{aligned} &\rho(\sup(z \cap \mu), y \cap \mu)(i) \\ &\qquad = \min(y - \sup(C_{y \cap (y \cap \mu)_i^{\sup(z \cap \mu)}} \cap \sup(z \cap \mu))) < \sup(z \cap \mu) \end{aligned}$$

for $i < l(\mu)$, since $\sup(y \cap \sup(z \cap \mu)) = \sup(z \cap \mu)$. Thus we can define

$$t(\mu) : l(\mu) \to z \cap \mu$$

by

$$t(\mu)(i) = \min(z - (\rho(\sup(z \cap \mu), y \cap \mu)(i) + 1)).$$

Then $y \in X_{t\delta}$ by $h(\delta) =^* \langle \sup(z \cap \mu) : \mu \in a \rangle$, $\{\sup(z \cap \mu) : \mu \in a\} \subset y$ and the choice of $t = \langle t(\mu) : \mu \in a \rangle$. Note that $\sup A_t = \lambda^+$: Otherwise we would have $\operatorname{ran} h(\xi) \subset g_t(\alpha) \subset \operatorname{cl}(d \cap (\alpha + 1)) \subset z$ for some $\sup A_t < \xi < \lambda^+$ and $\alpha \in d$. Then $\xi \leq \delta$, hence $\delta \notin A_t$, since $h(\xi) \leq^* \langle \sup(z \cap \mu) : \mu \in a \rangle =^* h(\delta)$. Thus $\sup\{\xi < \lambda^+ : \exists x \in X_{t\delta}(\operatorname{ran} h(\xi) \subset x)\} < \delta^*$ by $t \in \prod_{\mu \in a}(z \cap \mu)^{<\omega}$ and the choice of δ^*, contradicting $\operatorname{ran} h(\delta^*) \subset y \in X_{t\delta}$.

Set $\gamma = \gamma_t$. Then

$$h(\gamma) \leq^* \langle \sup(g_t(\xi) \cap \mu) : \mu \in a \rangle$$

for $\xi \in d$, since $g_t(\xi) \in X_{t\gamma}$. Hence

$$h(\gamma) <^* \langle \sup(z \cap \mu) : \mu \in a \rangle,$$

since

$$\sup(g_t(\xi) \cap \mu) \in \operatorname{cl}(d \cap (\xi + 1)) \cap \mu \subset \sup(z \cap \mu)$$

for $\mu \in a$ and $\xi \in d$. Thus $\gamma < \delta$ by $\langle \sup(z \cap \mu) : \mu \in a \rangle =^* h(\delta)$. We next have $\alpha \in d$ such that $C_{y \cap \sup(z \cap \mu)} \cap h(\alpha)(\mu) - h(\gamma)(\mu) \neq \emptyset$ for a.a. $\mu \in a$, since $\langle \sup(z \cap \mu) : \mu \in a \rangle = \langle \sup C_{y \cap \sup(z \cap \mu)} : \mu \in a \rangle$ is an exact upper bound of $h``\delta$. Set

$$x = g_t(\alpha) \in X_{t\gamma} \subset X.$$

Then

$$(x \cap \mu) \cup \{\sup(x \cap \mu)\} \subset z \cap \mu \subset y \cap \sup(z \cap \mu)$$

for $\mu \in a$, since $t \in \prod_{\mu \in a}(z \cap \mu)^{<\omega}$, $\sup A_t = \lambda^+$ and $\alpha \in d$. We claim that $f(x, y) = \delta$, as desired.

Note first

$$|\rho(\sup(z \cap \mu), y \cap \mu)| = n(\sup(z \cap \mu), y \cap \mu) = l(\mu)$$

and

$$\max \operatorname{ran} \rho(\sup(z \cap \mu), y \cap \mu) < \max \operatorname{ran} t(\mu) < h(\gamma)(\mu) \in x \cap \mu$$

for a.a. $\mu \in a$ by the choice of $t(\mu)$ and $\gamma_t = \gamma$, and $x \in X_{t\gamma}$. Hence

$$\max \rho(\sup(z \cap \mu), y \cap \mu)\text{``}l(\mu) < \sup(x \cap \mu) < \sup(z \cap \mu) \in y \cap \mu$$

for a.a. $\mu \in a$. By the basic facts $l(\mu) < n(\sup(x \cap \mu), y \cap \mu)$,

$$(y \cap \mu)^{\sup(x \cap \mu)}_{l(\mu)} = (y \cap \mu)^{\sup(z \cap \mu)}_{l(\mu)} = \sup(z \cap \mu)$$

and

$$\begin{aligned}\rho(\sup(x \cap \mu), y \cap \mu) \mid l(\mu) \\ = \rho(\sup(z \cap \mu), y \cap \mu) \mid l(\mu) = \rho(\sup(z \cap \mu), y \cap \mu)\end{aligned}$$

for a.a. $\mu \in a$. Also

$$\begin{aligned}\max \rho(\sup(x \cap \mu), y \cap \mu)\text{``}l(\mu) \\ = \max \operatorname{ran} \rho(\sup(z \cap \mu), y \cap \mu) < h(\gamma)(\mu) \in x\end{aligned}$$

for a.a. $\mu \in a$ by the claims above. Thus $\varepsilon(x, y, j)$ (exists and) $\leq \gamma$ for $j \leq^* l$.

Note next

$$n(h(\gamma)(\mu), x \cap \mu) = |\rho(h(\gamma)(\mu), x \cap \mu)| = |t(\mu)| = l(\mu)$$

and

$$\forall i < l(\mu)(\rho(h(\gamma)(\mu), x \cap \mu)(i) < t(\mu)(i))$$

for a.a. $\mu \in a$, since $x \in X_{t\gamma}$. Hence

$$\forall i < l(\mu)(\rho(h(\gamma)(\mu), x \cap \mu)(i) \leq \rho(\sup(z \cap \mu), y \cap \mu)(i))$$

for a.a. $\mu \in a$, since $\rho(h(\gamma)(\mu), x \cap \mu)(i) \in x \cap t(\mu)(i) \subset z \cap t(\mu)(i) \subset \rho(\sup(z \cap \mu), y \cap \mu)(i) + 1$ for $i < l(\mu)$ by the choice of $t(\mu)(i)$. Thus

$$\begin{aligned}\max \rho(h(\gamma)(\mu), x \cap \mu)\text{``}j(\mu) &\leq \max \rho(\sup(z \cap \mu), y \cap \mu)\text{``}j(\mu) \\ &= \max \rho(\sup(x \cap \mu), y \cap \mu)\text{``}j(\mu) \\ &< h(\varepsilon(x, y, j))(\mu) \leq h(\gamma)(\mu) \in x \cap \mu\end{aligned}$$

for a.a. $\mu \in a$ if $j \leq^* l$ by the claims above and the choice of $\varepsilon(x, y, j)$. Again by the basic facts

$$j \leq^* \langle n(h(\varepsilon(x, y, j))(\mu), x \cap \mu) \colon \mu \in a \rangle$$

and

$$\langle (x \cap \mu)_{j(\mu)}^{h(\varepsilon(x,y,j))(\mu)} \colon \mu \in a \rangle =^* \langle (x \cap \mu)_{j(\mu)}^{h(\gamma)(\mu)} \colon \mu \in a \rangle$$

for $j \leq^* l$.

Finally

$$\begin{aligned}(x \cap \mu)_{l(\mu)}^{h(\varepsilon(x,y,l))(\mu)} &= (x \cap \mu)_{l(\mu)}^{h(\gamma)(\mu)} \\ &= h(\gamma)(\mu) \\ &\leq \sup(C_{y \cap \sup(z \cap \mu)} \cap \sup(x \cap \mu)) \\ &= \sup(C_{y \cap (y \cap \mu)_{l(\mu)}^{\sup(x \cap \mu)}} \cap \sup(x \cap \mu)) \\ &\leq \min(y - \sup(C_{y \cap (y \cap \mu)_{l(\mu)}^{\sup(x \cap \mu)}} \cap \sup(x \cap \mu))) \\ &= \rho(\sup(x \cap \mu), y \cap \mu)(l(\mu))\end{aligned}$$

for a.a. $\mu \in a$, since

$$C_{y \cap \sup(z \cap \mu)} \cap \sup(x \cap \mu) - h(\gamma)(\mu) \neq \emptyset$$

by

$$C_{y \cap \sup(z \cap \mu)} \cap h(\alpha)(\mu) - h(\gamma)(\mu) \neq \emptyset$$

and

$$h(\alpha)(\mu) \in g_t(\alpha) \cap \mu = x \cap \mu$$

for a.a. $\mu \in a$. On the other hand

$$\begin{aligned}(x \cap \mu)_{j(\mu)}^{h(\varepsilon(x,y,j))(\mu)} &= (x \cap \mu)_{j(\mu)}^{h(\gamma)(\mu)} \\ &> (x \cap \mu)_{l(\mu)}^{h(\gamma)(\mu)} \\ &= h(\gamma)(\mu) \\ &> \rho(\sup(x \cap \mu), y \cap \mu)(j(\mu))\end{aligned}$$

for a.a. $\mu \in a$ if $j <^* l$ by the claims above. ⊣

§4. Remarks. Define

$$f : [\kappa^+]^2 \to \kappa^+$$

by

$$f(\alpha,\beta) = \max\Big\{\beta_j^\alpha \colon j < n(\alpha,\beta) \wedge \varepsilon(\alpha,\beta)(j) < \alpha \wedge j \leq n(\varepsilon(\alpha,\beta)(j),\alpha) \wedge \alpha_j^{\varepsilon(\alpha,\beta)(j)} \leq \rho(\alpha,\beta)(j)\Big\},$$

where $\varepsilon(\alpha,\beta)(j) = (\max \rho(\alpha,\beta)``j) + 1$. By our proof of Proposition in Section 2, for $X \subset \kappa^+$ unbounded there is a club $D \subset \kappa^+$ with $S^\kappa_{\kappa^+} \cap D \subset f``[X]^2$, which suffices to show $\kappa^+ \not\to [\kappa^+]^2_{\kappa^+}$. One might notice the difference between Shelah's coloring [4] and ours. This is because we could not work out the details of the penultimate paragraph in his proof. We could not verify either the corresponding uniqueness claim in the sentence (3) on page 240 of [2], even after replacing each occurrence of (ϵ,β) in the sentence by (ϵ,α). We nevertheless owe a great debt to both sources.

The rest of this section is devoted to Todorčević's original proof of Proposition as we understand it. We freely use the material in our previous proof.

Proof of Proposition. Fix $(x,y) \in [[\lambda]^\kappa]^2_\subset$ with $\sup x \in y$. Set

$$\begin{aligned} f(x,y) = \max\Big\{y_j^{\sup x} \colon\, & j < n(\sup x, y) \wedge \varepsilon(x,y,j) \text{ exists} \wedge \\ & j \leq n(\varepsilon(x,y,j),x) \wedge \\ & C_{y \cap y_j^{\sup x}} \cap \sup(x \cap \rho(\sup x, y)(j)) \\ & \quad - \sup(C_{x \cap x_j^{\varepsilon(x,y,j)}} \cap (\rho(\sup x,y)(j)+1)) \neq \emptyset\Big\}. \end{aligned}$$

We claim that f works.

Let $X \subset [\lambda]^\kappa$ be unbounded. Fix $s,t \in \lambda^l$ with $l < \omega$, and $\eta < \lambda$. Set

$$A_{st\eta} = \Big\{\xi < \lambda \colon \exists x \in X \exists \gamma \in x \Big(\rho(\gamma,x) \in \prod_{i<l}(t(i) - s(i)) \wedge \eta \in x \wedge \sup(C_{x\cap\gamma} \cap \xi) < \eta < \xi < \gamma\Big)\Big\}.$$

If $\sup A_{st\eta} = \lambda$, take $g_{st\eta}(\alpha) \in X$ for $\alpha < \lambda$ so that $g_{st\eta}(\alpha)$ and some $\gamma \in g_{st\eta}(\alpha)$ witness $\xi \in A_{st\eta}$ for some $\alpha < \xi < \gamma$. Let E be the club

$$\Big\{\delta < \lambda \colon \forall l < \omega \forall s,t \in \delta^l \forall \eta < \delta \Big(\sup A_{st\eta} < \delta \vee \Big(\sup A_{st\eta} = \lambda \wedge \forall \alpha < \delta(\sup g_{st\eta}(\alpha) < \delta)\Big)\Big)\Big\}.$$

We show that $\lim(S^\kappa_\lambda \cap E)$ is the desired club set.

Fix $\delta \in S^\kappa_\lambda \cap \lim(S^\kappa_\lambda \cap E)$. Build $z \in [\delta]^\kappa$ in κ stages so that

$$\sup(z \cap S^\kappa_\lambda \cap E) = \delta, \quad z \cap S^\kappa_\lambda \subset \lim z,$$
$$\{0\} \cup \{\beta + 1 \colon \beta \in z\} \cup \{\beta \in \lim z \colon \operatorname{cf}\beta < \kappa\} \subset z,$$

and

$$g_{st\eta}(\alpha) \cup \{\sup g_{st\eta}(\alpha)\} \subset z$$

for $s, t \in z^{<\omega}$, $\eta \in z$ with $\sup A_{st\eta} = \lambda$, and $\alpha \in z$. Then $\sup(z \cap \beta) \notin z$ for $\beta \in \delta - z$: Otherwise we would have

$$\sup(z \cap \beta) \in z \cap (\beta + 1) = z \cap \beta,$$

hence the desired contradiction $\sup(z \cap \beta) + 1 \in z \cap (\beta + 1) = z \cap \beta$ by the choice of z. Thus $\sup(z \cap \beta) \in \lim z - z \subset S^\kappa_\lambda$ for $\beta \in \delta - z$ by the choice of z. We have $z \cup \{\delta\} \subset y \in X$ with $\delta < \sup y$, since $X \subset [\lambda]^\kappa$ is unbounded.

Set

$$l = n(\delta, y) < \omega.$$

Then $l > 0$ and $\rho(\delta, y)(i) < \delta$ for $i < l$, hence we can define $t\colon l \to z$ by $t(i) = \min(z - (\rho(\delta, y)(i) + 1))$ as in the previous proof. We have $s\colon l \to z$ so that $s(i) \leq \rho(\delta, y)(i)$ and $C_{y \cap y^\delta_i} \cap \sup(z \cap \rho(\delta, y)(i)) - s(i) = \emptyset$: Set

$$s(i) = \rho(\delta, y)(i) \text{ if } \rho(\delta, y)(i) \in z.$$

Otherwise

$$\operatorname{cf} \sup(z \cap \rho(\delta, y)(i)) = \kappa$$

by the claim above. Hence

$$\sup(C_{y \cap y^\delta_i} \cap \sup(z \cap \rho(\delta, y)(i))) < \sup(z \cap \rho(\delta, y)(i))$$

by

$$\operatorname{ot}(C_{y \cap y^\delta_i} \cap \sup(z \cap \rho(\delta, y)(i))) < \kappa.$$

Thus we have

$$\sup(C_{y \cap y^\delta_i} \cap \sup(z \cap \rho(\delta, y)(i))) \leq s(i) \in z \cap \rho(\delta, y)(i)$$

as desired.

We have

$$\max \operatorname{ran} t < \delta' \in z \cap S^\kappa_\lambda \cap E,$$

since

$$\sup(z \cap S^\kappa_\lambda \cap E) = \delta.$$

Then

$$\zeta = \sup(C_{y \cap \delta} \cap \delta') < \delta'$$

by

$$\operatorname{ot}(C_{y \cap \delta} \cap \delta') < \kappa = \operatorname{cf} \delta'.$$

Hence

$$\sup(C_{y \cap \delta} \cap \delta') \leq \min(y - \sup(C_{y \cap \delta} \cap \delta')) \in C_{y \cap \delta} \cap \delta'$$

by the choice of $C_{y\cap\delta}$ and $\delta' \in z \cap S^\kappa_\lambda \subset \lim z \subset \lim y$. Thus

$$\zeta = \sup(C_{y\cap\delta} \cap \delta') \in C_{y\cap\delta} \cap \delta'.$$

Also $\eta = \min(z - (\zeta + 1)) < \delta'$ by $\zeta < \delta' \in \lim z$. Then $y \in X$ and $\delta \in y$ witness $\delta' \in A_{st\eta}$ by the choice of s and t, $\eta \in z \subset y$ and $\sup(C_{y\cap\delta} \cap \delta') < \eta < \delta' < \delta$. Note that $\sup A_{st\eta} = \lambda$: Otherwise we would have $\sup A_{st\eta} < \delta'$ by $\delta' \in E$, $s, t \in (\delta')^l$ and $\eta < \delta'$, contradicting $\delta' \in A_{st\eta}$.

We have

$$\delta' < \delta'' \in z \cap S^\kappa_\lambda \cap E$$

with

$$C_{y\cap\delta} \cap \delta'' - \delta' \neq \emptyset,$$

since

$$\sup C_{y\cap\delta} = \sup(z \cap S^\kappa_\lambda \cap E) = \delta.$$

Then

$$\delta' \leq \sup(C_{y\cap\delta} \cap \delta'') \in C_{y\cap\delta} \cap \delta''$$

as for

$$\zeta = \sup(C_{y\cap\delta} \cap \delta') \in C_{y\cap\delta} \cap \delta'.$$

We have

$$\sup(C_{y\cap\delta} \cap \delta'') < \alpha \in z \cap \delta'',$$

since

$$\delta'' \in z \cap S^\kappa_\lambda \subset \lim z.$$

Set

$$x = g_{st\eta}(\alpha) \in X.$$

Then

$$x \cup \{\sup x\} \subset z \cap \delta'' \subset y,$$

since $s, t \in (z \cap \delta'')^l$, $\eta \in z \cap \delta''$, $\sup A_{st\eta} = \lambda$, $\alpha \in z \cap \delta''$ and $\delta'' \in E$. Take $\gamma \in x$ and $\alpha < \xi < \gamma$ so that x and γ witness $\xi \in A_{st\eta}$. We claim that $f(x, y) = \delta$, as desired.

Note first

$$|\rho(\delta, y)| = n(\delta, y) = l$$

and

$$\max \operatorname{ran} \rho(\delta, y) < \max \operatorname{ran} t < \delta' \leq \sup(C_{y\cap\delta} \cap \delta'') < \alpha < \xi < \gamma \in x.$$

Hence

$$\max \rho(\delta, y)``l < \sup x < \delta \in y.$$

Thus $l < n(\sup x, y)$, $y_j^{\sup x} = y_j^\delta$ for $j \le l$, $\rho(\sup x, y) \mid l = \rho(\delta, y)$ and $\varepsilon(x, y, j) \le \gamma$ for $j \le l$ as in the previous proof. Also

$$\sup(C_{y\cap\delta} \cap \sup x) = \sup(C_{y\cap\delta} \cap \delta'') \in C_{y\cap\delta} \subset y$$

by

$$\sup(C_{y\cap\delta} \cap \delta'') < \sup x < \delta''.$$

Hence $\eta < \delta' \le \sup(C_{y\cap\delta} \cap \sup x) < \alpha < \xi$ by the claims above.

Note next

$$n(\gamma, x) = |\rho(\gamma, x)| = |s| = |t| = l$$

and $s(i) \le \rho(\gamma, x)(i) < t(i)$ for $i < l$, since x and γ witness $\xi \in A_{st\eta}$. Hence $\rho(\gamma, x)(i) \le \rho(\delta, y)(i)$ for $i < l$, $j \le n(\varepsilon(x, y, j), x)$ and $x_j^{\varepsilon(x,y,j)} = x_i^\gamma$ for $j \le l$ as in the previous proof. Recall that $\rho(\gamma, x)(i) \in C_{x\cap x_i^\gamma}$ for $i < l$. Thus

$$s(i) \le \rho(\gamma, x)(i) \le \sup(C_{x\cap x_i^\gamma} \cap (\rho(\delta, y)(i) + 1))$$

for $i < l$. Since x and γ witness $\xi \in A_{st\eta}$, $\sup(C_{x\cap\gamma} \cap \xi) < \eta \in x$. Hence $\sup(C_{x\cap\gamma} \cap \xi) \le \zeta$, since $C_{x\cap\gamma} \cap \xi \subset x \cap \eta \subset z \cap \eta \subset \zeta + 1$ by the choice of η.

Finally we have $\eta < \rho(\sup x, y)(l) < \xi$, since

$$\begin{aligned}\rho(\sup x, y)(l) &= \min(y - \sup(C_{y\cap y_l^{\sup x}} \cap \sup x)) \\ &= \min(y - \sup(C_{y\cap\delta} \cap \sup x)) \\ &= \sup(C_{y\cap\delta} \cap \sup x)\end{aligned}$$

by the claims above. Hence

$$\begin{aligned}\sup(C_{x\cap\gamma} \cap (\rho(\sup x, y)(l) + 1)) &\le \sup(C_{x\cap\gamma} \cap \xi) \\ &\le \zeta < \eta \le \sup(x \cap \rho(\sup x, y)(l))\end{aligned}$$

by the claims above. Thus

$$\begin{aligned}&\zeta \in C_{y\cap\delta} \cap \sup(x \cap \rho(\sup x, y)(l)) - \sup(C_{x\cap\gamma} \cap (\rho(\sup x, y)(l)+1)) \\ &\quad = C_{y\cap y_l^{\sup x}} \cap \sup(x \cap \rho(\sup x, y)(l)) - \sup(C_{x\cap x_l^{\varepsilon(x,y,l)}} \cap (\rho(\sup x, y)(l)+1)).\end{aligned}$$

On the other hand

$$\begin{aligned}&C_{y\cap y_i^{\sup x}} \cap \sup(x \cap \rho(\sup x, y)(i)) - \sup(C_{x\cap x_i^{\varepsilon(x,y,i)}} \cap (\rho(\sup x, y)(i) + 1)) \\ &= C_{y\cap y_i^\delta} \cap \sup(x \cap \rho(\delta, y)(i)) - \sup(C_{x\cap x_i^\gamma} \cap (\rho(\delta, y)(i) + 1)) \\ &\subset C_{y\cap y_i^\delta} \cap \sup(z \cap \rho(\delta, y)(i)) - s(i) \\ &= \emptyset\end{aligned}$$

for $i < l$ by the choice of $s(i)$ and the claims above. ⊣

In the case of coloring pairs of ordinals $<\kappa^+$, the above definition (and proof) can be simplified as in [9]: Set

$$f(\alpha,\beta) = \max\Big\{\beta_j^\alpha : j < n(\alpha,\beta) \wedge \varepsilon(\alpha,\beta)(j) < \alpha \wedge \\ j \leq n(\varepsilon(\alpha,\beta)(j),\alpha) \wedge \rho(\alpha,\beta)(j) \notin C_{\alpha_j^{\varepsilon(\alpha,\beta)(j)}}\Big\}.$$

The simplified coloring, however, does not appear to work in general.

REFERENCES

[1] Matthew Foreman and Menachem Magidor, *Mutually stationary sequences of sets and the non-saturation of the non-stationary ideal on $p_\kappa(\lambda)$*, ***Acta Mathematica***, vol. 186 (2001), pp. 271–300.

[2] András Hajnal and Peter Hamburger, ***Set theory***, London Mathematical Society Student Texts, vol. 48, Cambridge University Press, Cambridge, 1999.

[3] Akihiro Kanamori, ***The higher infinite***, Perspectives in Mathematical Logic, Springer, 1994.

[4] Saharon Shelah, *Was Sierpinski right? I*, ***Israel Journal of Mathematics***, vol. 62 (1988), pp. 355–380.

[5] ———, *Advances in cardinal arithmetic*, ***Finite and infinite combinatorics in sets and logic*** (N. Sauer etal, editor), NATO Adv. Sci. Inst. Ser. C Math. Phys. Sci., no. 411, Kluwer, Dordrecht, 1993, pp. 355–383.

[6] ———, *$\aleph_{\omega+1}$ has a jonsson algebra*, ***Cardinal arithmetic***, Oxford University Press, New York, 1994, pp. 34–116.

[7] ———, *There are jonsson algebras in many inaccessible cardinals*, ***Cardinal arithmetic***, Oxford University Press, New York, 1994, pp. 117–184.

[8] Masahiro Shioya, *Splitting $\mathcal{P}_\kappa\lambda$ into maximally many stationary sets*, ***Israel Journal of Mathematics***, vol. 114 (1999), pp. 347–357.

[9] Stevo Todorčević, *Partitioning pairs of countable ordinals*, ***Acta Mathematica***, vol. 159 (1987), pp. 261–294.

[10] ———, *Partitioning pairs of countable sets*, ***Proceedings of the Amererican Mathematical Society***, vol. 111 (1991), pp. 841–844.

[11] Daniel Velleman, *Partitioning pairs of countable sets of ordinals*, ***The Journal of Symbolic Logic***, vol. 55 (1990), pp. 1019–1021.

INSTITUTE OF MATHEMATICS
UNIVERSITY OF TSUKUBA
TSUKUBA, 305-8571, JAPAN
E-mail: shioya@math.tsukuba.ac.jp

ASPECTS OF THE TURING JUMP

THEODORE A. SLAMAN

§1. Introduction.

DEFINITION 1.1. The *Turing Jump* is the function which maps a set $X \subseteq \mathbb{N}$ to X', the halting problem relative to X. Fixing a recursive enumeration of all Turing machines,

$$X' = \{e\colon \text{The } e\text{th Turing machine with oracle } X \text{ halts.}\}$$

X' is the canonical example of a set which is definable from X but not recursive in X. The Turing degree of X' depends only on the Turing degree of X, so the jump induces an increasing function on the Turing degrees $\mathcal{D}$.

In this paper, we will discuss two aspects of the jump and its iterations. First, we will show that they are implicitly characterized by general properties of relative definability. Second, we will present the Shore and Slaman [1999] theorem that the function $x \mapsto x'$ is first order definable in the Turing degrees. Finally, we will pose analogous questions about the relation *y is recursively enumerable in x* and discuss what is known about them.

Our discussion will rest on two technical facts, which are generalizations of the following two theorems.

THEOREM 1.2 (Friedberg [1957]). *Suppose that $x \geq_T 0'$. Then there is a g such that $g' = x$.*

In other words, every sufficiently complicated degree x is the jump of some degree g.

THEOREM 1.3 (Posner and Robinson [1981]). *Suppose that $0 \not\geq_T x$. Then there is a g such that $x \vee g = g'$.*

Similarly, for every nontrivial degree x, there is a degree g such that by considering x relative to g, that is by considering $x \vee g$, one can view x as a jump.

Slaman was partially supported by National Science Foundation Grant DMS-9988644 and partially supported by the Mittag-Leffler Institute during the preparation of this paper. Slaman wishes to thank David Lippe for his comments on a preliminary draft of this paper.

Logic Colloquium 2000
Edited by R. Cori, A. Razborov, S. Todorčević, and C. Wood
Lecture Notes in Logic, 19

1.1. Hierarchies of definability. The arithmetic and hyperarithmetic hierarchies provide the context in which we can extend the inversion and join theorems.

The arithmetic hierarchy.

DEFINITION 1.4.

1. Σ^0_0 and Π^0_0 denote the set of bounded formulas in first order arithmetic.
2. Σ^0_{n+1} denotes the set of formulas $(\exists n_1, \dots, n_k)\psi$, where $\psi \in \Pi^0_n$.
3. Π^0_{n+1} denotes the set of formulas $(\forall n_1, \dots, n_k)\psi$, where $\psi \in \Sigma^0_n$.

DEFINITION 1.5.

1. $X \subseteq \mathbb{N}$ is Σ^0_n or Π^0_n if and only if it is definable in arithmetic by a formula of the corresponding type.
2. $X \subseteq \mathbb{N}$ is Δ^0_n if and only if it is both Σ^0_n and Π^0_n.

As is known very well, X is Δ^0_{n+1} if and only if $X \leq_T \emptyset^{(n)}$. Thus, the arithmetic sets are generated from the empty set by applying the jump and recursive functionals.

The hyperarithmetic hierarchy. Davis [1950] extended the arithmetic hierarchy into the transfinite by iterating the jump along recursive well-orderings of $\mathbb{N}$. At limits λ, he used the recursive presentation of λ to form the recursive join of the sets associated (by that presentation) to ordinals less than λ.

For example, $\emptyset^{(\omega)} = \{(n,m)\colon m \in \emptyset^{(n)}\}$ represents the ωth jump of the empty set.

Spector [1955] showed that any two sets associated with the same recursive ordinal have the same Turing degree. Consequently, we have degree invariant functions $X \mapsto X^{(\alpha)}$. Of course, we can define the hyperarithmetic hierarchy relative to a presentation of any countable ordinal, and then we obtain a degree invariant function $X \mapsto X^{(\alpha)}$ which is defined on those X's which compute that presentation of α.

1.1.1. *Jockusch–Shore REA-operators.*

DEFINITION 1.6 (Jockusch and Shore [1984]). An *REA-operator* is a function J from $2^{\mathbb{N}}$ to $2^{\mathbb{N}}$ such that there is an e such that for all X, $J(X)$ is the join of X with the eth set which is recursively enumerable relative to X (in a fixed recursive enumeration of all the relativized recursive enumerations).

An α-*REA*-operator is an α-length iteration of *REA*-operators, where the iteration is organized using a recursive presentation of α as in the hyperarithmetic hierarchy. For example, the canonical 2-*REA* operator is the map $A \mapsto A''$. But, note that not every 2-*REA* operator is degree invariant. The reader should consult Jockusch and Shore [1984] for a detailed presentation.

1.2. Extensions of the prototypes. Surprisingly, the inversion and join properties of the Turing jump are shared by all α-*REA* operators.

THEOREM 1.7 (Jockusch and Shore [1984]). *Suppose that $C \geq_T \emptyset^{(\alpha)}$ and J is an α-REA operator. There is a G such that $J(G) \equiv_T C$.*

Thus, every sufficiently complicated degree is in the range of J.

THEOREM 2.1 (Shore and Slaman [1999]). *Suppose that J is an α-REA operator and for all $\beta < \alpha$, $\emptyset^{(\beta)} \not\geq_T X$. There is a G such that $X \oplus G \equiv_T J(G)$.*

Similarly, every nontrivial set represents the value of J relative to some set. We will outline the proof of Theorem 2.1 in Section 2. We should note that there is a history behind it, with substantial preliminary results by Jockusch and Shore and by Cooper. A full account is given in Shore and Slaman [1999].

We can interpret Theorems 1.2 and 1.3, and their generalizations to Theorems 1.7 and 2.1, as asserting a fundamental role for the jump and its iterations. By Theorem 1.3, the only way by which a set can be not recursive is by being equivalent to the jump relative to some other set. And by Theorem 2.1, the same phenomenon is repeated through the transfinite. The applications which follow can be viewed as realizations of this interpretation.

1.3. Abstract closure operators. In our first application, we argue that the iterations of the Turing jump have a special role within relative definability. In the following definition, we are thinking of M as a function which maps X in $2^{\mathbb{N}}$ to the set of those Y's in $2^{\mathbb{N}}$ which are definable from X (in some specific way).

DEFINITION 1.8. A *closure operator* is a map $M : 2^{\mathbb{N}} \to 2^{2^{\mathbb{N}}}$ with the following properties.

1. For all $X \in 2^{\mathbb{N}}$, $X \in M(X)$.
2. For all X and for all Z, if Z is recursive in finitely many elements of $M(X)$, then $Z \in M(X)$. *$M(X)$ is closed under join and relative computation.*
3. For all X and Y in $2^{\mathbb{N}}$, if X is recursive in Y then $M(X) \subseteq M(Y)$. *M is monotone.*

The functions mapping X to the collection of subsets of $\mathbb{N}$ recursive in X, recursive in X', or arithmetically definable from X are closure operators.

Comparing closure operators.

DEFINITION 1.9. If M_1 and M_2 are closure operators, then $M_1 \leq M_2$ if and only if there is a set of natural numbers B, such that for all X, if B is recursive in X, then $M_1(X) \subseteq M_2(X)$.

In other words, $M_1 \leq M_2$ if and only if for all sufficiently complicated X, $M_1(X) \subseteq M_2(X)$. We say M_1 and M_2 are *equivalent* when for all sufficiently complicated X, $M_1(X) = M_2(X)$.

REMARK 1.10. This notion is a $2^{\mathbb{N}} \to 2^{2^{\mathbb{N}}}$ variation on Martin's ordering of degree invariant functions on reals. See Kechris and Moschovakis [1978] for further information on the Martin order and on Martin's conjectures concerning that order.

1.3.1. *Implicitly characterizing the hyperarithmetic hierarchy.* With the following theorem, we give a complete classification of the Borel closure operators, up to equivalence.

THEOREM 3.1. *If M is a closure operator such that the relation $Y \in M(X)$ is Borel, then one of the following conditions holds.*

1. *There is a countable ordinal α such that M is equivalent to the map*
$$X \mapsto \{Y\colon Y \text{ is recursive in } X^{(\alpha)}\}.$$
2. *There is a countable ordinal α such that M is equivalent to the map*
$$X \mapsto \{Y\colon (\exists \beta < \alpha)[Y \text{ is recursive in } X^{(\beta)}]\}.$$
3. *M is equivalent to the map $X \mapsto 2^{\mathbb{N}}$.*

In items 1 and 2, we fix a presentation of α and define the relevant iterations of the jump for all X relative to which that presentation is recursive.

Thus, every Borel closure operators is equivalent to one obtained by canonically iterating the Turing jump. Note, in the statement of Theorem 3.1, case 1 can be viewed as the successor ordinal case of case 2. We have chosen to state the theorem in this way to highlight the difference between the successor and limit cases. In the successor case, $M(X)$ has an element of greatest degree, and in the limit case, it does not.

We will prove Theorem 3.1 in Section 3. We note that both the theorem and its proof have precursors in Slaman and Steel's [1981] analysis of the Borel order preserving functions from $\mathcal{D}$ to $\mathcal{D}$.

It follows from Theorem 3.1 that the Borel closure operators are well-ordered. In analogy to Martin's conjectures concerning degree invariant functions, one can ask whether the Axiom of Determinacy implies that the set of all closure operators is well-ordered. The fact that we will use Martin's [1975] theorem that all Borel games are determined to prove Theorem 3.1 could be taken as evidence supporting an affirmative answer. On the other hand, this use of Borel determinacy may not be necessary. It is not known whether the proof of Theorem 3.1, like Borel Determinacy, requires the use of uncountably many iterations of the power set operation.

1.4. A degree theoretic definition of the jump. For our second application, we outline Shore and Slaman's [1999] proof that the jump is definable by a first order formula in $\mathcal{D}$, the partial order of the Turing degrees.

First, we invoke a theorem of Slaman and Woodin.

THEOREM 1.11 (Slaman and Woodin). *The function $x \mapsto x''$ is first order definable in $\mathcal{D}$.*

Theorem 1.11 is proved by applying the Slaman and Woodin analysis of automorphisms of the Turing degrees. This analysis involves a fair amount of metamathematics, but not much technical recursion theory.

Second, we apply Theorem 2.1 to the canonical 2-*REA* operator, the double-jump. And so, we define $0'$ in $\mathcal{D}$.

THEOREM 1.12. *$0'$ is defined within $\mathcal{D}$ as the greatest degree z such that there is no g such that $z \vee g$ is equal to g''.*

PROOF. For every degree g, $0' \vee g$ is recursive in g', and so no degree less than or equal to $0'$ can join any g to g''. Theorem 2.1 states that any degree not below $0'$ does join some g up to g''. Consequently, $0'$ is first order definable in terms of order, join and the double-jump. Order and join are clearly definable in $\mathcal{D}$ and, by Theorem 1.11, so is the double-jump. ⊣

We can relativize the above proof to obtain the following definition of the function mapping x to x' in $\mathcal{D}$.

THEOREM 1.13. *For any degree x, x' is definable from x within $\mathcal{D}$ as the greatest degree z such that $z \geq x$ and there is no g greater than or equal to x such that $z \vee g$ is equal to g''.*

We note that Cooper has claimed to define the jump by different means.

1.5. Recursive enumerability. Finally, we investigate whether the above properties of the jump have analogs for the relation of relative recursive enumerability.

To set the context, recall that recursive enumerability generates a finer hierarchy of degrees than the hyperarithmetic hierarchy.

DEFINITION 1.14 (Ershov, 1968). The *Difference Hierarchy*:

1. A set X is n-recursively enumerable if it has a recursive approximation
$$X(k) = \lim_{s \to \infty} \psi(k, s)$$
such that there are at most n many numbers s such that $\psi(k, s+1) \neq \psi(k, s)$.
2. For infinite α, X is α-recursively enumerable relative to a recursive system S of notations for ordinals if and only if there is a partial recursive function ψ such that for all k, $X(k)$ is equal to $\psi(k, z)$, where z is the S-least notation b for an ordinal less than α such that $\psi(k, b)$ converges.

Jockusch and Shore [1984] showed that every α-re set is α-*REA*. The converse fails, as every α-re set is Δ^0_2.

In fact, every Δ^0_2 set appears in the difference hierarchy, in the following precise sense.

THEOREM 1.15 (Ershov, 1968). *For every path S through $\mathcal{O}$, Kleene's complete set of notations for the recursive ordinals, and for every set $X \in \Delta^0_2$, there is a recursive ordinal α such that X is α-recursively enumerable relative to S.*

However, there is no analog for Spector's theorem that the Turing degree of the αth iterate of the Turing jump does not depend on which recursive presentation of α is used to form the iteration. The stratification of the Δ^0_2 sets by the difference hierarchy depends on S, the choice of notations for the recursive ordinals.

1.5.1. *Inversion and join.* We can consider the definition of an n-re or even α-re sets relative to another set X. By doing so, we obtain the n-re and α-re operators on sets. In particular, a 2-re operator is a function D from $2^{\mathbb{N}}$ to $2^{\mathbb{N}}$ such that there is an e such that for all X, $D(X)$ is the eth set which is 2-re relative to X (in a fixed recursive enumeration of all of the relativized 2-re definitions).

How do inversion and join apply to the operators which appear in the difference hierarchy? Since the difference hierarchy is contained the the *REA*-hierarchy, Theorem 1.7 applies to increasing α-re operators and provides an inversion theorem for them. We also suspect it provides the optimal result, although that is yet to be proven. The following example shows that Theorem 2.1 cannot be improved to reflect the resolution of the difference hierarchy.

THEOREM 4.1. *There is a 2-re operator* $D\colon Z \mapsto D(Z) \geq_T Z$ *and a 2-re set* X *with the following properties.*

1. X *is not of recursively enumerable degree.*
2. *For all* $G \subseteq \mathbb{N}$, $X \oplus G \not\equiv_T D(G)$.

We will present the proof of Theorem 4.1 in Section 4.

The existence of a counterexample to a sharp join theorem would lead us to believe that recursive enumerability and the difference hierarchy behave differently than the jump and the hyperarithmetic hierarchy. The following questions test this hypothesis.

A. Is there an implicit characterization for recursive enumerability and/or for the difference hierarchy?
B. Is there a degree theoretic definition of the relation *y is recursively enumerable relative to x*?

Of course, the second question is well known. See Rogers [1967].

1.5.2. Σ-*closure operators.* In the following, we are again thinking of $M(X)$ as the set of $Y \in 2^{\mathbb{N}}$ which are definable from X, but we are not assuming that the degrees of the definable sets are closed downward in $\mathcal{D}$.

DEFINITION 1.16. A Σ-*closure operator* is a map $M\colon 2^{\mathbb{N}} \to 2^{2^{\mathbb{N}}}$ with the following properties.

1. For all $X \in 2^{\mathbb{N}}$, $X \in M(X)$.
2. For all $X \in 2^{\mathbb{N}}$ and all $\{Z_1, \dots, Z_k\} \subseteq M(X)$, every set which is Turing equivalent to the join $\bigoplus_{i \leq k} Z_i$ is an element of $M(X)$.
3. For all X and Y in $2^{\mathbb{N}}$, if X is recursive in Y then $M(X) \subseteq M(Y)$.

Every closure operator is a Σ-closure operator, but there are others as well. For example, the maps sending X to the collection of sets with degree recursively enumerable in X or to the collection of sets with degree 2-re in X are Σ-closure operators which are not closure operators.

An implicit characterization of *REA*. The relation $Y \in REA(X)$ is an invariant of the class of Σ-closure operators.

THEOREM 5.2. *For any Borel Σ-closure operator M, if there is a cone of X's for which $M(X) \not\subseteq \Delta^0_1(X)$, then there is a cone of X's such that $M(X)$ contains all of the sets which are REA in X.*

In the sense of Theorem 5.2, recursive enumerability is an unavoidable consequence of nontriviality. Additionally, Theorem 5.2 provides operational limits on the possible uniformly definable in X divisions of the class $REA(X)$. We prove it in Section 5.

Despite Theorem 5.2, there is no classification of the Borel Σ-closure operators as simply presented as the one for closure operators given in Theorem 3.1. Shore has pointed out that the *REA* and difference hierarchies give different resolutions of Δ^0_2, so there are natural incomparable (under eventual pointwise inclusion) Σ-closure operators. Horowitz has shown that the Borel Σ-closure operators are not well-founded.

1.5.3. *Is relative recursive enumerability degree theoretically definable?* Cooper has claimed that the relation w *is recursively enumerable relative to* x is definable in $\mathcal{D}$, but his proof relied on a join principle for 2-re operators which is contradicted by Theorem 4.1. Cooper has since claimed to use additional properties of specific 2-re operators to circumvent this problem, but the full proof is not yet available.

§2. Proving the join theorem.

In this section, we outline the proof of the Shore and Slaman [1999] Join Theorem for α-*REA* operators.

THEOREM 2.1 (Shore and Slaman [1999]). *Suppose that J is an α-REA operator and for all $\beta < \alpha$, $\emptyset^{(\beta)} \not\geq_T X$. There is a G such that $X \oplus G \equiv_T J(G)$.*

We first show that it is sufficient to prove a weak join theorem for the α-*REA* operators $X \mapsto X^{(\alpha)}$. Namely, it is sufficient to show that if X and α are given so that α is a recursive ordinal and for all $\beta < \alpha$, $\emptyset^{(\beta)} \not\geq_T X$, then there is a G such that $X \oplus G \geq_T G^{(\alpha)}$.

Suppose that X and α are given as above and that J is an α-*REA* operator. Assuming the weak join theorem for the α-jump, fix G so that $X \oplus G \geq_T G^{(\alpha)}$. By Theorem 1.7 relative to G, there is an $H \geq_T G$ such that $J(H) \equiv X \oplus G$. But then $X \oplus H \geq_T X \oplus G \equiv_T J(H)$. Similarly, $J(H) \equiv_T X \oplus G \geq_T X$, since J is increasing in degree $J(H) \geq_T H$, and so $J(H) \geq_T X \oplus H$. Thus, $J(H) \equiv_T X \oplus H$, as required to verify the join theorem for J.

Turing functionals. We begin with a formalization of the way by which one set is recursive in another.

DEFINITION 2.2.

1. A *Turing functional* Φ is a set of sequences (x, y, σ) such that x and y are natural numbers and σ is a finite binary sequence. Further, for all x, for all y_1 and y_2, and for all compatible σ_1 and σ_2, if $(x, y_1, \sigma_1) \in \Phi$ and $(x, y_2, \sigma_2) \in \Phi$, then $y_1 = y_2$ and $\sigma_1 = \sigma_2$.
2. Φ is *use-monotone* if the following conditions hold.
   (a) For all (x_1, y_1, σ_1) and (x_2, y_2, σ_2) in Φ, if σ_1 is a proper initial segment of σ_2, then x_1 is less than x_2.
   (b) For all x_1 and x_2, y_2 and σ_2, if $x_2 > x_1$ and $(x_2, y_2, \sigma_2) \in \Phi$, then there are y_1 and σ_1 such that $\sigma_1 \subseteq \sigma_2$ and $(x_1, y_1, \sigma_1) \in \Phi$.
3. We write $\Phi(x, \sigma) = y$ to indicate that there is a τ such that τ is an initial segment of σ, possibly equal to σ, and $(x, y, \tau) \in \Phi$. If $X \subseteq \mathbb{N}$, we write $\Phi(x, X) = y$ to indicate that there is an ℓ such that $\Phi(x, X \upharpoonright \ell) = y$, and write $\Phi(X)$ for the function evaluated in this way. We say that $\Phi(X) = Y$ if and only if $\Phi(X)$ is equal to the characteristic function of Y.

In Definition 2.2, we do not require that Φ be recursively enumerable. Consequently, if Φ is a Turing functional and $X \subseteq \mathbb{N}$, then $\Phi(X)$ is recursive only in the join of Φ and X. Note also that in this formulation, a Turing functional is just a particular way to define a continuous function from $2^{\mathbb{N}}$ to $2^{\mathbb{N}}$.

In this language, $X \geq_T Y$ if and only if there is a recursive Turing functional Φ such that $\Phi(X) = Y$.

Kumabe–Slaman forcing. Kumabe and Slaman introduced the following notion of forcing in an earlier join theorem for the ω-jump.

1. A *condition* is a pair $p = (\Phi_p, \boldsymbol{Z}_p)$ consisting of a finite use-monotone Turing functional and a finite subset of $2^{\mathbb{N}}$.
2. If $p = (\Phi_p, \boldsymbol{Z}_p)$ and $q = (\Phi_q, \boldsymbol{Z}_q)$ are conditions, then $p \geq q$ if and only if
   (a) i. $\Phi_p \subseteq \Phi_q$ and
   ii. for all $(x_q, y_q, \sigma_q) \in \Phi_q \setminus \Phi_p$ and all $(x_p, y_p, \sigma_p) \in \Phi_p$, the length of σ_q is greater than the length σ_p,
   (b) $\boldsymbol{Z}_p \subseteq \boldsymbol{Z}_q$,
   (c) for every x, y, and $Z \in \boldsymbol{Z}_p$, if $\Phi_q(x, Z) = y$ then $\Phi_p(x, Z) = y$.

Hence, a condition p specifies all of the elements (x, y, σ) in Φ for which σ has length less than or equal to the maximum length occurring in Φ_p and specifies finitely many sets relative to which Φ is only defined at arguments where Φ_p is already defined.

2.1. An illuminating special case. We will only give the argument for the weak join theorem in case when $\alpha = 1$. The general proof involves a transfinite analysis of the Kumabe-Slaman forcing relation for atomic statements about $\Phi^{(\alpha)}$ and is available in Shore and Slaman [1999]. In this special case, we assume that X is not recursive and we build Φ so that $\Phi(X) \geq_T \Phi'$, thereby establishing the weak join property that $X \oplus \Phi \geq_T \Phi'$.

We will build Φ to have the following properties.

1. Every atomic statement about Φ' is decided by a Kumabe-Slaman condition on Φ.
2. $\Phi(X) = \Phi'$.

In our construction, we will first decide whether $n \in \Phi'$ using forcing and then we will extend our condition on Φ so that $\Phi(n, X)$ is equal to the value already decided for Φ'.

Deciding $\Sigma^0_1(\Phi)$ sentences. Our goal is to decide atomic statements about Φ' using conditions q while first maintaining $\Phi(X) = \Phi'$ and second maintainng $X \notin \boldsymbol{Z}_q$.

Given a condition $p = (\Phi_p; \boldsymbol{Z}_p)$ and a Σ^0_1-sentence $\psi(\Phi)$, one of the following conditions holds.

1. There is a $q = (\Phi_q, \boldsymbol{Z}_p)$ extending p such that $\Phi_p(X) = \Phi_q(X)$, and $\psi(\Phi_q)$ is satisfied by means of a witness less than the length of some σ such that there are x and y with $(x, y, \sigma) \in \Phi_q$.
2. For all q extending p, if $\psi(\Phi_q)$ then there is a $(x, y, \sigma) \in \Phi_q \setminus \Phi_p$ such that X extends σ.

In the first case, $(\Phi_q, \boldsymbol{Z}_p)$ is a condition extending p as required. Namely, the finite amount Φ_q specified about Φ is sufficient to establish a witness that Σ^0_1 statement ψ is true of Φ.

Otherwise, it is not possible to extend Φ_p to make ψ hold without adding a computation relative to X or one of the elements of $\boldsymbol{Z}_p$. We convert this situation into a definition.

DEFINITION 2.3. Fixing k, $\tau \in (2^{k_1})^k$ is *essential* if and only if for all finite Φ_q extending Φ_p, if $\psi(\Phi_q)$, then there is a $(x, y, \sigma) \subset \Phi_q \setminus \Phi_p$ such that σ is compatible with some component of τ.

The set of essential k-sequences form a finitely branching Π^0_1 tree T.

Now consider the second case in the analysis of ψ. Let k be the size of $\boldsymbol{Z}_p \cup \{X\}$. $\boldsymbol{Z}_p \cup \{X\}$ determines an infinite path through T.

Now, we use the hypothesis that X is not recursive. Since X is not recursive, T must have another path with coordinates $\boldsymbol{Y}$ such that $X \notin \boldsymbol{Y}$. But then, $q = (\Phi_p, \boldsymbol{Z}_p \cup \boldsymbol{Y})$ forces $\neg\psi(\Phi)$ and $X \notin \boldsymbol{Y}$ as desired.

Constructing Φ. By the previous paragraphs, we can start with one condition p and with a Σ^0_1 sentence ψ about Φ, and find a condition q extending p such that q decides ψ and such that $\Phi_q(X)$ is equal to $\Phi_p(X)$. In other words, we can decide an atomic statement about Φ' without adding any computations to Φ which apply to X. Now, we can proceed by recursion to build Φ

so that $\Phi(X) = \Phi'$ by alternating between deciding the next atomic statement ψ_n about Φ' and then extending Φ so that $\Phi(n, X)$ is equal to the value so decided. The statements forced along the way will be true of the functional Φ obtained in the limit, since whenever we decide that Σ^0_1 is supposed to hold, we actually fix the witness that makes it so.

§3. Proof of the hierarchy theorem.

THEOREM 3.1. *If M is a closure operator such that the relation $Y \in M(X)$ is Borel, then one of the following conditions holds.*

1. *There is a countable ordinal α such that M is equivalent to the map*

$$X \mapsto \{Y\colon Y \text{ is recursive in } X^{(\alpha)}\}.$$

2. *There is a countable ordinal α such that M is equivalent to the map*

$$X \mapsto \{Y\colon (\exists \beta < \alpha)[Y \text{ is recursive in } X^{(\beta)}]\}.$$

3. *M is equivalent to the map $X \mapsto 2^{\mathbb{N}}$.*

In items 1 and 2, we fix a presentation of α and define the relevant iterations of the jump for all X relative to which that presentation is recursive.

PROOF. In this argument, we apply some effective descriptive set theory. The reader may consult Sacks [1990] for the relevant background material.

First, suppose that there is a cone of X's such that $M(X)$ does not include all of the sets which are hyperarithmetic in X. Fix B so that B is the base of such a cone and so that the relation $Y \in M(X)$ is Δ^1_1 relative to B. Then, for all $X \geq_T B$, there is an $\alpha(X)$ such that $\omega_1^X > \alpha(X)$ and $X^{(\alpha(X))} \notin M(X)$. By Spector's [1955] Bounding Theorem, there is a single α such that $\omega_1^B > \alpha$ and for all $X \geq_T B$, $X^{(\alpha)} \notin M(X)$. Let α_0 be such an α.

By Martin's [1975] Borel Determinacy, for every β less than α_0, either there is a cone of degrees X such that $X^{(\beta)} \in M(X)$ or there is a cone of degrees X such that $X^{(\beta)} \notin M(X)$. Since every countable set of degrees has an upper bound and α_0 is countable, by increasing B and decreasing α_0 as needed, we may assume that for all β less than α_0 and all $X \geq_T B$, $X^{(\beta)} \in M(X)$ and that for all $X \geq_T B$, $X^{(\alpha_0)} \notin M(X)$.

Now, we claim that for all $X \geq_T B$, if $Y \in M(X)$ then there is a $\beta < \alpha_0$ such that $Y \leq_T X^{\beta}$. Suppose not, and let X and Y be a counterexample to the claim. By Theorem 2.1 relative to X, there is a G such that $G \geq_T X$ and $Y \oplus G \geq_T G^{(\alpha_0)}$. But then, since $G \geq_T X$, $M(X) \subseteq M(G)$ and so $Y \in M(G)$. Since M is a closure operator, $G \in M(G)$ and so $Y \oplus G \in M(G)$. Consequently, $G^{(\alpha_0)} \in M(G)$, which is a contradiction.

Thus for all $X \geq_T B$ and all Y, $Y \in M(X)$ if and only if there is a $\beta < \alpha_0$ such that Y is recursive in $X^{(\beta)}$, and the theorem is proven for this case.

Now, suppose that in every cone there is an X such that $M(X)$ includes all of the sets which are hyperarithmetic in X.

Fix B so that the relation $Y \in M(X)$ is Δ^1_1 relative to B and so that every set hyperarithmetic in B belongs to $M(B)$. Let $HYP(B)$ denote the collection of sets Y such that Y is hyperarithmetic in B. Since $HYP(B)$ is not Δ^1_1 relative to B, let Y be a set in $M(B) \setminus HYP(B)$. By a result of Woodin (unpublished), a join theorem for the hyperjump, there is a $G \geq_T B$ such that $Y \oplus G \equiv_T \mathcal{O}^G$, where $\mathcal{O}^G$ is the complete Π^1_1 subset of $\mathbb{N}$ relative to G. As above, $\mathcal{O}^G$ is an element of $M(G)$. But now consider the set $\{Y\colon Y \notin M(G)\}$. This set is Δ^1_1 relative to G. If it were nonempty, then it would have an element recursive in $\mathcal{O}^G$. But $\mathcal{O}^G \in M(G)$ implies that every set recursive in $\mathcal{O}^G$ belongs to $M(G)$. Consequently, $M(G) = 2^{\mathbb{N}}$. Finally, note that if $X \geq_T G$ then $M(G) \subseteq M(X)$. Thus, for every X, if $X \geq_T G$ then $M(X) = 2^{\mathbb{N}}$, and the theorem is proven in this case as well. $\dashv$

§4. A 2-re operator without the join property.

THEOREM 4.1. *There is a 2-re operator $D\colon A \mapsto D(A) \oplus A$ and a set X such that the following conditions hold.*

1. *The Turing degree of X is not recursively enumerable.*
2. *For all A, $D(A) \oplus A$ and $X \oplus A$ have different Turing degrees.*

We present the proof of Theorem 4.1 in two parts. In Section 4.1, we construct a 2-re set X. As stated above, we will ensure that the Turing degree of X is not recursively enumerable. We will also ensure that the recursive approximation to X is *self-restraining*, a dynamic feature which we explain in Definition 4.3. In Section 4.2, we start with any Δ^0_2 set X with a self-restraining recursive approximation, and we produce a 2-re increasing operator D such that for all A, $D(A)$ and $X \oplus A$ do not have the same Turing degree. Theorem 4.1 follows: apply the method of Section 4.2 to the set of Section 4.1 to produce D and X as required.

DEFINITION 4.2. For Φ a Turing functional, let ϕ be the functional such that $\phi(x, X) = \ell$ if and only if $(x, y, X \upharpoonright \ell) \in \Phi$.

In other words, $\phi(x, X)$ is the amount of X used to determine the value of $\Phi(x, X)$. Note that $\Phi(X)$ and $\phi(X)$ have the same domain.

We will assume that all recursive Turing functionals are use-monotone; see Definition 2.2. In particular, if Φ is a recursive Turing functional then, by our convention, for every X, $\phi(X)$ is a nondecreasing function. We do not lose any generality: if Y is recursive in X, then there is a use-monotone recursive Φ such that $\Phi(X) = Y$.

4.1. Constructing X. In the following, we will write $X(n)[s]$ to denote the value of a recursive binary predicate at the pair of arguments n and s. We say that this predicate approximates a set if for all n, $\lim_{s\to\infty} X(n)[s]$ exists and is equal to 0 or 1. In this case, we will write $\lim_{s\to\infty} X(n)[s] = X(n)$. We are anticipating the case in which $X(n)[s]$ is the approximation to $X(n)$ during stage s.

We will also approximate the application of a recursive functional Φ to sets which are also being approximated. We will use the suffix $[s]$ to indicate the approximation to the preceding expression during stage s. We adopt the usual convention that we will only approximate a functional's having a value at stage s when the use of that functional is less than s.

DEFINITION 4.3. Suppose that $\lim_{s\to\infty} X(n)[s]=X(n)$. We say that $X(n)[s]$ is a *self-restraining* approximation to X if and only if there is an increasing recursive function g from $\mathbb{N}$ to $\mathbb{N}$ such that for all ℓ and all s, if $X(\ell)[s] \neq X(\ell)[s+1]$ then there are less than $g(\ell)$ numbers $t > s$ such that $(\exists m \leq s)$ $\big[X(m)[t] \neq X(m)[t+1]\big]$.

The following construction is not original to this paper (cf. Jockusch and Shore [1984, Theorem 1.6]). We reproduce it here so that we can verify that the recursive approximation to the set constructed is self-restraining.

THEOREM 4.4. *There is recursive approximation $X(n)[s]$ to a 2-re set X such that the following conditions hold.*

1. *The Turing degree of X is not recursively enumerable.*
2. *The approximation $X(n)[s]$ is self-restraining.*

PROOF. We define $X(n)[s]$ by recursion on s, in the context of a finite injury construction. We begin by setting $X(n)[0]$ equal to 0. Equivalently, during stage 0, X is empty. In a later stage $s+1$, we may add n to X by setting $X(n)[s+1]$ equal to 1 when $X(n)[s]$ was equal to 0, or we may remove n from X by setting $X(n)[s+1]$ equal to 0 when $X(n)[s]$ was equal to 1. For each n, we will add n to X at most once and remove n from X at most once, and so we will construct a 2-re set.

A single requirement. Suppose that W is a recursively enumerable set and that Φ and Ψ are Turing functionals. We must satisfy the following requirement.

$$\Psi(W) \neq X \text{ or } \Phi(X) \neq W$$

Our strategy works as follows.

1. Choose a number n larger than the current stage and larger than any number ever mentioned in the construction prior to this point. Prohibit n from entering X until reaching a stage s during which $\Psi(n, W)[s] = 0$, predicting that n is not an element of X, and for all m less than or equal to $\psi(n, W)[s]$, $\Phi(m, X)[s]$ is equal to $W(m)[s]$. We can visualize this situation as in Figure 1.

   Upon reaching such a stage s_1, go to Step 2.
2. Put n into X upon entering Step 2 in stage s_1.

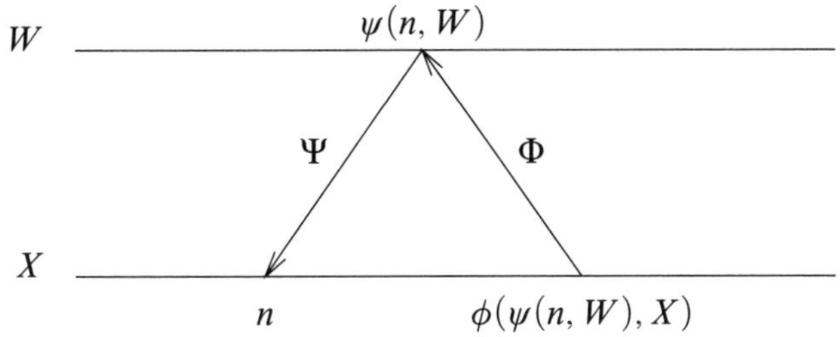

FIGURE 1. Configuring $\Phi(W) \neq X$ or $\Phi(X) \neq W$.

Wait for a later stage $s > s_1$ such that $\Psi(n, W)[s] = 1$. While waiting for that stage, prohibit any strategy of lower priority from changing X at any number less than or equal to s_1. Upon reaching such a stage s_2, go to Step 3.

3. Take n out of X, and prohibit any strategy of lower priority from changing X at any number less than or equal to s_2. Thereby, we set $X \upharpoonright s_1$ equal to $X[s_1] \upharpoonright s_1$ (as approximated before we put n into X).

The strategy has three possible outcomes. It could wait forever in one of the first two steps, in which case the requirement is clearly satisfied. Or the strategy could reach Step 3. In this last case, X is equal to $X[s_1]$ on all numbers less than or equal to s_1. But some number m less than $\psi(n, W)[s_1]$ must have entered W between stages s_1 and s_2 as $\Psi(n, W)[s_1] = 0$ and $\Psi(n, W)[s_2] = 1$. Consequently, if the strategy reaches Step 3, then for this m, $\Phi(m, X)$ is not equal to $W(m)$.

Priority construction. We define the recursive approximation $X(n)[s]$ to X by applying the finite injury priority method. We arrange the strategies in order type ω, with strategies later in the list having lower priority. During stage s, we identify the highest priority strategy which is supposed to take action, either by choosing its value for n or by going from one step to another, and follow the instructions for that strategy. We say that the lower priority strategies are injured, and we return them to the state in which they will begin Step 1 during the next stage.

Since each strategy acts only finitely often and for each strategy there are only finitely many others of higher priority, each strategy will eventually be implemented without injury and will satisfy its associated requirement.

Self-restraint. Suppose that at stage s we change our approximation to X at ℓ. Since only the first ℓ strategies can change the approximation to X at ℓ, each strategy of index greater than or equal to ℓ is injured during stage s and will not change X below s at any later stage. Similarly, each strategy of index less than ℓ can change our approximation to X at most twice below s before it is injured and then unable to change X below s. Consequently, we change

X at a number less than or equal to s at most 2ℓ many times after stage s. It follows that our construction is self-restraining.

This ends the proof of Theorem 4.4. ⊣

4.2. Constructing D. Suppose that X is a Δ^0_2 set which is not of recursively enumerable degree and has a self-restraining approximation $X(n)[s]$. It is safe to think of X as being the set constructed during the proof of Theorem 4.4.

We construct the required 2-re operator. In our presentation, we will assume that we are given the set A and will uniformly describe the set $D(A)$ so that it is 2-re relative to A.

We construct $D(A)$ by a finite injury construction similar to that in the previous section.

4.2.1. *A single requirement.* Suppose that Φ and Ψ are Turing functionals. We must satisfy the following requirement.

$$\Psi(X \oplus A) \neq D(A) \text{ or } \Phi(D(A) \oplus A) \neq X$$

We fix a recursive partition of $\mathbb{N}$ into infinitely many infinite subsets, and we allocate one of these subsets, R, to this requirement. In the following description of our strategy, the stage s approximation $D(A)[s]$ to $D(A)$ refers to the state of $D(A)$ at the beginning of stage s in the recursion being defined relative to A. In contrast, the stage s approximation $X[s]$ to X refers to the self-restraining recursive approximation which is given.

1. Choose a number $m_0 \in R$ larger than the current stage and larger than any number ever mentioned in the construction prior to this point. Prohibit any element of R which is greater than or equal to m_0 from entering $D(A)$ until reaching a stage s during which the following conditions hold.
   - $\Psi(m_0, X \oplus A)[s] = 0$, predicting that m_0 is not an element of $D(A)$. Let $\ell = \psi(m_0, X \oplus A)[s]$ be the length of the associated computation.
   - For each i less than or equal to $g(\ell) + 1$, $\Psi(m_i, X \oplus A)[s] = 0$, where $m_1, \dots, m_{g(\ell)+1}$ are the first $g(\ell) + 1$ elements of R which are strictly greater than m_0. Let ℓ^* be $\psi(m_{g(\ell)+1}, X \oplus A)[s]$. By Convention 4.2, ℓ^* is the least upper bound of $\psi(m_i, X \oplus A)[s]$ for i between 1 and $g(\ell) + 1$.
   - For all m less than or equal to ℓ^*, $\Phi(m, D(A) \oplus A)[s]$ is equal to $X(m)[s]$.

   We can visualize this situation as in Figure 2.

   Upon reaching such a stage s_0, let $j = 0$ and go to Step 2a.
2. (a) Put m_j into $D(A)$ upon entering Step 2a.

   Wait for a later stage $t > s_j$ such that $\Psi(m_j, X \oplus A)[t] = 1$. While waiting for that stage, prohibit any strategy of lower priority from changing $D(A)$ at any number less than or equal to s_0, thus we

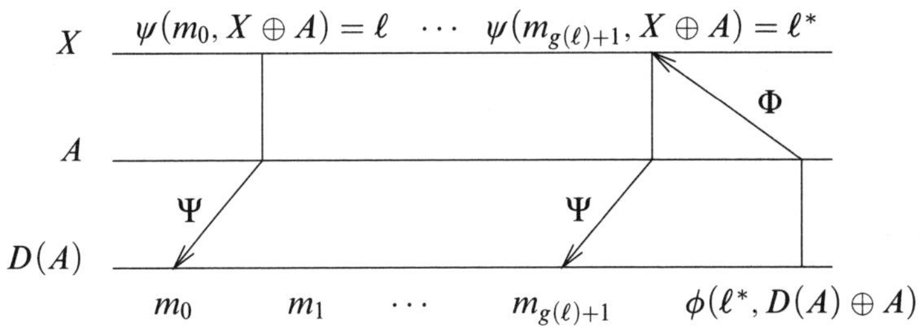

FIGURE 2. Configuring $\Psi(X \oplus A) \neq D(A)$ or $\Phi(D(A) \oplus A) \neq X$.

maintain the viability of the computations seen during stage s_0. Upon reaching such a stage t_j, go to Step 2b.

(b) Remove m_j from $D(A)$ upon entering Step 2b, thereby returning the value of $D(A) \upharpoonright s_0$ to $D(A)[s_0] \upharpoonright s_0$.

Wait for a later stage $s > t_j$ such that $\Phi(m_j, D(A) \oplus A)[s] = X[s]$ on all numbers less than or equal to ℓ^*. While waiting for that stage, prohibit any strategy of lower priority from changing $D(A)$ at any number less than or equal to s_0. Upon reaching such a stage s, increase the value of j by 1, define s_j to be equal to s, and go to Step 2a with these values for j and s_j.

This strategy could wait forever in the Step 1, in which case the requirement is clearly satisfied. Once the strategy leaves Step 1, it has defined ℓ and has $2(g(\ell)+1)$ possible outcomes. If for some fixed $j_0 \leq g(\ell)+1$, it waits forever in Step 2a with $j = j_0$ then the requirement is again clearly satisfied. Similarly, the requirement is satisfied if for some fixed $j_0 < g(\ell) + 1$, it waits forever in Step 2b with $j = j_0$.

The last possibility is for the strategy to reach Step 2b with $j = g(\ell) + 1$. In particular, it went from Step 2a to Step 2b during some stage t_0 after s_0. But then, $\Psi(m_0, X \oplus A)[s_0] = 0$ and $\Psi(m_0, X \oplus A)[t_0] = 1$. It can only be that the approximation to X changed during some stage s between s_0 and t_0 at some number ℓ_0 less than or equal to $\psi(m_0, X[s] \oplus A)[s_0] = \ell$. By the assumption that the approximation to X is self-restraining, the approximation to X changes no more than $g(\ell_0)$ times at numbers less than or equal to s during stages after s. Now we can check some inequalities. By the terms of Definition 4.3 g is increasing, and so the approximation to X changes no more than $g(\ell)$ times at numbers less than or equal to s during stages after s. Since $\psi(m_{g(\ell)+1}, X[s_0] \oplus A) = \ell^*$ is defined at stage s_0, it's value is less than s_0. Consequently, the approximation to $X \upharpoonright \ell^*$ changes no more than $g(\ell)$ times during stages after s. Finally, s is less than t_0, and so the approximation to $X \upharpoonright \ell^*$ changes changes no more than $g(\ell)$ times during stages after t_0.

But each time the strategy went from one step to the next, it must be that the approximation to $X \restriction \ell^*$ changed between the stage when the strategy entered that step and the one when it went to the next step. Either, in reaction to our adding m_j to $D(A)$, the approximation to $X \restriction \ell^*$ changed to make $\Psi(m_j, X \oplus A)[s] = 1$, and therefore it became incompatible with $X[s_0] \restriction \ell^*$. Or, in reaction to our returning the value of $D(A) \restriction \ell^*$ to $D(A) \restriction \ell^*[s_0]$, the approximation to X below ℓ^* returned to $X[s_0] \restriction \ell^*$ in order to agree with $\Phi(m_j, D(A) \oplus A)[s_0] \restriction \ell^*$. But then the approximation to $X \restriction \ell^*$ changed at least $2(g(\ell) + 1) - 1 = 2g(\ell) + 1$ times after stage t_0. Notice that $g(\ell) < 2g(\ell) + 1$, and that we have a contradiction.

It follows that the strategy cannot reach Step 2b for $j = \ell + 1$ and that the requirement is satisfied.

4.2.2. *Priority Construction.* As in the previous section, we organize our strategies in a finite injury priority construction. Once the strategies of higher priority have stopped acting, the next one begins in Step 1 and ensures that its associated requirement is satisfied.

This completes the proof of Theorem 4.1 ⊣

§5. An implicit characterization of *REA*. In this section, we prove that every nontrivial Σ-closure operator eventually extends the map $X \mapsto REA(X)$. Our proof makes use of the Shoenfield Jump Inversion Theorem.

THEOREM 5.1 (Shoenfield [1959]). *Suppose that W is REA relative to $\emptyset'$. Then there is a set W_0 such that $\emptyset' \geq_T W_0$ and $W_0' \equiv_T W$.*

THEOREM 5.2. *For any Borel Σ-closure operator M, if there is a cone of X's for which $M(X) \not\subseteq \Delta^0_1(X)$, then there is a cone of X's such that $M(X)$ contains all of the sets which are REA in X.*

PROOF. Suppose that M is a Borel Σ-closure operator and that there is a cone of X's for which $M(X) \not\subseteq \Delta^0_1(X)$. Choose B_1 so that for all X, if $X \geq_T B_1$ then $M(X) \not\subseteq \Delta^0_1(X)$.

We first apply the argument from Section 3 to conclude that there is a cone of X's for which $X' \in M(X)$. Suppose that $X \geq_T B_1$. Let A be an element of $M(X)$ such that $X \not\geq_T A$. By Theorem 1.3 relativized to X, choose G so that $G \geq_T X$ and $A \oplus G \equiv_T G'$. Then, $A \in M(X) \subseteq M(G)$ and $G \in M(G)$, and so $G' \in M(G)$, since it is equivalent to the join of elements of $M(G)$.

By the previous paragraph, we may conclude that for every X, there is a G such that $G \geq_T X$ and $G' \in M(G)$. Martin's [1975] Borel Determinacy implies that there is a cone of G's such that $G' \in M(G)$. Choose B_2 so that for every G, if $G \geq_T B_2$ then $G' \in M(G)$.

We will now argue that B_2' is the base of a cone as required. The situation that we will describe is the one indicated in Figure 3. Suppose that H is given with $H \geq_T B_2'$. Let W be a set which is *REA* relative to H. By Theorem 1.2 relativized to B_2, fix H_0 so that $H_0 \geq_T B_2$ and $H_0' \equiv_T H$. Finally, by

Theorem 5.1 relativized to H_0, choose W_0 so that $H \geq_T W_0 \geq_T H_0$ and $W_0' \equiv_T W$. We use solid lines to indicate the Turing order with the higher set being $\geq_T$ the lower one, and we use dotted lines to indicate the Turing jump with the higher set having the same Turing degree as the jump of the lower one.

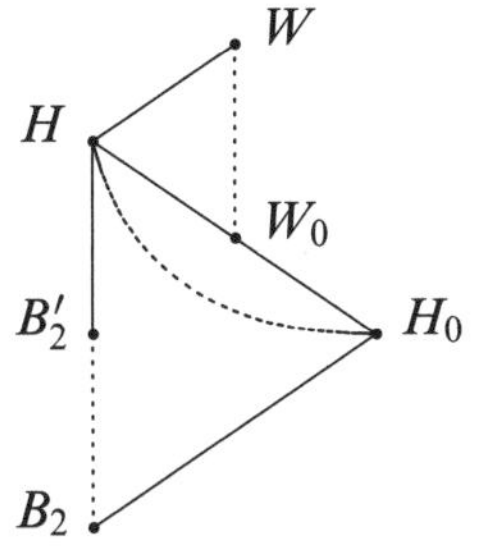

FIGURE 3. Relationships between B_2, B_2', the H's, and the W's.

Now, $W_0 \geq_T B_2$ implies that $W_0' \in M(W_0)$. Since $H \geq_T W_0$, $M(W_0) \subseteq M(H)$. Consequently, $W_0' \in M(H)$. Since $W_0' \equiv_T W$, $W \in M(H)$. Since W was an arbitrary set *REA* relative to H, every set *REA* relative to H belongs to $M(H)$. Since H was an arbitrary element of the cone above B_2', for every set H in the cone above B_2', all of the sets which are *REA* relative to H belong to $M(H)$, as required. $\dashv$

As we mentioned in the introduction, the Σ-closure operators are not as well behaved as the closure operators are. Even so, the function $X \mapsto REA(X)$ is intrinsic to the notion of Σ-closure. It is open whether the function mapping X to the set of reals which are 2-re in X and above X is also intrinsic. It is open whether there is a Borel Σ-closure operator M with the following property: there is a cone of X's, such that

1. there is a set in $M(X)$ whose degree is not recursively enumerable relative to X
2. and there is a set $D \geq_T X$ which is 2-re in X and not in $M(X)$.

REFERENCES

M. DAVIS [1950], *On the theory of recursive unsolvability*, ***Ph.D. thesis***, Princeton University.

R. M. FRIEDBERG [1957], *A criterion for completeness of degrees of unsolvability*, ***The Journal of Symbolic Logic***, vol. 22, pp. 159–160.

C. G. JOCKUSCH, JR. AND R. A. SHORE [1984], *Pseudo-jump operators II: Transfinite iterations, hierarchies, and minimal covers*, ***The Journal of Symbolic Logic***, vol. 49, pp. 1205–1236.

A. S. Kechris and Y. N. Moschovakis [1978], *The Victoria Delfino problems*, ***Cabal seminar 76–77*** (A. S. Kechris and Y. N. Moschovakis, editors), Lecture Notes in Mathematics, vol. 689, Springer-Verlag, Heidelberg, pp. 279–282.

D. A. Martin [1975], *Borel determinacy*, ***Annals of Mathematics (2)***, vol. 102, pp. 363–371.

D. B. Posner and R. W. Robinson [1981], *Degrees joining to* $\mathbf{0}'$, ***The Journal of Symbolic Logic***, vol. 46, no. 4, pp. 714–722.

H. Rogers, Jr. [1967], *Some problems of definability in recursive function theory*, ***Sets models and recursion theory, Proceedings of the Summer School in Mathematical Logic and Tenth Logic Colloquium, Lercester, August–September 1965*** (J. N. Crossley, editor), North-Holland Publishing Co.

G. E. Sacks [1990], ***Higher recursion theory***, Springer-Verlag, Berlin.

J. R. Shoenfield [1959], *On degrees of unsolvability*, ***Annals of Mathematics***, vol. 69, pp. 644–653.

R. A. Shore and T. A. Slaman [1999], *Defining the Turing jump*, ***Mathematical Research Letters***, vol. 6, pp. 711–722.

T. A. Slaman and J. R. Steel [1981], *Complementation in the Turing degrees*, ***The Journal of Symbolic Logic***, vol. 54, no. 1, pp. 160–176.

C. Spector [1955], *Recursive well-orderings*, ***The Journal of Symbolic Logic***, vol. 20, pp. 151–163.

DEPARTMENT OF MATHEMATICS
UNIVERSITY OF CALIFORNIA, BERKELEY
BERKELEY, CA 94720-3840, USA
E-mail: slaman@math.berkeley.edu

LIOUVILLE FUNCTIONS

A. J. WILKIE

Introduction. The purpose of this note is to report on an (as yet, unsuccessful) attempt to realise Zilber's axioms for a "generic" function on an algebraically closed field (see [4]) by an entire function on the complex numbers. These axioms are complete and ω-stable, and so if my program could be carried out it would provide the first example, as far as I know, of an ω-stable expansion of the complex field by a transcendental, entire function.

Zilber's argument is an interesting application of Hrushovski's amalgamation technique using, as predimension function, $\delta_f^K(X) := \text{t.d.}(X \cup f[X]) - |X|$ (where K is an algebraically closed field, $\text{t.d.}(Y)$ denotes the transcendence degree of the field generated by Y over the prime subfield of K, $f : K \to K$ is any function and X is a finite subset of K which, for convenience in this paper, satisfies $0 \notin X$). One immediately restricts attention to the (clearly axiomatizable) class consisting of those structures $\langle K, f \rangle$ for which δ_f^K is everywhere non-negative, and then shows that the collection of existentially closed[1] structures within this class is, indeed, axiomatizable.

Of course, no account is (nor, in the presence of ω-stability, could be) taken here of any metric structure that might naturally exist on the field K and so, in the case $K = \mathbb{C}$, we cannot expect to construct entire functions this way.[2] So my method in this note is simply to write down a suitable candidate, $H : \mathbb{C} \to \mathbb{C}$ say, and verify that Zilber's axioms hold for it. This amounts to showing thc following:

(A) suppose that $n \geq 1$ and that $\theta_1, \ldots, \theta_n$ are pairwise distinct, non-zero, complex numbers. Then $\text{t.d.}\,\mathbb{Q}(\theta_1, \ldots, \theta_n, H(\theta_1), \ldots, H(\theta_n)) \geq n$ (i.e., $\delta_H^{\mathbb{C}}$ is everywhere non-negative), and

(B) suppose that $F : \mathbb{C}^n \times \mathbb{C}^n \to \mathbb{C}^n$ is a polynomial map, defined over some (countable) subfield $k \subseteq \mathbb{C}$, having a non-singular zero which is both *balanced* and *free over* k. Then the map $F^* : \mathbb{C}^n \to \mathbb{C}^n$ given by $F^*(z_1, \ldots, z_n) := F(z_1, \ldots, z_n, H(z_1), \ldots, H(z_1))$ has a non-singular zero with pairwise distinct, nonzero coordinates.

[1]Strictly speaking, existentially closed with respect to so-called *strong* embeddings: we do not get a model complete theory here.

[2]See, however, the discussion concluding the introductory section of [5].

Logic Colloquium 2000
Edited by R. Cori, A. Razborov, S. Todorčević, and C. Wood
Lecture Notes in Logic, 19

The italicized terms in the "existential closedness" condition (B) will be explained below, where I also show that they are necessary conditions for the conclusion to hold. I should also mention that (B) is not Zilber's formulation of his existential closedness axioms (he makes no reference in [4] to differentiability, whereas the conclusion of (B) only makes sense for differentiable H) and my original proof of the equivalence of the two versions was cumbersome. I am grateful to Pascal Koiran for pointing out that this equivalence follows easily from a result in [2] (Lemma 2).

Liouville functions. A Liouville *number* is a real, irrational number ξ having the property that for every $k \geq 1$, there exists a rational number p/q $(q > 0)$ with $|\xi - p/q| < 1/q^k$. Such numbers must be transcendental because of Liouville's theorem on the (lack of good) rational approximation of algebraic numbers. This suggests candidates for H which, at least, should satisfy (A). I call an entire function $H : \mathbb{C} \to \mathbb{C}$ a *Liouville function* if it is well approximated by polynomials with rational coefficients in the sense that H has a Taylor series of the form

$$H(z) = \sum_{i=1}^{\infty} a_i^{-1} \cdot z^i \tag{1}$$

where the a_i are non-zero integers satisfying

$$\text{for every } l \geq 1,\ |a_{i+1}| > |a_i|^{i^l} \text{ for all sufficiently large } i. \tag{2}$$

It is almost immediate that the value of a Liouville function at a non-zero rational number is a Liouville number, and hence transcendental. Thus (A) holds for $n = 1$ and θ_1 rational. In fact, we have

THEOREM 1. Let H be a Liouville function. Then (A) holds.

I strongly conjecture that (B) also holds for Liouville functions. Notice that, as well as the consequence stated in the introduction, this would also imply that all expansions of the complex field by a Liouville function have the same theory. (Indeed, Zilber conjectures that they are all isomorphic.) Unfortunately, at the time of writing I can only prove

THEOREM 2. Suppose that $F : \mathbb{C}^n \times \mathbb{C}^n \to \mathbb{C}^n$ is a polynomial map defined over $\mathbb{Q}$. Suppose that F has a balanced non-singular zero. Then the map $F^* : \mathbb{C}^n \to \mathbb{C}^n$ given by $F^*(z_1, \ldots, z_n) := F(z_1, \ldots, z_n, H(z_1), \ldots, H(z_n))$ has a non-singular zero with pairwise distinct, non-zero coordinates.

The main lemmas. The proofs of Theorems 1 and 2 rely on two propositions. These provide the necessary effective estimates and explain why the growth rate (2) for the coefficients of a Liouville function is as it is.

PROPOSITION 1. Let m, n, k be positive integers. Then there exists $c = c(n, m, k) > 0$ such that for any positive integers h, d the following holds:—if, for each $i = 1, \ldots, k$, $Q_i(y_1, \ldots, y_m, x_1, \ldots, x_n)$ is a polynomial (over $\mathbb{Z}$) of height at most h and degree at most d, then the formula (in the language of rings)

$$\exists y_1, \ldots, y_m \bigwedge_{i=1}^{k} Q_i(y_1, \ldots, y_m, x_1, \ldots, x_n) = 0 \tag{3}$$

is equivalent (in the theory ACF_0) to a quantifier-free formula involving (equalities and inequalities between) polynomials (over $\mathbb{Z}$) of height at most h^{d^c} and degree at most d^c only. Further, if $x_1, \ldots, x_n \in \mathbb{C}$ are given of modulus less than 1 such that (3) is true (in the field $\mathbb{C}$), then witnesses $y_1, \ldots, y_m \in \mathbb{C}$ can be found of absolute value bounded by $h^{d^c} \cdot |q(x_1, \ldots, x_n)|^{-1}$, for some polynomial q (over $\mathbb{Z}$) of height at most h^{d^c} and degree at most d^c, satisfying $q(x_1, \ldots, x_n) \neq 0$.

[The *height* of a polynomial over $\mathbb{Z}$ is, by definition, the sum of the absolute values of its coefficients.]

Proposition 2. Let n, k be positive integers. Then there exists $c = c(n,k) > 0$ such that for any positive integers h, d the following holds:— if $Q_1(x_1, \ldots, x_n), \ldots, Q_k(x_1, \ldots, x_n)$ are polynomials (over $\mathbb{Z}$), of height at most h and degree at most d, having no common zero (in $\mathbb{C}^n$) then $\forall R \geq 1$,

$$\inf \left\{ \max_{1 \leq i \leq k} |Q_i(\omega_1, \ldots, \omega_n)| \colon \omega_i \in \mathbb{C}, |\omega_i| \leq R, i = 1, \ldots, k \right\} > h^{-Rd^c}. \tag{4}$$

Both Proposition 1 and Proposition 2 follow from effective versions of the Hilbert Nullstellensatz (see, e.g., [1] p. 92 for a very precise, explicit statement of (4)).

Notation. For $\bar{\eta}$ an n-tuple $\langle \eta_1, \ldots, \eta_n \rangle$ of complex numbers (or variables) and f a unary function, $f(\bar{\eta})$ denotes the n-tuple $\langle f(\eta_1), \ldots, f(\eta_n) \rangle$.

Proof of Theorem 1. Fix $n \geq 1$ and let U denote the open subset

$$\{\bar{z} \in \mathbb{C}^n \colon z_i \neq 0 \text{ and } z_i \neq z_j \text{ for all } i, j = 1, \ldots, n \text{ with } i \neq j\}$$

of $\mathbb{C}^n$.

Suppose, for a contradiction, that $\bar{\theta} \in U$ and $\text{t.d.}\,\mathbb{Q}(\bar{\theta}, H(\bar{\theta})) \leq n - 1$. Then there are polynomials (which we may take to be over $\mathbb{Z}$), $P_1(\bar{x}, \bar{y}), \ldots, P_{n+1}(\bar{x}, \bar{y})$ such that

$$\langle \bar{\theta}, H(\bar{\theta}) \rangle \in V, \text{ and} \tag{5}$$

$$\dim((U \times \mathbb{C}^n) \cap V) \leq n - 1, \tag{6}$$

where

$$V := \{\langle \bar{z}, \bar{w} \rangle \in \mathbb{C}^n \times \mathbb{C}^n \colon P_i(\bar{z}, \bar{w}) = 0 \text{ for } i = 1, \ldots, n+1\}.$$

For each $d \geq 1$, let $H_d(z)$ denote the partial sum $\sum_{i=1}^{d} a_i^{-1} \cdot z^i$ of $H(z)$ and define the function $\mu_d : \mathbb{C} \times \mathbb{C}^n \to \mathbb{C}$ by

$$\mu_d(z, t_1, \ldots, t_n) = H_d(z) + z^{d+1} \cdot \sum_{j=1}^{n} t_j \cdot z^{j-1}. \tag{7}$$

Now consider the map $\Phi_d : U \times \mathbb{C}^n \to U \times \mathbb{C}^n$ defined by

$$\Phi_d(\bar{z}, \bar{t}) = \langle \bar{z}, \mu_d(z_1, \bar{t}), \ldots, \mu_d(z_n, \bar{t}) \rangle. \tag{8}$$

Notice that for each $\bar{z} \in U$, $\Phi_d(\bar{z}, \bar{t})$ is linear in $\bar{t}$ with determinant (of the homogeneous part) equal to $\Delta'(\bar{z}) := (z_1 \cdots z_n)^{d+1} \cdot \Delta(\bar{z})$, where $\Delta(\bar{z})$ is the usual Van der Monde determinant given by

$$\begin{vmatrix} 1 & z_1 & \cdots & z_1^{n-1} \\ 1 & z_2 & \cdots & z_2^{n-1} \\ \vdots & \vdots & & \vdots \\ 1 & z_n & \cdots & z_n^{n-1} \end{vmatrix} = \prod_{1 \leq i < j \leq n} (z_j - z_i).$$

So, by the definition of U, $\Delta'(\bar{z})$ is non-vanishing throughout U and hence Φ_d is a bijection. Since Φ_d is, obviously, also algebraic, it follows from (6) that $\dim(\Phi_d^{-1}((U \times \mathbb{C}^n) \cap V)) \leq n-1$ and hence the projection of $\Phi_d^{-1}((U \times \mathbb{C}^n) \cap V)$ onto the second n coordinates also has dimension at most $n-1$. In other words, the subset W of $\mathbb{C}^n$ defined by the formula

$$\bar{t} \in W \Leftrightarrow \exists \bar{z} \in U \bigwedge_{i=1}^{n+1} P_i\left(\bar{z}, \mu_d\left(z_1, \bar{t}\right), \ldots, \mu_d\left(z_n, \bar{t}\right)\right) = 0 \tag{9}$$

has dimension at most $n-1$.

Now by clearing denominators in the P_i's and removing, by the usual tricks, the inequalities in the definition of U (e.g., replace $z_1 \neq z_2$ by $\exists w (z_1 - z_2) \cdot w - 1 = 0$, etc.), we see that W has a definition of the form (3) with $k = O(1)$, $m = O(1)$, $\deg(Q_i) = O(d)$ and $\operatorname{height}(Q_i) = (a_1 \cdots a_d)^{O(1)}$ (for $i = 1, \ldots, k$), where the implied constants depend only on the original polynomials $P_1, \ldots, P_{n+1}$ and not on d. (We shall be considering $d \to \infty$.)

Hence, by (2) and Proposition 1, it follows that, for large d, W can be defined by a quantifier-free formula involving polynomials of height $a_d^{d^{O(1)}}$ and degree $d^{O(1)}$ only. But as W has dimension at most $n-1$ and $W \subseteq \mathbb{C}^n$, it follows that W must be contained in a finite union of zero sets of (non-trivial) such polynomials. Now an easy calculation (using the growth rate condition (2)) shows that, for large d, the n-tuple $\langle a_{d+1}^{-1}, \ldots, a_{d+n}^{-1} \rangle$ cannot be a zero of any non-trivial polynomial of height $a_d^{d^{O(1)}}$ and degree $d^{O(1)}$, so it follows that $\langle a_{d+1}^{-1}, \dots, a_{d+n}^{-1} \rangle \notin W$. Since, by (7), $\mu_d(z, a_{d+1}^{-1}, \dots, a_{d+n}^{-1}) = H_{d+n}(z)$, we obtain from (9) that $\forall \bar{z} \in U \bigvee_{i=1}^{n+1} P_i(\bar{z}, H_{d+n}(\bar{z})) \neq 0$.

Thus the polynomials (in $\bar{z}, z_{n+1}$), $P_1(\bar{z}, H_{d+n}(\bar{z})), \dots, P_{n+1}(\bar{z}, H_{d+n}(\bar{z}))$, $\Delta'(\bar{z}) \cdot z_{n+1} - 1$ have no common zero in $\mathbb{C}^{n+1}$. After clearing denominators, setting $R = \max\{|\theta_1|, \dots, |\theta_n|, |\Delta'(\bar{\theta})|^{-1}\}$ (recall that $\bar{\theta} \in U$, so $\Delta'(\bar{\theta}) \neq 0$), performing height and degree calculations as above and setting $z_{n+1} = |\Delta'(\bar{\theta})|^{-1}$, we see from Proposition 2 that for all large d, there is some $i(d) \in \{1, \dots, n+1\}$ such that

$$\left| P_{i(d)}\left(\bar{\theta}, H_{d+n}(\bar{\theta})\right)\right| \geq a_{d+n}^{-Rd^{O(1)}} = a_{d+n}^{-d^{O(1)}} \text{ (as } R \text{ is fixed).} \tag{10}$$

However, by (1) and (2), $|H_{d+n}(\bar{\theta}) - H(\bar{\theta})| = O(a_{d+n+1}^{-1} \cdot R^{d+n+1}) = O(a_{d+n+1}^{-1/2})$. This implies (by the mean value theorem) that $|P_i(\bar{\theta}, H_{d+n}(\bar{\theta})) - P_i(\bar{\theta}, H(\bar{\theta}))| = O(a_{d+n+1}^{-1/2})$ (for each $i = 1, \dots, n+1$) and hence, by (5), that

$$\left| P_i\left(\bar{\theta}, H_{d+n}(\bar{\theta})\right)\right| = O\left(a_{d+n+1}^{-1/2}\right) \text{ (for each } i = 1, \dots, n+1).$$

However, this contradicts (10) and (2) for large d, and proves Theorem 1. ⊣

Balanced, non-singular zeros. Recall that if $F\colon \mathbb{C}^n \to \mathbb{C}^m$ (where $n \geq m \geq 1$) is a holomorphic function, then a point $\bar{\omega} \in \mathbb{C}^n$ is called a non-singular zero of F if $F(\bar{\omega}) = 0$, and, further, the Jacobian matrix of F:

$$\frac{\partial F}{\partial \bar{z}} = \frac{\partial(F_1, \dots, F_m)}{\partial(z_1, \dots, z_n)} = \begin{pmatrix} \frac{\partial F_1}{\partial z_1} & \cdots & \frac{\partial F_1}{\partial z_n} \\ \vdots & & \vdots \\ \frac{\partial F_m}{\partial z_1} & \cdots & \frac{\partial F_m}{\partial z_n} \end{pmatrix}$$

(where $F(\bar{z}) = \langle F_1(\bar{z}), \dots, F_m(\bar{z}) \rangle$) has maximum rank m when the entries are evaluated at $\bar{\omega}$.

NOTATION. For $I \subseteq \{1, \dots, n\}$, say $I = \{i_1, \dots, i_r\}$ (in increasing order), and $\bar{z} = \langle z_1, \dots, z_n \rangle$ an n-tuple of complex numbers (or variables), write $\bar{z}_I$ for $\langle z_{i_1}, \dots, z_{i_r} \rangle$.

Thus, by elementary algebra, a zero $\bar{\omega}$ of F (as above) is non-singular if and only if $\det\left(\frac{\partial F}{\partial \bar{z}_I}\right)(\bar{\omega}) \neq 0$ for some $I \subseteq \{1, \dots, n\}$ with $|I| = m$.

We shall be interested in holomorphic (in fact polynomial) maps $F\colon \mathbb{C}^{2n} = \mathbb{C}^n \times \mathbb{C}^n \to \mathbb{C}^n$, $\langle \bar{z}, \bar{w} \rangle \mapsto F(\bar{z}, \bar{w})$. A zero $\langle \bar{\alpha}, \bar{\beta} \rangle$ of F is non-singular, therefore, if $\det\left(\frac{\partial F}{\partial(\bar{z}_I, \bar{w}_J)}\right)(\bar{\alpha}, \bar{\beta}) \neq 0$ for some $I, J \subseteq \{1, \dots, n\}$ with $|I| + |J| = n$.

If we can choose I, J here partitioning $\{1, \dots, n\}$ (i.e., such that $I \cap J = \emptyset$ and (hence) $I \cup J = \{1, \dots, n\}$) and, further, the $\alpha_1, \dots, \alpha_n$ are all non-zero and pairwise distinct, then we call $\langle \bar{\alpha}, \bar{\beta} \rangle$ a *balanced*, non-singular zero of F.

Now consider the situation of Theorem 2 (except that we allow $H: \mathbb{C} \to \mathbb{C}$ to be any entire function) and suppose that the map $F^*: \mathbb{C}^n \to \mathbb{C}^n$ does, in fact, have a non-singular zero $\bar{\alpha}$ with the properties stated. Set $\beta_i = H(\alpha_i)$ for $i = 1, \ldots, n$. Then the Jacobian of F^* at $\bar{\alpha}$ is given by

$$\frac{\partial F^*}{\partial \bar{z}}(\bar{\alpha}) = \left(\frac{\partial F_i}{\partial z_j}(\bar{\alpha}, \bar{\beta}) + \frac{\partial F_i}{\partial w_j}(\bar{\alpha}, \bar{\beta}) \cdot H'(\alpha_j) \right)_{1 \le i,j \le n}. \tag{11}$$

Hence, by the multilinearity of the determinant,

$$\det\left(\frac{\partial F^*}{\partial \bar{z}}(\bar{\alpha}) \right) = \sum_{I,J} \left(\frac{\partial F}{\partial(\bar{z}_I, \bar{w}_J)}(\bar{\alpha}, \bar{\beta}) \cdot \prod_{j \in J} H'(\alpha_j) \right), \tag{12}$$

where the sum is taken over all partitions I, J of $\{1, \ldots, n\}$.

Since the left hand side of (12) is non-zero (by assumption), at least one summand on the right must be non-zero and so $\langle \bar{\alpha}, \bar{\beta} \rangle$ is a balanced, non-singular zero of F. Thus the condition on F in Theorem 2 is, indeed, necessary.

As a partial converse to this, suppose that $\langle \bar{\alpha}, \bar{\beta} \rangle$ is a balanced, non-singular zero of F, witnessed by the partition I_0, J_0 of $\{1, \ldots, n\}$ say. Suppose $H: \mathbb{C} \to \mathbb{C}$ is an entire function satisfying

$$H(\alpha_i) = \beta_i \text{ for } i = 1, \ldots, n, \tag{13}$$

$$H'(\alpha_i) = \begin{cases} 1 & \text{for } i \in J_0, \\ 0 & \text{for } i \in I_0. \end{cases}$$

(Such an H exists since $\alpha_1, \ldots, \alpha_n$ are pairwise distinct.)

Then $F^*(\bar{\alpha}) = F(\bar{\alpha}, \bar{\beta}) = 0$, and the only non-zero term on the right hand side of (12) is the one corresponding to I_0, J_0. Thus $\det\left(\frac{\partial F^*}{\partial \bar{z}}(\bar{\alpha})\right) = \frac{\partial F}{\partial(\bar{z}_{I_0}, \bar{w}_{J_0})}(\bar{\alpha}, \bar{\beta}) \neq 0$ (by hypothesis), and so $\bar{\alpha}$ is a non-singular zero of F^*.

Of course, in order to prove Theorem 2 we must work with a *given* (Liouville) function H — we are not free to construct it! However, we shall see that the partial sums of the given H can be modified (via μ_d functions — see the proof of Theorem 1) so that equations (13) are solvable, and the following lemma concerns the linear map that arises.

LEMMA. Fix $n \ge 1$ and let $\Delta^*(\bar{z})$ denote the determinant of the $2n \times 2n$ matrix

$$\begin{pmatrix} 1 & z_1 & z_1^2 & \cdots & z_1^{2n-1} \\ \vdots & \vdots & \vdots & & \vdots \\ 1 & z_n & z_n^2 & \cdots & z_n^{2n-1} \\ 0 & 1 & 2z_1 & \cdots & (2n-1)z_1^{2n-2} \\ \vdots & \vdots & \vdots & & \vdots \\ 0 & 1 & 2z_n & \cdots & (2n-1)z_n^{2n-2} \end{pmatrix}.$$

(So the $(n+i)$th row is the derivative w.r.t. z_i of the ith row, for $i = 1, \dots, n$.)
Then $\Delta^*(\bar{z}) = \pm\Delta(\bar{z})^2$, so $\Delta^*(\bar{z}) \neq 0$ for $z_1, \dots, z_n$ pairwise distinct.

PROOF. Exercise. [Hint: First show that for fixed $\bar{z}$, $\Delta^*(z_1 + t, \dots, z_n + t)$ is independent of t. Then set $t = -z_1$ and use induction on n.] ⊣

COROLLARY. Suppose $n \geq 1$ and that $F: \mathbb{C}^n \times \mathbb{C}^n \to \mathbb{C}^n$ is a holomorphic map with a balanced, non-singular zero, $\langle \bar{\alpha}, \bar{\beta} \rangle$ say. Suppose further that $f: \mathbb{C} \to \mathbb{C}$ is an entire function and that d is a non-negative integer. Then there exist complex numbers $\gamma_0, \dots, \gamma_{2n-1}$ such that, setting

$$R(z) = f(z) + z^{d+1} \cdot \sum_{j=0}^{2n-1} \gamma_j \cdot z^j \quad \text{and} \quad F^*(\bar{z}) = F(\bar{z}, R(\bar{z})),$$

$\bar{\alpha}$ is a non-singular zero of F^* and $R(\bar{\alpha}) = \bar{\beta}$.

PROOF. By the discussion preceding the lemma, it is sufficient to find $\gamma_0, \dots, \gamma_{2n-1} \in \mathbb{C}$ so that the corresponding function $R(z)$ satisfies equations (13) (with R in place of H). where, as there, I_0, J_0 is a partition of $\{1, \dots, n\}$ witnessing the fact that $\langle \bar{\alpha}, \bar{\beta} \rangle$ is a balanced, non-singular zero of F. These equations are

$$\sum_{j=0}^{2n-1} X_j \cdot \alpha_i^{d+1+j} = \beta_i - f(\alpha_i) \text{ for } i = 1, \dots, n,$$

$$\sum_{j=0}^{2n-1} (d+1+j) X_j \cdot \alpha_i^{d+j} = \begin{cases} 1 - f'(\alpha_i) & \text{for } i \in J_0, \\ -f'(\alpha_i) & \text{for } i \in I_0 \end{cases}$$

and they need to be solved for $X_0, \dots, X_{2n-1}$.
So we must show that the matrix

$$\begin{pmatrix} \alpha_1^{d+1} & \cdots & \alpha_1^{d+2n} \\ \vdots & & \vdots \\ \alpha_n^{d+1} & \cdots & \alpha_n^{d+2n} \\ (d+1)\alpha_1^d & \cdots & (d+2n)\alpha_1^{d+2n-1} \\ \vdots & & \vdots \\ (\alpha+1)\alpha_n^d & \cdots & (d+2n)\alpha_n^{d+2n-1} \end{pmatrix}$$

is invertible.
However, its determinant is $(\alpha_1 \cdots \alpha_n)^{2d+1} \Delta^*(\bar{\alpha})$ (as can be seen by taking out the initial factor and then subtracting $(d+1) \cdot (i\text{th row})$ from the $(n+i)$th row, for each $i = 1, \dots, n$), which is non-zero by the (second part of the) balanced property of $\langle \bar{\alpha}, \bar{\beta} \rangle$ and the lemma. ⊣

PROOF OF THEOREM 2. Let H and F be as in the statement of Theorem 2 (H being given by (1)) and let $\langle \bar{\alpha}, \bar{\beta} \rangle$ be a balanced nonsingular zero of F.

For each $d \geq 1$, define the function $R_d \colon \mathbb{C} \times \mathbb{C}^{2n} \to \mathbb{C}$ by

$$R_d(z, t_0, \ldots, t_{2n-1}) = H_d(z) + z^{d+1} \cdot \sum_{j=0}^{2n-1} t_j \cdot z^j \tag{14}$$

where $H_d(z)$ is the partial sum $\sum_{i=1}^{d} a_i^{-1} \cdot z^i$ of $H(z)$.

By the corollary, there exist values of the variables $\tilde{t} := \langle t_0, \ldots, t_{2n-1} \rangle$ such that the map $\bar{z} \mapsto F(\bar{z}, R_d(\bar{z}, \tilde{t}))$ has a nonsingular zero.

Now, after clearing denominators, write the ith coordinate function of $F(\bar{z}, R_d(\bar{z}, \tilde{t}))$ as $F_{d,i}(\bar{z}, \tilde{t})$. Then each $F_{d,i}$ is clearly a polynomial over $\mathbb{Z}$ of degree $O(d)$ and height $a_d^{O(1)}$ (just as in the proof of Theorem 1).

Set $J_d(\bar{z}, \tilde{t}) := \det\left(\frac{\partial(F_{d,1}, \ldots, F_{d,n})}{\partial(z_1, \ldots, z_n)}\right)$, so that J_d is also a polynomial over $\mathbb{Z}$ of degree $O(d)$ and height $a_d^{O(1)}$. Consider the formula

$$\exists \bar{z}, z_{n+1} \left(\bigwedge_{i=1}^{n} F_{d,i}(\bar{z}, \tilde{t}) = 0 \wedge J_d(\bar{z}, \tilde{t}) \cdot z_{n+1} - 1 = 0 \right). \tag{15}$$

By Proposition 1, (15) defines a subset, X say, of $\mathbb{C}^{2n}$ ($=$ the $\tilde{t}$-space) which can also be defined by a quantifier-free formula involving equalities and inequalities between polynomials of degree $d^{O(1)}$ and height $a_d^{d^{O(1)}}$ only.

Now it easily follows from the implicit function theorem that X is open in $\mathbb{C}^{2n}$ (for the usual topology) and, from the remark following (14) above, that X is non-empty. Hence $\mathbb{C}^{2n} \setminus X$ has dimension strictly less than $2n$ and so is contained in a finite union of sets of the form $\{\tilde{t} \in \mathbb{C}^{2n} \colon \rho(\tilde{t}) = 0\}$, where ρ is a non-trivial polynomial (over $\mathbb{Z}$) of degree $d^{O(1)}$ and height $a_d^{d^{O(1)}}$.

Now it easily follows from (2) that, for sufficiently large d, the point $\langle a_{d+1}^{-1}, \ldots, a_{d+2n}^{-1} \rangle$ of $\mathbb{C}^{2n}$ cannot be a zero of such a ρ, and hence that $\langle a_{d+1}^{-1}, \ldots, a_{d+2n}^{-1} \rangle \in X$. Further, by the last part of Proposition 1, witnesses for (15) (for $\tilde{t} = \langle a_{d+1}^{-1}, \ldots, a_{d+2n}^{-1} \rangle$) can be found satifying $|z_1|, \ldots, |z_n|, |z_{n+1}| < a_{d+2n}^{d^{O(1)}}$. (It is at this point, by the way, that the proof breaks down for F defined over $\mathbb{C}$, rather than $\mathbb{Z}$.)

To sum up, the map $\bar{z} \mapsto F(\bar{z}, H_{d+2n}(\bar{z}))$ has a zero, $\bar{\alpha}$ say, whose coordinates are bounded in absolute value by $a_{d+2n}^{d^{O(1)}}$ and at which the Jacobian of F is bounded below in absolute value by $a_{d+2n}^{-d^{O(1)}}$.

Now arguing as in the proof of Theorem 1, we see that (for large d) each coordinate of $F(\bar{\alpha}, H(\bar{\alpha}))$ is bounded in absolute value by $a_{d+2n+1}^{-1/2}$ (since $\|H(\bar{\alpha}) - H_{d+2n}(\bar{\alpha})\|$ is of this order, up to multiplication by terms of order $a_{d+2n}^{dO(1)}$). As this is small compared with the norm of the inverse of the derivative of the map $\bar{z} \mapsto F(\bar{z}, H(\bar{z}))(= F^*(\bar{z}))$ and with the norm of the second derivative of this map (over a sufficiently large neighbourhood of $\bar{\alpha}$) it follows from the Newton approximation method (as described, e.g., in [3] p. 270) that F^* has a nonsingular zero (close to $\bar{\alpha}$), as required. ⊣

Back to (B). Theorem 2 cannot be extended to maps F defined over arbitrary subfields of $\mathbb{C}$. For example, let k be the subfield of $\mathbb{C}$ generated (over $\mathbb{Q}$) by the complex number $H(1)$, and consider the polynomial map $\langle F_1, F_2\rangle : \mathbb{C}^2 \times \mathbb{C}^2 \to \mathbb{C}^2$ defined (over k) by $F_1(z_1, z_2, w_1, w_2) = z_1 - 1$ and $F_2(z_1, z_2, w_1, w_2) = (w_1 - H(1)) \cdot z_2 - 1$. Let $\bar{\alpha} = \langle 1, 1/2\rangle$, $\bar{\beta} = \langle H(1) + 2, 0\rangle$. Then $F_1(\bar{\alpha}, \bar{\beta}) = F_2(\bar{\alpha}, \bar{\beta}) = 0$ and

$$\begin{pmatrix} \frac{\partial F_1}{\partial z_1} & \frac{\partial F_1}{\partial z_2} \\ \frac{\partial F_2}{\partial z_1} & \frac{\partial F_2}{\partial z_2} \end{pmatrix} = \begin{pmatrix} 1 & 0 \\ 0 & 2 \end{pmatrix}$$

which has determinant $2 (\neq 0)$ at $\langle \bar{\alpha}, \bar{\beta}\rangle$. Thus $\langle \bar{\alpha}, \bar{\beta}\rangle$ is a balanced, nonsingular zero of $\langle F_1, F_2\rangle$. But $\langle F_1, F_2\rangle^*(z_1, z_2) = \langle z_1 - 1, (H(z_1) - H(1)) \cdot z_2 - 1\rangle$, so $\langle F_1, F_2\rangle^*$ has no zeros at all.

Such an example is, however, ruled out by the second hypothesis in (B): a point $\langle \bar{\alpha}, \bar{\beta}\rangle \in \mathbb{C}^n \times \mathbb{C}^n$ is said to be *free* over the subfield k of $\mathbb{C}$ if each coordinate of $\bar{\alpha}$ is transcendental over k.

Added in proof. Pascal Koiran has now established (B), and so Liouville functions do indeed satisfy all the Zilber axioms for a generic function. In fact, Koiran's method also shows that this theory is precisely the "limit theory" of a generic polynomial (with zero constant term) as its degree goes to infinity. His results may be found in: P. Koiran, *The theory of Liouville functions*, LIP Research Report 2002-13, Lyon, and ***The Journal of Symbolic Logic***, vol. 68 (2003), pp. 353–365.

REFERENCES

[1] W. D. Brownawell, *Aspects of the Hilbert Nullstellensatz*, ***New advances in transcendence theory*** (Alan Baker, editor), CUP, 1988, pp. 90–101.

[2] O. Chapuis, E. Hrushovski, P. Koiran, and B. Poizat, *La limit des théories de courbes génériques*, ***The Journal of Symbolic Logic***, vol. 67 (2002), pp. 24–34.

[3] J. Dieudonné, ***Foundations of modern analysis***, Academic Press, 1960.

[4] B. Zilber, *A theory of absolute Schanuel function*, preprint (but now contained in [5]).

[5] ———, *Analytic and pseudo-analytic structures*, ***Logic colloquium 2000*** (R. Cori et al., editors), Lecture Notes in Logic, vol. 19, AK Peters, 2005, this volume, pp. 392–408.

MATHEMATICAL INSTITUTE
UNIVERSITY OF OXFORD
24-29 ST GILES, OXFORD OX1 3LB, UK
E-mail: wilkie@maths.ox.ac.uk

ANALYTIC AND PSEUDO-ANALYTIC STRUCTURES

BORIS ZILBER

One of the questions frequently asked nowadays about model theory is whether it is still logic. The reason for asking the question is mainly that more and more of model theoretic research focuses on concrete mathematical fields, uses extensively their tools and attacks their inner problems. Nevertheless the logical roots in the case of model theoretic geometric stability theory are not only clear but also remain very important in all its applications.

This line of research started with the notion of a κ-categorical first order theory, which soon mutated into the more algebraic and less logical notion of a κ-categorical structure.

A structure $\mathbf{M}$ in a first order language L is said to be **categorical in cardinality** κ if there is exactly one structure of cardinality κ (up to isomorphism) satisfying the L-theory of $\mathbf{M}$.

In other words, if we add to $\mathrm{Th}(\mathbf{M})$ the (non first-order) statement that the cardinality of the domain of the structure is κ, the description becomes categorical.

The principal breakthrough, in the mid-sixties, from which stability theory started was the answer to J. Los' problem.

THE MORLEY THEOREM. *A countable theory which is categorical in one uncountable cardinality is categorical in all uncountable cardinalities.*

The basic examples of uncountably categorical structures in a countable language are:

(1) Trivial structures (the language allows only equality);

(2) Abelian divisible torsion-free groups; Abelian groups of prime exponent (the language allows $+, =$); Vector spaces over a (countable) division ring;

(3) Algebraically closed fields in language $(+, \cdot, =)$.

Also, any structure definable in one of the above is uncountably categorical in the language which witnesses the interpretation.

The structures definable in algebraically closed fields, for example, are effectively objects of algebraic geometry.

As a matter of fact the main logical problem after answering the question of J. Los was *what properties of* $\mathbf{M}$ *make it* κ*-categorical for uncountable* κ?

Logic Colloquium 2000
Edited by R. Cori, A. Razborov, S. Todorčević, and C. Wood
Lecture Notes in Logic, 19

The answer is now reasonably clear:

The key factors are measurability by a dimension and high homogeneity of the structure.

This gave rise to (Geometric) Stability Theory, the theory studying structures with good dimensional and geometric properties (see [Bu] and [P]). When applied to fields, the stability theoretic approach is in many respects very close to Algebraic Geometry.

The abstract dimension notion for finite $X \subset \mathbf{M}$ mentioned above is best understood by examples:

(1a) Trivial structures: **size of** X;

(2a) Abelian divisible torsion-free groups; Abelian groups of prime exponent; Vector spaces over a division ring: **linear dimension of** X;

(3a) Algebraically closed fields: **transcendence degree** tr.d.(X).

Dually, one can classically define another type of dimension using the initial one:

$$\dim V = \max\{\text{tr.d.}\,(\bar{x}) \mid \bar{x} \in V\}$$

for $V \subseteq \mathbf{M}^n$ an algebraic variety. In model theory the latter type of dimension notion is called the **Morley rank**.

The last example can also serve as a good illustration of the significance of the homogeneity of the structures. So, in general, the transcendence degree makes good sense in any field, and there is quite a reasonable dimension theory for algebraic varieties over a field. But the dimension theory in arbitrary fields fails if we want to consider it for wider classes of definable subsets, e.g., the images of varieties under algebraic mappings. In algebraically closed fields any definable subset is a boolean combination of varieties, by elimination of quantifiers, which is the eventual consequence of the fact that algebraically closed fields are existentially closed in the class of fields. The latter effectively means high homogeneity, as an existentially closed structure absorbs any amalgam with another member of the class.

One of the achievements of stability theory is the establishing of some hierarchy of types of structures that allows to say which ones are more 'analysable' (see [Sh]).

The next natural question to ask is *whether there are 'very good' stable structures which are not reducible to* (1)–(3) *above*?

The initial hope of the present author in [Z1], that any uncountably categorical structure comes from the classical context (the trichotomy conjecture), was based on the general belief that logically perfect structures could not be overlooked in the natural progression of mathematics. Allowing some philosophical licence here, this was also a belief in a strong logical predetermination of basic mathematical structures.

As a matter of fact this turned out to be true in many cases. Specifically it is true for *Zariski geometries*, which are defined as structures with a good dimension theory and nice topological properties, similar to the Zariski topology on algebraic varieties (see [HZ]).

Another situation where this principle works is the context of o-minimal structures (see [PS]).

Powerful applications of the result on Zariski geometries and of the underlying methodology were found by Hrushovski [H3], [H4]. This not only led to new and independent solutions to some Diophantine problems, the Manin–Mumford and Mordell–Lang (the function field) conjectures, but also a new geometric vision of these.

Yet the trichotomy conjecture proved to be false in general as Hrushovski found a source of a great variety of counterexamples.

We analyse below the Hrushovski construction, purporting to answer the question of whether the counterexamples it provides dramatically overhaul the trichotomy conjecture or if there is a way to save at least the spirit of it. As the reader will find below the author is inclined to stick to the second alternative.

§1. Hrushovski construction of new structures. The main steps:

Suppose we have a, usually elementary, class of structures $\mathcal{H}$ with a good dimension notion $d(X)$ for finite subsets of the structures. We want to introduce a new function or relation on $\mathbf{M} \in \mathcal{H}$ so that the new structure gets a good dimension notion.

The main principle, which Hrushovski found will allow us to do this, is that of the free fusion. That is, the new function should be related to the old structure in as free a way as possible. At the same time we want the structure to be homogeneous. He then found an effective way of writing down the condition: *the number of explicit dependencies in X in the new structure must not be greater than the size (the cardinality) of X.*

The explicit L-dependencies on X can be counted as L-codimension, $\text{size}(X) - d(X)$. The explicit dependencies coming with a new relation or function are the ones given by simplest 'equations', basic formulas.

So, for example, if we want a *new unary function f on a field* (implicit in [H2]), the condition should be

$$\text{tr.d.}(X \cup f(X)) - \text{size}(X) \geq 0, \tag{1}$$

since in the set $Y = X \cup f(X)$ the number of explicit field dependencies is $\text{size}(Y) - \text{tr.d.}(Y)$, and the number of explicit dependencies in terms of f is $\text{size}(X)$.

If we want, e.g., to put a *new ternary relation R on a field*, then the condition would be

$$\text{tr.d.}(X) - r(x) \geq 0, \tag{2}$$

where $r(X)$ is the number of triples in X satisfying R.

The very first of Hrushovski's examples (see [H1]) introduces just a *new structure of a ternary relation*, which effectively means putting new relation on the trivial structure. So then we have

$$\text{size}(X) - r(X) \geq 0. \tag{3}$$

If we similarly introduce an automorphism σ on the field (*difference fields*, [CH]), then we have to count

$$\text{tr.d.}(X \cup \sigma(X)) - \text{tr.d.}(X) \geq 0, \tag{4}$$

and the inequality here always holds.

Similarly for *differential fields* with the differentiation operator D (see [Ma]), where we always have

$$\text{tr.d.}(X \cup D(X)) - \text{tr.d.}(X) \geq 0. \tag{5}$$

The left hand side in each of the inequalities (1)–(5), denote it $\delta(X)$, is a counting function, which is called **predimension**, as it satisfies some of the basic properties of the dimension notion.

At this point we have carried out the first step of the Hrushovski construction, that is:

(Dim) we introduced the class $\mathcal{H}_\delta$ of the structures with a new function or relation, and the extra condition

$$\text{(GS)} \qquad \delta(X) \geq 0 \quad \text{for all finite } X.$$

(GS) here stands for 'Generalised Schanuel', the reason for which will be given below. The condition (GS) allows us to introduce another counting function with respect to a given structure $\mathbf{M} \in \mathcal{H}_\delta$

$$\partial_M(X) = \min\{\delta(Y) \colon X \subseteq Y \subseteq_{\mathit{fin}} M\}.$$

We also need to adjust the notion of embedding in the class for further purposes. This is the **strong embedding**, $\mathbf{M} \leq \mathbf{L}$, meaning that $\partial_M(X) = \partial_L(X)$ for every $X \subseteq_{\mathit{fin}} \mathbf{M}$.

The next step is:

(EC) Using the inductiveness of the class construct an existentially closed structure in $(\mathcal{H}_\delta, \leq)$.

If the class has the amalgamation property, then the existentially closed structures are sufficiently homogeneous. Also, *for* $\mathbf{M}$ *existentially closed* $\partial_M(X)$ *becomes a dimension notion.*

So, if the class $\mathcal{EC}$ of existentially closed structures is also axiomatisable, one can rather easily check that the existentially closed structures are ω-stable. This is the case for examples (1)–(3) and (5) above.

In more general situations the e.c. structures may be unstable, but still with reasonably good model-theoretic properties.

Notice that though condition (GS) is trivial in examples (4)–(5), the derived dimension notion ∂ is non-trivial. In both examples $\partial(x) > 0$ iff the corresponding rank of x is infinite (which is the SU-rank in algebraically closed difference fields and the Morley rank, in differentially closed fields).

Notice that the dimension notion ∂ for finite subsets, similarly to the example (3a), gives rise to a dual dimension notion for definable subsets $S \subseteq \mathbf{M}^n$ over a finite set of parameters C:

$$\dim(S) = \max\{\partial(\{x_1, \dots, x_n\}/C): \langle x_1, \dots, x_n\rangle \in S\}.$$

(Mu) This stage originally had been considered prior to (EC), but as one easily sees, it can be equivalently introduced afterwards.

We want to find now a finite Morley rank structure as a substructure (maybe non-elementary) of a structure $\mathbf{M} \in \mathcal{EC}$. In fact an existentially closed $\mathbf{M}$ would be of finite Morley rank, if '$\dim(S) = 0$' is equivalent to 'S is finite'. But in general $\dim(S)$ may be zero for some infinite definable subsets S, e.g., the set $S = \{x \in \mathbf{M}: f(x) = 0\}$ is one such in example (1): 'some equations have too many solutions'.

To eliminate the redundant solutions Hrushovski introduces a counting function μ for the maximal allowed size of potentially Morley rank 0 subsets. Then $\mathcal{H}_{\delta,\mu}$ is the subclass of structures of $\mathcal{H}_\delta$ satisfying the bounds given by μ. Equivalently, since existentially closed structures are universal for structures of $\mathcal{H}_\delta$, we see that $\mathcal{H}_{\delta,\mu}$ is the class of substructures of existentially closed structures $\mathbf{M}$, satisfying the bound by μ.

Inside this class we can just as well carry out the construction of existentially closed structures $\mathbf{M}_\mu$. Again, if the subclass has the amalgamation property and is first order definable, then an existentially closed substructure M_μ of this subclass is of finite Morley rank, in fact strongly minimal in cases (1)–(3). It is also important for the further discussion that $\mathbf{M}_\mu \subseteq \mathbf{M}$.

The infinite dimensional structures emerging after step (EC) in natural classes we call *natural Hrushovski structures*. Some but not all of them lead after step (Mu) to finite Morley rank structures.

It follows immediately from the construction, that the class of natural Hrushovski structures is singled out in $\mathcal{H}$ by three properties: the generalised Schanuel property (GS), the property of existentially closedness (EC) and the property (ID), stating the existence of n-dimensional subsets for all n.

It takes a bit more model theoretic analysis, as is done in [H1], to prove that in examples (1)–(3), and in many others, (GS), (EC) and (ID) form a complete set of axioms.

Since Hrushovski found the counterexamples the main question that has arisen is whether the pathological structures demonstrate the failure of the general principle or if there is a classical context which explains the counterexamples.

We now want to try and find grounds for the latter.

We start with one more example of Hrushovski construction.

§2. **Pseudo-exponentiation.** Suppose we want to put a new function ex on a field K of characteristic zero, so that ex is a homomorphism between the additive and the multiplicative groups of the field:

$$\mathrm{ex}(x_1 + x_2) = \mathrm{ex}(x_1) \cdot \mathrm{ex}(x_2).$$

Then the corresponding predimension on new structures $\mathbf{K}_{\mathrm{ex}} = (K, +, \cdot, \mathrm{ex})$ must be

$$\delta(X) = \mathrm{tr.d.}(X \cup \mathrm{ex}(X)) - \mathrm{lin.d.}(X) \geq 0, \qquad \text{(GS)}$$

where lin.d.(X) is the linear dimension of X over $\mathbb{Q}$.

Equivalently (GS) can be stated:

$$\text{assuming that } X \text{ is linearly independent over } \mathbb{Q},$$
$$\mathrm{tr.d.}(X \cup \mathrm{ex}(X)) \geq \mathrm{size}(X),$$

which in case $\mathbf{K}$ is the field of complex numbers and ex $=$ exp is known as the *Schanuel conjecture* (see [L]).

Now start with the class $\mathcal{H}(\mathrm{ex}\,/\,\mathrm{st})$ consisting of structures $\mathbf{K}_{\mathrm{ex}}$ satisfying (GS) and with the additional property that the kernel ker $= \{x \in K\colon \mathrm{ex}(x) = 1\}$ is a cyclic subgroup of the additive group of the field K, which we call a **standard kernel** (see [Z2]). This class is non-empty and can be described as a subclass of an elementary class defined by omitting countably many types.

The subclass $\mathcal{EC}(\mathrm{ex}\,/\,\mathrm{st})$ of existentially closed substructures of $\mathcal{H}(\mathrm{ex}\,/\,\mathrm{st})$ is first order axiomatisable inside $\mathcal{H}(\mathrm{ex}\,/\,\mathrm{st})$ (see [Z3] for the proof about a similar structure).

By the obvious analogy with the structure $\mathbb{C}_{\mathrm{exp}} = (\mathbb{C}, +, \cdot, \mathrm{exp})$ on the complex numbers we conjecture that $\mathbb{C}_{\mathrm{exp}}$ is one of the structures in $\mathcal{EC}(\mathrm{ex}/\,\mathrm{st})$. And we want to find the condition that might single out the isomorphism type of $\mathbb{C}_{\mathrm{exp}}$ among the other structures in the class. We do this by introducing an extra step to the Hrushovski construction, when it is possible, which comes after (EC). This is applicable in a wide variety of classes:

(P) Consider a ∂-independent set C of cardinality κ (embeddable in an existentially closed structure) and let $E(C)$ be a structure **prime over** C **in class** $\mathcal{EC}$.

When a prime structure over ∂-independent C, card $C = \kappa$, exists and is unique over C, we call $E(C)$ **the** κ**-canonical structure**.

The nice thing about having κ-canonical structures is that we get a link with the logical background again: the notion is a good analogue of κ-categoricity, or rather the strong minimality concepts (see also [Le] for recent developments in this direction).

We prove:

If the κ*-canonical structure* $E(C)$ *exists and the language is countable, we have* card $E(C) = \kappa$ *and* $E(C)$ *is* $\aleph_0$*-quasi-minimal, i.e., any definable subset* *of $E(C)$ is either countable, or the complement of a countable. Moreover, when $\kappa > \aleph_0$, for a definable $S \subseteq E(C)^n$ S is countable iff* $\dim(S) = 0$.

It follows from a well-known theorem of Shelah (see [Sh] and [Bu]), that canonical structures exist if the class $\mathcal{EC}$ is first order axiomatisable, complete and ω-stable, e.g., in examples (1)–(3) and (5). There is no prime structure in the class of algebraically closed difference fields (example (4)) due to the fact that the theory is not complete.

In spite of the fact that $\mathcal{EC}(\text{ex} / \text{st})$ is not axiomatisable and even interprets the ring of integers we managed to prove in [Z2] that:

There exists a weaker (non unique) version of κ-canonical structure for any infinite cardinal κ in $\mathcal{H}(\text{ex} / \text{st})$ which is $\aleph_0$-quasi-minimal.[1]

This finally brings us to

CONJECTURE. *$\mathbb{C}_{\exp}$ is the canonical structure of cardinality $2^{\aleph_0}$ in $\mathcal{H}(\text{ex} / \text{st})$ if we put* ex *to be* exp.

The conjecture (and even the one stating that $\mathbb{C}_{\exp} \in \mathcal{EC}(\text{ex} / \text{st})$) is obviously stronger than the Schanuel conjecture. Another consequence of the conjecture is the fact that $\mathbb{C}_{\exp}$ is existentially closed. We show in [Z2] that this is equivalent to the statement:

EC(exp): *Any non obviously-contradictory system of equations over $\mathbb{C}$ in terms of $+, \cdot$ and* exp *has a solution in $\mathbb{C}$.*

The definition of **non obviously-contradictory system** is quite similar to ones (implicitly) formulated for other classes in the form of axiom schemes, e.g., ACFA(iii) in [CH] for algebraically closed difference fields, and see also examples in section 3. We are not going to give the definition here, but a good example of such a system is an equation of the form $t(x) = 0$, where $t(x)$ is a term in $+, \cdot, \exp$ over $\mathbb{C}$ which is not of the form $\exp(s(x))$ for some other term $s(x)$. Such an equation by the conjecture should have a solution in $\mathbb{C}$. I have learned, while writing this paper, that such was exactly a conjecture by S. Schanuel which was proved by W. Henson and L. Rubel using Nevanlinna theory (see [HR]). A. Wilkie gave a rather simple proof of the solvability of an equation of the form $\sum_{i \leq N} q_i(x) e^{p_i(x)} = 0$, for q_i and p_i polynomials in one variable, $N > 1$, based on the rate of growth argument. The general case is open, but some research suggesting the truthfulness of the conjecture can be traced in the literature. See also the discussion below.

Notice also that one can easily replace the exponentiation by other classical functions and observe similar effects, including corresponding versions of the Schanuel conjecture.

Based on the analysis of pseudo-exponentiation one would like to conclude hypothetically that

---

[1] *Added in proof.* We have now proved that there is an $L_{\omega_1,\omega}(Q)$-sentence axiomatizing a subclass of $\mathcal{H}(\text{ex} / \text{st})$ with unique model in every infinite cardinal, and the models are $\aleph_0$-quasi-minimal. We don't know whether the models are canonical in the above sense.

1. Basic Hrushovski structures have analytic prototypes.

2. The statement of the Schanuel conjecture along with its analogs is an intrinsic property of classical analytic functions, probably responsible for a good dimension theory of the corresponding structures on the complex numbers.

3. Another basic property of classical analytic structures is the EC-property: *Any non obviously-contradictory system of equations over the structure has a solution.*

Of course, we don't possess technical means to check the truthfulness of the conjectures. But the general picture drawn above in view of the conjectures can be tested in simplest examples.

§3. Analytic interpretations.

New ternary relation. Let $g\colon \mathbb{C}^3 \to \mathbb{C}$ be an entire function with the properties:

(GS[R]) If a system of $n+1$ equations of the form $g(v_{i_k}, v_{j_k}, v_{l_k}) = 0$ $(k = 1, \dots, n+1)$, with $v_{i_k}, v_{j_k}, v_{l_k} \in \{v_1, \dots, v_n\}$, has a solution $\langle a_1, \dots, a_n\rangle$ in $\mathbb{C}$, then at least two of the triples $\langle a_{i_k}, a_{j_k}, a_{l_k}\rangle$ coincide.

(EC[R]) Let $\{\langle x_j, y_j, z_j\rangle\colon j \le n+m\}$ be distinct triples, where each of x_j, y_j, z_j is either a complex constant or one of variables $v_1, \dots, v_n$, but not all three of them constants. Then the system

$$\{g(x_i, y_i, z_i) = 0\colon i \le n\} \cup \{g(x_i, y_i, z_i) \neq 0\colon n < i \le n+m\}$$

of n equations and m inequalities has a solution in $\mathbb{C}$, provided no k of the n equations have less than k explicit variables.

It is easy to see that if we interpret the ternary relation $R(v_1, v_2, v_3)$ as $g(v_1, v_2, v_3) = 0$, by GS[R] we get condition (3) for δ, and by EC[R] the existential closedness is satisfied. Then ID[R] follows from the fact that ∂-closure of a finite set is countable, since n independent analytic equations in the complex n-space have only countably many non-singular solutions.

Problem 1. (i) Construct an entire function g such that GS[R] and EC[R] hold.

(ii) Prove that $(\mathbb{C}, R)$ is canonical in this case, i.e., the structure is prime over its basis (of cardinality $2^{\aleph_0}$).

Remark. One can construct g satisfying (GS[R]), using an argument of A. Wilkie, as

$$g(v_1, v_2, v_3) = \sum_{i_1, i_2, i_3 \in \mathbb{N}} a_{i_1, i_2, i_3} v_1^{i_1} v_2^{i_2} v_3^{i_3}$$

with complex coefficients a_{i_1, i_2, i_3} algebraically independent and very rapidly decreasing.

New functions on a field. This is another class mentioned above. One can easily write down in a first order way the following two schemes of axioms:

(GS[f]) Let $V \subseteq K^{2n}$ be a variety over $\mathbb{Q}$ in variables $x_1, \dots, x_n, y_1, \dots, y_n$. If $\dim V < n$, then there is no point $\langle x_1, \dots, x_n, y_1, \dots, y_n \rangle$ in V with $y_i = \mathrm{f}(x_i)$ and $x_i \neq x_j$ for all distinct $i, j \in \{1, \dots, n\}$.

(EC[f]) Let $V \subseteq K^{2n}$ be an irreducible variety over K in variables $x_1, \dots, x_n, y_1, \dots, y_n$ such that

(i) V is not contained in a hyperplane given by an equation of the form $x_i = x_j$ for $i < j \leq n$, or $x_i = c$, for $c \in K$;

(ii) for any $0 < i_1 < \dots < i_k \leq n$ the dimension of $V_{i_1,\dots,i_k}$, the projection of V onto $(x_{i_1}, \dots, x_{i_k}, y_{i_1}, \dots, y_{i_k})$-space, is not less than k.

Then there is a point $\langle x_1, \dots, x_n, y_1, \dots, y_n \rangle$ in V with $y_i = \mathrm{f}(x_i)$, for all $i \in \{1, \dots, n\}$.

It has been proved in [Z6] that:

GS[f] *and* EC[f] *along with* ID[f], *the axiom of infinite* ∂-*dimensionality, determine a complete* ω-*stable theory* (*of Morley rank* ω). *The theory has canonical model* $\mathbf{K}_\mathrm{f}$ *in every cardinality* κ.

Problem 2. (i) Construct an entire holomorphic function $\mathrm{f}\colon \mathbb{C} \to \mathbb{C}$ satisfying GS[f] and EC[f] for $K = \mathbb{C}$.

(ii) Prove that $\mathbb{C}_\mathrm{f}$ is isomorphic to canonical $\mathbf{K}_\mathrm{f}$ of cardinality $2^{\aleph_0}$.

Notice that we get basically the same theory, with minor changes, if we change GS[f] and, correspondingly, the counting function δ in (1) to:

(GS′[f]) $\mathrm{tr.d.}(X \cup \mathrm{f}(X)) - \mathrm{size}(X) \geq 0$, provided X does not contain certain elements, say, 0.

A. Wilkie in [W] proves that an entire analytic function given as

$$\mathrm{f}(x) = \sum_{i \geq 0} \frac{x^i}{a_i}$$

with a_i very rapidly increasing integers, satisfies (GS′[f]) and (EC[f]), if in the latter V is defined over $\mathbb{Q}$.[2]

New differentiable functions on a field. To get better understanding of the 'analytic' nature of previous examples we consider the class $\mathcal{H}(F)$ of fields K of characteristic zero with a collection $F = \{\mathrm{f}^{(i)} : i \in \mathbb{Z}\}$ of unary functions on K.

We first introduce, for a finite $I \subseteq \mathbb{Z}$

$$\delta_I(X) = \mathrm{tr.d.}\left(X \cup \bigcup_{i \in I} f^{(i)}(X)\right) - \mathrm{size}(X) \cdot \mathrm{size}(I),$$

the I**-predimension** of $X \subseteq_{fin} K$.

[2] *Added in proof.* Problem 2 has now been solved. P. Koiran in [K] proves the full (EC[f]) for Wilkie's functions. We can also show that $\mathbb{C}_f$ satisfies the *countable closure property* of [Z8], which implies (ii) of Problem 2 by the main result of [Z8].

Then the predimension of $X \subseteq_{fin} K$ is

$$\delta(X) = \min\{\delta_I(X) \colon I \subseteq_{fin} \mathbb{Z}\}.$$

We then, as usual, introduce the subclass satisfying first order definable condition

$$(\mathrm{GS}[F]) \qquad \delta(X) \geq 0,$$

and after going through the construction stage (EC) find out that the resulting structures satisfy first order axiom scheme EC[F], similar to EC[f].

If we add the corresponding ID[F], *the first order theory defined by*

$$\mathrm{GS}[F] \cup \mathrm{EC}[F] \cup \mathrm{ID}[F]$$

is ω-stable of rank ω, and the reduct of the theory to $(+, \cdot, \mathrm{f})$, where $\mathrm{f} = \mathrm{f}^{(0)}$, is just the theory in the previous example.

We then consider for each $i \in \mathbb{Z}$ a definable function

$$g^{(i)}(x_1, x_2) = \begin{cases} \frac{\mathrm{f}^{(i)}(x_1) - \mathrm{f}^{(i)}(x_2)}{x_1 - x_2}, & \text{if } x_1 \neq x_2, \\ \mathrm{f}^{(i+1)}(x_1), & \text{if } x_1 = x_2. \end{cases}$$

We want to introduce on $\mathbf{K}_F$, a model of the theory, a Zariski type topology τ. Consider a formal 'completion' $\bar{K} = K \cup \{\infty\}$, which can be viewed as the projective line over the field K. Define **basic τ-closed subsets of $\bar{K}^n$** to be

(i) all Zariski-closed subsets of $\bar{K}^n$ for every $n \in \mathbb{N}$,

(ii) for $n = 2$ the 'closures' of graphs of $\mathrm{f}^{(i)}$:

$$\left\{\langle x, y\rangle \in \bar{K}^2 \colon \left(x \in K \wedge y = \mathrm{f}^{(i)}(x)\right) \vee x = \infty\right\}$$

and

(iii) for $n = 3$ the 'closures' of graphs of $g^{(i)}$:

$$\left\{\langle x_1, x_2, y\rangle \in \bar{K}^3 \colon \left(x_1, x_2 \in K \wedge y = g^{(i)}(x_1, x_2)\right) \vee x_1 = \infty \vee x_2 = \infty\right\}.$$

Notice that (ii) and (iii) would actually be the closures of the graphs in the complex topology, if $K = \mathbb{C}$ and $f^{(i)}$ were holomorphic functions.

Define now the family of **τ-closed sets** to be the minimal family of subsets of $\bar{K}^n$, all n, containing the τ-closed subsets and closed under intersections, finite unions and projections $\bar{K}^{n+1} \to \bar{K}^n$. We prove in [Z6] that:

The topology τ is compact in the sense that any family with the finite intersection property of τ-closed sets has a non-empty intersection, and the projection of a closed set is closed (but not Hausdorff).

Notice that there is no DCC for closed sets.

We can then view K as a locally compact space. By definition, $f^{(i)}$ and $g^{(i)}$ are continuous functions on K, in the sense of the topology, and $f^{(i+1)}$ satisfies the definition of the derivative of $f^{(i)}$. Moreover, one can then carry out complex style analysis and 'non-standard analysis' on K, as shown in [Z4].

In particular, using the notion of *infinitesimal elements* in $K^* \succ K$, in the case char $K = 0$, one gets the Taylor formula for any definable function h: *given $x \in K$, $n \in \mathbb{N}$ and an infinitesimal α, there is an infinitesimal β such that*

$$h(x+\alpha) = \sum_{0 \le k \le n} \frac{h^{(k)}(x)}{k!} \cdot \alpha^k + \alpha^n \beta,$$

where $h^{(k)}$ are the derivatives of h. In case of positive characteristic p it holds only for $n < p$.

REMARK. It is worth mentioning that in the example of *new function on a field* one could consider f to be a one-to-one function with the same predimension formula (1), which is in fact the original Hrushovski example in [H2] in the case of an equi-characteristic pair of fields. In this case *we cannot* expand the structure to include derivatives under a reasonable topology as above, and this agrees very well with the fact that there is no transcendental analytic one-to-one function. But any one-to-one Hrushovski structure, by the way of construction, is a substructure of a natural Hrushovski structure above.

§4. Mixed characteristics structures. There are examples of Hrushovski structures which have no analytic prototype. Such an example is *the fusion of two fields*:

Let $\mathcal{H}$ be the class of two-sorted structures (L, R) with both sorts fields, char $L = p$, char $R = q$, in the language of fields for both sorts and an extra binary relation between the two sorts, interpreted as a bijective function $\mathrm{f}\colon L \to R$.

One then introduces the predimension function for $X \subseteq_{fin} L$

$$\delta(X) = \text{tr.d.}(X) + \text{tr.d.}(\mathrm{f}(X)) - \text{size}(X) \ge 0. \tag{6}$$

After going through steps (Dim) and (EC) one gets an ω-stable theory of rank ω which is called the **ω-stable fusion of two fields**. Notice that the pull-back of the field structure of R to L gives a second field structure on L, which was the original Hrushovski example. More importantly, Hrushovski showed that after carrying out step (Mu) one gets a strongly minimal fusion of two fields.

Let $\mathcal{H}$ again be the class of two-sorted structures with both sorts fields, char $L = 0$, char $R = p$, with the condition, that L is a valued field with the residue field R and the valuation group $\mathbb{Z}$. So there are also definable mappings $v\colon L \to \mathbb{Z}$ and $\rho\colon L_0 \to R$, where v is the valuation, $L_0 = \{x \in L\colon v(x) \ge 0\}$, the valuation ring, and ρ the residue ring homomorphism.

Following the freeness principle one comes to the following predimension function, analogous to (4) and (5), for the class:

Let $v(p)=1$, and for $X \subseteq_{fin} L$, $[X]=\{p^z x : x \in X, z \in \mathbb{Z} \text{ and } v(x)=-z\}$. Then

$$\delta(X) = \text{tr.d.}(X) - \text{tr.d.}(\rho[X]).$$

Like in (4) and (5), $\delta(X) \geq 0$ holds automatically in this class, so $\mathcal{H}_\delta = \mathcal{H}$ and every embedding is strong.

Then, completing step (EC), one gets the subclass of existentially closed structures, which are elementary equivalent to *maximal unramified extensions of the p-adic field* $\mathbb{Q}_p$. A more complex structure, with a definable automorphism σ, is studied in [BM] and is identified there as *the field of Witt vectors with an automorphism.*

Another class of mixed characteristic structures, *algebraically closed valued fields*, with the stability flavor is studied in [HHM], and there is a strong evidence that this example is of the pattern under consideration.

We present here a series of examples of mixed characteristics of a different kind.

Let A be a 1-dimensional torus or an elliptic curve defined over a field F_0 of characteristic p, by which we mean just the group scheme, so for every field F containing F_0 there is an algebraic group $\mathcal{A}(F)$ (written multiplicatively) of F-points of $\mathcal{A}$. Let $\mathcal{H}(\mathcal{A})$ be a class of two sorted structures (D, A), where $A = \mathcal{A}(F)$ is as above, and with the language of Zariski closed relations on A, D is a field in characteristic zero considered with family of relations and operations Σ, all definable by polynomial equations (with parameters) in the field and, at least, containing the additive operation, so D is a group with an extra structure.

Notice, that both sorts are then strongly minimal structures and so have corresponding dimension notions $\dim_D$ and $\dim_A$.

Let also the language of the class to contain a function symbol e, which is interpreted as a surjective homomorphism

$$e\colon D \to A.$$

We can do the Hrushovski construction for this class putting for $X \subseteq_{fin} D$

$$\delta(X) = \dim_D(X) + \dim_A(e(X)) - \text{lin.d.}(X),$$

where $\text{lin.d.}(X)$ is the dimension in the additive group, i.e., the linear dimension over $\mathbb{Q}$.

It is easy to see that the resulting structure will have the D and A existentially closed, so F is an algebraically closed field. Also, the kernel $\ker = \{x \in D\colon e(x) = 1\}$ is an additive subgroup of D, and by algebraic reasons ker is elementarily equivalent to:

$\mathbb{Z}[1/p]$, in case A is the multiplicative group of the field;

or to

$\mathbb{Z}[1/p]^{2-i} \times \mathbb{Z}^i$, in case A is an elliptic curve with Hasse invariant $i \in \{0, 1\}$.

Here $\mathbb{Z}[1/p]$ is the additive subgroup of rationals with denominators p^n, $n \in \mathbb{N}$, if $p > 0$, and is just the group of integers, if $p = 0$.

(See also about elliptic curves in [Ha] and the characterisation of elementary equivalence for abelian groups in [EF]).

It is well known that in the given language $A(F)$ is biinterpretable with the algebraically closed field F, perhaps expanded by finitely many constants. Thus for $Y \subseteq_{fin} A$ we have $\dim_A(Y) = \text{tr.d.}(Y/C)$, with the transcendence degree on the right calculated by means of the biinterpretation, and C is the set of constants.

Because of trichotomy results for definability in algebraically closed fields (see [R]), following also from later work [HZ], there are essentially two possibilities to consider for D:

(i) D is an algebraically closed field;

(ii) D is a vector space over a field K of characteristic zero.

Thus, in case (i) the inequality for δ takes form

$$\text{tr.d.}(X) + \text{tr.d.}(e(X)) - \text{lin.d.}(X) + d \geq 0, \tag{7}$$

where d is a non-negative integer depending on the constants needed for the interpretation.

In case (ii) we have

$$\text{lin.d.}_K(X) + \text{tr.d.}(e(X)) - \text{lin.d.}(X) + d \geq 0, \tag{8}$$

where $\text{lin.d.}_K(X)$ is the dimension in the sense of the K-vector space.

The existentially closed structures in the classes bear rather close similarity to *universal coverings of corresponding algebraic varieties* A, and the corresponding kernel ker plays the role of *the fundamental group of the variety*, denoted usually $\pi_1(A)$. Both model theoretically and algebraic geometrically the following two types of the kernel are especially interesting:

the case of minimal group satisfying the condition above, i.e., precisely the group

$$\mathbb{Z}[1/p] \quad \text{or} \quad \mathbb{Z}[1/p]^{2-i} \times \mathbb{Z}^i,$$

depending on A, which we call **standard kernel**;

the case of minimal *algebraically compact group* (see [EF]) satisfying the condition above, i.e., precisely the group

$$\prod_{l \neq p \text{ prime}} \mathbb{Z}_l \quad \text{or} \quad \prod_{l \neq p \text{ prime}} \mathbb{Z}_l^{2-i} \times \prod_{l \text{ prime}} \mathbb{Z}_l^i,$$

depending on A, which we call **compact kernel**. (Here $\mathbb{Z}_l$ stands for the additive group of l-adic numbers).

The standard kernel corresponds to the, so-called, *topological* $\pi_1(A)$, and the compact kernel corresponds to the, so-called, *algebraic* $\pi_1(A)$ [M].

So even in the mixed characteristic case the construction results in something which, even if not analytic, has a definite analytic flavour. But anyway, we hardly know what is the right generalisation of analyticity in many cases and there is hardly a satisfactory theory in the case of non-Archimedean valued fields.

The case corresponding to formula (7) with $p = 0$ is quite similar to the case of the field with pseudoexponentiation.

The case corresponding to formula (8) is studied in [Z3] mainly under assumption that the characteristic of F is zero. Notice that the corresponding structure is quite rich in expressive power, in particular, in sort A we can 'raise to powers $k \in K$' by putting for $x \in A$

$$x^k = e(k \ln(x)),$$

where $\ln(x)$ is an e-pull-back of x in D, so this is a 'multivalued operation'. We call $\mathcal{EC}$-structures corresponding to this case *the group A with 'raising to K-powers'*. It is also interesting that, given $k_1, \ldots, k_n \in K$,

$$H = k_1 \ker + \ldots + k_n \ker \quad \text{and} \quad e(H) = e(k_1 \ker) \cdot \ldots \cdot e(k_n \ker) \qquad (9)$$

are definable subgroups of D and A, correspondingly, and when the kernel is standard these are finite group-rank subgroups. If $n = 1$ and $k_1 = N^{-1}$, where $N \in \mathcal{Z}$ is a non-standard integer, the group

$$e\left(\frac{1}{N} \ker\right)$$

is a torsion subgroup of A.

We can prove the existence of the canonical structure in all the classes, including pseudo-exponentiation, with fixed compact kernel (see [Z2]). But the following problem remains open

PROBLEM 3. Prove the existence of κ-canonical structures for the classes above with standard kernel.

Notice that when $K = \mathbb{Q}$, formula (8) with $d = 0$ holds for any structure in the class, and the class $\mathcal{EC}$ (with all possible kernels) is axiomatisable, complete and superstable. We call the structures in the class **group covers of $\mathcal{A}$**. Even for this class Problem 3 is non trivial and has been proved for the case $\mathcal{A}$ is the multiplicative group of a field of characteristic zero very recently in [Z7] and [Z8] using rather subtle field arithmetic results and some model theoretic techniques of Shelah's.

We believe that the answer for group covers with standard kernel is crucial for understanding the general case.

Further model theoretic properties and a Diophantine conjecture. Any deeper model theory of most of the above mentioned classes depends on the following conjecture about intersections in semi-Abelian varieties (characteristic zero case):

Given a semi-Abelian variety B over $\mathbb{C}$ and an algebraic subvariety $W \subseteq B$ over $\mathbb{Q}$ there is a finite collection $\tau(W, B)$ of proper semi-Abelian subvarieties of B such that: for any algebraic subgroup $S \leq B$ and any irreducible component U of $S \cap W$

$$\textit{either } \dim U = \dim W + \dim S - \dim B,$$
$$\textit{or } U \subseteq T \textit{ for some } T \in \tau(W, B).$$

The conjecture, as a matter of fact, is of Diophantine type:

The conjecture on intersections in semi-Abelian varieties implies the Mordell–Lang (and Manin–Mumford) conjecture for number fields.

This can be proved *ab initio* as in [Z5], but a nicer, model theoretical way is to see directly as in [Z3] that:
under the assumption that the conjecture is true, the structure A with raising to K-powers is superstable, and the definable finitely generated or torsion subgroups of A are locally modular.

In fact, the proof of this theorem goes via elimination of quantifiers to the level of existential formulas (*near model completeness*, as typical for Hrushovski structures), which yields local modularity of the kernel. Hence one gets local modularity of subgroups in (9), which by the definition of local modularity implies the Mordell–Lang statement. It remains to notice that any finitely generated or torsion subgroup of A is embeddable in a group $e(H)$ of (9) for a right choice of K and $k_1, \ldots, k_n$.

The conjecture effects the fields with pseudo-exponentiation:

Assume the conjecture on intersection in semi-Abelian varieties, the case $B = (\mathbb{C}^*)^n$. Then on any canonical model $\mathbf{K}_{\mathrm{ex}}$ in $\mathcal{EC}(\mathrm{ex}\,/\,\mathrm{st})$ there is a (non-Hausdorff) locally compact topology under which ex *is continuous and, moreover, in infinitesimal neighborhoods of any point $a \in \mathbf{K}$ the function* ex *can be represented by the Taylor expansion*

$$\mathrm{ex}(a + x) = \mathrm{ex}(a) \cdot \sum_n \frac{x^n}{n!}.$$

Also, under the conjecture the structure of genuine exponentiation (or rather raising to powers) on the complex numbers becomes very clear.

(See [Z5]) *Assume the Schanuel conjecture and the conjecture on intersections in semi-Abelian varieties (the case $B = (\mathbb{C}^*)^n$). Then the structure of complex numbers with the (multivalued) operations $y = \exp(r \ln x) = x^r$ of raising to real powers satisfies* [EC], *i.e., any non-obviously contradictory system of equations of the form*

$$\sum_{r_1,\ldots,r_n} a_{r_1,\ldots,r_n} x_1^{r_1} \cdot \ldots \cdot x_n^{r_n} = 0,$$

with $a_{r_1,\ldots,r_n} \in \mathbb{C}$ and $r_1, \ldots, r_n \in \mathbb{R}$, has a solution in $\mathbb{C}$.

The proof is based on a theory developed by D. Bernstein, A. Kushnirenko, B. Kazarnovski and A. Khovanski (see [Kh]).

It follows then from the results of preceding sections that

(See [Z3]) *Assume the Schanuel conjecture and the conjecture on intersections in semi-Abelian varieties (the case* $B = (\mathbb{C}^*)^n$*). Then the structure of complex numbers with the (multivalued) operations of raising to real powers is superstable.*

It is also interesting to notice that the Diophantine conjecture is equivalent to Hrushovski–Schanuel type inequality (8) with $d = 0$ and $K = \mathcal{Q}$, a non-standard model of the field of rationals. In this form we can suggest the corresponding conjecture for case $p > 0$:

$$\text{lin.d.}_{\mathcal{Q}}(X) + \text{tr.d.}(e(X)) - \text{lin.d.}_{Q(\mathcal{P})}(X) \geq 0, \tag{10}$$

where

$$\mathcal{P} = \{p^N : N \in \mathcal{Z}\}$$

is the subgroup of the multiplicative group of $\mathcal{Q}$ inducing automorphisms on A, and $Q(\mathcal{P})$ is the subfield of $\mathcal{Q}$ generated by $\mathcal{P}$. Thus the corresponding class $\mathcal{EC}$ generalises the class of algebraically closed difference fields, with commuting generic automorphisms. So, most probably the theory of the class is *simple*.

One can easily rewrite the conjecture in (10) in a more standard form as follows

Given a semi-Abelian variety A over an algebraically closed field F of characteristic p and an algebraic subvariety $W \subseteq A^n$ there are finite collections $\lambda(W, A^n)$ of constants and $\tau(W, A^n)$ of n-tuples $\langle f_1(v_1, \ldots, v_m), \ldots, f_n(v_1, \ldots, v_m)\rangle$ of polynomials with integer coefficients with the property:

for any algebraic subgroup $S \leq A^n$ and any irreducible component U of $S \cap W$ there are a constant $c \in \lambda(W, A^n)$, an n-tuple $\langle f_1, \ldots, f_n\rangle \in \tau(W, A^n)$ and non-negative integers $t_1, \ldots, t_m$ such that

$$S \subseteq \left\{ \langle x_1, \ldots, x_n\rangle \in A^n : x_1^{f_1(p^{t_1}, \ldots, p^{t_m})} \cdot \ldots \cdot x_n^{f_n(p^{t_1}, \ldots, p^{t_m})} = c \right\}.$$

REFERENCES

[BM] L. Belair and A Macintyre, *L'automorphisme de Frobenius des vecteurs de Witt*, ***Comptes Rendus de l'Académie des Sciences de Paris***, tome 331, Série 1, (2000), pp. 1–4.

[Bu] S. Buechler, ***Essential stability theory***, Springer-Verlag, Berlin, 1996.

[CH] Z. Chatzidakis and E. Hrushovski, *Model theory of difference fields*, preprint, 1995.

[EF] P. Eklof and E. Fisher, *The elementary theory of abelian groups*, ***Annals of Mathematical Logic***, vol. 4 (1972), pp. 115–171.

[Ha] R. Hartshorne, ***Algebraic geometry***, Springer-Verlag, Berlin-Heidelberg-New York, 1977.

[HHM] D. Haskell, E. Hrushovski, and D. Macpherson, *Independence and elimination of imaginaries in algebraically closed valued fields*, preprint, 2000.

[HR] W. Henson and L. Rubel, *Some applications of Nevanlinna theory to mathematical logic: identities of exponential functions*, ***Transactions of the American Mathematical Society***, vol. 282 (1984), pp. 1–32, correction in vol. 294 (1986), p. 381.

[H1] E. HRUSHOVSKI, *A new strongly minimal set*, ***Annals of Pure and Applied Logic***, vol. 62 (1993), pp. 147–166.

[H2] ———, *Strongly minimal expansions of algebraically closed fields*, ***Israel Journal of Mathematics***, vol. 79 (1992), pp. 129–151.

[H3] ———, *The Mordell–Lang conjecture for function fields*, ***Journal of the American Mathematical Society***, vol. 9 (1996), no. 3, pp. 667–690.

[H4] ———, *Difference fields and the Manin-Mumford conjecture*, preprint, 1996.

[HZ] E. HRUSHOVSKI and B. ZILBER, *Zariski geometries*, ***Journal of the American Mathematical Society***, vol. 9 (1996), pp. 1–56.

[Kh] A. KHOVANSKI, ***Fewnomials***, Fazis, Moscow, 1997, in Russian.

[K] P. KOIRAN, *The theory of Liouville functions*, ***The Journal of Symbolic Logic***, vol. 68 (2003), no. 2, pp. 353–365.

[L] S. LANG, ***Introduction to transcendental numbers***, Addison-Wesley, Reading, Massachusetts, 1966.

[Le] O. LESSMANN, ***Dependence relation in non-elementary classes***, Ph.D. thesis, Carnegie Mellon University, 1998.

[Ma] D. MARKER, *Model theory of differential fields*, ***Model theory of fields***, Lecture Notes in Logic, Springer-Verlag, 1996.

[M] D. MUMFORD, ***Abelian varieties***, Tata Institute, Bombay, 1968.

[PS] Y. PETERZIL and S. STARCHENKO, *A trichotomy theorem for o-minimal structures*, ***Proceedings of the London Mathematical Society***, vol. 77 (1998), no. 3, pp. 481–523.

[P] A. PILLAY, ***Geometric stability theory***, Clarendon Press, Oxford, 1996.

[R] E. RABINOVICH, ***Interpreting a field in a sufficiently rich incidence system***, QMW Press, 1993.

[Sh] S. SHELAH, ***Classification theory***, North-Holland, Amsterdam, 1990, revised edition.

[Sh1] ———, *Finite diagrams stable in power*, ***Annals of Mathematical Logic***, vol. 2 (1970), pp. 69–118.

[W] A. WILKIE, *Liouville functions*, ***Logic colloquium 2000*** (R. Cori et al., editors), Lecture Notes in Logic, vol. 19, AK Peters, 2005, this volume, pp. 383–391.

[Z1] B. ZILBER, *The structure of models of uncountably categorical theories*, ***Proceedings of the International Congress of Mathematicians,*** 1983, Warszawa, vol. 1, PWN—North-Holland Publishing Company, Amsterdam—New York—Oxford, 1984, pp. 359–368.

[Z2] ———, *Fields with pseudo-exponentiation*, e-archive: `math.LO/0012023`.

[Z3] ———, *Raising to powers in algebraically closed fields*, ***Journal of Mathematical Logic***, vol. 3 (2003), pp. 217–238.

[Z4] ———, *Quasi-Riemann surfaces*, ***Logic: from foundations to applications, European Logic Colloquium*** (W. Hodges, M. Hyland, C. Steinhorn, and J. Truss, editors), Clarendon Press, Oxford, 1996, pp. 515–536.

[Z5] ———, *Exponential sums equations and the Schanuel conjecture*, ***Journal of the London Mathematical Society***, vol. 65 (2002), no. 2, pp. 27–44.

[Z6] ———, *A theory of a generic function with derivations*, ***Logic and Algebra*** (Yi Zhang, editor), Contemporary Mathematics AMS series, vol. 302, AMS, 2002, pp. 85–100, available on `www.maths.ox.ac.uk/~zilber`.

[Z7] ———, *Covers of the multiplicative group of an algebraically closed field of characteristic zero*, available on `www.maths.ox.ac.uk/~zilber`, 2001.

[Z8] ———, *A categoricity theorem for quasi-minimal excellent classes*, to appear, available on `www.maths.ox.ac.uk/~zilber`.

MATHEMATICAL INSTITUTE, OXFORD UNIVERSITY
24–29 ST. GILES, OXFORD, OX5 1SS, UK
E-mail: zilber@maths.ox.ac.uk

Lecture Notes in Logic

General Remarks

This series is intended to serve researchers, teachers, and students in the field of symbolic logic, broadly interpreted. The aim of the series is to bring publications to the logic community with the least possible delay and to provide rapid dissemination of the latest research. Scientific quality is the overriding criterion by which submissions are evaluated.

Books in the Lecture Notes in Logic series are printed by photo-offset from master copy prepared using LaTeX and the ASL style files. For this purpose the Association for Symbolic Logic provides technical instructions to authors. Careful preparation of manuscripts will help keep production time short, reduce costs, and ensure quality of appearance of the finished book. Authors receive 50 free copies of their book. No royalty is paid on LNL volumes.

Commitment to publish may be made by letter of intent rather than by signing a formal contract, at the discretion of the ASL Publisher. The Association for Symbolic Logic secures the copyright for each volume.

The editors prefer email contact and encourage electronic submissions.

Editorial Board

Editorial Policy

1. Submissions are invited in the following categories:
i) Research monographs
ii) Lecture and seminar notes
iii) Reports of meetings
iv) Texts which are out of print
Those considering a project which might be suitable for the series are strongly advised to contact the publisher or the series editors at an early stage.

2. Categories i) and ii). These categories will be emphasized by Lecture Notes in Logic and are normally reserved for works written by one or two authors. The goal is to report new developments quickly, informally, and in a way that will make them accessible to non-specialists. Books in these categories should include
– at least 100 pages of text;
– a table of contents and a subject index;
– an informative introduction, perhaps with some historical remarks, which should be accessible to readers unfamiliar with the topic treated;

In the evaluation of submissions, timeliness of the work is an important criterion. Texts should be well-rounded and reasonably self-contained. In most cases the work will contain results of others as well as those of the authors. In each case, the author(s) should provide sufficient motivation, examples, and applications. Ph.D. theses will be suitable for this series only when they are of exceptional interest and of high expository quality.

Proposals in these categories should be submitted (preferably in duplicate) to one of the series editors, and will be refereed. A provisional judgment on the acceptability of a project can be based on partial information about the work: a first draft, or a detailed outline describing the contents of each chapter, the estimated length, a bibliography, and one or two sample chapters. A final decision whether to accept will rest on an evaluation of the completed work.

3. Category iii). Reports of meetings will be considered for publication provided that they are of lasting interest. In exceptional cases, other multi-authored volumes may be considered in this category. One or more expert participant(s) will act as the scientific editor(s) of the volume. They select the papers which are suitable for inclusion and have them individually refereed as for a journal. Organizers should contact the Managing Editor of Lecture Notes in Logic in the early planning stages.

4. Category iv). This category provides an avenue to provide out-of-print books that are still in demand to a new generation of logicians.

5. Format. Works in English are preferred. After the manuscript is accepted in its final form, an electronic copy in LaTeX format will be appreciated and will advance considerably the publication date of the book. Authors are strongly urged to seek typesetting instructions from the Association for Symbolic Logic at an early stage of manuscript preparation.

Lecture Notes in Logic

From 1993 to 1999 this series was published under an agreement between the Association for Symbolic Logic and Springer-Verlag. Since 1999 the ASL is Publisher and A K Peters, Ltd. is Co-publisher. The ASL is committed to keeping all books in the series in print.

Current information may be found at `http://www.aslonline.org`, the ASL Web site. Editorial and submission policies and the list of Editors may also be found above.

Previously published books in the *Lecture Notes in Logic* are:

1. *Recursion theory.* J. R. Shoenfield. (1993, reprinted 2001; 84 pp.)
2. *Logic Colloquium '90; Proceedings of the Annual European Summer Meeting of the Association for Symbolic Logic, held in Helsinki, Finland, July 15–22, 1990.* Eds. J. Oikkonen and J. Väänänen. (1993, reprinted 2001; 305 pp.)
3. *Fine structure and iteration trees.* W. Mitchell and J. Steel. (1994; 130 pp.)
4. *Descriptive set theory and forcing: how to prove theorems about Borel sets the hard way.* A. W. Miller. (1995; 130 pp.)
5. *Model theory of fields.* D. Marker, M. Messmer, and A. Pillay. (1996; 154 pp.)
6. *Gödel '96; Logical foundations of mathematics, computer science and physics; Kurt Gödel's legacy. Brno, Czech Republic, August 1996, Proceedings.* Ed. P. Hajek. (1996, reprinted 2001; 322 pp.)
7. *A general algebraic semantics for sentential objects.* J. M. Font and R. Jansana. (1996; 135 pp.)
8. *The core model iterability problem.* J. Steel. (1997; 112 pp.)
9. *Bounded variable logics and counting.* M. Otto. (1997; 183 pp.)
10. *Aspects of incompleteness.* P. Lindstrom. (1997, 2nd ed. 2003; 163 pp.)
11. *Logic Colloquium '95; Proceedings of the Annual European Summer Meeting of the Association for Symbolic Logic, held in Haifa, Israel, August 9–18, 1995.* Eds. J. A. Makowsky and E. V. Ravve. (1998; 364 pp.)
12. *Logic Colloquium '96; Proceedings of the Colloquium held in San Sebastian, Spain, July 9–15, 1996.* Eds. J. M. Larrazabal, D. Lascar, and G. Mints. (1998; 268 pp.)
13. *Logic Colloquium '98; Proceedings of the Annual European Summer Meeting of the Association for Symbolic Logic, held in Prague, Czech Republic, August 9–15, 1998.* Eds. S. R. Buss, P. Hájek, and P. Pudlák. (2000; 541 pp.)
14. *Model Theory of Stochastic Processes.* S. Fajardo and H. J. Keisler. (2002; 136 pp.)
15. *Reflections on the Foundations of Mathematics; Essays in honor of Solomon Feferman.* Eds. W. Seig, R. Sommer, and C. Talcott. (2002; 444 pp.)

16. *Inexhaustibility; a non-exhaustive treatment.* T. Franzén. (2004; 255 pp.)

17. *Logic Colloquium '99; Proceedings of the Annual European Summer Meeting of the Association for Symbolic Logic, held in Utrecht, Netherlands, August 1–6, 1999.* Eds. J. van Eijck, V. van Oostrom, and A. Visser. (2004; 208 pp.)

18. *The Notre Dame Lectures.* Ed. P. Cholak. (2005, 185 pp.)

19. *Logic Colloquium 2000; Proceedings of the Annual European Summer Meeting of the Association for Symbolic Logic, held in Paris, France, July 23–31, 2000.* Eds. R. Cori, A. Razborov, S. Todorčević, and C. Wood. (2005; 408 pp.)

20. *Logic Colloquium '01; Proceedings of the Annual European Summer Meeting of the Association for Symbolic Logic, held in Vienna, Austria, August 1–6, 2001.* Eds. M. Baaz, S. Friedman, and J. Krajíček. (2005, 486 pp.)